Solving Problems in Multiply Connected Domains

CBMS-NSF REGIONAL CONFERENCE SERIES IN APPLIED MATHEMATICS

A series of lectures on topics of current research interest in applied mathematics under the direction of the Conference Board of the Mathematical Sciences, supported by the National Science Foundation and published by SIAM.

GARRETT BIRKHOFF, *The Numerical Solution of Elliptic Equations*
D. V. LINDLEY, *Bayesian Statistics, A Review*
R. S. VARGA, *Functional Analysis and Approximation Theory in Numerical Analysis*
R. R. BAHADUR, *Some Limit Theorems in Statistics*
PATRICK BILLINGSLEY, *Weak Convergence of Measures: Applications in Probability*
J. L. LIONS, *Some Aspects of the Optimal Control of Distributed Parameter Systems*
ROGER PENROSE, *Techniques of Differential Topology in Relativity*
HERMAN CHERNOFF, *Sequential Analysis and Optimal Design*
J. DURBIN, *Distribution Theory for Tests Based on the Sample Distribution Function*
SOL I. RUBINOW, *Mathematical Problems in the Biological Sciences*
P. D. LAX, *Hyperbolic Systems of Conservation Laws and the Mathematical Theory of Shock Waves*
I. J. SCHOENBERG, *Cardinal Spline Interpolation*
IVAN SINGER, *The Theory of Best Approximation and Functional Analysis*
WERNER C. RHEINBOLDT, *Methods of Solving Systems of Nonlinear Equations*
HANS F. WEINBERGER, *Variational Methods for Eigenvalue Approximation*
R. TYRRELL ROCKAFELLAR, *Conjugate Duality and Optimization*
SIR JAMES LIGHTHILL, *Mathematical Biofluiddynamics*
GERARD SALTON, *Theory of Indexing*
CATHLEEN S. MORAWETZ, *Notes on Time Decay and Scattering for Some Hyperbolic Problems*
F. HOPPENSTEADT, *Mathematical Theories of Populations: Demographics, Genetics and Epidemics*
RICHARD ASKEY, *Orthogonal Polynomials and Special Functions*
L. E. PAYNE, *Improperly Posed Problems in Partial Differential Equations*
S. ROSEN, *Lectures on the Measurement and Evaluation of the Performance of Computing Systems*
HERBERT B. KELLER, *Numerical Solution of Two Point Boundary Value Problems*
J. P. LASALLE, *The Stability of Dynamical Systems*
D. GOTTLIEB AND S. A. ORSZAG, *Numerical Analysis of Spectral Methods: Theory and Applications*
PETER J. HUBER, *Robust Statistical Procedures*
HERBERT SOLOMON, *Geometric Probability*
FRED S. ROBERTS, *Graph Theory and Its Applications to Problems of Society*
JURIS HARTMANIS, *Feasible Computations and Provable Complexity Properties*
ZOHAR MANNA, *Lectures on the Logic of Computer Programming*
ELLIS L. JOHNSON, *Integer Programming: Facets, Subadditivity, and Duality for Group and Semi-group Problems*
SHMUEL WINOGRAD, *Arithmetic Complexity of Computations*
J. F. C. KINGMAN, *Mathematics of Genetic Diversity*
MORTON E. GURTIN, *Topics in Finite Elasticity*
THOMAS G. KURTZ, *Approximation of Population Processes*
JERROLD E. MARSDEN, *Lectures on Geometric Methods in Mathematical Physics*
BRADLEY EFRON, *The Jackknife, the Bootstrap, and Other Resampling Plans*
M. WOODROOFE, *Nonlinear Renewal Theory in Sequential Analysis*
D. H. SATTINGER, *Branching in the Presence of Symmetry*
R. TEMAM, *Navier–Stokes Equations and Nonlinear Functional Analysis*
MIKLÓS CSÖRG, *Quantile Processes with Statistical Applications*
J. D. BUCKMASTER AND G. S. S. LUDFORD, *Lectures on Mathematical Combustion*
R. E. TARJAN, *Data Structures and Network Algorithms*
PAUL WALTMAN, *Competition Models in Population Biology*
S. R. S. VARADHAN, *Large Deviations and Applications*
KIYOSI ITÔ, *Foundations of Stochastic Differential Equations in Infinite Dimensional Spaces*
ALAN C. NEWELL, *Solitons in Mathematics and Physics*

Pranab Kumar Sen, *Theory and Applications of Sequential Nonparametrics*
László Lovász, *An Algorithmic Theory of Numbers, Graphs and Convexity*
E. W. Cheney, *Multivariate Approximation Theory: Selected Topics*
Joel Spencer, *Ten Lectures on the Probabilistic Method*
Paul C. Fife, *Dynamics of Internal Layers and Diffusive Interfaces*
Charles K. Chui, *Multivariate Splines*
Herbert S. Wilf, *Combinatorial Algorithms: An Update*
Henry C. Tuckwell, *Stochastic Processes in the Neurosciences*
Frank H. Clarke, *Methods of Dynamic and Nonsmooth Optimization*
Robert B. Gardner, *The Method of Equivalence and Its Applications*
Grace Wahba, *Spline Models for Observational Data*
Richard S. Varga, *Scientific Computation on Mathematical Problems and Conjectures*
Ingrid Daubechies, *Ten Lectures on Wavelets*
Stephen F. McCormick, *Multilevel Projection Methods for Partial Differential Equations*
Harald Niederreiter, *Random Number Generation and Quasi-Monte Carlo Methods*
Joel Spencer, *Ten Lectures on the Probabilistic Method, Second Edition*
Charles A. Micchelli, *Mathematical Aspects of Geometric Modeling*
Roger Temam, *Navier–Stokes Equations and Nonlinear Functional Analysis, Second Edition*
Glenn Shafer, *Probabilistic Expert Systems*
Peter J. Huber, *Robust Statistical Procedures, Second Edition*
J. Michael Steele, *Probability Theory and Combinatorial Optimization*
Werner C. Rheinboldt, *Methods for Solving Systems of Nonlinear Equations, Second Edition*
J. M. Cushing, *An Introduction to Structured Population Dynamics*
Tai-Ping Liu, *Hyperbolic and Viscous Conservation Laws*
Michael Renardy, *Mathematical Analysis of Viscoelastic Flows*
Gérard Cornuéjols, *Combinatorial Optimization: Packing and Covering*
Irena Lasiecka, *Mathematical Control Theory of Coupled PDEs*
J. K. Shaw, *Mathematical Principles of Optical Fiber Communications*
Zhangxin Chen, *Reservoir Simulation: Mathematical Techniques in Oil Recovery*
Athanassios S. Fokas, *A Unified Approach to Boundary Value Problems*
Margaret Cheney and Brett Borden, *Fundamentals of Radar Imaging*
Fioralba Cakoni, David Colton, and Peter Monk, *The Linear Sampling Method in Inverse Electromagnetic Scattering*
Adrian Constantin, *Nonlinear Water Waves with Applications to Wave-Current Interactions and Tsunamis*
Wei-Ming Ni, *The Mathematics of Diffusion*
Arnulf Jentzen and Peter E. Kloeden, *Taylor Approximations for Stochastic Partial Differential Equations*
Fred Brauer and Carlos Castillo-Chavez, *Mathematical Models for Communicable Diseases*
Peter Kuchment, *The Radon Transform and Medical Imaging*
Roland Glowinski, *Variational Methods for the Numerical Solution of Nonlinear Elliptic Problems*
Bengt Fornberg and Natasha Flyer, *A Primer on Radial Basis Functions with Applications to the Geosciences*
Fioralba Cakoni, David Colton, and Houssem Haddar, *Inverse Scattering Theory and Transmission Eigenvalues*
Mike Steel, *Phylogeny: Discrete and Random Processes in Evolution*
Peter Constantin, *Analysis of Hydrodynamic Models*
Donald G. Saari, *Mathematics Motivated by the Social and Behavioral Sciences*
Yuji Kodama, *Solitons in Two-Dimensional Shallow Water*
Douglas N. Arnold, *Finite Element Exterior Calculus*
Qiang Du, *Nonlocal Modeling, Analysis, and Computation*
Alain Miranville, *The Cahn–Hilliard Equation: Recent Advances and Applications*
Per-Gunnar Martinsson, *Fast Direct Solvers for Elliptic PDEs*
Darren Crowdy, *Solving Problems in Multiply Connected Domains*

Darren Crowdy
Imperial College London
London, United Kingdom

Solving Problems in Multiply Connected Domains

SOCIETY FOR INDUSTRIAL AND APPLIED MATHEMATICS
PHILADELPHIA

10 9 8 7 6 5 4 3 2 1

Publications Director Kivmars H. Bowling
Executive Editor Elizabeth Greenspan
Acquisitions Editor Paula Callaghan
Developmental Editor Mellisa Pascale
Managing Editor Kelly Thomas
Production Editor Ann Manning Allen
Copy Editor Julia Cochrane
Production Manager Donna Witzleben
Production Coordinator Cally A. Shrader
Compositor Cheryl Hufnagle
Graphic Designer Doug Smock

Library of Congress Cataloging-in-Publication Data

Names: Crowdy, Darren, author.
Title: Solving problems in multiply connected domains / Darren Crowdy, Imperial College London, London, United Kingdom.
Description: Philadelphia : Society for Industrial and Applied Mathematics, [2020] | Series: CBMS-NSF regional conference series in applied math ; 97 | Includes bibliographical references and index. | Summary: "The aim of this monograph is to present a mathematical framework which makes solving problems in multiply connected domains a very natural generalization of solving them in simply connected ones"-- Provided by publisher.
Identifiers: LCCN 2019052802 (print) | LCCN 2019052803 (ebook) | ISBN 9781611976144 (paperback) | ISBN 9781611976151 (ebook)
Subjects: LCSH: Problem solving.
Classification: LCC QA63 .C76 2020 (print) | LCC QA63 (ebook) | DDC 510--dc23
LC record available at *https://lccn.loc.gov/2019052802*
LC ebook record available at *https://lccn.loc.gov/2019052803*

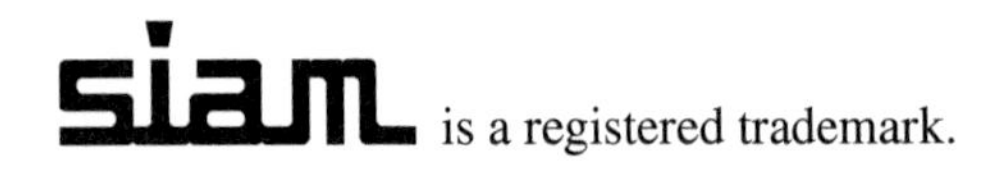

Contents

Preface

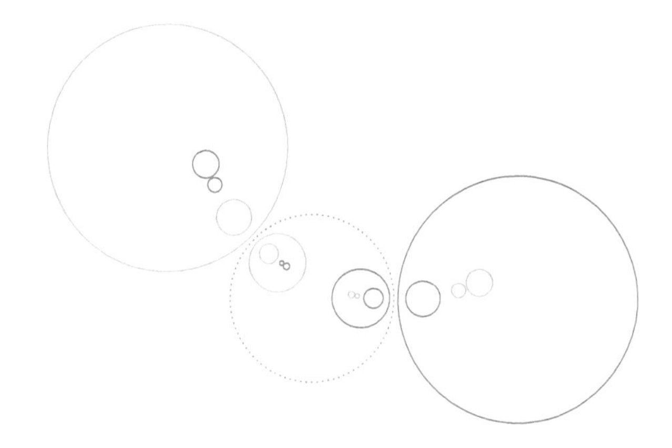

> *The universe, of course, is based upon the principle of the circle within the circle. At the moment, I am in an inner circle. Of course, smaller circles within this circle are also possible.*
>
> *A Confederacy of Dunces*
> John Kennedy Toole

Whenever two or more objects, or entities, be they bubbles, vortices, black holes, magnets, colloidal particles, microorganisms, swimming bacteria, Brownian random walkers, airfoils, turbine blades, electrified drops, magnetized particles, dislocations, cracks, or heterogeneities in an elastic solid, interact in some ambient medium, those objects or entities make "holes" in that medium. At least, that is how a mathematician would see it. Mathematicians call the holey regions containing a collection of interacting entities multiply connected.

In every field of the applied and natural sciences there is a well-known solution of the "single-entity" problem and usually another for a "singly periodic array of entities." These solutions are well known not only because they provide important baseline studies but often too because they can be expressed analytically in terms of recognized special functions. Between these two cases—when two, three, or some other finite number of entities interact—the literature in every field becomes much more sparse. Most commonly studies of two or more interacting entities become numerical in nature, or rely on some asymptotic "diluteness" or "weakly interacting" approximation so that the single-entity analysis can be used in an approximate sense. For two interacting objects it is common for elliptic function theory to appear in the analysis, and the function theories of Weierstrass or Jacobi.

The aim of this monograph is to present a mathematical framework which makes solving problems in multiply connected domains a very natural generalization of solving them in simply connected ones. For simply connected domains, where an integer $M = 0$, the framework reproduces the well-known single-entity and periodic-array solutions just mentioned. In the doubly connected case, or when $M = 1$ corresponding to two objects, the framework reproduces, in a different form, the results available using elliptic function theory. But, importantly, the framework works in the same basic way for any $M \geq 0$ meaning that a mathematical description of one, two, three, or any finite number of objects is captured in a unified fashion.

For applied and natural scientists this mathematical framework turns out to provide a flexible route to solving all kinds of problems involving interacting entities. Building this framework has been a key theme of the author's research over the last two decades, and this monograph is a distillation of the main outcomes of that work. By now, the framework is well developed, and the aim in writing this monograph is to make the ideas accessible to a wider range of scientists and to foster broader application of these versatile mathematical techniques. No single mathematical framework can reasonably claim to solve *every* problem in a multiply connected domain, but the framework described here is relevant to a sufficiently wide range of applications that it deserves to be better known.

The two analytic functions of a complex variable z,

$$\omega(z,a) = z - a \qquad \text{and} \qquad \mathcal{G}_0(z,a) = \frac{1}{2\pi \mathrm{i}} \log\left(\frac{z-a}{|a|(z-1/\overline{a})}\right), \tag{1}$$

thinking of a as a parameter, will be familiar to many scientists, and they can be used to illustrate the foundation on which our mathematical framework is built. The function $\omega(z,a)$ is a simple monomial. As such, its significance is easy to overlook. It happens to be much more important than is generally appreciated: it is the simplest example of a *prime function.*

The function $\mathcal{G}_0(z,a)$, on the other hand, is the analytic extension of the Green's function for Laplace's equation in a unit disc $|z| < 1$. Mathematicians, physicists, and engineers come across this function early in their scientific lives and they appreciate its fundamental significance. For mathematicians it is a basic object in complex analysis and potential theory; for applied scientists it encodes, inter alia, the electric field potential generated by a point charge at position $z = a$ situated outside a grounded circular conductor of unit radius.

Many texts have been written about Green's functions. This book is novel in being all about the *other* function—the prime function—and its natural generalizations. We argue that the prime function is at least as fundamental as the Green's function, perhaps more so, even if use of the Green's function turns out to be the easiest way to establish the existence of the prime function, as we will see here.

The alleged importance of the prime function may seem surprising given that most scientists have not heard of it. But a clue to its central role follows from the simple observation that the analytic extension of the Green's function above can be rewritten as

$$\mathcal{G}_0(z,a) = \frac{1}{2\pi \mathrm{i}} \log\left(\frac{\omega(z,a)}{|a|\omega(z,1/\overline{a})}\right), \tag{2}$$

meaning that *two* instances of the prime function are required to write down $\mathcal{G}_0(z,a)$. From this observation alone one might argue that the prime function is the more basic building block.

Given $\omega(z,a)$, the analytic extension of the Green's function $\mathcal{G}_0(z,a)$ is given by (2). Conversely, given $\mathcal{G}_0(z,a)$, there is an interesting way to derive the prime function $\omega(z,a)$.

First notice, from (1), that

$$2\pi\mathrm{i}\left[\frac{\partial\mathcal{G}_0(z,a)}{\partial a}-\frac{\partial\mathcal{G}_0(w,a)}{\partial a}\right]=2\pi\mathrm{i}\left[-\frac{1}{z-a}+\frac{1}{w-a}\right], \tag{3}$$

which, unlike $\mathcal{G}_0(z,a)$ itself, is an analytic function of a. Now introduce the integral of this analytic function of a:

$$\Pi_{a,b}^{z,w}=2\pi\mathrm{i}\int_b^a\left[-\frac{1}{z-a'}+\frac{1}{w-a'}\right]da', \tag{4}$$

which gives a result that is analytic in all variables z, a, w, and b. The value of $\Pi_{a,b}^{z,w}$ is only determined up to integer multiples of $2\pi\mathrm{i}$ depending on which contours are taken joining b to a. However its exponential $e^{-\Pi_{a,b}^{z,w}}$ is well defined, and we can therefore consider the double limit

$$X(a,b)\equiv\lim_{\substack{z\to a\\ w\to b}}\left[-(z-a)(w-b)e^{-\Pi_{a,b}^{z,w}}\right]. \tag{5}$$

This limiting process brings us back down to a function of just two variables, a and b. It is a simple exercise to establish that

$$\Pi_{a,b}^{z,w}=\log\left[\frac{(z-a)(w-b)}{(z-b)(w-a)}\right]+2\pi\mathrm{i}n,\qquad n\in\mathbb{Z}, \tag{6}$$

and consequently that

$$X(a,b)\equiv\lim_{\substack{z\to a\\ w\to b}}\left[-(z-a)(w-b)e^{-\Pi_{a,b}^{z,w}}\right]=(a-b)^2. \tag{7}$$

After taking the square root we arrive at the prime function $\omega(a,b)=(a-b)$.

These more or less elementary mathematical steps can be generalized, in a natural way, to provide a framework—indeed, we will go so far as to use the word "calculus"—for solving problems in multiply connected domains using the prime function as the essential mathematical tool. Demonstrating this is one of the goals of this monograph. A benefit of this approach is theoretical unification: solving problems in multiply connected domains becomes a natural generalization of solving them in simply connected ones, with the prime function furnishing the theoretical linchpin.

The basic notion of a prime function, or a "prime form," on a Riemann surface goes back well over a hundred years, to work by the likes of Weierstrass, Poincaré, Schottky, Klein, and Osgood (among others). An excellent summary of the history of the development of the idea of prime functions (and prime forms), together with all the relevant classical references, can be found in the monograph by Bottazzini and Gray [71], and we refer the reader there for a comprehensive history. The present monograph adopts the spirit of these general ideas for the specific goal of solving problems in multiply connected planar domains; such an endeavor was not a focus of attention for any of these classical authors, who had the much more general matter of uniformization on their minds. Nor does the program of ideas laid out here appear to have been developed since by other authors. In this author's view, the importance of this function for the applied sciences has been overlooked for too long.[1]

[1]The lucid treatment by Hejhal [90], which includes some discussion of the prime function in the context of domain functionals, has been a particular inspiration to the author but has very different objectives than the present book, which is more constructive in its spirit and intent.

An array of classical results, familiar from standard references in complex analysis and relevant only to simply connected domains, will be generalized, using the prime function, to domains of any other finite connectivity. The familiar results will be just a special case that corresponds to setting the prime function to be $\omega(z, a) = z - a$. For higher connected domains, the relevant prime function is no longer elementary—as must be expected given the additional topological complexity—yet, in terms of it, many formulas for mathematical objects and functions that are useful in applications are identical to, or closely similar to, their counterparts in the simply connected case. This is appealing theoretically and aesthetically. And it points to the fundamental importance of the prime function.

The overarching spirit of this monograph is to give a geometrical view of the function theory. Thinking of a function geometrically, as a conformal mapping between planes, is a fundamental idea for us. And it will steer our development.

The monograph is divided in two: Part I and Part II.

Part I gives the mathematical foundations and lays out the theoretical framework. Part I can be read independently of Part II.

In Part I the exigencies of mathematical rigor have been forgone in order to make the main ideas as accessible as possible to applied scientists. At the same time, the essential mathematical arguments are laid out quite fully. Each chapter in Part I ends with a set of exercises, which include other information, ideas, methodologies, or connections with other approaches which, although relevant and of interest, are deemed by the author to be nevertheless tangential to the key tenets of the framework. Earlier review articles advocating the same general approach expounded in more detail in this monograph have been given by the author in [19, 13, 10].

Chapter 1 deals with simply connected domains but casts the material in a new light. A central role is given to the prime function, something not typically seen in complex analysis texts. With the prime function as the building block, this chapter systematically constructs an increasingly sophisticated suite of analytic functions and interprets them geometrically as conformal mappings. The developmental template set out in Chapter 1 is then emulated in Chapters 5 and 6. This is meant to underline how this approach to the multiply connected situation is a natural development of the simply connected case.

Chapters 2–4 lay out the theoretical underpinnings for our approach. They define what the prime function is, where it comes from, and what its properties are. With a focus on problems in multiply connected domains, we use the canonical class of multiply connected circular domains as a basis for the mathematical framework; it is known that every planar multiply connected domain is conformally equivalent to a domain of this type. We then establish the existence of, and properties of, the prime function in this canonical class by assuming the existence of the first-type Green's function in these same domains, something applied scientists will be comfortable with since it is familiar to them and proofs of its existence are well known and readily available. The analytic extensions of the so-called harmonic measures associated with these multiply connected domains are then used to nullify the periods of the (analytic extension of the) first-type Green's function around the holes and to build an appropriate *modified* Green's function. Then, much as we just did above for the case without any holes, a double limit involving the analytic extension of the latter function defines the prime function and allows us to establish its function theoretic properties.

Having demonstrated the existence of a prime function associated with any multiply connected circular domain, Chapters 5 and 6 use it as a basic building block for the mathematical framework. A diverse array of other useful functions are then constructed using nothing other than the existence of the prime function and its function theoretic properties.

We can do all this without even telling the reader how to actually calculate values of the prime function. This is left until the very end of Part I.

If the reader is happy to believe that the prime function exists and has the properties claimed, then it is feasible to omit Chapters 2–4 on first reading and to jump from Chapter 1 (the simply connected case) straight to Chapter 5 (the doubly connected case) and then on to Chapter 6 (triply, and all higher, connected cases). When read successively, these chapters aim to exhibit the naturalness of the approach. Further comments on this are made in a "Note to Instructors" to follow. To preserve the expository integrity of the book, however, it seemed necessary to interrupt the natural flow of Chapters 1, 5, and 6 with the foundational Chapters 2–4, which explain the prime function and its properties.

Chapters 7–13 showcase more sophisticated things that can be done with the prime function, with the constructs of earlier chapters used as basic tools. We argue that the basic conformal slit mappings introduced in Chapters 1–6 should be viewed as the simplest instances of Schwarz–Christoffel mappings and polycircular arc mappings, and, in Chapters 7 and 10, we show how to use these basic slit mappings to construct the conformal mappings from circular domains to *any* multiply connected polygonal or polycircular arc domain. In this way, the function theory builds up naturally, motivated by our geometrical demands on it. Already this represents a significant departure from standard texts, where such matters are usually not discussed for the multiply connected case, even though multiply connected polygons and polycircular arc domains are ubiquitous in applications.

To tackle polycircular arc domains we need some elements of automorphic function theory, and these are given in Chapter 9. That chapter follows the gentle introduction in Chapter 8 to loxodromic functions, a special subclass of automorphic functions relevant to the concentric annulus (the doubly connected case).

In Chapter 11 we further broaden the class of domains amenable to our framework by studying quadrature domains. Again, these are not usually encountered in standard texts. Following in the general spirit of this monograph, we give a geometrical view of them. Just as we showed that the conformal mappings to general multiply connected polygons and polycircular arc domains can be built up functionally from simple slit and circular slit mappings, we show that quadrature domains can likewise be built up by a geometrical process we refer to as the "merging" of simple circular discs. This gives us some valuable geometrical intuition about this class of domains. The notion of the Bergman kernel of a multiply connected planar domain comes in useful here and is briefly discussed.

Armed with this even larger toolbox of techniques, and a now very broad class of multiply connected domains describable with our methods, the number of problems amenable to the framework increases dramatically. Chapter 12 introduces the Cauchy transform of a domain since it proves to be a useful concept in several free boundary problems considered in Part II. In Chapter 13 we consider generalized Poisson kernels for multiply connected circular domains and show how to solve standard boundary value problems in multiply connected domains, all with the prime function playing a central role. All these concepts arise in various physical problems involving multiply connected domains in Part II.

All matters concerning the numerical evaluation of the prime function are relegated to the final chapter of Part I, Chapter 14, since, within our special function approach, this is a secondary concern, albeit one of great importance for practical application of the theory. Unfortunately, owing to the fact discussed earlier that the prime function and its significance have not to date been fully appreciated by the scientific community, the special function packages in popular platforms such as MATLAB, Mathematica, and Maple do not currently contain any routines to evaluate it (except perhaps in disguise in the doubly connected case where connections with more recognized special functions can be made—see Chapter 14). However, it is important to point out that, with a view to popularizing the use of the prime function, the author and coworkers have made available fast and accurate numerical routines in MATLAB to evaluate them; these routines are freely available at the website `github.com/ACCA-Imperial`.[2] Readers are encouraged not only to use these codes to evaluate the prime function in their work but also to contribute to the ongoing development of these resources. Chapter 14 contains more details.

Part II is a compendium of essays on a curated selection of physical science applications where aspects of the general theory of Part I are showcased in action. Each chapter can be read independently since no prior expertise in the various applications is assumed. The chapters in Part II focus on how the general mathematical theory of Part I is being used rather than on what new can be learned from the analysis in each application area. For that, readers should refer to the original references.

To highlight the essentially constructive nature of the framework Part II contains many figures, contour plots, and graphs, all computed using the methods described in Chapter 14 of Part I. The author's aim in Part II is to inspire readers to deploy the methods of Part I to solve problems in their own areas of interest. The survey of particular applications in Part II, which is necessarily skewed towards the author's own research interests and limited by his lack of knowledge of other areas, is just the tip of the iceberg of all the possible applications of the general theory.

An epilogue closes the monograph with a discussion of connections with other approaches and other areas of mathematics.

[2]Should this link become inactive readers should refer to the author's institutional website for information.

Note to instructors

This book is really a treatise on how to extend calculus beyond the Riemann sphere, where we know it best, to spheres with "handles" (think "holes"). For applied scientists, this amounts to taking familiar results for the single-object case and extending it to any number of objects.

The book has been written as a research monograph, but with more than a nod to pedagogy in the many exercises at the end of each chapter of Part I, and in the series of essays on applications making up Part II. As such, the book provides ample fodder for advancing the subject matter in existing courses, for extended coursework, and for student projects, as well as for research projects at the undergraduate level, the graduate level, and beyond.

A few comments on the author's suggested strategy for integrating this material into existing courses, or using it as an educational resource, may be helpful to instructors. Incidentally, the following remarks are a good entry point for any reader of the book.

Chapter 1 is indispensable and offers a novel take on basic complex analysis and geometric function theory, with the prime function taking center stage. It offers students new perspectives on ideas they already encounter in university courses even before multiple objects/entities/holes are introduced. This chapter alone provides a useful counterpoint to the more traditional treatments.

A jump to Chapter 5 is then possible, at least on first reading, and much of it should be understandable. It treats the two-body problem, or problems in doubly connected domains. As emphasized in the text, the concentric annulus

$$\rho < |z| < 1 \qquad (0 < \rho < 1)$$

is all that is needed for the two-body problem, and it provides the simplest example of the general paradigm the monograph aims to convey. Pedagogically it can be helpful to direct students to this chapter immediately after Chapter 1, without necessarily burdening them, yet, with the foundational Chapters 2–4. Instead, students can be shown that it is possible to *define* the prime function for the concentric annulus using the infinite-product expansion (14.59) and (14.60) from which all the identities (5.4), (5.5), (5.6), and (5.7) satisfied by this prime function can easily be established by simple exercises using this infinite-product definition. Exercise 14.3 elucidates this possible approach. Then, a good portion of the results of Chapter 5 can subsequently be established too, just using these identities. And the simple limit $\rho \to 0$ retrieves everything seen in Chapter 1.

The advantage of this approach is that the student gets down to calculations right away and has at hand a way of evaluating the prime function for this case, simply by truncating the infinite-product expansion (14.59) and (14.60). Students can write themselves a simple numerical algorithm to do this and then calculate and plot all the many other functions discussed in Chapter 5, and in other chapters, the restriction being that only the doubly connected case is within reach.

Of course, the price to pay is a loss of the global picture this monograph seeks to give in the foundational Chapters 2–4, where the prime function is defined more generally for higher connected domains. But they can be read later.

Acknowledgments

I would like to thank Stefan Llewellyn Smith, Bernard DeConinck, and Thomas Trogdon for suggesting me as a lecturer in the NSF-CBMS workshop series. The week at UC Irvine in June 2018 was a delight. It was a pleasure to share these ideas with so many fantastic participants, mathematicians and engineers, of all genders and career stages, and from all over the world. Thanks to all for making it such a wonderful experience, and to the NSF and the CBMS for the support. I also acknowledge longstanding support from the Engineering and Physical Sciences Research Council, the Leverhulme Trust, and the Royal Society in the UK. I have benefited greatly from their various grant and fellowship schemes over the years, and this monograph stands, I hope, as a testament to what can be achieved under their auspices.

This book is an outgrowth of the 10 lectures I gave at Irvine, but it encapsulates almost two decades of mathematical thought and endeavor. It contains much more material than I was able to cover in those lectures but is identical in spirit. The topic here—solving problems in multiply connected domains—has been a constant source of fascination for me since my PhD work at Caltech in the mid- to late 1990s. Why did these problems have the reputation for being so difficult? Why did mathematical results that worked for simply connected, and sometimes doubly connected, domains seem to simply stop working if you added another hole? I felt that mathematics was failing me. I spotted early in my research career that, historically, this has been a largely neglected topic, one for which the best way forward was far from clear, and for which there was an urgent need for something to be done, especially for use in the applied sciences. There were some techniques out there, but, speaking personally, I found them unsatisfactory, theoretically unappealing, and unnatural. As I started out on this scientific journey, putting all the pieces of this puzzle together, I could not have foreseen that such a beautiful and coherent framework lay waiting to be uncovered. I have tried to tell the story here, as I see it now, in one place, and with emphasis on how naturally all the ideas fit together.

The scientific journey would have been a lonely one, perhaps a failed one, had it not been for the constant support of my colleagues and students. I cannot explain the satisfaction, even joy, I feel now when I see colleagues tackling their problems using the prime function as the essential tool. Thanks to all my valued colleagues who appreciated the potential of the ideas and embraced them early on, and with such verve. I will not include a list of names, lest I miss somebody: you know who you are! Thanks also go to the colleagues who challenged me constructively, offered differing viewpoints, and forced me to think more deeply. This book is all the stronger for it.

I offer an explicit thank-you to my former doctoral student Jonathan Marshall, who has been a stalwart companion in these endeavors from the early 2000s; I cannot even estimate how much you have contributed to the existence of this book. Sincere thanks also go to Everett Kropf and Rhodri Nelson for all their contributions, their enthusiasm, and especially for all their work on the GitHub codes.

This book took almost three years to write. It was as arduous as it was fulfilling. A final thanks, a special thanks, goes to Laurence, who was there with unerring love, encouragement, and support just when I needed it, and every step of the way.

Part I

Mathematical framework

Dr Sloper's opposition was the unknown quantity in the problem he had to work out. . .; but in mathematics there are many short cuts. . . .

Washington Square
Henry James

Chapter 1

Function theory and the prime function

1.1 ▪ The prime function

Every reader will have encountered the simple function

$$\omega(z, a) \equiv (z - a), \tag{1.1}$$

where $z = x + \mathrm{i}y$ is a complex variable with real and imaginary parts that can be viewed geometrically as coordinates in a two-dimensional (x, y) plane; this complex plane will be denoted by $\mathbb{C}$ in the usual way. For now, we will think of a as a complex-valued parameter, that is, a fixed point in this complex plane. Later, however, we will find it convenient to think of a not as a parameter but also as another complex variable moving around in the complex plane. This is the reason for writing $\omega(z, a)$ as a function of the two variables z and a.

The function (1.1) is an analytic function of both variables z and a with a simple zero when both arguments coincide at $z = a$ and with a simple pole at $z = \infty$. We will call it a *prime function*. More precisely, (1.1) is the prime function on the Riemann sphere or, as we prefer to say, the prime function of the unit disc. We will see in Chapter 3, by a natural "doubling" procedure, how to associate the unit disc to the extended complex plane or Riemann sphere.

There are, as will be shown, many prime functions, of which (1.1) is just the simplest instance.

Why call $\omega(z, a)$ a *prime* function? The fundamental theorem of algebra states that any polynomial of degree N, where $N \geq 1$ is an integer, with complex-valued coefficients has precisely N roots in the complex plane $\mathbb{C}$. The roots are counted by multiplicity. From the point of view of function theory this means that we can exchange the familiar sum representation

$$z^N + c_{N-1}z^{N-1} + c_{N-2}z^{N-2} + \cdots + c_1 z + c_0 \tag{1.2}$$

of a polynomial of degree N, which involves the N coefficients $\{c_j | j = 0, 1, \ldots, N-1\}$, for a product representation of the form

$$(z - a_1) \times (z - a_2) \times \cdots \times (z - a_N), \tag{1.3}$$

or, now that we have introduced the prime function (1.1), this can be written

$$\omega(z, a_1) \times \omega(z, a_2) \times \cdots \times \omega(z, a_N). \tag{1.4}$$

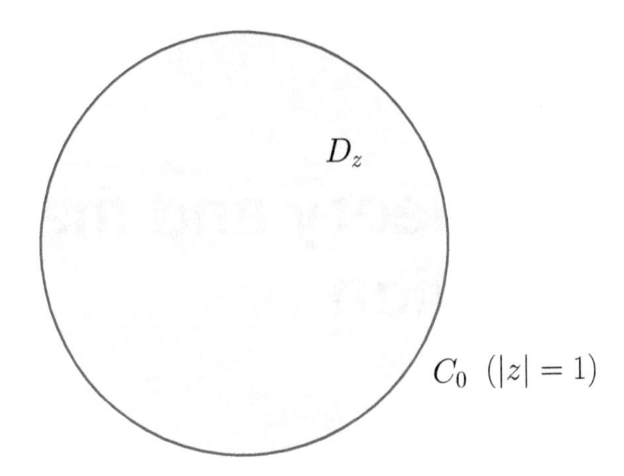

Figure 1.1. *The unit disc D_z with boundary given by the unit circle $|z| = 1$ denoted by C_0.*

The fundamental theorem of algebra can now be stated as follows: any polynomial of degree N with complex coefficients can be factorized into a product of N prime functions, each having a simple zero at one of the N roots of the polynomial in the complex plane. This statement has the spirit of the fundamental theorem of arithmetic—where the word "prime" is more familiar—that any integer can be uniquely factorized into a product of prime numbers.

While (1.2) is written in terms of the set of coefficients $\{c_j | j = 0, 1, \ldots, c_{N-1}\}$, the alternative representation (1.4) involves $\{a_j | j = 1, \ldots, N\}$, which is the set of N roots of the polynomial (1.2). Displaying the roots of a polynomial in a complex plane provides a more *geometrical* picture of the polynomial function than the sum representation (1.2). In the spirit of geometric function theory [93] this monograph will emphasize throughout the geometrical aspects of the functions being constructed, often by viewing them as conformal mappings between two complex planes.[3]

An important concept for us will be the *Schwarz conjugate* $\overline{h}(z)$ of a given analytic function $h(z)$. It is defined by

$$\overline{h}(z) \equiv \overline{h(\overline{z})}. \tag{1.5}$$

If $h(z) = e^{iz}$, say, then its Schwarz conjugate is e^{-iz}, which, we note, is a function different from the original $h(z)$. For a function of two variables, such as the prime function (1.1), the Schwarz conjugate function is defined by

$$\overline{\omega}(z, a) \equiv \overline{\omega(\overline{z}, \overline{a})}. \tag{1.6}$$

Thus the idea is to evaluate the function at the complex conjugate of *all* the variables on which that function depends and then take a complex conjugate of the result.

Let D_z denote the unit disc $|z| < 1$, as shown in Figure 1.1, and let its boundary, the unit circle $|z| = 1$, be called C_0.

The prime function (1.1), which we will associate with D_z, is characterized by several features: its simple zero at $z = a$, its simple pole at $z = \infty$, and the fact that it and its

[3]Nehari [104] is an excellent treatise on conformal mapping that we recommend as a general reference. We will, however, build up from scratch our own compendium of important mappings in a novel fashion using the prime function as a basic building block. Many of our formulas are not to be found in [104], or indeed in any other standard references.

Schwarz conjugate function satisfy the following two functional identities:

$$\omega(a,z) = -\omega(z,a), \tag{1.7}$$

$$\overline{\omega}(1/z,1/a) = -\frac{1}{za}\omega(z,a). \tag{1.8}$$

These identities are easily verified using the definition (1.1).

As it happens, the Schwarz conjugate of the unit disc D_z satisfies the additional functional identity

$$\overline{\omega}(z,a) = \omega(z,a). \tag{1.9}$$

This is listed separately because not all the prime functions to be introduced subsequently will satisfy this one. Indeed, we will come to understand (1.9) as a consequence of the unit disc D_z being reflectionally symmetric about the real z axis. Later, when we introduce other prime functions associated with more general classes of domains, all of them will satisfy (1.7) and (1.8), but only those associated with domains that are reflectionally symmetric about the real axis will also satisfy (1.9).

The aim now is to develop a view of function theory in the disc using the prime function (1.1) as a basic building block. As natural as this sounds, this is not something traditionally done in complex analysis texts, but it provides a valuable perspective. Since our emphasis is on the *geometry* of the analytic functions we build, we will often think of a function $f(z)$ as providing a *conformal mapping* between regions in two complex planes, i.e., between a complex z plane and a complex w plane:

$$z \mapsto w = f(z). \tag{1.10}$$

This approach has the added bonus that many of the conformal mappings we build along the way will have direct relevance in a wide range of important physical applications considered in Part II.

1.2 ▪ A ratio of prime functions

The prime function $\omega(z,a)$ defined in (1.1) has a simple zero at $z = a$ and a simple pole at $z = \infty$. Since, by changing the parameter a, the simple zero of this function is both visible and movable, it is natural to arrange for the pole to share this feature. This leads us to consider the ratio of prime functions

$$R(z;a,b) = \frac{\omega(z,a)}{\omega(z,b)}, \tag{1.11}$$

where a and b are distinct arbitrary points. This is an analytic function of z with a simple zero at a and a simple pole at b instead of at infinity. We can now move the zero by altering a and move the pole by altering b.

It is helpful to find an expression for the complex conjugate of the ratio $R(z;a,b)$ on C_0 where $z = 1/\overline{z}$. The reader is encouraged to study carefully the simple mathematical steps to follow since similar manipulations will be used extensively throughout the book.

For $z \in C_0$, where $z = 1/\overline{z}$, we can write

$$R(z;a,b) = \frac{\omega(1/\overline{z},a)}{\omega(1/\overline{z},b)}. \tag{1.12}$$

On taking a complex conjugate we find

$$\overline{R(z;a,b)} = \frac{\overline{\omega}(1/z,\overline{a})}{\overline{\omega}(1/z,\overline{b})}. \tag{1.13}$$

Assuming that $a \neq 0$ and $b \neq 0$, use of the functional relation (1.8) then produces

$$\overline{R(z;a,b)} = \frac{\overline{a}\,\omega(z,1/\overline{a})}{\overline{b}\,\omega(z,1/\overline{b})}. \tag{1.14}$$

We now recognize the function we started with on the right-hand side, albeit with modified parameters. The conclusion is that, for z on C_0, and for any distinct nonzero a and b,

$$\overline{R(z;a,b)} = \frac{\overline{a}}{\overline{b}} R(z;1/\overline{a},1/\overline{b}), \qquad z \in C_0. \tag{1.15}$$

This will prove to be a useful result.

When $\omega(z,a) = (z-a)$, as we assume in this chapter, this ratio of prime functions $R(z;a,b)$ is important theoretically, and, when interpreted as a conformal mapping, it has a special name: a *Möbius map*. Other common designations for the same object are *linear fractional maps* or *bilinear functions*.[4]

1.3 ▪ Green's function in the unit disc

Consider the function defined, for any nonzero $a \in D_z$, by

$$\mathcal{G}_0(z,a) = \frac{1}{2\pi \mathrm{i}} \log\left(\frac{1}{|a|} R(z;a,1/\overline{a})\right) = \frac{1}{2\pi \mathrm{i}} \log\left(\frac{\omega(z,a)}{|a|\omega(z,1/\overline{a})}\right). \tag{1.16}$$

It follows from (1.15), with $b = 1/\overline{a}$, that, for $z \in C_0$,

$$\overline{R(z;a,1/\overline{a})} = |a|^2 R(z;1/\overline{a},a) = \frac{|a|^2}{R(z;a,1/\overline{a})}, \tag{1.17}$$

which, after multiplying by $R(z;a,1/\overline{a})$ and taking the square root, means that

$$|R(z;a,1/\overline{a})| = |a| \tag{1.18}$$

or, equivalently,

$$\left|\frac{1}{|a|} R(z;a,1/\overline{a})\right| = \left|\frac{\omega(z,a)}{|a|\omega(z,1/\overline{a})}\right| = 1, \qquad z \in C_0. \tag{1.19}$$

If we define

$$G_0(z,a) = \mathrm{Im}[\mathcal{G}_0(z,a)], \tag{1.20}$$

then

$$G_0(z,a) = -\frac{1}{2\pi} \log\left|\frac{\omega(z,a)}{|a|\omega(z,1/\overline{a})}\right|. \tag{1.21}$$

It follows from (1.19) that, for $z \in C_0$,

$$G_0(z,a) = 0, \qquad z \in C_0. \tag{1.22}$$

[4]Needham [103] gives a fascinating treatment of the geometrical properties of Möbius maps.

It is also clear that $\mathcal{G}_0(z,a)$ is analytic everywhere in D_z except for a single logarithmic singularity at $z = a$ of the form

$$\mathcal{G}_0(z,a) = \frac{1}{2\pi \mathrm{i}} \log(z-a) + \text{a locally analytic function.} \tag{1.23}$$

The only other singularity of $\mathcal{G}_0(z,a)$ is at $z = 1/\overline{a}$, which is outside the unit disc D_z.

The real function $G_0(z,a)$ satisfies the boundary value problem

$$\nabla^2 G_0(z,a) = -\delta(z-a), \qquad z \in D_z, \tag{1.24}$$

for some fixed point a in D_z with boundary condition

$$G_0 = 0 \qquad z \in C_0. \tag{1.25}$$

$G_0(z,a)$ is the *Green's function* of the unit disc. The function $\mathcal{G}_0(z,a)$ is the *analytic extension* of $G_0(z,a)$ as a function of z in D_z.

From the residue theorem it is easy to deduce that

$$\int_{C_0} d\mathcal{G}_0 = 1, \tag{1.26}$$

where the integral around a closed contour of this kind is always taken to be one that traverses the contour in an anticlockwise sense.

Remark: The assumption that $a \neq 0$ is not a limitation, just a case that requires special treatment. Indeed, the Green's function for the unit disc with singularity at the origin is easily verified to be simply

$$G_0(z,0) = -\frac{1}{2\pi} \log|z|. \tag{1.27}$$

This can be seen by taking a limit of (1.21) as $|a| \to 0$.

Let us give a geometrical interpretation of the analytic function $\mathcal{G}_0(z,a)$ by defining a new complex variable w that is related to z via

$$w = \mathcal{G}_0(z,a). \tag{1.28}$$

This functional identification between two complex variables allows us to think of the function $\mathcal{G}_0(z,a)$ as a *conformal mapping* $z \mapsto w = f(z)$ [104, 61]. For definiteness, we pick the parameter a to be real with $-1 < a < 0$. This function has two logarithmic branch points at a and $1/a$, and it is useful to choose a branch of the multivalued function by introducing a branch cut joining a to $1/a$ along the negative real axis. The function (1.28) then transplants the unit disc D_z in the complex z plane to a semistrip of unit width in the complex w plane with

$$0 < \mathrm{Im}[w] < \infty, \tag{1.29}$$

as shown in Figure 1.2. The two vertical edges of the semistrip in the w plane correspond to the images under the mapping (1.28) of the two sides of the portion of the branch cut sitting inside the unit disc. Since we have chosen the branch cut to be a straight line along the negative real axis, the two vertical edges of the semistrip are also straight lines.

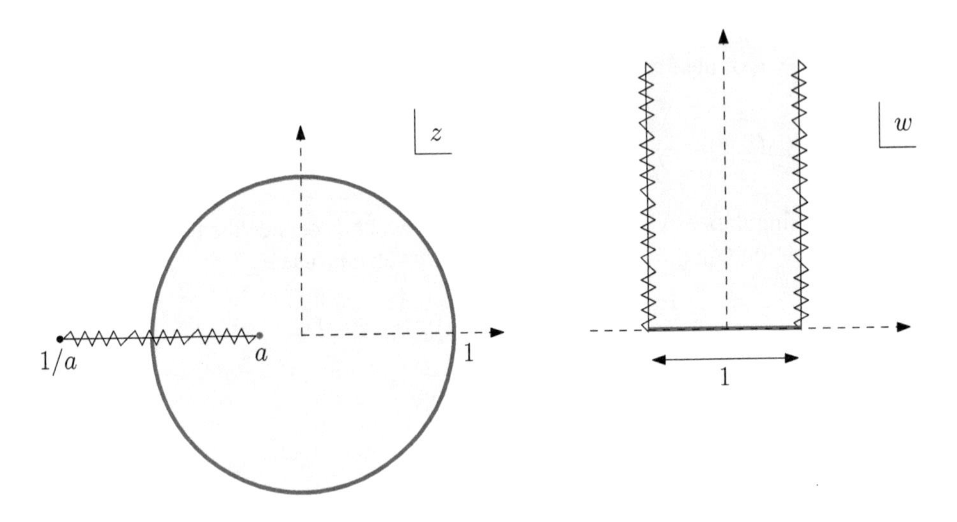

Figure 1.2. *Image in the w plane of the cut unit disc D_z under the action of $\mathcal{G}_0(z,a)$ (1.28) when viewed as a conformal map with $-1 < a < 0$. The two sides of a branch cut joining a to $1/\overline{a}$ inside D_z correspond to the two sides of a period window.*

1.4 ▪ Disc automorphisms

Another conformal mapping can be formed by taking the exponential of $\mathcal{G}_0(z,a)$:

$$w = e^{2\pi \mathrm{i} \mathcal{G}_0(z,a)}, \tag{1.30}$$

where a is again taken to be a general nonzero point in D_z. In terms of the prime function this can be written as

$$w = \frac{1}{|a|} R(z; a, 1/\overline{a}) = \frac{1}{|a|} \frac{\omega(z,a)}{\omega(z,1/\overline{a})}. \tag{1.31}$$

We notice that this is a Möbius map; indeed, it is an important one. Since the prime function has a simple zero at $z = a$, this point is transplanted to $w = 0$. To find the image of the unit circle C_0 we use the fact that the imaginary part of $\mathcal{G}_0(z,a)$ vanishes on C_0:

$$\overline{w} = e^{-2\pi \mathrm{i} \overline{\mathcal{G}_0(z,a)}} = e^{-2\pi \mathrm{i} \mathcal{G}_0(z,a)} = \frac{1}{w}, \qquad z \in C_0. \tag{1.32}$$

The image of C_0 in the z plane therefore certainly lies on the unit circle $|w| = 1$. Does it cover the *whole* of this unit circle? To establish this we must examine the change in the argument of w as z traverses C_0. This is given by

$$[\arg[w]]_{C_0} = \frac{1}{\mathrm{i}} \int_{C_0} d \log w = 2\pi \int_{C_0} d\mathcal{G}_0(z,a) = 2\pi, \tag{1.33}$$

where the square bracket notation $[.]_{C_0}$ denotes the change in the enclosed quantity as z traverses C_0 once in an anticlockwise direction. We have used (1.26) in the final equality. The conclusion is that the image of C_0 is the entire unit circle $|w| = 1$. Because of this property the conformal maps (1.31) are called *automorphisms of the unit disc* because they transplant the unit disc in the z plane to the unit disc in the w plane. Provided $a \neq 0$, it is, nevertheless, a nontrivial mapping with all the points inside and on the boundary of the

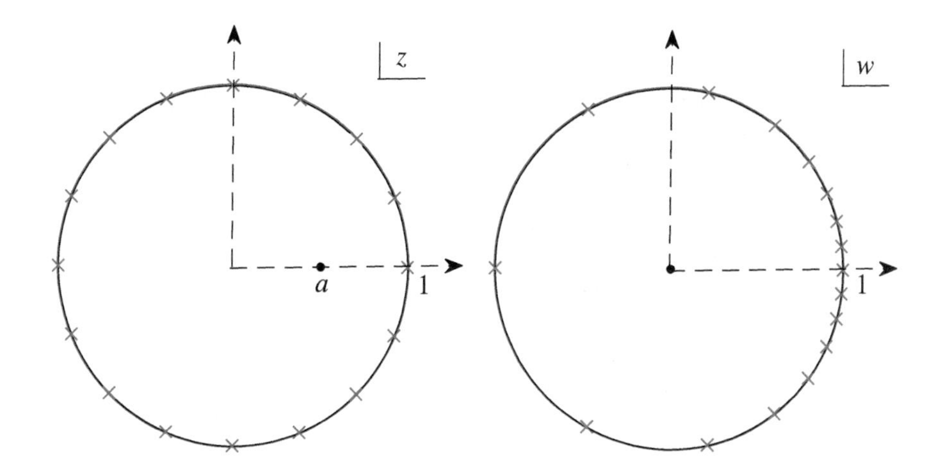

Figure 1.3. *The effect of a disc automorphism (1.31) is to transplant the unit z disc to the unit w disc while shifting $z = a$ to $w = 0$. A uniform distribution of points on the boundary (blue crosses) is displaced to a nonuniform distribution (red crosses).*

disc being rearranged and with $z = a$ being transplanted to $w = 0$, as indicated in Figure 1.3.

The argument principle [61] can be used to show that such an automorphism is a univalent, or one-to-one, mapping of the unit disc in the z plane to the unit disc in the w plane. This principle says that if $f(z)$ is a meromorphic function inside a domain D_z, then the change $\Delta\arg[f]_{D_z}$ in the argument of f as the boundary ∂D_z is traversed in a direction keeping the domain D_z to the left is given by

$$\frac{1}{2\pi}\Delta\arg[f]_{D_z} = N_{\text{zeros}}(f; D_z) - N_{\text{poles}}(f; D_z), \tag{1.34}$$

where $N_{\text{zeros}}(f; D_z)$ is the number of zeros of f and $N_{\text{poles}}(f; D_z)$ is the number of poles of f inside D_z, both counted by multiplicity. To show that the mapping (1.31) to the unit w disc is one-to-one we make the choice

$$f = \frac{1}{|a|}R(z; a, 1/\overline{a}) - \alpha = \frac{1}{|a|}\frac{\omega(z, a)}{\omega(z, 1/\overline{a})} - \alpha, \tag{1.35}$$

where $\alpha \in \mathbb{C}$ is a parameter. Since we know that the mapping (1.31) completely traverses $|w| = 1$, then the left-hand side of (1.34) equals 1 for any point α with modulus less than unity. But f has no poles in D_z, so we conclude from (1.34) that the function f has just one zero in D_z when $|\alpha| < 1$. But a zero of f means that the mapping (1.31) attains the value α. Hence we have established that the mapping (1.31) attains the value α precisely once if α is inside the unit w disc.

The argument principle can be used in a similar way to establish the univalency of all the conformal mappings to be introduced in the remainder of this chapter; see Exercise 1.11.

The most general automorphism of the disc is, in fact,

$$w = e^{\mathrm{i}\phi}e^{2\pi\mathrm{i}\mathcal{G}_0(z,a)} = \frac{e^{\mathrm{i}\phi}}{|a|}\frac{\omega(z, a)}{\omega(z, 1/\overline{a})} \tag{1.36}$$

for some real ϕ. The multiplicative factor $e^{i\phi}$ merely rotates the unit disc through the angle ϕ.

This disc automorphism (1.36) depends on *three* real parameters: $\mathrm{Re}[a]$, $\mathrm{Im}[a]$, and ϕ. These are precisely the three degrees of freedom in the statement of the Riemann mapping theorem for simply connected domains [61], which guarantees the existence of a conformal mapping to any given planar domain from the unit disc. This conformal mapping is unique up to specification of these three degrees of freedom in the disc automorphism corresponding to mapping the unit disc back to itself.

Remark: We will come to recognize (1.31) as the formula for a *bounded circular slit map*. This designation seems out of place here for the simply connected case because there are no slits! But slits generated by the same functional form of mapping (1.31) in terms of the prime function will appear in the doubly and higher connected cases, as will be seen in §5.6 and §6.6. The thing to remember is that in the simply connected case the "bounded circular slit map" is just a disc automorphism.

1.5 ▪ Unbounded circular slit map

Other interesting conformal maps can be built from $\mathcal{G}_0(z,a)$. Let

$$w = e^{2\pi i(\mathcal{G}_0(z,a) - \mathcal{G}_0(z,b))}, \tag{1.37}$$

where a and b are two distinct nonzero points in D_z. From (1.16) this can be written in terms of the prime function as

$$w = \frac{|b|}{|a|} \frac{\omega(z,a)\omega(z,1/\overline{b})}{\omega(z,b)\omega(z,1/\overline{a})}. \tag{1.38}$$

This is no longer a Möbius map. Under this map the point $z = a$ is transplanted to $w = 0$ and $z = b$ is taken to $w = \infty$, so the image domain is unbounded. Since $\mathcal{G}_0(z,a)$ and $\mathcal{G}_0(z,b)$ are real on C_0, C_0 is transplanted to $|w| = 1$. Again we ask: does this image make a complete traversal of this unit circle? The answer is no. The change in the argument of w as z traverses C_0 is

$$[\arg[w]]_{C_0} = \frac{1}{i}\int_{C_0} d\log w = 2\pi \int_{C_0} d\mathcal{G}_0(z,a) - d\mathcal{G}_0(z,b) = 0. \tag{1.39}$$

Hence the image of C_0 is a finite-length circular arc slit lying on the unit circle in the w plane, as shown in Figure 1.4. This is an example of an *unbounded circular slit map*.

There is a special case of this class of maps that turns out to be especially useful in applications. It corresponds to setting

$$a = ir, \qquad b = \overline{a}, \qquad 0 < r < 1, \tag{1.40}$$

and premultiplying by -1 so that (1.38) becomes

$$w = -\frac{\omega(z,a)\omega(z,1/a)}{\omega(z,\overline{a})\omega(z,1/\overline{a})}. \tag{1.41}$$

For this choice of parameters it is easy to verify that points on the real diameter $-1 < z < 1$ have the property that the corresponding images in the w plane satisfy

$$\overline{w} = \frac{1}{w}, \tag{1.42}$$

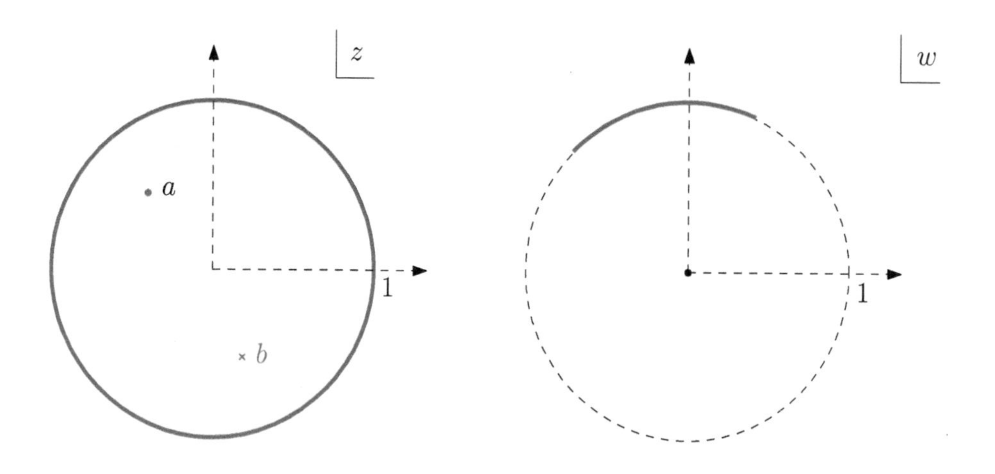

Figure 1.4. *Unbounded circular slit map: the image of the unit disc D_z under the map (1.38) is the entire w plane exterior to a circular arc slit on $|w| = 1$. A point $z = a$ maps to $w = 0$, and $z = b$ maps to infinity.*

where we have used the additional identity (1.9) satisfied by the prime function. This means that the real diameter of the unit disc *also* maps to the unit circle $|w| = 1$. Indeed, it maps to that arc of the unit circle that is outside the image of C_0, as indicated in Figure 1.5. It can be verified that the boundary of the upper *half* disc D_z is now transplanted to the interior of the unit disc in the w plane (Exercise 1.13).

Another interesting mapping, to a complex χ plane say, is formed by taking a logarithm of this unbounded circular slit map (1.41) to the w plane, producing the following sequence of mappings from the z plane to the w plane and on to the χ plane:

$$\chi = \frac{1}{2\pi \mathrm{i}} \log w = \frac{1}{2\pi \mathrm{i}} \log \left[-\frac{\omega(z,a)\omega(z,1/a)}{\omega(z,\overline{a})\omega(z,1/\overline{a})} \right], \qquad a = \mathrm{i}r. \tag{1.43}$$

As a function of w, this function has logarithmic branch points at $w = 0, \infty$. We can, for example, choose a branch cut along the negative real axis in the w plane, as illustrated in Figure 1.5. Then, on composing these two mappings, the image of the unit disc D_z outside this cut maps to an infinite strip parallel to the imaginary axis with the unit circle mapping to a finite-length slit along the real χ axis. The upper half unit disc D_z outside this cut maps to the upper half strip in the w plane; see Figure 1.5.

1.6 ▪ Parallel slit map

A useful class of conformal maps can be formed by taking parametric derivatives of $\mathcal{G}_0(z,a)$. Let

$$a = a_x + \mathrm{i}a_y, \tag{1.44}$$

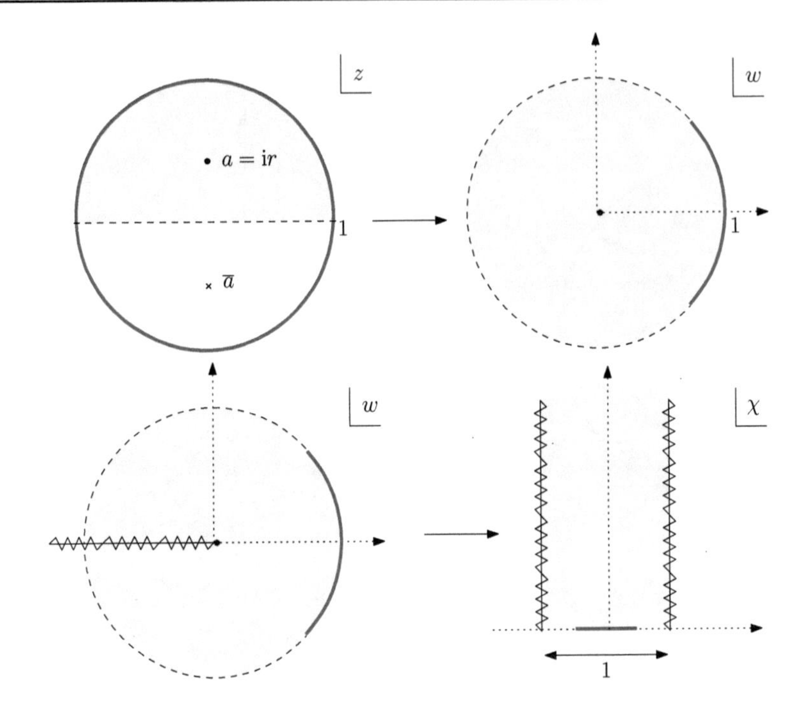

Figure 1.5. *Conformal map (1.41) from the upper half unit disc in the z plane to the unit disc in a w plane with $|z| = 1$ being transplanted to a circular slit on $|w| = 1$ with the real diameter $-1 < z < +1$ mapping to the remainder of $|w| = 1$. The lower two schematics show the logarithm of this map: the image of the upper half unit disc is a semistrip $0 < \mathrm{Im}[\chi] < \infty$ with a slit along its lower edge on the real χ axis.*

where a is any nonzero point inside D_z, and consider the two maps

$$w_1 = -2\pi\mathrm{i}\frac{\partial \mathcal{G}_0(z,a)}{\partial(\mathrm{i}a_y)}, \qquad w_2 = -2\pi\mathrm{i}\frac{\partial \mathcal{G}_0(z,a)}{\partial a_x}. \tag{1.45}$$

Since

$$\frac{\partial}{\partial a_x} = \frac{\partial}{\partial a} + \frac{\partial}{\partial \overline{a}}, \qquad \frac{\partial}{\partial(\mathrm{i}a_y)} = \frac{\partial}{\partial a} - \frac{\partial}{\partial \overline{a}}, \tag{1.46}$$

it is clear that both parametric derivatives of $-2\pi\mathrm{i}\mathcal{G}_0(z,a)$ produce a simple pole, with unit residue at the point $z = a$ when viewed as a function of z.

Since $-2\pi\mathrm{i}\mathcal{G}_0(z,a)$ has constant real part (in fact, equal to zero) on C_0, its differentiation with respect to the purely imaginary parameter $\mathrm{i}a_y$ means that the imaginary part of w_1 will be constant (in fact, equal to zero) on C_0 ; thus D_z is transplanted to the entire w_1 plane exterior to a slit parallel to the real axis (in fact, on the real axis).

On the other hand, differentiation of $-2\pi\mathrm{i}\mathcal{G}_0(z,a)$ with respect to the real parameter a_x means that the real part of w_2 will be constant (equal to zero) on C_0; thus the interior of the unit z disc to the entire w_2 plane is exterior to a slit parallel to the imaginary axis (on the imaginary axis).

It follows by direct calculation that

$$w_1 = -\frac{1}{a}\mathcal{K}(z,a) + \frac{1}{\overline{a}}\mathcal{K}(z,1/\overline{a}) + \frac{1}{2a} - \frac{1}{2\overline{a}}, \tag{1.47}$$

$$w_2 = -\frac{1}{a}\mathcal{K}(z,a) - \frac{1}{\overline{a}}\mathcal{K}(z,1/\overline{a}) + \frac{1}{2a} + \frac{1}{2\overline{a}}, \tag{1.48}$$

where we introduce

$$\mathcal{K}(z,a) \equiv a\frac{\partial \log \omega(z,a)}{\partial a}, \tag{1.49}$$

a function that will turn out to be of great utility.

It is clear from (1.47)–(1.48) why we insisted earlier that $a \neq 0$. Actually, formulas (1.48) still hold for $a = 0$ except that they must be regularized. This matter is explored in Exercise 1.17.

The pure functions of a in (1.47)–(1.48) do not affect the properties of the mapping; they just provide an additive shift, so we can drop them and introduce

$$\phi_0(z,a) = -\frac{1}{a}\mathcal{K}(z,a) + \frac{1}{\overline{a}}\mathcal{K}(z,1/\overline{a}), \tag{1.50}$$

$$\phi_{\pi/2}(z,a) = -\frac{1}{a}\mathcal{K}(z,a) - \frac{1}{\overline{a}}\mathcal{K}(z,1/\overline{a}), \tag{1.51}$$

which represent conformal maps to the unbounded region exterior to a slit parallel to the real and imaginary axes, respectively.

Even more results are available. The linear combination

$$w = \phi_\theta(z,a) = e^{\mathrm{i}\theta}\left[\cos\theta\phi_0(z,a) - \mathrm{i}\sin\theta\phi_{\pi/2}(z,a)\right], \qquad 0 \le \theta < \pi, \tag{1.52}$$

also has a simple pole at $z = a$ with residue 1; this is easy to check. Moreover, it satisfies

$$\mathrm{Im}[e^{-\mathrm{i}\theta}\phi_\theta(z,a)] = 0, \qquad z \in C_0. \tag{1.53}$$

The image of C_0 is therefore a slit turned at angle θ to the positive real w axis. The generalized definition (1.52) is consistent with the notation in (1.50)–(1.51) when $\theta = 0$ and $\pi/2$.

We refer to (1.52) as the formula for a *parallel slit map*. This designation seems out of place here because there is only one slit, so it is not "parallel" to any other slit. However, when multiply connected domains are considered in §5.9 and §6.9, the very same formulas in terms of the prime function will produce conformal mappings to domains with multiple slits, all of which are parallel to each other.

1.7 ▪ Cayley map

All the functions developed so far have been derivable in a straightforward way from the analytic extension $\mathcal{G}_0(z,a)$ of the Green's function, which was, in turn, built using the prime function, as seen in (1.16). However many other functions not readily expressible in terms of $\mathcal{G}_0(z,a)$ can be constructed using the prime function $\omega(z,a)$ itself as the basic building block.

Since, from (1.16), it is known that $\mathcal{G}_0(z,a)$ can be constructed from the ratio of prime functions $R(z;a,b)$ in the special case $b = 1/\overline{a}$, it is natural to relax this condition and ask about

$$w = R(z;a,b), \tag{1.54}$$

where $a, b \in \mathbb{C}$ are arbitrary points. This is a Möbius map. Such functions have been extensively studied in the literature [103, 61], so we will not explore them in detail here; instead several of the exercises review some of their important properties.

There is, however, a special case that will be of particular interest to us: the situation in which a and b are both distinct points on the boundary circle C_0 of D_z. In such a case, the ratio of prime functions $R(z;a,b)$ has a special name: the *Cayley map*.

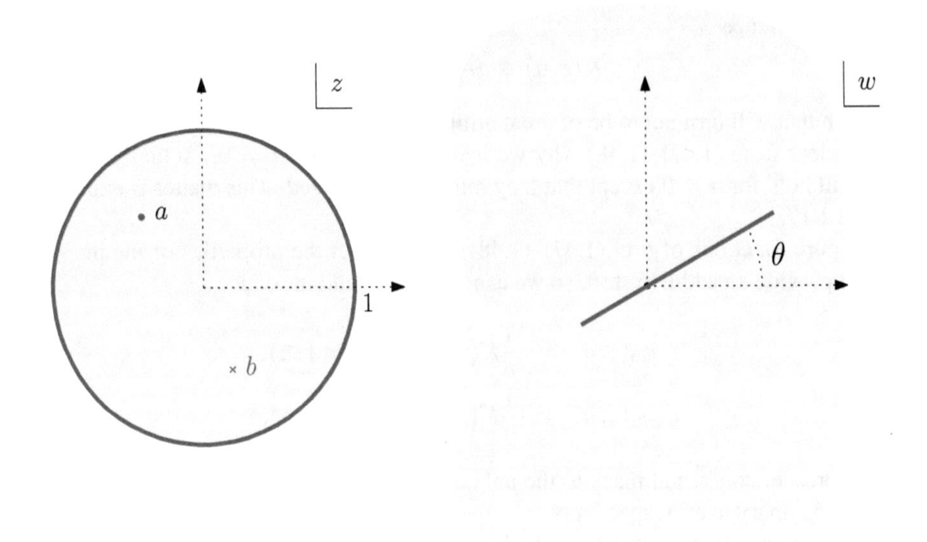

Figure 1.6. *Parallel slit map. Under the map (1.52) the unit disc D_z is mapped to the unbounded plane exterior to a finite-length slit making angle θ to the real axis. The point $z = a$ maps to the origin $w = 0$; $z = b$ is the preimage of the point at infinity in the w plane.*

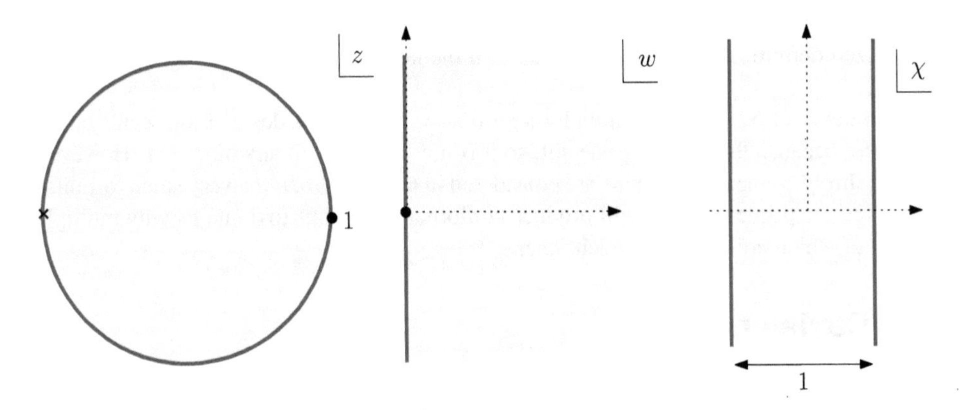

Figure 1.7. *Cayley map and a logarithm of the Cayley map: under the map (1.58) the unit disc D_z is mapped to the right half w plane. Under a further logarithmic transformation (1.60), the right half w plane is transplanted to a unit-width strip, or channel, with $-\infty < \mathrm{Im}[\chi] < \infty$.*

If a and b are on C_0, then $1/\overline{a} = a$ and $1/\overline{b} = b$, so that, from (1.15), for z on C_0,

$$\overline{R(z;a,b)} = \frac{b}{a} R(z;a,b), \qquad z \in C_0, \tag{1.55}$$

and, hence,

$$\frac{R(z;a,b)}{\overline{R(z;a,b)}} = e^{2\mathrm{i}\arg[R(z;a,b)]} = \frac{a}{b}, \qquad z \in C_0. \tag{1.56}$$

This implies that

$$\arg[R(z;a,b))] = \text{constant}, \qquad z \in C_0. \tag{1.57}$$

An interesting example is to set $a = 1, b = -1$ and to add a convenient additional prefactor of -1 so that

$$w = -\frac{\omega(z,+1)}{\omega(z,-1)}. \tag{1.58}$$

As a conformal mapping from a complex z plane to a complex w plane it transplants $z = 1$ to $w = 0$ and $z = -1$ to $w = \infty$. To find the image of C_0 we notice that (1.55) implies

$$\overline{w} = -w. \tag{1.59}$$

Hence C_0 maps to the imaginary axis in the w plane where $w + \overline{w} = 0$. Since $z = 0$ corresponds to $w = 1$, we conclude that the Cayley map (1.58) transplants the unit disc to the right-half w plane $\text{Re}[w] > 0$. If the additional prefactor of -1 is omitted, a map to the left half plane is produced.

It is easy to confirm that this is a self-inverse mapping (Exercise 1.10).

We have seen how useful it can be to take a logarithm of the mappings we have constructed. In this case, on taking a logarithm of the Cayley map to produce a mapping to a complex χ plane we find

$$\chi = \frac{1}{\pi \mathrm{i}} \log w = \frac{1}{\pi \mathrm{i}} \log \left[-\frac{\omega(z,+1)}{\omega(z,-1)} \right]. \tag{1.60}$$

This transplants the right half w plane to the infinite strip of unit width. We note that (1.60) can be rewritten in terms of the $\tan^{-1}$ function if desired (Exercise 1.12).

The map (1.60) from the unit disc to an infinite strip, or "channel," is useful in applications.

1.8 ▪ Radial slit map

The functional form (1.38) of the unbounded circular slit map leads us to inquire about the modification of it given by

$$w = R(z;a,b)R(z;1/\overline{a},1/\overline{b}) = \frac{\omega(z,a)\omega(z,1/\overline{a})}{\omega(z,b)\omega(z,1/\overline{b})}, \tag{1.61}$$

where a and b are two distinct nonzero points inside D_z. This is similar to formula (1.38), but the prime functions $\omega(z,1/\overline{a})$ and $\omega(z,1/\overline{b})$ appearing there have been swapped around. The point $z = a$ still maps to $w = 0$, and $z = b$ still maps to $w = \infty$. But the properties of the mapping for $z \in C_0$ are different. From (1.15) we deduce that

$$\overline{w} = \frac{\overline{a}b}{a\overline{b}} w, \qquad z \in C_0. \tag{1.62}$$

This means that w has constant argument for z on C_0. We conclude that (1.61) maps the unit disc D_z to the exterior of a finite-length slit in the w plane with constant argument. This is a *radial slit map.*

Just as we did for the unbounded circular slit maps, we consider the special case where we let

$$a = \mathrm{i}r, \qquad b = \overline{a}, \qquad 0 < r < 1. \tag{1.63}$$

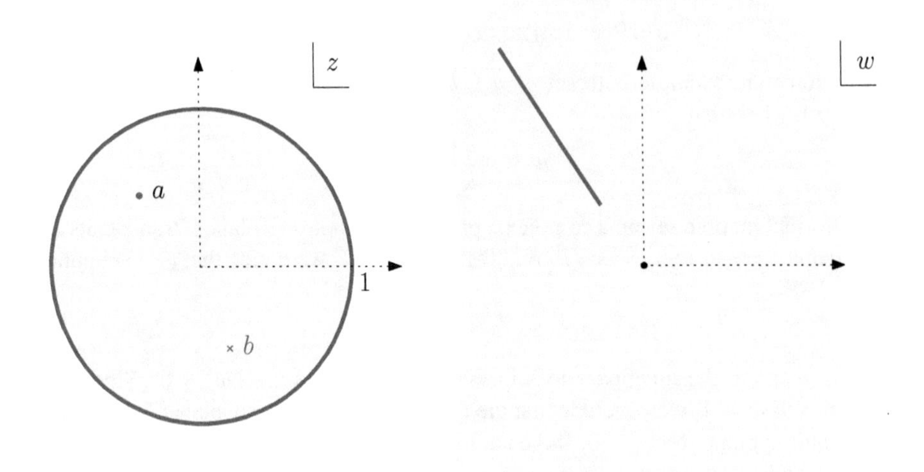

Figure 1.8. *Radial slit map. Under the map (1.61) the unit disc D_z is transplanted to the unbounded region exterior to a finite-length slit on a ray emanating from the origin $w = 0$. The point $z = a$ is the preimage of $w = 0$.*

We also premultiply by -1 so that (1.61) becomes

$$w = -\frac{\omega(z,a)\omega(z,1/\overline{a})}{\omega(z,\overline{a})\omega(z,1/a)}. \tag{1.64}$$

Then (1.62) tells us

$$\arg[w] = 0, \qquad z \in C_0. \tag{1.65}$$

It is easy to check that, for points z on the real axis $-1 < z < 1$,

$$\overline{w} = \frac{1}{w}, \tag{1.66}$$

where we have used the additional identity (1.9). Hence the real diameter of the unit disc maps to the unit circle $|w| = 1$. The upper two schematics in Figure 1.9 show the effect of this special form (1.64) of the radial slit mapping.

We have restricted ourselves to the cases where a and b are both nonzero; this is because the quantities $1/\overline{a}$ and $1/\overline{b}$ appear in formula (1.61) and these are undefined as $a, b \to 0$. However, consider the function

$$w = \frac{\omega(z,0)}{\omega(z,b)\omega(z,1/\overline{b})}. \tag{1.67}$$

On C_0 it can be verified that

$$\overline{w} = \frac{b}{\overline{b}} w, \tag{1.68}$$

implying that

$$\arg[w] = \arg[\overline{b}], \qquad z \in C_0. \tag{1.69}$$

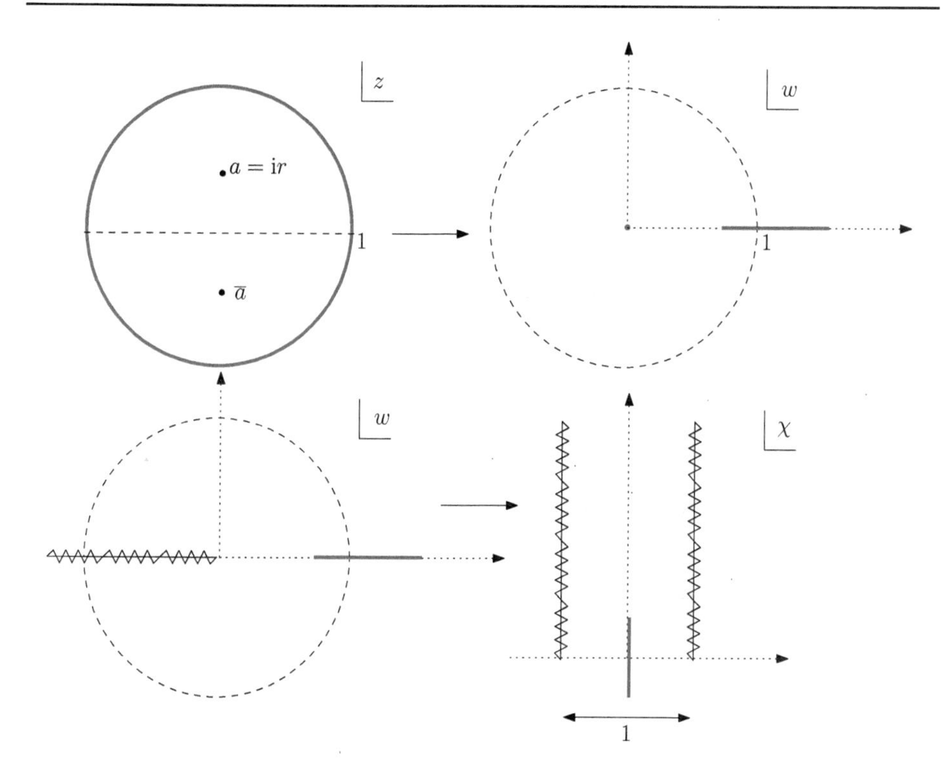

Figure 1.9. *Conformal map (1.64) from the upper half unit disc in the z plane to the unit disc in a w plane with $|z| = 1$ being transplanted to a radial slit passing through $|w| = 1$ with the real diameter $-1 < z < +1$ mapping to the remainder of $|w| = 1$. The lower schematics show the logarithm of this map (1.70): the interior of the unit w disc is transplanted to the upper half of a unit-width semistrip $0 < \mathrm{Im}[\chi] < \infty$ in the χ plane with a finite-length slit parallel to the imaginary χ axis emanating from the real χ axis.*

Therefore (1.67) is also a radial slit map. It is the relevant form of the radial slit map in the limit $a \to 0$. A similar result pertains when $b \to 0$ (a and b must be distinct, so they are never both zero).

We can go on to take a logarithm of (1.64) and consider the mapping

$$\chi = \frac{1}{2\pi\mathrm{i}} \log w = \frac{1}{2\pi\mathrm{i}} \log \left[-\frac{\omega(z,a)\omega(z,1/\overline{a})}{\omega(z,\overline{a})\omega(z,1/a)} \right]. \tag{1.70}$$

As a function of w this has two logarithmic branch points at $w = 0, \infty$, so we can define a single-valued branch of the function by selecting a branch cut along the negative real w axis. The interior of the unit w disc is transplanted by this branch of the function to the upper half of a unit-width semistrip $0 < \mathrm{Im}[\chi] < \infty$ in the χ plane with a finite-length slit parallel to the imaginary χ axis emanating from the real χ axis, as shown in Figure 1.9.

1.9 ▪ The Schwarz function of a circle

The *Schwarz function* of an analytic curve C [76] is defined to be the function $S(z)$, analytic in an annular neighborhood of the curve, satisfying the condition that

$$S(z) = \overline{z}, \qquad z \in C. \tag{1.71}$$

Consider the circle with center $\delta \in \mathbb{C}$ and radius $q \in \mathbb{R}$ defined by

$$|z-\delta|^2 = q^2. \tag{1.72}$$

It follows that the Schwarz function of this circle is

$$\overline{z} = \overline{\delta} + \frac{q^2}{z-\delta} = S(z). \tag{1.73}$$

This can be rewritten in terms of the ratio of prime functions $R(z;a,b)$ as

$$S(z) = \overline{\delta}\, R(z; \delta - q^2/\overline{\delta}), \delta). \tag{1.74}$$

These Schwarz functions of a circle, which are simple Möbius maps, will play an important theoretical role in the framework to be developed in subsequent chapters.

1.10 ▪ Reflection in a circle

A circle is a closed curve of constant curvature. The real axis is a simple example having zero curvature. The Schwarz function $S_{\mathbb{R}}(z)$, say, of the real axis in the z plane is

$$\overline{z} = S_{\mathbb{R}}(z) = z. \tag{1.75}$$

It is well known that, geometrically, taking the complex conjugate of a point z produces its mirror "reflection" in the real axis. Taking this complex conjugate can also be written as effecting the (antiholomorphic) transformation

$$z \mapsto \overline{S_{\mathbb{R}}(z)}. \tag{1.76}$$

It is natural to extend this notion of the reflection of a point with respect to a more general circle C, of nonzero curvature, by defining it to be given by

$$z \mapsto \overline{S(z)}, \tag{1.77}$$

where, now, $S(z)$ is the Schwarz function of the circle C introduced in §1.9. In this way, reflection in the unit circle C_0 corresponds to the antiholomorphic mapping

$$z \mapsto \frac{1}{\overline{z}}. \tag{1.78}$$

1.11 ▪ The functions $\mathcal{K}(z,a)$ and $K(z,a)$

When considering parallel slit maps, we encountered in (1.49) the logarithmic derivative $\mathcal{K}(z,a)$ of the prime function with respect to the parameter a, namely, the function

$$\mathcal{K}(z,a) \equiv a \frac{\partial \log \omega(z,a)}{\partial a}. \tag{1.79}$$

An important feature of this function is that it has a simple pole when $z=a$. The identity (1.8) satisfied by the prime function can be used to establish that

$$\overline{\mathcal{K}}(1/z, 1/a) = 1 - \mathcal{K}(z,a). \tag{1.80}$$

In fact, the extra identity (1.9)—associated with the fact that the disc is reflectionally symmetric about the real axis—means that we can write (1.80) as

$$\mathcal{K}(1/z, 1/a) = 1 - \mathcal{K}(z,a). \tag{1.81}$$

Since $\omega(z,a)$ is an analytic function of two variables, z and a, we can also introduce its logarithmic derivative with respect to z:

$$K(z,a) \equiv z\frac{\partial \log \omega(z,a)}{\partial z}. \tag{1.82}$$

This also has a simple pole when $z = a$. Indeed, the functions $\mathcal{K}(z,a)$ and $K(z,a)$ are intimately related. The antisymmetry of the prime function (1.7) in its arguments z and a implies that these functions are related by

$$K(z,a) = \mathcal{K}(a,z). \tag{1.83}$$

From the identity (1.8) we can deduce that

$$\overline{K}(1/z,1/a) = 1 - K(z,a) \tag{1.84}$$

or, in view of the extra identity (1.9),

$$K(1/z,1/a) = 1 - K(z,a). \tag{1.85}$$

A simple direct calculation with $\omega(z,a) = z - a$ (Exercise 1.16) shows that $\mathcal{K}(z,a)$ and $K(z,a)$ satisfy the functional relation

$$\mathcal{K}(z,a) + K(z,a) = 1. \tag{1.86}$$

1.12 ▪ The cross-ratio

Consider the conformal mapping from a complex z plane to a complex χ plane given, in terms of the prime function, by

$$\chi = p(z,w,a,b) = \frac{\omega(z,a)\omega(w,b)}{\omega(z,b)\omega(w,a)}, \tag{1.87}$$

where a, w, and b can be viewed as complex-valued parameters. As a function of z it is a Möbius map, and it is clear that

$$z = a \mapsto \chi = 0, \qquad z = w \mapsto \chi = 1, \qquad z = b \mapsto \chi = \infty. \tag{1.88}$$

This Möbius mapping can be used to transplant three arbitrary points $\{a,w,b\}$ in the z plane to the special points $\{0,1,\infty\}$ in the χ plane. This is often useful in applications. The function $p(z,w,a,b)$ is known as the *cross-ratio*.

Interestingly, the function (1.87) is also a Möbius map when considered as a function of a or w or b.

An important observation for us, with which we conclude this chapter, is that

$$\lim_{\substack{z\to a\\ w\to b}} \left[-(z-a)(w-b)p(z;w;a;b)^{-1}\right] = \omega(a,b)^2. \tag{1.89}$$

Exercise 1.22 asks the reader to verify this. We see that the prime function $\omega(a,b)$—or, more precisely, its square $\omega(a,b)^2$—can be retrieved from this double limit associated with the cross-ratio function.

Exercises

1.1. **(Mapping properties of inversion)** Consider the inversion

$$w = \frac{1}{z}. \tag{1.90}$$

Show that, when viewed as a conformal map, the circle

$$|z - \delta|^2 = q^2 \tag{1.91}$$

is transplanted to the circle

$$|w - \Delta|^2 = Q^2, \tag{1.92}$$

where

$$\Delta = \frac{\overline{\delta}}{|\delta|^2 - q^2}, \qquad Q = \frac{q}{||\delta|^2 - q^2|}. \tag{1.93}$$

1.2. **(Möbius maps take circles to circles)** Consider the ratio of two prime functions as a conformal map:

$$w = A\, R(z; a, b), \qquad a, b, A \in \mathbb{C}. \tag{1.94}$$

When $\omega(z, a) = (z - a)$ this is a Möbius map.

(a) Show that (1.94) can be written as

$$w = A\, \frac{\omega(z, a)}{\omega(z, b)} = A + A\frac{\omega(b, a)}{\omega(z, b)}. \tag{1.95}$$

(b) Now argue that, geometrically, $z \mapsto w$ is equivalent to the composition of this sequence of three simpler maps:

$$z \mapsto w_1 = z - b, \qquad w_1 \mapsto w_2 = \frac{1}{w_1}, \qquad w_2 \mapsto w = A + A\omega(b, a)w_2. \tag{1.96}$$

(c) Hence show that a Möbius map transplants any circle in the z plane to a circle in the w plane.

1.3. **(Inverse of the Schwarz function of a circle)** Show that the inverse of the Schwarz function of a circle is its Schwarz conjugate function:

$$S^{-1}(z) = \overline{S}(z). \tag{1.97}$$

1.4. **(Reflection of a circle in C_0 is also a circle)** Let C be some chosen circle contained inside the unit disc D_z. Verify that the reflection of C in the unit circle C_0 is itself a circle.

1.5. **(Reflection in C_0 as a Möbius map)** Consider the circle C inside the unit disc D_z defined by

$$|z - \delta|^2 = q^2 \tag{1.98}$$

for some $\delta \in \mathbb{C}$ and $q \in \mathbb{R}$. Let C' denote the image of C under reflection in the unit circle C_0.

(a) Let $S(z)$ denote the Schwarz function of the circle C and let $S_0(z)$ denote the Schwarz function of C_0. Show that the Möbius map defined by

$$\theta(z) \equiv \overline{S}(S_0(z)) \tag{1.99}$$

is the analytic function that transplants each point on C' to its corresponding point on C obtained by reflection of C' in C_0.

(b) Verify that

$$\theta^{-1}(z) = \frac{1}{\overline{\theta}(1/z)} \tag{1.100}$$

and that this map takes points on C to their corresponding points on C' obtained by reflection of C in C_0.

1.6. **(Mapping of an eccentric to a concentric annulus)** It is required to map the eccentric annulus with boundaries

$$|z| = 1, \qquad |z - \delta| = q, \tag{1.101}$$

where $\delta, q > 0$ are positive real parameters, to a concentric annulus $\rho < |w| < 1$ for some real positive ρ. Show that the required map is

$$w = -\frac{z - a}{|a|(z - a^{-1})}, \tag{1.102}$$

where

$$a = \frac{1 + \delta^2 - q^2 - [(1 + \delta^2 - q^2)^2 - 4\delta^2]^{1/2}}{2\delta} \tag{1.103}$$

and

$$\rho = \frac{1 - \delta^2 + q^2 - [(1 - \delta^2 + q^2)^2 - 4q^2]^{1/2}}{2q}, \tag{1.104}$$

and where the minus sign in (1.102) ensures that $z = 1$ maps to $w = 1$.

1.7. **(Mapping effect of a disc automorphism)** Let C be a circle inside D_z defined by

$$|z - \tilde{\delta}| = \tilde{q}. \tag{1.105}$$

Show that the image of C under the disc automorphism

$$\eta = \frac{z - a}{1 - z\overline{a}}, \tag{1.106}$$

where a is some interior point in the unit z disc D_z (i.e., $|a| < 1$), is the circle

$$|\eta - \delta| = q, \tag{1.107}$$

where

$$\delta = \frac{(\tilde{\delta} - a)(1 - a\overline{\tilde{\delta}}) + a\tilde{q}^2}{|1 - \tilde{\delta}\overline{a}|^2 - \tilde{q}^2|a|^2}, \qquad q = \tilde{q}\frac{1 - |a|^2}{\big||1 - \tilde{\delta}\overline{a}|^2 - \tilde{q}^2|a|^2\big|}. \tag{1.108}$$

1.8. **(Blaschke products)** It is required to find a function $f(z)$ with the following properties:

(i) $f(z)$ is analytic everywhere in the unit disc D_z;

(ii) $f(z)$ has precisely N zeros inside D_z at given locations $\{a_j | j = 1, \ldots, N\}$;

(iii) the modulus of $f(z)$ is unity on the boundary $|z| = 1$ of D_z.

By thinking about the automorphism of the unit disc (1.31) construct a function $f(z)$ having properties (i)–(iii).

1.9. **(A self-inverse disc automorphism)** Show that a disc automorphism (1.36) with $\phi = 0$ and a real and positive is self-inverse.

1.10. **(A self-inverse Cayley map)** Show that the Cayley map (1.58) is self-inverse.

1.11. **(Univalency and the argument principle)** Just as was done for the automorphism of the disc, use the argument principle to establish that the interior of the disc D_z is transplanted in a one-to-one fashion to the various target domains for the unbounded circular slit map, the parallel slit map, the Cayley map, and the radial slit map. *Hint:* Note that the Cayley map can be viewed as having "half a pole" inside D_z.

1.12. **(On the mapping to a strip)** Show that the inverse of the function (1.60) is

$$z = -\mathrm{i}\tan\left(\frac{\pi\chi}{2}\right). \tag{1.109}$$

1.13. **(On the special unbounded circular slit mapping)** This exercise concerns the special case of the unbounded circular slit map (1.41).

(a) Verify that this mapping transplants the *upper* half of the unit disc D_z to the interior $|w| < 1$ of the unit disc in the w plane.

(b) Confirm also that the lower half of the unit disc D_z is transplanted to the unbounded region $|w| > 1$.

(c) Show that the mapping (1.41) maps the unit circle $|z| = 1$ to a circular slit on the unit w circle with

$$-\phi < \arg[w] < +\phi, \tag{1.110}$$

where

$$r = \tan\left(\frac{\phi}{4}\right). \tag{1.111}$$

Find the inverse mapping.

1.14. **(Symmetric triply connected domain)** Consider the domain D_z comprising the unit disc with two smaller circular discs with boundaries C_1 and C_2 excised. Let δ_1 and δ_2 be the centers of C_1 and C_2, let q_1 and q_2 be their radii, and suppose that

$$\delta_1 = -\delta_2 = \delta \in \mathbb{R}, \qquad q_1 = q_2 = q, \qquad 0 < q < \delta < 1. \tag{1.112}$$

Thus, the domain has reflectional symmetry about both the real and imaginary axes. Use the result of Exercise 1.6 to show that, under the disc automorphism

$$z \mapsto f(z) = -\frac{z-a}{|a|(z-1/\overline{a})} \tag{1.113}$$

with

$$a = \frac{(1+\delta^2-q^2) - [(1+\delta^2-q^2)^2 - 4\delta^2]^{1/2}}{2\delta}, \tag{1.114}$$

the domain D_z is mapped to the circular domain $\tilde{D}_z$ for which the two interior circles $\tilde{C}_1$ and $\tilde{C}_2$ are

$$|z| = \rho, \qquad |z - D| = Q, \tag{1.115}$$

where

$$\rho = \frac{1-\delta^2+q^2 - [(1-\delta^2+q^2)^2 - 4q^2]^{1/2}}{2q}, \tag{1.116}$$

and

$$D = \frac{1}{2}\left[f(-\delta+q) + f(-\delta-q)\right], \qquad Q = \frac{1}{2}\left[f(-\delta+q) - f(-\delta-q)\right]. \tag{1.117}$$

1.15. **(Two views of a slit map)** Consider a slit along the real axis in a complex w plane between $[r, 1]$ for $r > 0$. We will construct two different, but equivalent, representations of the conformal mapping from the unit z disc D_z to the region exterior to this slit. There are three real degrees of freedom in the mapping theorem, deriving from the three real degrees of freedom in the automorphisms of the disc. We will fix two of these by insisting that $z = b \in \mathbb{R}$ is the preimage of infinity; we fix the remaining real degree of freedom by insisting that $z = 1$ is the preimage of the end of the slit at $w = 1$.

(a) One approach is to construct this map as an instance of a radial slit map. Show that the required mapping is

$$w = f_1(z) = \frac{\omega(1,b)\omega(1,1/b)\omega(z,a)\omega(z,1/a)}{\omega(1,a)\omega(1,1/a)\omega(z,b)\omega(z,1/b)}, \tag{1.118}$$

where $a \in \mathbb{R}$ determines the position of the endpoint r.

(b) This same mapping can be constructed as an instance of a parallel slit mapping. Show that the required map is

$$w = f_2(z) = \frac{\mathcal{K}(z,b) - \mathcal{K}(z,1/b) - (\mathcal{K}(a,b) - \mathcal{K}(a,1/b))}{\mathcal{K}(1,b) - \mathcal{K}(1,1/b) - (\mathcal{K}(a,b) - \mathcal{K}(a,1/b))}. \tag{1.119}$$

The function $f_2(z)$ must be identical to $f_1(z)$. Indeed, $f_2(z)$ is simply the partial fraction decomposition of the rational function $f_1(z)$.

(c) Show that in the limit $b \to 0$, both (1.118) and (1.119) become

$$w = A\left[\frac{1}{z} + z - B\right], \qquad A = \frac{1}{2-(a+1/a)}, \quad B = a + \frac{1}{a}. \tag{1.120}$$

The combination

$$\frac{1}{z} + z \tag{1.121}$$

is the well-known form of the *Joukowski map*.

1.16. **(An identity between $\mathcal{K}(z,a)$ and $K(z,a)$)** By direct substitution of $\omega(z,a) = (z-a)$, verify that, for this prime function, the functional relation (1.86) holds.

1.17. **(Regularization of parallel slit maps for $a = 0$)** When $a \to 0$, the formulas (1.50)–(1.51) for the parallel slit maps become unbounded—this is because of the appearance in them of $1/\overline{a}$—but they can be regularized.

(a) Use the identities found in §1.11 to show that, after subtraction of any constants that become unbounded as $a \to 0$, the parallel slit maps in (1.50)–(1.51) can be written as

$$\phi_0(z,0) = -\frac{\partial}{\partial a}\log\omega(z,a)\bigg|_{a=0} - \frac{\partial}{\partial a}\log\overline{\omega}(1/z,a)\bigg|_{a=0}, \tag{1.122}$$

$$\phi_{\pi/2}(z,0) = -\frac{\partial}{\partial a}\log\omega(z,a)\bigg|_{a=0} + \frac{\partial}{\partial a}\log\overline{\omega}(1/z,a)\bigg|_{a=0}. \tag{1.123}$$

Note that, on the right-hand sides of these expressions, the logarithmic derivative is taken with respect to a and *then* a is set equal to zero.

(b) Verify that formula (1.122) again retrieves the Joukowski map (1.121).

1.18. **(Möbius invariance of cross-ratio)** Show that the cross-ratio function $p(z, w, a, b)$ is invariant under the transformations

$$z \mapsto M(z), \qquad w \mapsto M(w), \qquad a \mapsto M(a), \qquad b \mapsto M(b), \tag{1.124}$$

where M is any Möbius map.

1.19. **(Cross-ratio identity)** Show that

$$p(z, w, a, b) + p(z, a, w, b) = 1. \tag{1.125}$$

1.20. **(Van der Pauw method)** A physical problem in a unit disc involves four points a, b, c, and d on the unit circle $|z| = 1$. There are two conjugate harmonic potentials Φ and Ψ associated with the physical problem, so that one can introduce the complex potential

$$h(z) = \Phi(z, \overline{z}) + \mathrm{i}\Psi(z, \overline{z}). \tag{1.126}$$

Applying a forcing at points a and b leads to the complex potential

$$h^{(a,b)}(z) = \frac{m}{\pi} \log \left[\frac{\omega(z, a)}{\omega(z, b)} \right] = \Phi^{(a,b)}(z, \overline{z}) + \mathrm{i}\Psi^{(a,b)}(z, \overline{z}). \tag{1.127}$$

Notice that this is the logarithm of a Cayley map.

(a) Verify that Ψ is constant for all points with $|z| = 1$ except at $z = a, b$, where it is undefined.

(b) Let $R^{(a,b)}_{(c,d)}$ denote the difference in the values of Φ between points c and d, i.e.,

$$R^{(a,b)}_{(c,d)} \equiv \Phi^{(a,b)}(c, \overline{c}) - \Phi^{(a,b)}(d, \overline{d}). \tag{1.128}$$

Verify that

$$e^{\pi R^{(a,b)}_{(c,d)}/m} + e^{\pi R^{(a,c)}_{(b,d)}/m} = 1. \tag{1.129}$$

1.21. **(On boundary normal derivatives)** For points z on the unit circle C_0, verify the formula

$$\frac{\partial G_0(z, a)}{\partial n_z} ds_z = -d\mathcal{G}_0(z, a), \qquad z \in C_0, \tag{1.130}$$

where n_z denotes the normal to C_0 pointing out of the unit disc D_z and ds_z denotes an element of arclength along C_0. By extension, show that for any function $H(z)$ that is the imaginary part of a function $\mathcal{H}(z)$ that is analytic in an annular neighborhood of some closed curve C_1—such as the circle C_1 shown in Figure 1.10—so that

$$H(z) = \mathrm{Im}[\mathcal{H}(z)] \qquad \text{for } z \in C_1, \tag{1.131}$$

and where $H(z)$ is constant on C_1, i.e.,

$$H(z) = \text{constant} \qquad \text{for } z \in C_1, \tag{1.132}$$

then

$$\frac{\partial H(z)}{\partial n_z} ds_z = -d\mathcal{H}(z) \qquad \text{for } z \in C_1. \tag{1.133}$$

1.22. **(Prime function as a double limit)** Verify the identity

$$\lim_{\substack{z \to a \\ w \to b}} \left[-(z - a)(w - b) p(z, w, a, b)^{-1} \right] = \omega(a, b)^2, \tag{1.134}$$

where $\omega(z, a) = z - a$.

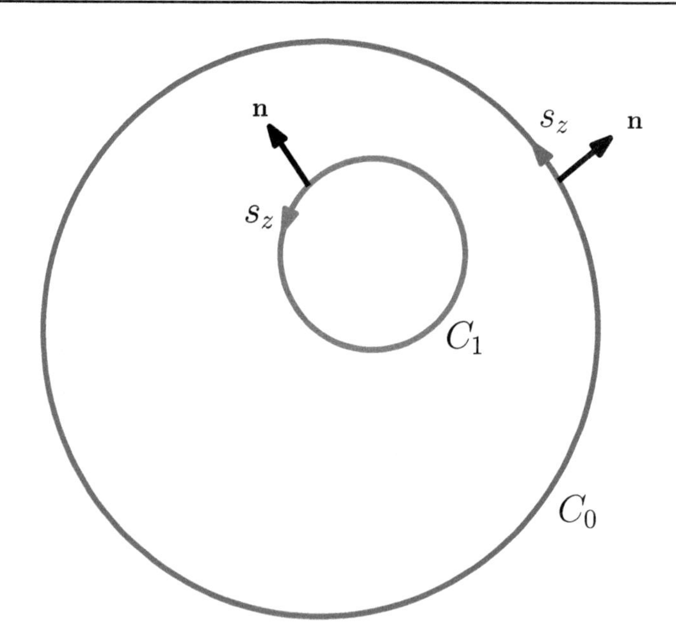

Figure 1.10. *Definition of normal vectors (Exercise 1.21). Arclength increases as the curves C_0 and C_1 are traversed in an anticlockwise sense.*

1.23. **(Cross-ratio from the Green's function)** Consider the function $\Pi_{a,b}^{z,w}$ defined by the double integral

$$\Pi_{a,b}^{z,w} = 2\pi\mathrm{i} \int_b^a \int_w^z \frac{\partial^2 \mathcal{G}_0(z', a')}{\partial z' \partial a'} dz' da', \tag{1.135}$$

where the integration contour between w and z and that between b and a are chosen arbitrarily within the unit disc D_z.

(a) Show that

$$2\pi\mathrm{i} \frac{\partial^2 \mathcal{G}_0(z, a)}{\partial z \partial a} = \frac{1}{(z-a)^2}. \tag{1.136}$$

(b) Show that

$$\Pi_{a,b}^{z,w} = \log p(z, w, a, b) + 2\pi\mathrm{i}n \tag{1.137}$$

for some integer $n \in \mathbb{Z}$ where $p(z, w, a, b)$ is the cross-ratio.

(c) Use your answer to Exercise 1.22 to show that

$$\lim_{\substack{z \to a \\ w \to b}} \left[-(z-a)(w-b) e^{-\Pi_{a,b}^{z,w}} \right] = \omega(a, b)^2. \tag{1.138}$$

Chapter 2

Function theory in multiply connected circular domains

2.1 ▪ Towards new prime functions

In Chapter 1 the function $\omega(z,a) = (z-a)$ associated with function theory in the unit disc was written down at the outset and a diverse array of functions was then built from it. We called $\omega(z,a)$ a prime function. In this case the prime function is elementary, so simple in fact that it hardly seems to merit having any special designation. Among the functions built from the prime function in Chapter 1 was $\mathcal{G}_0(z,a)$, the analytic extension of the first-type Green's function in the unit disc.

Our objective now is to go the other way. We will *start* with the first-type Green's function in a given domain and ask how we might arrive at a prime function. This will not just be done for the unit disc, however. Instead we will start with a large class of multiply connected domains and examine the properties of the first-type Green's function associated with them. That is the subject of this chapter. Then, once this has been done, in Chapter 4 we will use these properties to define the prime function associated with these multiply connected domains. Those generalized prime functions will be much more interesting.

2.2 ▪ Multiply connected circular domains

The class of multiply connected planar domains in which we will build the theoretical framework is the class of *circular domains*. Such a domain will be denoted by D_z and taken to have connectivity $M+1$ where the integer $M \geq 0$. Its boundary is denoted by ∂D_z. An example circular domain with $M=2$ is shown in Figure 2.1. The domain D_z is characterized as the unit disc $|z| < 1$ in a complex z plane with M smaller circular discs excised. The outer boundary of D_z, the unit circle $|z| = 1$, will be called C_0, and the inner circular boundaries of D_z will be denoted by $\{C_j | j = 1, \ldots, M\}$.

The important fact—to be exploited later—is that this class of domains is *canonical* in that an extension of Riemann's mapping theorem, due to Koebe [82], states that any planar multiply connected domain D is conformally equivalent to such a circular domain D_z. To within an automorphism of the unit disc, the geometry of such a conformally equivalent circular domain D_z will be dictated by the choice of D. We will see examples of how this works later.

From a practical point of view this means that there is much to be gained from developing a function theory in this canonical class of circular domains since, by virtue of their

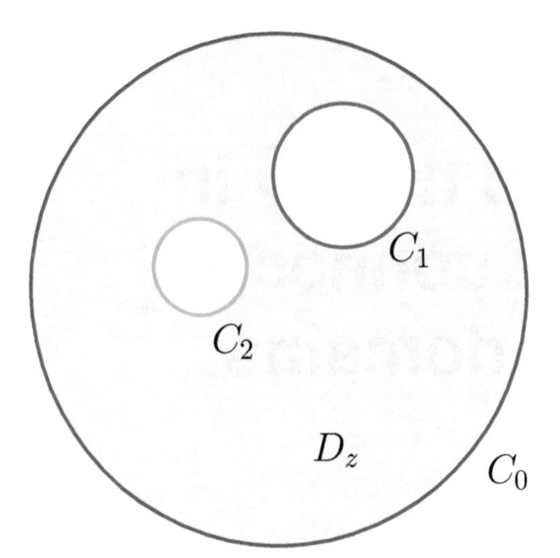

Figure 2.1. *A typical triply connected circular domain D_z corresponding to $M = 2$. The unit disc, whose boundary is the unit circle C_0, has two smaller circular discs excised with boundaries C_1 and C_2.*

conformal equivalence to every other multiply connected planar domain, that theory finds relevance in a great many problems.

2.3 ▪ First-type Green's function

In Chapter 1 a key role was played by the Green's function $G_0(z,a)$ of the unit disc. When D_z is a multiply connected circular domain we have to be more careful about what we mean by a Green's function since there are now Green's functions of various kinds. This is due to the presence of the holes.

An important rule of thumb when studying analytic functions in multiply connected domains is always to ask about the *periods* of the functions around the holes. Roughly speaking, by what amount do these functions alter their values on analytic continuation around a continuous loop encircling any given hole? This consideration will emerge as a key feature of this chapter.

Let $G(z,a)$ be the so-called *first-type Green's function* in a multiply connected circular domain D_z. This is the real-valued function satisfying

$$\nabla^2 G(z,a) = -\delta(z-a), \qquad z \in D_z, \tag{2.1}$$

with the boundary conditions

$$G = 0, \qquad z \in \partial D_z. \tag{2.2}$$

$G(z,a)$ therefore vanishes on *all* the circular boundary components of D_z.

An important fact that we will use repeatedly is that if some harmonic function ϕ in the multiply connected domain D_z is zero on all its boundaries, then, by the uniqueness of the solution of the Dirichlet problem in D_z [82], the function ϕ must be identically zero.[5]

Take two distinct points a and b inside D_z. It follows from Green's second identity that

$$\int\int_{D_z} \left[G(z,b)\nabla^2 G(z,a) - G(z,a)\nabla^2 G(z,b)\right] dA_z \tag{2.3}$$

[5]A harmonic function is defined as a function which vanishes when acted on by the Laplacian operator.

$$= \int_{\partial D_z} G(z,b)\frac{\partial G(z,a)}{\partial n_z}ds_z - G(z,a)\frac{\partial G(z,b)}{\partial n_z}ds_z = 0, \tag{2.4}$$

where the subscript z on n_z and ds_z is added to emphasize that z is the integration variable. On use of (2.1) this can be seen to imply

$$G(a,b) = G(b,a), \tag{2.5}$$

or on reassigning the variable b as z,

$$G(a,z) = G(z,a). \tag{2.6}$$

The first-type Green's function is therefore symmetric with respect to swapping its two arguments. From this we conclude that $G(z,a)$ is a harmonic function of *both* complex variables z and a, except of course when $z = a$.

We define $\mathcal{G}(z,a)$ to be the analytic extension of $G(z,a)$ in D_z with $G(z,a)$ being its imaginary part, namely,

$$G(z,a) = \text{Im}[\mathcal{G}(z,a)]. \tag{2.7}$$

Since D_z is multiply connected, we ask about the periods of the analytic function $\mathcal{G}(z,a)$ about each of the holes in D_z. These periods, which in this case must be functions of the second variable a, are defined by

$$\sigma_j(a) \equiv -\int_{C_j} d\mathcal{G}(z,a), \qquad j = 1,\ldots,M, \tag{2.8}$$

where the integration is with respect to the z variable. From this definition we infer three things.

First, since the imaginary part of $\mathcal{G}(z,a)$ vanishes on each C_j for $j = 1,\ldots,M$, then $d\mathcal{G}(z,a)$ is real on these circles and, consequently, $\sigma_j(a)$ is a real function of a.

Second, since $\mathcal{G}(z,a)$ is a harmonic function of a in D_z, then so is $\sigma(a)$ since it is defined as an integral, as a function of the variable z, of a differential which is harmonic as a function of a.

Finally, we also infer that

$$\sigma_j(a) = \begin{cases} 0, & z \in C_0, \\ \delta_{jk}, & z \in C_k,\ k = 1,\ldots,M, \end{cases} \tag{2.9}$$

where δ_{jk} is the Kronecker delta.[6] This result follows on consideration of Green's second identity for an arbitrary function Φ that is regular in D_z, namely,

$$\iint_{D_z} \left[\Phi(z)\nabla^2 G(z,a) - G(z,a)\nabla^2\Phi\right] dA \tag{2.10}$$

$$= \int_{\partial D_z} \Phi(z)\frac{\partial G(z,a)}{\partial n_z}ds_z - G(z,a)\frac{\partial \Phi(z)}{\partial n_z}ds_z. \tag{2.11}$$

Making the specific choice $\Phi = \sigma_j(z)$, which we know is harmonic in D_z, (2.11) leads to the result

$$-\sigma_j(a) = \int_{\partial D_z} \sigma_j(z)\frac{\partial G(z,a)}{\partial n_z}ds_z \tag{2.12}$$

$$= \int_{C_0} \sigma_j(z)\frac{\partial G(z,a)}{\partial n_z}ds_z - \sum_{k=1}^{M}\int_{C_k} \sigma_j(z)\frac{\partial G(z,a)}{\partial n_z}ds_z, \tag{2.13}$$

[6] $\sigma_j(a)$ is sometimes called the *harmonic measure* of the boundary C_j with respect to the point a [75].

where here, and throughout this book, the notation

$$\int_{C_k} \tag{2.14}$$

denotes a contour integral traversing a closed contour C_k in an anticlockwise sense and with s_z and the normals n_z chosen as in Figure 1.10. Now on any boundary component of D_z,

$$\frac{\partial G}{\partial n_z} ds_z = \text{Re}\left[2\frac{\partial G}{\partial z}\left(-\text{i}\frac{dz}{ds_z}\right)\right] ds_z = \text{Re}[-d\mathcal{G}(z,a)] = -d\mathcal{G}(z,a), \tag{2.15}$$

where the last equality follows from the fact that the imaginary part of $\mathcal{G}(z,a)$ is constant on ∂D_z; see Exercise 1.21. Therefore (2.13) can be written as

$$-\sigma_j(a) = -\int_{C_0} \sigma_j(z) d\mathcal{G}(z,a) + \sum_{k=1}^{M} \int_{C_k} \sigma_j(z) d\mathcal{G}(z,a), \ \ j = 1, \dots, M. \tag{2.16}$$

Conditions (2.9) are deduced from a comparison of (2.8) and (2.16).

Each function $\sigma_j(a)$, being harmonic for a in D_z, has its own analytic extension as a function of the variable a which will be denoted by $\hat{v}_j(a)$ with

$$\sigma_j(a) = \text{Im}[\hat{v}_j(a)], \qquad j = 1, \dots, M. \tag{2.17}$$

It is immediate on inspection of (2.9) and (2.17) that each analytic function $\hat{v}_j(a)$ has constant imaginary part on all the boundary circles of D_z.

2.4 ▪ The period matrix τ_{jk}

Each of the analytic functions $\hat{v}_j(a)$ of a for $j = 1, \dots, M$ will also have periods about the holes of D_z. This time, the periods will comprise a set of constants which we collect in a matrix:

$$\mathcal{P}_{jk} \equiv \int_{C_k} d\hat{v}_j(a). \tag{2.18}$$

Since the imaginary part of each $\hat{v}_j(a)$ for $j = 1, \dots, M$ is constant on C_k for $k = 1, \dots, M$, we can write, on use of the Cauchy–Riemann equations,

$$\mathcal{P}_{jk} = \int_{C_k} d\hat{v}_j(a) = -\int_{C_k} \frac{\partial \sigma_j(a)}{\partial n_a} ds_a, \tag{2.19}$$

where the subscript a on n_a and ds_a reflects the fact that a is now the integration variable. In deriving this we have used the fact that, on any boundary component of D_z,

$$\frac{\partial \sigma_j}{\partial n_a} ds_a = \text{Re}\left[2\frac{\partial \sigma_j}{\partial a}\left(-\text{i}\frac{da}{ds_a}\right)\right] ds_a = \text{Re}[-d\hat{v}_j(a)] = -d\hat{v}_j(a), \tag{2.20}$$

where the last equality follows from the fact that the imaginary part of $\hat{v}_j(a)$ is constant on ∂D_z; see Exercise 1.21 again. Similarly (2.8) can be written as

$$\sigma_j(a) = -\int_{C_j} d\mathcal{G}(z,a) = \int_{C_j} \frac{\partial G(z,a)}{\partial n_z} ds_z, \qquad j = 1, \dots, M. \tag{2.21}$$

On substitution of (2.21) into (2.19), we find

$$\mathcal{P}_{jk} \equiv \int_{C_k} d\hat{v}_j(z) = -\int_{C_k}\int_{C_j} \frac{\partial^2 G(z,a)}{\partial n_a \partial n_z} ds_z ds_a. \tag{2.22}$$

It can be shown from this representation, and on use of (2.6), that the matrix $\mathcal{P}_{jk}$ is symmetric, namely,

$$\mathcal{P}_{jk} = \mathcal{P}_{kj}. \tag{2.23}$$

It is also invertible. To see this, consider a linear combination of the harmonic measures given by

$$\Psi(z) = \sum_{j=1}^{M} a_j \sigma_j(z), \tag{2.24}$$

where $\{a_j | j = 1, \ldots, M\}$ is a set of real constants. It follows from (2.9) that

$$\Psi(z) = \begin{cases} 0, & z \text{ on } C_0, \\ a_j, & z \text{ on } C_j, \qquad j = 1, \ldots, M, \end{cases} \tag{2.25}$$

implying that, from the divergence theorem,

$$\int\int_{D_z} |\nabla \Psi|^2 \, dA = \int\int_{D_z} \nabla.(\nabla\Psi) \, dA = \int_{\partial D_z} \Psi \frac{\partial \Psi}{\partial n_z} ds_z \tag{2.26}$$

$$= -\sum_{k=1}^{M} a_k \int_{C_k} \sum_{j=1}^{M} a_j \frac{\partial \sigma_j(z)}{\partial n_z} ds_z. \tag{2.27}$$

On use of (2.19), this quantity is

$$\sum_{j,k=1}^{M} a_k \mathcal{P}_{jk} a_j. \tag{2.28}$$

Since the integral on the left-hand side of (2.27) is nonnegative, we conclude that

$$\sum_{j,k=1}^{M} a_k \mathcal{P}_{jk} a_j \geq 0. \tag{2.29}$$

Indeed, the only way the integral on the left-hand side of (2.27), and hence the quantity (2.28), can vanish is if $\nabla\Psi$ vanishes, meaning that Ψ is a constant in D_z. But the harmonic function Ψ vanishes on C_0. Hence, Ψ must vanish identically, meaning that $a_1 = a_2 = \cdots = a_k = 0$. The conclusion is that $\mathcal{P}_{jk}$ is a positive definite symmetric matrix. As such it is nonsingular and invertible. The matrix τ having elements

$$\tau_{jk} = 2\mathrm{i}[\mathcal{P}^{-1}]_{jk} \tag{2.30}$$

therefore exists and is well defined. It too is symmetric. By the deductions just made it has a positive definite imaginary part. It has a special role to play in the subsequent development, so we call this the *period matrix* of the domain D_z.

2.5 ▪ The functions $\{v_j(z)|j = 1, \ldots, M\}$

Armed with the matrix $\mathcal{P}_{jk}$ and its inverse, we can also define, for each $j = 1, \ldots, M$, the analytic function

$$v_j(z) = \sum_{k=1}^{M} [\mathcal{P}^{-1}]_{jk} \hat{v}_k(z), \tag{2.31}$$

which, being a linear combination of the set $\{\hat{v}_j(z)|j = 1, \ldots, M\}$, has constant imaginary part on each boundary circle of D_z. Indeed,

$$\mathrm{Im}[v_j(z)] = \begin{cases} 0, & z \in C_0, \\ [\mathcal{P}^{-1}]_{jk}, & z \in C_k. \end{cases} \tag{2.32}$$

Again, we can ask about the periods of *these* functions around the holes of D_z. For any $j, n = 1, \ldots, M$,

$$\int_{C_n} dv_j(z) = \sum_{k=1}^{M} [\mathcal{P}^{-1}]_{jk} \int_{C_n} d\hat{v}_k(z) = \sum_{k=1}^{M} [\mathcal{P}^{-1}]_{jk} \mathcal{P}_{kn} = \delta_{jn}. \tag{2.33}$$

This result is important: it shows that $v_j(z)$ is single-valued as z makes a continuous anticlockwise circuit of any of the circles C_n for $n \neq j$ for $n = 1, \ldots, M$. It is not single-valued, however, as z makes a continuous anticlockwise circuit of C_j; indeed, it jumps by unity.

If we introduce the harmonic function

$$\Sigma_j(z) = \mathrm{Im}[v_j(z)], \ j = 1, \ldots, M \tag{2.34}$$

then, by Exercise 1.21, we know that

$$\int_{C_k} \frac{\partial \Sigma_j(z)}{\partial n_z} ds_z = -\int_{C_k} dv_j, \qquad j = 1, \ldots, M, \ k = 0, 1, \ldots, M. \tag{2.35}$$

By the divergence theorem and the fact that $\Sigma_j(z)$ is harmonic in D_z,

$$0 = \int\int_{D_z} \nabla^2 \Sigma_j(z) \, dA_z = \int_{C_0} \frac{\partial \Sigma_j(z)}{\partial n_z} ds_z - \sum_{k=1}^{M} \int_{C_k} \frac{\partial \Sigma_j(z)}{\partial n_z} ds_z \tag{2.36}$$

$$= \int_{C_0} \frac{\partial \Sigma_j(z)}{\partial n_z} ds_z - \int_{C_j} \frac{\partial \Sigma_j(z)}{\partial n_z} ds_z, \tag{2.37}$$

where, in the final equality, we have used (2.35) and (2.33). We conclude from (2.37) and (2.35) that

$$\int_{C_0} dv_j = \int_{C_j} dv_j = 1. \tag{2.38}$$

This means that $v_j(z)$ also jumps by unity as z makes a continuous anticlockwise circuit of C_0.

For any $j = 1, \ldots, M$ a single-valued branch of the analytic function $v_j(z)$ in D_z can be defined by drawing a "barrier" between a point on C_j and a point z_0 on C_0 and insisting that any continuous deformation of the argument z of $v_j(z)$ must not cross this barrier. Figure 2.2 shows a schematic for $M = 2$ where the barrier chosen to make $v_1(z)$ single-valued is shown in red and that chosen to render $v_2(z)$ single-valued is shown in green. There is another barrier shown in black that will be discussed in the next section.

A final observation is that each $v_j(z)$ for $j = 1, \ldots, M$ has only been specified up to the addition of an arbitrary real constant.

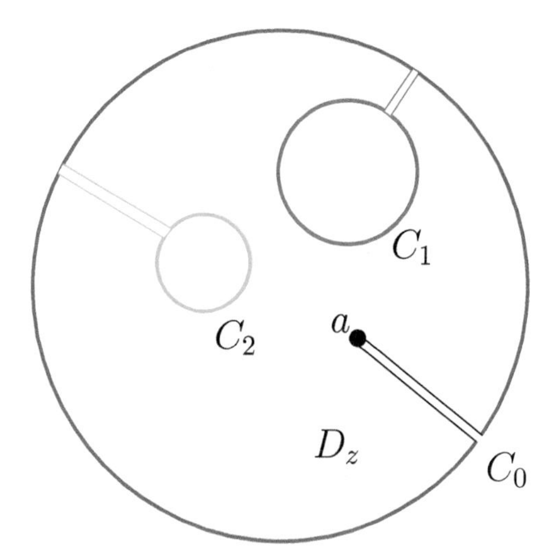

Figure 2.2. *Barriers that can be introduced in D_z to define single-valued branches of the analytic functions $\mathcal{G}_0(z, a)$ (the barrier joining a to C_0) and each function $v_j(z)$ for $j = 1, 2$ in this case of $M = 2$ (the barrier joining C_j to C_0).*

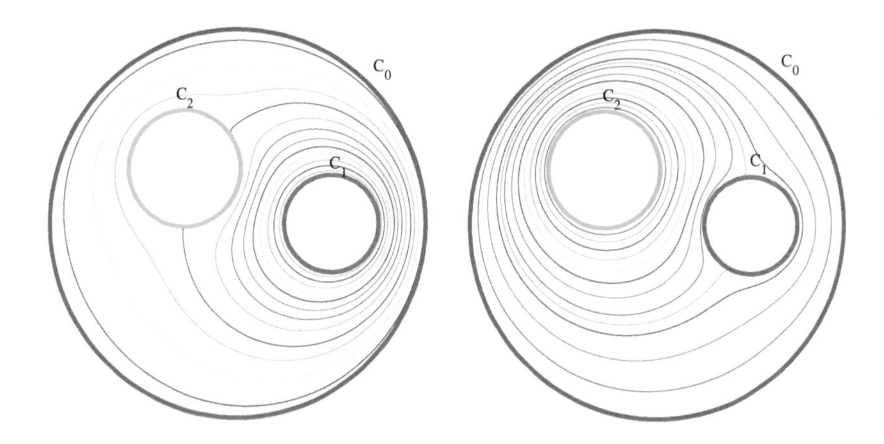

Figure 2.3. *Contours of $\Sigma_1(z) = \mathrm{Im}[v_1(z)]$ (left) and $\Sigma_2(z) = \mathrm{Im}[v_2(z)]$ (right) for $M = 2$. For $\Sigma_1(z)$ closed contours encircle C_1 (red), and for $\Sigma_2(z)$ closed contours encircle C_2 (green).*

2.6 ▪ The modified Green's function $G_0(z, a)$

Equipped with the collection of M analytic functions $\{v_j(z)|j = 1, \ldots, M\}$ it turns out that there are just enough of them to nullify all the periods of $\mathcal{G}(z, a)$ around the M holes in D_z. Consider the analytic function defined by

$$\mathcal{G}_0(z, a) = \mathcal{G}(z, a) + \sum_{j=1}^{M} \sigma_j(a) v_j(z), \tag{2.39}$$

where the reader should by now recognize all the functions on the right-hand side. Why combine all these functions in this way? The reason is that, for any $n = 1, \ldots, M$,

$$\int_{C_n} d\mathcal{G}_0(z,a) = \int_{C_n} d\mathcal{G}(z,a) + \int_{C_n} \sum_{j=1}^{M} \sigma_j(a) dv_j(z) \tag{2.40}$$

$$= \int_{C_n} d\mathcal{G}(z,a) + \sum_{j=1}^{M} \sigma_j(a)\delta_{jn} = 0, \tag{2.41}$$

where we have used (2.33) and (2.8).

We now define

$$G_0(z,a) = \text{Im}[\mathcal{G}_0(z,a)]. \tag{2.42}$$

It is clear from its definition (2.39) that

$$\nabla^2 G_0(z,a) = -\delta(z-a), \qquad z \in D_z, \tag{2.43}$$

a feature it shares with the first-type Green's function $G(z,a)$. It follows from (2.31) that

$$\hat{v}_j(a) = \sum_{n=1}^{M} \mathcal{P}_{jn} v_n(a). \tag{2.44}$$

On substitution of (2.17) and (2.44) into (2.39) and on taking the imaginary part we find

$$G_0(z,a) = G(z,a) + \sum_{j,n=1}^{M} \mathcal{P}_{jn} \text{Im}[v_n(a)] \text{Im}[v_j(z)]. \tag{2.45}$$

It is clear from this formula that

$$G_0(z,a) = G_0(a,z). \tag{2.46}$$

This feature of symmetry in the two variables z and a is another feature that $G_0(z,a)$ shares with the first-type Green's function; cf. (2.6).

It is worth noting that we can also write

$$G_0(z,a) = G(z,a) + \sum_{j,n=1}^{M} \mathcal{P}_{jn} \text{Im}[v_n(z)] \text{Im}[v_j(a)], \tag{2.47}$$

or indeed

$$G_0(z,a) = G(z,a) + \sum_{j=1}^{M} \sigma_j(z) \text{Im}[v_j(a)]. \tag{2.48}$$

See Exercise 2.3. These are all just alternative representations of the same function.

In summary $G_0(z,a)$ satisfies the partial differential equation

$$\nabla^2 G_0(z,a) = -\delta(z-a), \qquad z \in D_z, \tag{2.49}$$

and, as is perhaps seen most clearly from the representation (2.48),

$$G_0(z,a) = \begin{cases} 0, & z \in C_0, \\ \text{Im}[v_j(a)], & z \in C_j,\ j = 1, \ldots, M. \end{cases} \tag{2.50}$$

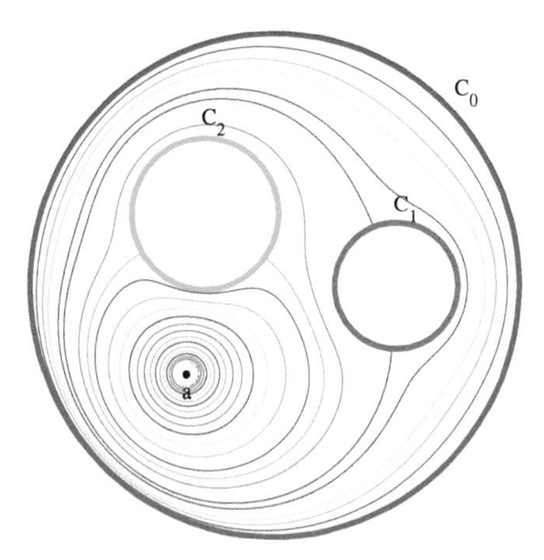

Figure 2.4. *Contours of $G_0(z,a)$ in D_z for $M = 2$. There are no closed contours around either C_1 (red) or C_2 (green). There are, however, closed contours near C_0. There are closed contours around the point a.*

We call $G_0(z,a)$ a *modified Green's function* because it satisfies the same partial differential equation (2.1) as the first-type Green's function $G(z,a)$ but it assumes generally nonzero constant values on the boundaries of D_z. Indeed, these constants take values that are precisely such that (2.41) holds, i.e., that the periods of $\mathcal{G}_0(z,a)$ around the holes $\{C_j | j = 1, \ldots, M\}$ all vanish.

By the divergence theorem we have

$$-1 = \int\int_{D(z)} \nabla^2 G_0(z,a) dA_z = \int_{C_0} \frac{\partial G_0(z,a)}{\partial n_z} ds_z = -\int_{C_0} d\mathcal{G}_0(z,a), \qquad (2.51)$$

where, in the second equation, we have used the fact that the periods of $\mathcal{G}_0(z,a)$ around the interior holes vanish. Collecting all these results together means that we can write

$$\int_{C_k} d\mathcal{G}_0(z,a) = \delta_{0k}, \qquad k = 0, 1, \ldots, M. \qquad (2.52)$$

This relation is important since it shows that the analytic function $\mathcal{G}_0(z,a)$ is single-valued as z encircles any of the interior boundaries $\{C_j | j = 1, \ldots, M\}$, but it is not, however, a single-valued function in D_z. This multivaluedness is associated with the logarithmic singularity of $\mathcal{G}_0(z,a)$ at $z = a$.

A single-valued branch of $\mathcal{G}_0(z,a)$ in D_z can be defined by drawing a "barrier" between a and any point on C_0 and insisting that any continuous deformation of the point z must not cross this barrier if we want to stay on a single-valued branch of $\mathcal{G}_0(z,a)$. Such a barrier between a and C_0 is drawn in black in Figure 2.2.

The function $\mathcal{G}_0(z,a)$ will play a crucial role in constructing the prime function in Chapter 4.

Remark on terminology: Throughout this book we will use the word "analytic" to describe functions that are holomorphic in some variable except for isolated or branch point singularities. Thus we will describe $\mathcal{G}_0(z,a)$ as being an analytic function of $z \in D_z$ even though it has a logarithmic branch point at $z = a$.

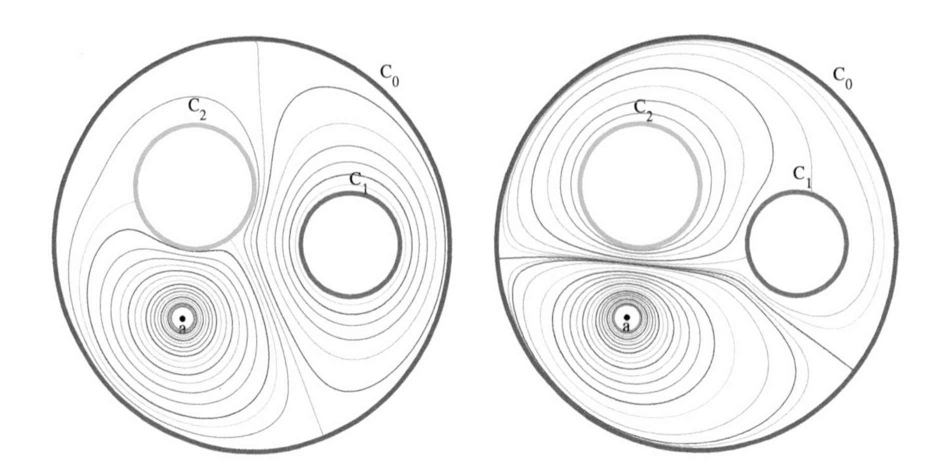

Figure 2.5. *Contours of the two "other" modified Green's functions, $G_1(z,a)$ and $G_2(z,a)$, for the $M = 2$ case shown in Figure 2.4. For $G_1(z,a)$ closed contours encircle C_1 (red), and for $G_2(z,a)$ they encircle C_2 (green). Both exhibit closed contours around the point a.*

2.7 ▪ Other modified Green's functions $\{G_j(z,a)\}$

It is useful to summarize what we just did. The construction of $G_0(z,a)$ made use of the functions $\{v_j(z)|j = 1,\ldots,M\}$ to nullify the periods of the analytic extension of the first-type Green's function $\mathcal{G}(z,a)$ about all the holes with boundaries $C_1,\ldots,C_M$, its period around C_0 remaining untouched. The circle C_0 therefore appears to have been singled out for special treatment. It is for this reason that we have included the subscript "0" in the notation $G_0(z,a)$.

What if we had picked a different circular boundary, that is, one of the circles $\{C_j|j = 1,\ldots,M\}$, for this special treatment instead?

To explore this we pick some value of j between 1 and M and, by a natural extension of our notation, define the analytic function $\mathcal{G}_j(z,a)$ for $j = 1,\ldots,M$ by

$$\mathcal{G}_j(z,a) \equiv \mathcal{G}_0(z,a) - v_j(z) - v_j(a) + \mathrm{i}[\mathcal{P}^{-1}]_{jj} \tag{2.53}$$

and then set

$$G_j(z,a) = \mathrm{Im}[\mathcal{G}_j(z,a)]. \tag{2.54}$$

It is not difficult to see that

$$\nabla^2 G_j(z,a) = -\delta(z-a), \qquad z \in D_z, \tag{2.55}$$

which is a property it shares with both $G(z,a)$ and $G_0(z,a)$; see (2.1) and (2.49). Since both functions $G_0(z,a)$ and $v_j(z)$ are constant on all the boundaries of D_z, so indeed is $G_j(z,a)$. Exercise 2.8 asks the reader to determine these constant values. Moreover, as a result of adding the term $-v_j(z)$ in (2.53), which is analytic in D_z, the function $\mathcal{G}_j(z,a)$ satisfies

$$\int_{C_k} d\mathcal{G}_j(z,a) = -\delta_{jk}, \qquad k = 0,1,\ldots,M. \tag{2.56}$$

In other words, the term $-v_j(z)$ has canceled out the unit period around C_0 associated with $\mathcal{G}_0(z,a)$ at the price of introducing a nonzero period -1 around C_j. It is the circle C_j

that has now been singled out for special treatment—hence the subscript j in the notation $G_j(z,a)$.

The term $-v_j(a)$ in (2.53), which is independent of z and therefore has no effect on the periods of the function, has been included to preserve the symmetry property that

$$G_j(z,a) = G_j(a,z). \tag{2.57}$$

The constant $\mathrm{i}[\mathcal{P}^{-1}]_{jj}$ has been included to enforce the normalization

$$G_j(z,a) = 0, \qquad z \in C_j. \tag{2.58}$$

In applications the last two properties (2.57) and (2.58) are not always necessary.

Each modified Green's function $G_j(z,a)$ satisfies the differential equation (2.55) with the subscript j conveniently conveying around which boundary circle of D_z the analytic extension of the function has a nonzero period.

Exercises

2.1. **(Symmetry of τ_{jk})** This exercise is about establishing the symmetry of the period matrix τ_{jk} defined in (2.30).

(a) On use of the definition (2.22), verify that the matrix $\mathcal{P}_{jk}$, and hence also the period matrix τ_{jk}, is symmetric as in (2.23).

(b) The harmonic functions $\Sigma_j(z)$ for $j = 1,\dots,M$ are defined in (2.34). Use Green's second identity with the two functions Σ_j and Σ_k to deduce that

$$\Sigma_j\Big|_{C_k} = \Sigma_k\Big|_{C_j}. \tag{2.59}$$

Use this result to argue that the period matrix τ_{jk} is symmetric.

2.2. **(Analytic functions in D_z)** Show (using Green's second identity, for example) that any function $H(z)$ that is analytic in D_z with constant imaginary part on ∂D_z is a linear combination of the functions $\{v_j(z)|j = 1,\dots,M\}$.

2.3. **(Alternative representations of $G_0(z,a)$)** Verify that (2.47) and (2.48) are valid alternative representations of the modified Green's function $G_0(z,a)$.

2.4. **(A Liouville-type theorem)** By using Green's second identity, together with the functions $\{v_j(z)|j = 1,\dots,M\}$, show that any function $H(z)$ that is analytic *and* single-valued in D_z so that

$$\int_{C_j} dH(z) = 0, \qquad j = 1,\dots,M, \tag{2.60}$$

and with constant imaginary part on ∂D_z is necessarily constant.

2.5. **(Concentric annulus)** Verify that if D_z is the concentric annulus

$$\rho < |z| < 1 \tag{2.61}$$

for some $0 < \rho < 1$, then the associated functions as introduced in this chapter are

$$\sigma_1(z) = \frac{\log|z|}{\log\rho}, \qquad \hat{v}_1(z) = \mathrm{i}\frac{\log z}{\log\rho}, \qquad v_1(z) = \frac{1}{2\pi\mathrm{i}}\log z, \tag{2.62}$$

while the period matrix has just the single entry

$$\mathcal{P}_{11} = \oint_{C_1} d\hat{v}_1 = \oint_{|z|=\rho} \frac{\mathrm{i}}{\log \rho}\frac{dz}{z} = -\frac{2\pi}{\log \rho}. \tag{2.63}$$

Note that in (2.62) we have chosen a real additive degree of freedom associated with $v_1(z)$ to be zero.

2.6. Suppose the real function ϕ is harmonic in some $(M+1)$-connected circular domain D_z and satisfies the boundary conditions

$$\phi = \begin{cases} 0, & z \in C_0, \\ \Phi_j, & z \in C_j,\ j = 1, \dots, M, \end{cases} \tag{2.64}$$

where $\{\Phi_j | j = 1, \dots, M\}$ is a set of constants. Define the M quantities

$$\Gamma_j = \int_{C_j} \frac{\partial \phi}{\partial n} ds, \qquad j = 1, \dots, M. \tag{2.65}$$

Show that

$$\Phi_j = [\mathcal{P}^{-1}]_{jk}\Gamma_k, \tag{2.66}$$

where the summation convention is assumed.

2.7. **(Conformal invariance of harmonic measures)** Let D_w be a bounded multiply connected domain in a complex w plane that is the image of some preimage circular domain D_z under a conformal mapping

$$w = f(z). \tag{2.67}$$

Let ∂D_j denote the jth boundary of D_w for $j = 0, 1, \dots, M$ and suppose that ∂D_j is the image of the boundary C_j of D_z. Let $\tilde{\sigma}_j(w)$ satisfy

$$\nabla^2 \tilde{\sigma}_j = 0, \qquad w \in D_w, \tag{2.68}$$

with

$$\tilde{\sigma}_j(w) = \begin{cases} 0, & w \in \partial D_0, \\ 1, & w \in \partial D_j, \\ 0, & w \in \partial D_k,\ k \neq 0, j, \end{cases} \tag{2.69}$$

and let the period matrix be

$$\tilde{\mathcal{P}}_{jk} = -\int_{\partial D_k} \frac{\partial \tilde{\sigma}_j}{\partial n_w} ds_w. \tag{2.70}$$

Verify that

$$\sigma_j(z) = \tilde{\sigma}_j(f(z)), \qquad \tilde{\mathcal{P}}_{jk} = \mathcal{P}_{jk}, \tag{2.71}$$

where $\{\sigma_j | j = 1, \dots, M\}$ are the functions (2.8) associated with the circular domain D_z and $\mathcal{P}_{jk}$ is its period matrix.

2.8. **(Boundary values of $\{G_j(z,a) | j = 1, \dots, M\}$)** The constant boundary values of the modified Green's function $G_0(z,a)$ are given in (2.50). The modified Green's functions $\{G_j(z,a) | j = 1, \dots, M\}$ as defined in (2.53)–(2.54) also assume constant values on the boundary circles $\{C_j | j = 0, 1, \dots, M\}$ of D_z. Show that, for $j = 1, \dots, M$,

$$G_j(z,a) = \begin{cases} -\mathrm{Im}[v_j(a)] + [\mathcal{P}^{-1}]_{jj}, & z \in C_0, \\ +\mathrm{Im}[v_k(a)] - [\mathcal{P}^{-1}]_{jk} - \mathrm{Im}[v_j(a)] + [\mathcal{P}^{-1}]_{jj}, & z \in C_k, \end{cases} \tag{2.72}$$

where $k = 1, \dots, M$.

Chapter 3

The Schottky double

3.1 ▪ Doubling the domain D_z

In order to develop the function theory it turns out to be useful to "double up" the multiply connected circular domain D_z. This doubling of the domain is done by a process of reflection of D_z in its outer boundary C_0. We will then work in a domain, to be called F, comprising two copies of D_z.[7] The idea behind the doubling procedure is that it generates a second set of circular boundaries that can, in a sense to be made precise in this chapter, be "glued" together in pairs. Mathematically, this gluing is done using a holomorphic identification of circular boundaries by means of Möbius maps.

The notion of "reflection in a circle" was discussed in §1.10 of Chapter 1. Let D_z' denote the reflection of D_z in the unit circle C_0, with C_j' used to denote the circle bounding D_z' that is the reflection of C_j in C_0. Exercise 1.4 asks the reader to confirm that this reflection of C_j in C_0 is indeed a circle. Consequently, it is easy to check that the reflected domain D_z' is itself a multiply connected circular domain with the same connectivity as D_z. Let F be the union of the original circular domain D_z and its reflection D_z' in C_0: $F = D_z \cup D_z'$. The region F, which is itself a circular domain, can be either bounded or unbounded, depending on whether the origin is in the domain D_z.

Figure 3.1 illustrates this doubling process applied to the triply connected circular domain D_z in Figure 2.1 of Chapter 2. In Figure 3.1 the origin is in D_z so that F is unbounded since it contains the reflection of the origin in C_0, which is the point at infinity. The boundaries of F are the $2M$ circles $\{C_j, C_j' | j = 1, \ldots, M\}$; in the $M = 2$ case shown these are the 4 circles C_1, C_1', C_2, and C_2'. The circle C_0, which we have used as the basis for carrying out this doubling procedure, is not a boundary of F.

This chapter outlines the theory for the case where F is unbounded, as in Figure 3.1. In Chapter 5, which deals with the case where D_z is the concentric annulus $\rho < |z| < 1$, the corresponding region F is the "doubled annulus" $\rho < |z| < 1/\rho$, which is bounded. All the details of this chapter can be generalized to the case of bounded F without difficulty. The same goes for the marginal case where the origin lies on the boundary of one of the circles $\{C_j | j = 1, \ldots, M\}$.

[7] This is a model of the so-called *Schottky double* of D_z.

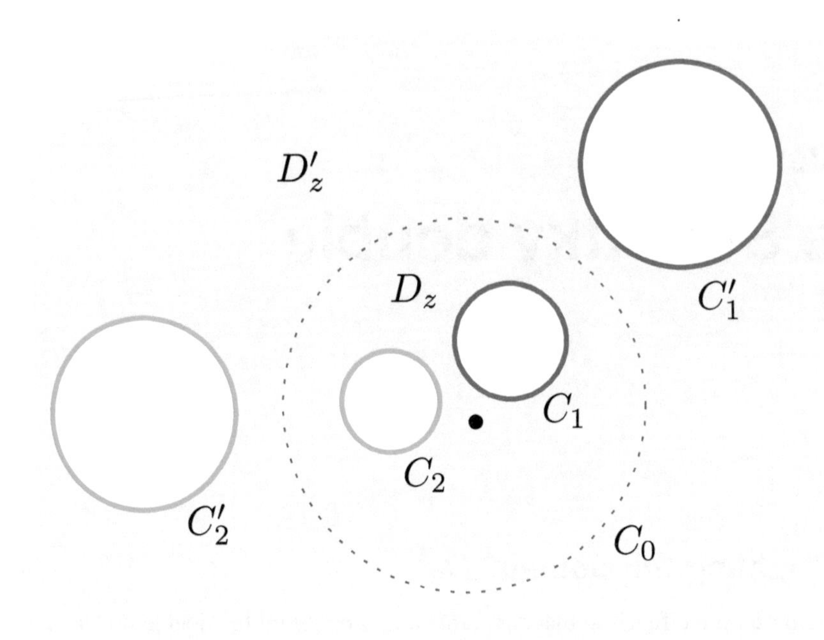

Figure 3.1. *An unbounded doubled domain $F = D_z \cup D'_z$ associated with the triply connected circular domain D_z of Figure 2.1. Since the origin is in D_z, then F is the unbounded (light blue) region exterior to the 4 circles C_1, C_2, C'_1, and C'_2. The circle C_0, shown as a dashed line, is not a boundary of F.*

3.2 ▪ Analytic continuation of $\mathcal{G}_0(z,a)$ to the doubled domain

The reason for doubling up the domain in this way becomes apparent when we start thinking about the analytic continuation of the function $\mathcal{G}_0(z,a)$ outside D_z. This analytic continuation can be effected by making use of the first boundary condition in (2.50), which implies that, for z on C_0,

$$\mathcal{G}_0(z,a) = \overline{\mathcal{G}_0(1/\overline{z},a)}. \tag{3.1}$$

This is a relation between analytic functions valid on C_0, and therefore also off C_0 by a process of analytic continuation. By the Schwarz reflection principle [76, 61], (3.1) allows us to analytically continue $\mathcal{G}_0(z,a)$ into D'_z.

In Chapter 2 we discussed how a barrier joining the point a to C_0 can be introduced to define a single-valued branch of $\mathcal{G}_0(z,a)$. Such a barrier was shown in Figure 2.2. It is natural, in view of (3.1), to extend this barrier into D'_z by the same process of reflection in C_0, resulting in a longer barrier extending into D'_z now joining a to $1/\overline{a}$. If this barrier meets C_0 at right angles, then this longer barrier extended by reflection will be smooth; Figure 3.2 shows a typical barrier drawn between a and its reflection $1/\overline{a}$ in C_0. By the Schwarz reflection principle we can use (3.1) to analytically continue $\mathcal{G}_0(z,a)$ into the doubled domain F and define a single-valued branch of it by allowing the argument z to move continuously throughout F but without crossing the barrier joining a to $1/\overline{a}$. Note that since $\mathcal{G}_0(z,a)$ is analytic everywhere in D_z except for a logarithmic singularity of

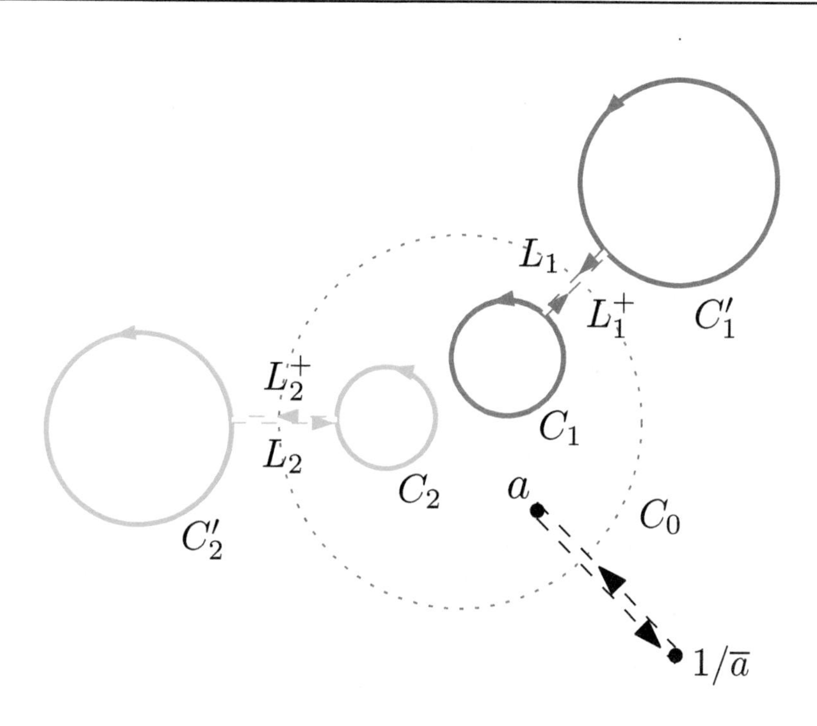

Figure 3.2. *The barrier joining the logarithmic branch points at a and $1/\overline{a}$ is shown in black. The barriers $\{(L_j, L_j^+)|j = 1,\ldots,M\}$ are introduced to define single-valued branches of the analytic functions $\{v_j(z)|j = 1,\ldots,M\}$. L_j starts at a point on C_j' and ends at its reflection in C_0 on C_j; L_j^+ is the same contour traversed in the opposite direction. For each j, the 4 contours C_j', L_j, C_j, L_j^+, with C_j', C_j traversed anticlockwise, form a dumbbell-shaped loop.*

strength $1/(2\pi\mathrm{i})$ at $z = a$, relation (3.1) allows us to deduce that it can be analytically continued everywhere into D_z' (off the barrier just discussed) except for a logarithmic singularity of strength $-1/(2\pi\mathrm{i})$ at $1/\overline{z} = a$, that is, when $z = 1/\overline{a}$. These two logarithmic singularities of opposite strength at a and $1/\overline{a}$ can be connected by the extended barrier (or "branch cut") shown in Figure 3.2.

Other useful consequences for $\mathcal{G}_0(z,a)$ follow from the other boundary conditions in (2.50) on the circles $\{C_j|j = 1,\ldots,M\}$.

Suppose z moves continuously from a point on C_0 into the domain D_z' until it reaches C_j' without crossing the barrier joining a to $1/\overline{a}$. Since if z lies on C_j' then $1/\overline{z}$ lies on C_j, (2.50) and (3.1) tell us that

$$\mathrm{Im}[\mathcal{G}_0(z,a)] = -\mathrm{Im}[v_j(a)], \qquad z \in C_j', \tag{3.2}$$

where we have used the elementary fact that the imaginary parts of complex conjugate quantities differ by a minus sign.

We combine all this information about the analytic continuation of $\mathcal{G}_0(z,a)$ into D_z' to conclude that, for $z \in F$, it can be written as

$$\mathcal{G}_0(z,a) = \frac{1}{2\pi\mathrm{i}} \log\left[\frac{(z-a)}{|a|(z-1/\overline{a})}\right] + \tilde{\mathcal{G}}_0(z,a), \tag{3.3}$$

where $\tilde{\mathcal{G}}_0(z,a)$ is analytic everywhere in F and must be chosen so that $\mathcal{G}_0(z,a)$ satisfies the boundary conditions

$$\mathrm{Im}[\mathcal{G}_0(z,a)] = \begin{cases} +\mathrm{Im}[v_j(a)], & z \in C_j,\ j = 1,\dots,M, \\ -\mathrm{Im}[v_j(a)], & z \in C_j',\ j = 1,\dots,M. \end{cases} \tag{3.4}$$

Other inferences concerning the periods of $\mathcal{G}_0(z,a)$ around the M holes with boundaries $\{C_j'|j = 1,\dots,M\}$ in D_z' can be made from (3.1). For $k = 1,\dots,M$, and on use of (3.1),

$$\int_{C_k'} d\mathcal{G}_0(z,a) = \int_{C_k'} d\overline{\mathcal{G}_0(1/\overline{z},a)}, \qquad k = 1,\dots,M. \tag{3.5}$$

On making a change of integration variable $z = 1/\overline{z}$, the right-hand side becomes

$$-\int_{C_k} d\overline{\mathcal{G}_0(z,a)} = -\int_{C_k} d\mathcal{G}_0(z,a) = 0, \qquad k = 1,\dots,M, \tag{3.6}$$

where the minus sign is introduced because the sense of integration is reversed by the change of variable; the first equality follows because, as seen in (3.4), the imaginary part of $\mathcal{G}_0(z,a)$ is constant on any circle C_k for $k = 1,\dots,M$ so that $d\mathcal{G}_0(z,a)$ is real there; and the second equality in (3.6) follows from (2.52). The conclusion is that

$$\int_{C_k'} d\mathcal{G}_0(z,a) = 0, \qquad k = 1,\dots,M. \tag{3.7}$$

This means that $\mathcal{G}_0(z,a)$ has vanishing periods around *all* $2M$ circular holes $\{C_j, C_j'|j = 1,\dots,M\}$ in F. $\mathcal{G}_0(z,a)$ is therefore single-valued as z continuously encircles *any* of the $2M$ boundaries $\{C_j, C_j'|j = 1,\dots,M\}$ of the doubled domain F, although its value changes by $+1$ if z (once) encircles the point a continuously or by -1 if z (once) encircles the point $1/\overline{a}$ continuously.

3.3 ▪ Analytic continuation of $v_j(z)$ to the doubled domain

In the same spirit, it is natural to ask about the analytic continuation into the doubled domain F of the set of M functions $\{v_j(z)|j = 1,\dots,M\}$ introduced in Chapter 2.

For each $j = 1,\dots,M$, the first boundary condition in (2.32) on C_0 implies that

$$v_j(z) = \overline{v_j(1/\overline{z})}, \qquad j = 1,\dots,M, \qquad z \in C_0. \tag{3.8}$$

This allows, again by the Schwarz reflection principle, for each $v_j(z)$ to be analytically continued into D_z'. In Chapter 2 a barrier was introduced joining C_j to C_0 in order to define a single-valued branch of $v_j(z)$. It is natural, in view of (3.8), to extend this barrier into D_z' by reflection in C_0, resulting in a longer barrier now joining C_j to C_j'. This is done for each $j = 1,\dots,M$. We will think of these barriers as having two sides, denoted by the pair (L_j, L_j^+): L_j starts at a point on C_j' and ends at its reflection in C_0 on C_j; L_j^+ is the same barrier traversed in the opposite direction, but, since they are "barriers," it is understood that a point on L_j can only reach the corresponding point on the "other side" L_j^+ by making a continuous anticlockwise traversal of C_j or a continuous clockwise traversal of C_j'. Figure 3.2 shows a schematic illustrating how M such barriers L_j and L_j^+ linking C_j to C_j' can be introduced; the case $M = 2$ is shown. On use of (3.8) we deduce from the Schwarz reflection principle that each $v_j(z)$ for $j = 1,\dots,M$ is analytic in F with a jump along the barrier (L_j, L_j^+).

Other useful consequences for $v_j(z)$ follow from the other boundary conditions in (2.32) on the circles $\{C_j|j=1,\ldots,M\}$.

Pick a value of j and suppose z moves continuously from a point on C_0 into the domain D'_z until it reaches C'_j without crossing the barrier (L_j, L_j^+). Since if z lies on C'_j then $1/\overline{z}$ lies on C_j, (2.32) and (3.8) tell us that

$$\mathrm{Im}[v_j(z)] = -\left[\mathcal{P}^{-1}\right]_{jk}, \qquad z \in C_k, \tag{3.9}$$

where we have used the fact that the imaginary parts of complex conjugate quantities differ by a minus sign. We therefore conclude that on the $2M$ boundaries of F, the function $v_j(z)$ satisfies

$$\mathrm{Im}[v_j(z)] = \begin{cases} \left[\mathcal{P}^{-1}\right]_{jk}, & z \in C_k,\ k=1,\ldots,M, \\ -\left[\mathcal{P}^{-1}\right]_{jk}, & z \in C'_k,\ k=1,\ldots,M. \end{cases} \tag{3.10}$$

Relation (3.8) also provides information concerning the periods of each $v_j(z)$ around the M holes with boundaries $\{C'_k|k=1,\ldots,M\}$ in D'_z. We find from (3.8) that

$$\int_{C'_k} dv_j(z) = \int_{C'_k} d\overline{v_j(1/\overline{z})}. \tag{3.11}$$

On making a change of integration variable $z = 1/\overline{z}$, the right-hand side becomes

$$-\int_{C_k} d\overline{v_j(z)} = -\int_{C_k} dv_j(z) = -\delta_{jk}, \tag{3.12}$$

where the minus sign is introduced because the sense of integration is reversed by the change of variable; the first equality follows because each $v_j(z)$ has constant imaginary part on each circle C_k for $k=1,\ldots,M$, meaning that $dv_j(z)$ is real there; and the second equality follows on use of (2.33). The conclusion is that

$$\int_{C'_k} dv_j(z) = -\delta_{jk}, \qquad j,k=1,\ldots,M. \tag{3.13}$$

3.4 ▪ A parametric identity for $\mathcal{G}_0(z,a)$

The function $\mathcal{G}_0(z,a)$ and the set of analytic functions $\{v_j(z)|j=1,\ldots,M\}$ have now been analytically continued, as functions of z, from D_z to the doubled domain F. We originally defined $\mathcal{G}_0(z,a)$ as the analytic extension in D_z of the modified Green's function $G_0(z,a)$ in D_z; the latter function is the solution of the boundary value problem

$$\nabla^2 G_0(z,a) = -\delta(z-a), \qquad z \in D_z, \tag{3.14}$$

with boundary values

$$G_0(z,a) = \begin{cases} 0, & z \in C_0, \\ +\mathrm{Im}[v_j(a)], & z \in C_j,\ j=1,\ldots,M. \end{cases} \tag{3.15}$$

In view of the analytic extension of $\mathcal{G}_0(z,a)$ to F just carried out we can now think of the same $G_0(z,a)$ as being the solution of the following boundary value problem in the doubled domain F:

$$\nabla^2 G_0(z,a) = -\delta(z-a) + \delta(z-1/\overline{a}), \qquad z \in F, \tag{3.16}$$

with boundary values

$$G_0(z,a) = \begin{cases} +\mathrm{Im}[v_j(a)], & z \in C_j,\ j = 1,\dots,M, \\ -\mathrm{Im}[v_j(a)], & z \in C_j',\ j = 1,\dots,M. \end{cases} \tag{3.17}$$

On use of the identities (3.8) satisfied by the functions $\{v_j(z)|j = 1,\dots,M\}$ the boundary conditions (3.17) can be rewritten as

$$G_0(z,a) = \begin{cases} \dfrac{1}{2}\left[\mathrm{Im}[v_j(a)] - \mathrm{Im}[v_j(1/\overline{a})]\right], & z \in C_j,\ j = 1,\dots,M, \\ -\dfrac{1}{2}\left[\mathrm{Im}[v_j(a)] - \mathrm{Im}[v_j(1/\overline{a})]\right], & z \in C_j',\ j = 1,\dots,M. \end{cases} \tag{3.18}$$

On setting $a \mapsto 1/\overline{a}$ in the statement (3.16) and (3.18) of the boundary value problem satisfied by $G_0(z,a)$ in F we notice that it produces an *almost* identical boundary value problem, the only difference being a minus sign on the right-hand sides of (3.16) and (3.18). It can be argued that its unique solution, which it is natural to denote by $G_0(z,1/\overline{a})$ since we have exchanged a for $1/\overline{a}$, is

$$G_0(z,1/\overline{a}) = -G_0(z,a). \tag{3.19}$$

The analytic extensions, as functions of z, of both sides of (3.19) for $z \in F$ imply the following parametric identity for $\mathcal{G}_0$:

$$\mathcal{G}_0(z,1/\overline{a}) = -\mathcal{G}_0(z,a). \tag{3.20}$$

This relation is very natural; since $\mathcal{G}_0(z,a)$ is characterized by a logarithmic singularity of strength $+1/(2\pi\mathrm{i})$ as $z = a \in D_z$ and one of strength $-1/(2\pi\mathrm{i})$ at its reflection in C_0 at $1/\overline{a} \in D_z'$, if a migrates into D_z' so that now there is a logarithmic singularity of strength $+1/(2\pi\mathrm{i})$ at $a \in D_z'$ and one of strength $-1/(2\pi\mathrm{i})$ at $1/\overline{a} \in D_z$, then $\mathcal{G}_0(z,a)$ must be just the negative of $\mathcal{G}_0(z,1/\overline{a})$.

This identity (3.20) is important: if $a \in D_z$, as has been assumed up to now, then $1/\overline{a} \in D_z'$ so the identity (3.20) provides a means of defining the function $\mathcal{G}_0(z,a)$ for any choice of the parameter a lying in the doubled domain F. It also shows that $\mathcal{G}_0(z,a)$ is harmonic in a for any $a \in F$.

3.5 ▪ The conformal moduli $\{\delta_j, q_j | j = 1,\dots,M\}$

The multiply connected circular domains D_z introduced in Chapter 2 are geometrically simple. They are characterized by a finite set of parameters: the centers and radii of the boundaries $\{C_j|j = 1,\dots,M\}$ of the excised circular discs.

Let $\delta_j \in \mathbb{C}$ be the center of circle C_j, and let $q_j \in \mathbb{R}$ be its radius. We therefore associate with any $(M+1)$-connected circular domain D_z the set of circle centers $\{\delta_j \in \mathbb{C}|j = 1,\dots,M\}$ and their radii $\{q_j \in \mathbb{R}|j = 1,\dots,M\}$. This geometrical data will be referred to as the *conformal moduli* of D_z. It constitutes a set of $3M$ real parameters (see Figure 3.3).

3.6 ▪ The Möbius maps $\{\theta_j(z)|j = 1,\dots,M\}$

For points z on circle C_j for $j = 1,\dots,M$ we can write

$$\overline{z} = S_j(z), \qquad z = \overline{S_j(z)}, \tag{3.21}$$

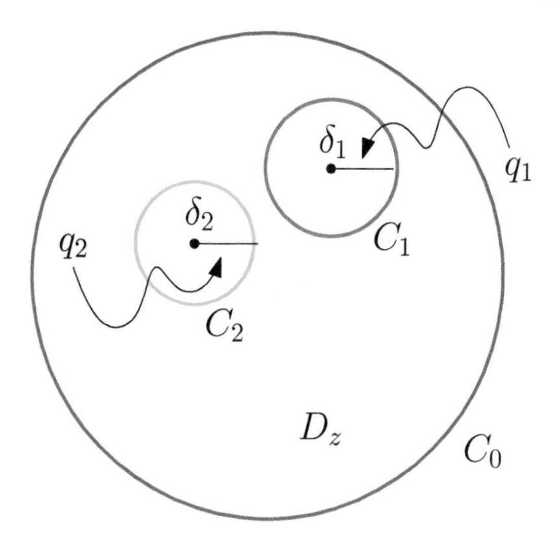

Figure 3.3. *Conformal moduli: these are the centers and radii* $\{\delta_j, q_j | j = 1, \ldots, M\}$ *of the circles* $\{C_j | j = 1, \ldots, M\}$*. The case* $M = 2$ *is shown.*

where $S_j(z)$ is the Schwarz function of the circle C_j introduced in §1.9. Since the equation for C_j is

$$|z - \delta_j|^2 = (z - \delta_j)(\overline{z} - \overline{\delta_j}) = q_j^2, \tag{3.22}$$

it is easy to rearrange this to find

$$\overline{z} = S_j(z) = \overline{\delta_j} + \frac{q_j^2}{z - \delta_j}, \qquad j = 1, \ldots, M. \tag{3.23}$$

It is expedient to introduce the new set of Möbius maps given by

$$\theta_j(z) \equiv \overline{S_j(1/\overline{z})} = \delta_j + \frac{q_j^2 z}{1 - \overline{\delta_j} z}, \qquad j = 1, \ldots, M. \tag{3.24}$$

This set will play an important role in the theoretical development and will arise naturally in the considerations of the next subsection. These Möbius maps have some interesting geometrical properties which it is useful to bear in mind.

First, it can be shown that the Möbius map $\theta_j(z)$ will transplant any point on C_j' to its corresponding point on C_j obtained by reflection in the unit circle C_0. This is shown schematically in Figure 3.4.

It can also be shown that if all points in the interior of F are subjected to the mapping $\theta_j(z)$, then the image will lie inside circle C_j.

Another feature is that if all points in the interior of F are subjected to the mapping $\theta_j^{-1}(z)$, then the image will lie inside circle C_j'.

Exercises 1.5 and 3.1 explore some of these geometrical properties of the Möbius maps $\{\theta_j(z) | j = 1, \ldots, M\}$.

3.7 ▪ Analytic continuations outside F

The Möbius maps $\{\theta_j(z) | j = 1, \ldots, M\}$ turn out to be useful in finding the analytic continuations of the functions $\mathcal{G}_0(z, a)$ and $\{v_j(z) | j = 1, \ldots, M\}$ outside the domain F and into other regions of the complex z plane.

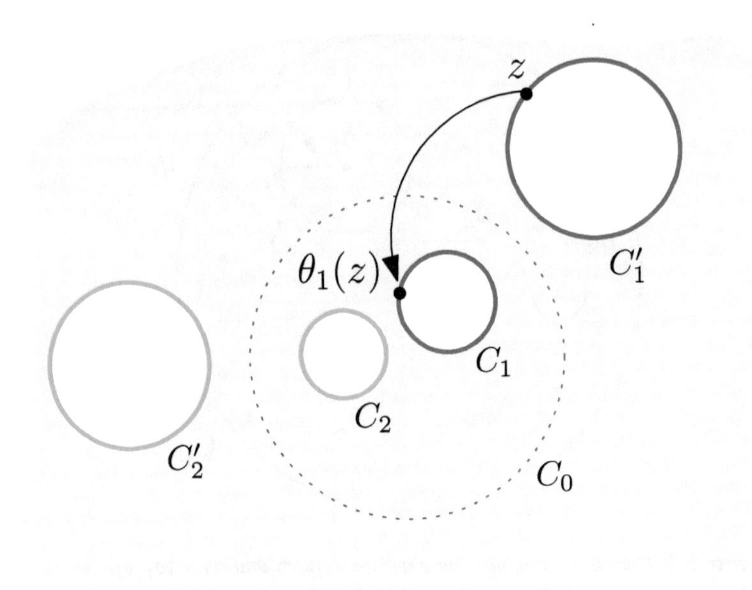

Figure 3.4. *Geometry of the Möbius map $\theta_1(z)$: it transplants a point z on C_1' to the corresponding point $1/\overline{z}$ on C_1 obtained by reflection in C_0.*

Pick a value of j between 1 and M and suppose that a point z starts on C_0 and moves continuously into D_z' until it reaches C_j' and without crossing the barrier between a and $1/\overline{a}$. Then $1/\overline{z}$ will lie on C_j and (3.1) implies that

$$\mathcal{G}_0(z,a) = \overline{\mathcal{G}_0(1/\overline{z},a)}. \tag{3.25}$$

But we know from (3.4) that

$$\mathrm{Im}[\mathcal{G}_0(1/\overline{z},a)] = \mathrm{Im}[v_j(a)], \qquad 1/\overline{z} \in C_j, \tag{3.26}$$

or, equivalently,

$$\mathcal{G}_0(1/\overline{z},a) - \overline{\mathcal{G}_0(1/\overline{z},a)} = 2\mathrm{i}\mathrm{Im}[v_j(a)], \qquad 1/\overline{z} \in C_j. \tag{3.27}$$

But if $1/\overline{z}$ is on C_j, then, from (3.21),

$$\frac{1}{\overline{z}} = \overline{S_j(1/\overline{z})} = \theta_j(z), \tag{3.28}$$

where S_j is the Schwarz function of circle C_j and, in the second equality, we have used the definition (3.24). Combining (3.28) and (3.27) with (3.25) leads to

$$\mathcal{G}_0(\theta_j(z),a) - \mathcal{G}_0(z,a) = 2\mathrm{i}\mathrm{Im}[v_j(a)], \qquad z \in C_j'. \tag{3.29}$$

It is important to note that while (3.29) holds for $z \in C_j'$, as a relation between analytic functions, it can be analytically continued off this circle. Writing it as

$$\mathcal{G}_0(\theta_j(z),a) = \mathcal{G}_0(z,a) + 2\mathrm{i}\mathrm{Im}[v_j(a)], \qquad j = 1,\ldots,M, \tag{3.30}$$

it becomes clear that we can use this to analytically continue $\mathcal{G}_0(z,a)$ into the regions of the complex z plane that are images of F under the Möbius maps $\{\theta_j | j = 1, \dots, M\}$.

Exercise 3.6 explores what can happen to relation (3.30) if z is dragged continuously to $\theta_j(z)$ and allowed to cross the barrier between a and $1/\overline{a}$ (or any of its images under Θ).

To arrive at (3.30) we took a point on z and continuously deformed it into D'_z, without crossing the barrier joining a to $1/\overline{a}$, until it reached C'_j; however, we might, on the other hand, have continuously dragged the point z into D_z—again without crossing the barrier joining a to $1/\overline{a}$—until it reached C_j. Similar steps to those carried out above would then lead to

$$\mathcal{G}_0(\theta_j^{-1}(z), a) = \mathcal{G}_0(z,a) - 2\mathrm{i}\mathrm{Im}[v_j(a)], \qquad j = 1, \dots, M. \tag{3.31}$$

The same relation also follows by letting $z \mapsto \theta_j^{-1}(z)$ in (3.30).

(3.30) and (3.31) hold provided we do not cross the barrier between a and $1/\overline{a}$. In view of (3.30) and (3.31), which provide analytic continuations of $\mathcal{G}_0(z,a)$ to the "images" of F in other parts of the complex plane, it is natural to introduce the "images" of the chosen barrier under the action of the Möbius maps $\{\theta_k(z) | k = 1, \dots, M\}$ and their inverses. These new barriers, and for $M \geq 1$ there is an infinite number of them, will lie in the images of F under the action of these Möbius maps. Relations (3.30) and (3.31) will hold provided z is continuously deformed to the point $\theta_k(z)$ without crossing any of these barriers.

By extension, we note that the analytic relations (3.30) and (3.31) are valid for any $j = 1, \dots, M$ and we might consider setting, for example, $z \mapsto \theta_k(z)$ in (3.30) for $k \neq j$:

$$\mathcal{G}_0(\theta_j(\theta_k(z)), a) = \mathcal{G}_0(\theta_k(z), a) + 2\mathrm{i}\mathrm{Im}[v_j(a)] \tag{3.32}$$

$$= \mathcal{G}_0(z,a) + 2\mathrm{i}\mathrm{Im}[v_k(a)] + 2\mathrm{i}\mathrm{Im}[v_j(a)], \tag{3.33}$$

where, in the second equality, we have used (3.30) again. Such a relation is clearly just one of infinitely many that can be obtained from the basic relations (3.30) and (3.31) by repeated action of the set of Möbius maps given by

$$\{\theta_j(z), \theta_j^{-1}(z) | j = 1, \dots, M\}. \tag{3.34}$$

This set of Möbius maps can be viewed as "generators" of these functional relations. Indeed, as shown in Exercise 3.1, the infinite set of Möbius maps generated by the free action of the basic Möbius maps $\{\theta_j(z), \theta_j^{-1}(z) | j = 1, \dots, M\}$ form what is called a *Schottky group*, which we here denote by Θ.

We will not need to say anything more about the Schottky group Θ until Chapter 14, where it may be used as a basis of a representation—one of several possible choices of representation—of the prime function.

On taking a partial derivative of (3.30) with respect to a we find

$$\frac{\partial \mathcal{G}_0(\theta_j(z), a)}{\partial a} - \frac{\partial \mathcal{G}_0(z,a)}{\partial a} = \frac{dv_j(a)}{da}, \qquad z \in C'_j. \tag{3.35}$$

This functional relation will be important in Chapter 4. It can be written as

$$\frac{\partial \mathcal{G}_0(\theta_j(z), a)}{\partial a} = \frac{\partial \mathcal{G}_0(z,a)}{\partial a} + \frac{dv_j(a)}{da}, \qquad z \in C'_j, \tag{3.36}$$

and, as a relation between analytic functions of z, this too can be used to analytically continue $\partial \mathcal{G}_0(z,a)/\partial a$ into the regions of the complex z plane that are images of F under the Möbius maps $\{\theta_j | j = 1, \dots, M\}$.

Relation (3.8) can similarly be used to find useful relations involving the functions $\{v_j(z)|j=1,\ldots,M\}$.

Pick a value of j between 1 and M and suppose that a point z starts on C_0 and is dragged continuously into D'_z until it reaches C'_k and without crossing the barrier (L_j, L_j^+). Then $1/\overline{z}$ will lie on C_k and (3.8) implies that

$$v_j(z) = \overline{v_j(1/\overline{z})}, \qquad j=1,\ldots,M. \tag{3.37}$$

But we know from (3.10) that

$$\mathrm{Im}[v_j(1/\overline{z})] = \left[\mathcal{P}^{-1}\right]_{jk} = \frac{\tau_{jk}}{2\mathrm{i}}, \qquad 1/\overline{z} \in C_k, \tag{3.38}$$

or, equivalently,

$$v_j(1/\overline{z}) - \overline{v_j(1/\overline{z})} = \tau_{jk}, \qquad 1/\overline{z} \in C_k. \tag{3.39}$$

After rearrangement, we find

$$\overline{v_j(1/\overline{z})} = v_j(1/\overline{z}) - \tau_{jk}, \qquad 1/\overline{z} \in C_k. \tag{3.40}$$

But if $1/\overline{z}$ is on C_k, then

$$\frac{1}{\overline{z}} = \overline{S_k(1/\overline{z})} = \theta_k(z), \tag{3.41}$$

where $S_k(z)$ is the Schwarz function of circle C_k. Combining (3.40) and (3.41) with (3.37) leads to

$$v_j(\theta_k(z)) - v_j(z) = \tau_{jk}, \qquad j,k=1,\ldots,M, \ \text{ for } z \in C'_k. \tag{3.42}$$

On rearrangement, we find

$$v_j(\theta_k(z)) = v_j(z) + \tau_{jk}, \qquad k=1,\ldots,M, \ \text{ for } z \in C'_k, \tag{3.43}$$

which, as a relation between analytic functions, furnishes the analytic continuation of $v_j(z)$ for z off C'_k and outside the fundamental region F, indeed, into the image of F under the mapping $\theta_k(z)$ which lies inside C_k. On setting $z \mapsto \theta_k^{-1}(z)$ in this relation we find

$$v_j(\theta_k^{-1}(z)) = v_j(z) - \tau_{jk}, \qquad k=1,\ldots,M. \tag{3.44}$$

This relation also follows by picking a point z on C_0 and dragging it continuously into D_z until it reaches C_k without crossing the barrier (L_j, L_j^+).

Earlier we introduced a barrier (L_j, L_j^+) between C_j and C'_j in F in order to define a single-valued branch of $v_j(z)$ there. In view of (3.43) and (3.44) it is natural to introduce additional barriers given by all the "images" of (L_j, L_j^+) under the action of the Schottky group Θ introduced earlier; for $M \geq 1$ there will be an infinite number of such image barriers. Relation (3.42) will hold provided z is continuously deformed to the point $\theta_k(z)$ without crossing any of these barriers. Figure 3.5 shows the images of the circles C_1, C'_1, C_2, C'_2 and typical barriers (L_1, L_1^+) and (L_2, L_2^+) under the action of the level 1 Möbius maps $\theta_1, \theta_2, \theta_1^{-1}, \theta_2^{-1}$. There are infinitely more images of these circles and barriers under the action of other elements of the infinite Schottky group Θ that are not shown in Figure 3.5.[8]

[8]The geometry of the images of circles, and points, under repeated Möbius transformations is a whole subject unto itself, but we will not need to use these ideas to develop our framework. The monograph by Mumford, Series, and Wright [101] is recommended to readers interested in these aspects.

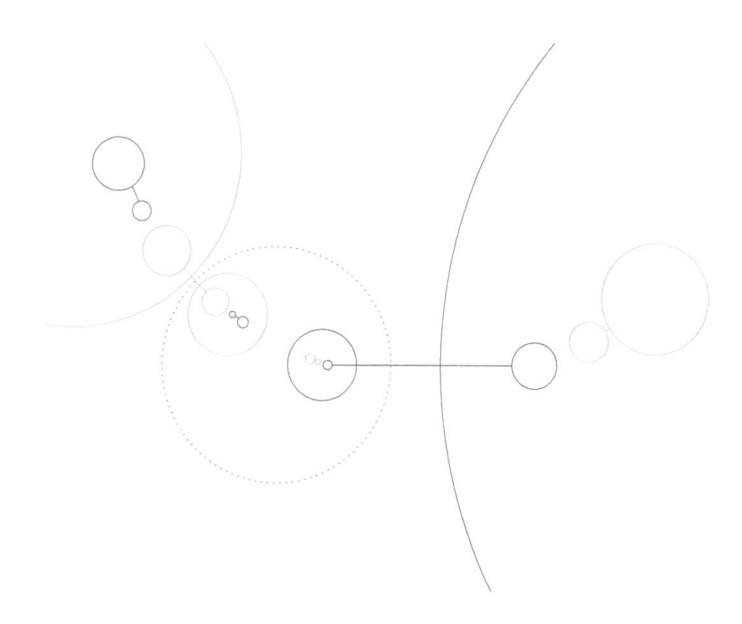

Figure 3.5. *Superposed images of the circles C_1, C_1', C_2, C_2' and typical barriers (L_1, L_1^+) and (L_2, L_2^+) under the action of the Möbius maps $\theta_1, \theta_2, \theta_1^{-1}, \theta_2^{-1}$. All images of C_1, C_1', and (L_1, L_1^+) are shown in red; the images of C_2, C_2', and (L_2, L_2^+) are shown in green. C_0 is shown as a blue dashed line. These are just the "first level" in an infinite hierarchy of images.*

Exercises

3.1. **(Geometry of the Möbius maps $\{\theta_j(z)\}$)** This exercise explores geometrical properties of the Möbius maps $\{\theta_j(z)\}$.

(a) Verify that if z is a point on C_j', then $\theta_j(z)$ lies on C_j for $j = 1, \ldots, M$; indeed

$$\theta_j(z) = \frac{1}{\overline{z}}. \tag{3.45}$$

(b) Verify that if z is a point on C_j, then

$$z = \theta_j(1/\overline{z}). \tag{3.46}$$

(c) Verify that if z is an arbitrary point strictly inside the fundamental region $F = D_z \cup D_z'$, then $\theta_j(z)$ lies inside the circle C_j.

(d) Show that each $\theta_j(z)$ for $j = 1, \ldots, M$ has precisely two *fixed points*: these are defined to be points z_* such that

$$\theta_j(z_*) = z_*. \tag{3.47}$$

Determine the location of these fixed points.

3.2. **(Analytical properties of the Möbius maps $\{\theta_j(z)\}$)** Suppose

$$\theta_j(\zeta) = \frac{a_j\zeta + b_j}{c_j\zeta + d_j}. \tag{3.48}$$

Show that, for any a, we have

$$\theta_j(\zeta) - a = (a_j - c_j a)\left(\frac{\zeta - \theta_j^{-1}(a)}{c_j\zeta + d_j}\right). \tag{3.49}$$

3.3. **(Schottky groups)** Verify the following facts:

(a) The identity map $\theta(z) = z$ is a Möbius map.

(b) The inverse of a Möbius map is another Möbius map.

(c) The composition of any two Möbius maps is another Möbius map.

(d) The set Θ made up of all compositions of the $2M$ Möbius maps

$$\{\theta_j(z), \theta_j^{-1}(z) | j = 1, \ldots, M\} \tag{3.50}$$

forms an infinite *group* under the action of composition.

The infinite set Θ is known as a free *Schottky group* on $2M$ generators.

3.4. **(Centers and radii of C_j')** Show that if q_j' denotes the radius of the circle C_j' and δ_j' denotes its center, then

$$q_j' = \frac{q_j}{||\delta_j|^2 - q_j^2|}, \qquad \delta_j' = \frac{\delta_j}{|\delta_j|^2 - q_j^2}, \tag{3.51}$$

where q_j is the radius of C_j and δ_j is its center.

3.5. By taking the imaginary part of (3.30), verify that it is consistent with relation (3.17).

3.6. **(What happens if you cross a barrier?)** Suppose that some point z on C_1' is dragged continuously to the point $\theta_1(z)$ on C_1 along the contour I_1, as shown in Figure 3.6. Then (3.30) holds and implies that

$$\mathcal{G}_0(\theta_1(z), a) = \mathcal{G}_0(z, a) + 2\mathrm{i}\mathrm{Im}[v_1(a)], \tag{3.52}$$

and (3.42) holds and implies the following relations:

$$v_1(\theta_1(z)) - v_1(z) = \tau_{11}, \qquad v_2(\theta_1(z)) - v_2(z) = \tau_{21}, \tag{3.53}$$

$$v_1(\theta_2(z)) - v_1(z) = \tau_{12}, \qquad v_2(\theta_2(z)) - v_2(z) = \tau_{22}. \tag{3.54}$$

Find what happens to all these relations if, instead, z is dragged continuously to the same point $\theta_1(z)$ but now along the contour I_2 also shown in Figure 3.6 and which crosses both the barrier (L_2, L_2^+) and the barrier joining a to $1/\overline{a}$.

3.7. **(Reflectionally symmetric domains)** Suppose all the circles $C_0, C_1, \ldots, C_M$ of a multiply connected circular domain D_z are centered on the real axis. Such a domain D_z is reflectionally symmetric about the real z axis. This additional symmetry has important ramifications for the function theory in such domains.

(a) The functions $\{\sigma_j(z) | j = 1, \ldots, M\}$ are real harmonic functions in D_z satisfying the boundary conditions (2.9); their analytic extensions are denoted by $\{\hat{v}_j(z) | j = 1, \ldots, M\}$. By considering the effect on the domain D_z of the coordinate transformation

$$z \mapsto \overline{z} \tag{3.55}$$

and considering the function

$$H(z) \equiv \hat{v}_j(z) + \overline{\hat{v}_j(\overline{z})}, \tag{3.56}$$

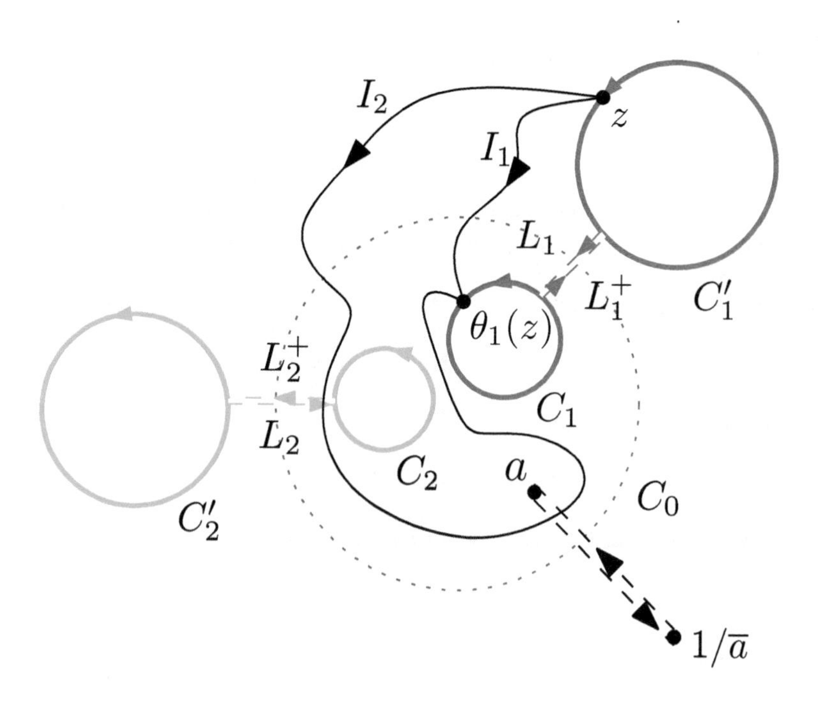

Figure 3.6. *The two contours I_1 and I_2 considered in Exercise 3.6.*

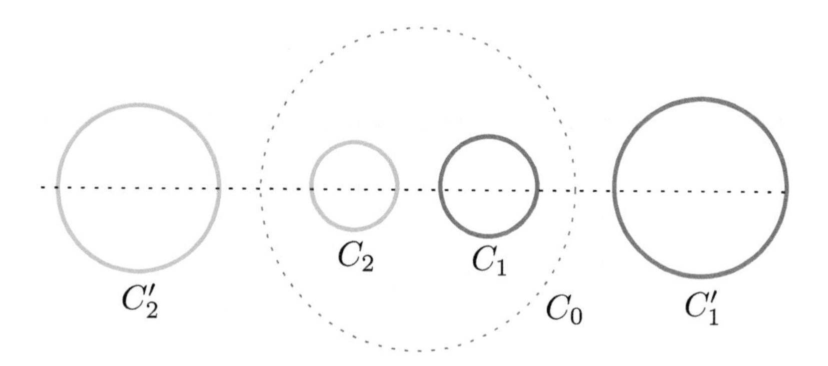

Figure 3.7. *A circular domain D_z with $\delta_j \in \mathbb{R}$ for all $j = 1, \ldots, M$ has, in addition to a reflectional symmetry with respect to C_0, an additional reflectional symmetry with respect to the real axis. The doubled domain for the case $M = 2$ is shown.*

show that for such reflectionally symmetric domains D_z the functions $\{v_j(z)|j = 1,\ldots,M\}$ and their Schwarz conjugate functions are related by

$$\overline{v_j}(z) = -v_j(z) + d_j, \qquad j = 1,\ldots,M, \tag{3.57}$$

for some set of real constants $\{d_j|j = 1,\ldots,M\}$.

(b) Verify that the real constants $\{d_j|j = 1,\ldots,M\}$ can all be set equal to zero without affecting any of the properties of the functions $\{v_j|j = 1,\ldots,M\}$ outlined in Chapters 2 and 3.

(c) By considering the effect on the domain D_z of the coordinate transformation

$$z \mapsto \overline{z}, \qquad a \mapsto \overline{a} \tag{3.58}$$

and considering the function

$$\tilde{H}(z,a) \equiv \mathcal{G}_0(z,a) + \overline{\mathcal{G}_0(\overline{z},\overline{a})}, \tag{3.59}$$

show that, for such reflectionally symmetric domains,

$$\mathcal{G}_0(z,a) = -\overline{\mathcal{G}_0(\overline{z},\overline{a})} + d_0, \tag{3.60}$$

where d_0 is a real constant.

(d) Verify that d_0 can be set equal to zero without affecting any of the properties of $\mathcal{G}_0(z,a)$ outlined in Chapters 2 and 3.

Chapter 4

What is the prime function?

4.1 ▪ The function $\Pi_{a,b}^{z,w}$

We are now in a position to introduce the *prime function* associated with a multiply connected circular domain D_z. It is the central object on which our framework is based. By the end of this chapter we will have retrieved, in a novel way, the simple prime function $\omega(z,a) = z - a$ for the unit disc. But, in addition, we will have defined broad new classes of prime functions associated with any other multiply connected circular domain D_z of the kind discussed in Chapter 2.

Let us introduce the following analytic function of four complex variables, z, w, a, and b, taking values in the doubled domain F and defined by

$$\Pi_{a,b}^{z,w} \equiv \int_b^a d\Pi^{z,w}(a'), \tag{4.1}$$

where

$$d\Pi^{z,w}(a') \equiv 2\pi \mathrm{i} \left[\frac{\partial \mathcal{G}_0(z,a')}{\partial a'} - \frac{\partial \mathcal{G}_0(w,a')}{\partial a'} \right] da'. \tag{4.2}$$

$\mathcal{G}_0(.,.)$ is the analytic extension of the modified Green's function $G_0(.,.)$ introduced in Chapter 2; consequently, the integrand in (4.1) is an analytic function of both z and w in F. Since $\mathcal{G}_0(z,a')$ is harmonic as a function of a', the integrand in (4.1), involving partial derivatives with respect to a', will also be an analytic function of a' in F. This integrand has isolated singularities, however. Near $a' = z$, $d\Pi^{z,w}(a')$ has the singular behavior

$$d\Pi^{z,w}(a') = \left[+\frac{1}{a'-z} + \text{a locally analytic function of } a' \right] da', \tag{4.3}$$

and near $a' = w$ it has the singular behavior

$$d\Pi^{z,w}(a') = \left[-\frac{1}{a'-w} + \text{a locally analytic function of } a' \right] da'. \tag{4.4}$$

Since $\Pi_{a,b}^{z,w}$ is defined by integration of an analytic function of a' between $a' = b$ and $a' = a$, its value must be expected to depend on the choice of integration contour between these two points. Now consider evaluating the function $\Pi_{a,b}^{z,w}$ using the two distinct choices of contour I_1 and I_2 shown in the upper schematic of Figure 4.1. Contour I_2 differs from

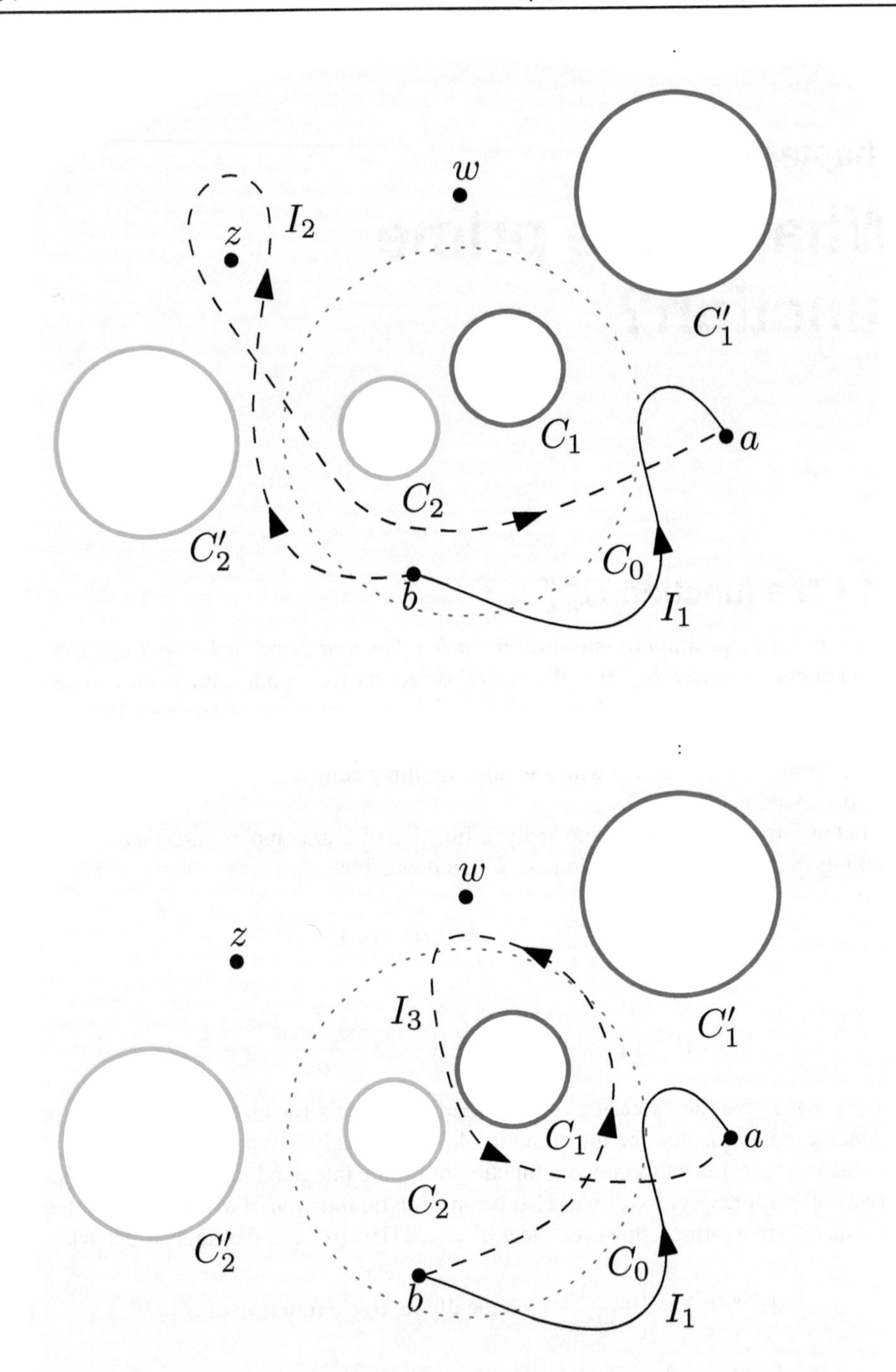

Figure 4.1. *Sources of ambiguity in the value of* $\Pi_{a,b}^{z,w}$ *depending on the choice of contour between* b *and* a *in* F. *The contour can encircle (any number of times) the fixed point* z *or* w, *or any of the* $2M$ *circles* $\{C_j, C_j' | j = 1, \ldots, M\}$. *By construction, however, it turns out that these different choices can only change the value of* $\Pi_{a,b}^{z,w}$ *by an integer multiple of* $2\pi i$.

contour I_1 by encircling the point z once in an anticlockwise direction. Consequently, on use of (4.3), the value of the integral along I_2 will equal the value of the integral along the contour I_1 plus $2\pi\mathrm{i}$, which is $2\pi\mathrm{i}$ times the residue contribution $+1$ at the point z.

If, on the other hand, I_2 had encircled w in an anticlockwise direction, on use of (4.4), the value of the integral along I_2 would equal the value of the integral along the contour I_1 minus $2\pi\mathrm{i}$, which is $2\pi\mathrm{i}$ times the residue contribution -1 at the point w.

It is clear that, depending on how many times the integration contour between b and a encircles the points z and w, the value of $\Pi_{a,b}^{z,w}$ can differ by integer multiples of $2\pi\mathrm{i}$.

But there is a second possible source of ambiguity in the value of the integral (4.1). This is illustrated in the lower schematic in Figure 4.1 by the two contours I_1 and I_3. Contour I_3 differs from I_1 by encircling, once in an anticlockwise direction, the hole in F with boundary C_1. The value of $\Pi_{a,b}^{z,w}$ using I_3 as the integration contour is then equal to the value obtained using I_1 as the contour plus the quantity

$$\int_{C_1} d\Pi^{z,w}(a'). \tag{4.5}$$

If instead I_3 had encircled C_1' once in an anticlockwise direction, then the value of $\Pi_{a,b}^{z,w}$ obtained using it would equal the value obtained using I_1 as the contour plus the quantity

$$\int_{C_1'} d\Pi^{z,w}(a'). \tag{4.6}$$

The value of the quantities (4.5) and (4.6), as well as analogous quantities deriving from all other circles $\{C_j, C_j' | j = 2, \ldots, M\}$, is therefore of considerable interest.

It turns out that

$$\int_{C_j} d\Pi^{z,w}(a') = \int_{C_j'} d\Pi^{z,w}(a') = 0, \qquad j = 1, \ldots, M. \tag{4.7}$$

This property, or *normalization*, of the differential (4.1) means that the difference in the value of $\Pi_{a,b}^{z,w}$ resulting from different choices of contour between b and a in F can *only* be integer multiples of $2\pi\mathrm{i}$. This is an important fact that will be needed later. Establishing the key result (4.7) that leads to this circumstance is one of the principal objectives of this chapter. First we need to understand more about $\Pi_{a,b}^{z,w}$.

4.2 ▪ Properties of the differential $d\Pi^{z,w}(a)$

It is necessary to show that the differential $d\Pi^{z,w}(a)$ has the following property:

$$d\Pi^{z,w}(\theta_j(a)) = d\Pi^{z,w}(a), \qquad j = 1, \ldots, M. \tag{4.8}$$

We therefore need to show that it is invariant under the action of the Schottky group Θ generated by the Möbius maps $\{\theta_j | j = 1, \ldots, M\}$ and their inverses, as introduced in Chapter 3.

First note that on swapping a and z in (3.29) we find the following relation between analytic functions of a:

$$\mathcal{G}_0(\theta_j(a), z) - \mathcal{G}_0(a, z) = 2\mathrm{i}\mathrm{Im}[v_j(z)]. \tag{4.9}$$

On partial differentiation of (4.9) with respect to a, we find

$$\mathcal{G}_0'(a, z) = \mathcal{G}_0'(\theta_j(a), z)\theta_j'(a), \tag{4.10}$$

where we have introduced the convenient notation $\mathcal{G}_0'(a,z)$ to denote the derivative of $\mathcal{G}_0(a,z)$ with respect to its first argument. (4.10) can be written as the equivalence of two differentials

$$\mathcal{G}_0'(a,z)da = \mathcal{G}_0'(\theta_j(a),z)\, d(\theta_j(a)). \tag{4.11}$$

The symmetry condition (2.46), which we know holds when both z and a are in D_z, can be written in terms of $\mathcal{G}_0(z,a)$ as

$$\mathcal{G}_0(z,a) - \overline{\mathcal{G}_0(z,a)} = \mathcal{G}_0(a,z) - \overline{\mathcal{G}_0(a,z)}. \tag{4.12}$$

On taking a partial derivative with respect to a we find

$$\frac{\partial}{\partial a}\left[\mathcal{G}_0(z,a) - \overline{\mathcal{G}_0(z,a)}\right] = \mathcal{G}_0'(a,z), \tag{4.13}$$

where we have used the fact that $\mathcal{G}_0(a,z)$ is analytic in a. We know the right-hand side of (4.13) to be analytic as a function of a for $a \in F$; so too therefore is the left-hand side of (4.13). Use of (4.13) on both sides of relation (4.11) leads to

$$\frac{\partial}{\partial a}\left[\mathcal{G}_0(z,a) - \overline{\mathcal{G}_0(z,a)}\right] da = \frac{\partial}{\partial \theta_j(a)}\left[\mathcal{G}_0(z,\theta_j(a)) - \overline{\mathcal{G}_0(z,\theta_j(a))}\right] d(\theta_j(a)). \tag{4.14}$$

On equating the parts of (4.14) that are analytic in z we deduce that

$$da\frac{\partial}{\partial a}\mathcal{G}_0(z,a) = d(\theta_j(a))\frac{\partial}{\partial \theta_j(a)}\mathcal{G}_0(z,\theta_j(a)) + df_j(a), \tag{4.15}$$

where $df_j(a)$ is a differential that is independent of z, but which must be expected to depend on a. We derived this assuming $z \in D_z$ but, by analytic continuation, it must hold for $z \in F$ (provided $z \neq a$). We can write down the same expression (4.15) with z replaced by a different point $w \in F$:

$$da\frac{\partial}{\partial a}\mathcal{G}_0(w,a) = d(\theta_j(a))\frac{\partial}{\partial \theta_j(a)}\mathcal{G}_0(w,\theta_j(a)) + df_j(a). \tag{4.16}$$

Subtraction of (4.16) from (4.15) leads to

$$d\Pi^{z,w}(\theta_j(a)) = d\Pi^{z,w}(a), \qquad j = 1,\ldots,M, \tag{4.17}$$

which is precisely (4.8).

One important consequence of (4.17) is the following. By a change of variable in the integrals, and the fact that, as a conformal mapping, θ_j takes points on C_j' to points on C_j produced by their reflection in C_0, as illustrated in Figure 3.4,

$$\int_{C_j} d\Pi^{z,w}(a) = -\int_{C_j'} d\Pi^{z,w}(\theta_j(a)), \qquad j = 1,\ldots,M, \tag{4.18}$$

where the minus sign arises because a must traverse C_j' in a clockwise sense for $\theta_j(a)$ to traverse C_j in an anticlockwise sense. Hence, by (4.17), (4.18) becomes

$$\int_{C_j} d\Pi^{z,w}(a) = -\int_{C_j'} d\Pi^{z,w}(a), \qquad j = 1,\ldots,M. \tag{4.19}$$

This means that, to establish the $2M$ equations (4.7), it is enough to establish the M equations

$$\int_{C_j'} d\Pi^{z,w}(a) = 0, \qquad j = 1,\ldots,M. \tag{4.20}$$

4.3 ▪ Properties of $\Pi_{a,b}^{z,w}$

Let us now examine the singularities of $\Pi_{a,b}^{z,w}$. The following four results follow immediately from the known form of the simple pole singularities of the integrand in (4.1). Near $z = a$,

$$\Pi_{a,b}^{z,w} = \log(a - z) + \text{a locally analytic function of } z, w, a, b, \tag{4.21}$$

and near $z = b$,

$$\Pi_{a,b}^{z,w} = -\log(b - z) + \text{a locally analytic function of } z, w, a, b. \tag{4.22}$$

Similarly, near $w = a$,

$$\Pi_{a,b}^{z,w} = -\log(a - w) + \text{a locally analytic function of } z, w, a, b, \tag{4.23}$$

and near $w = b$,

$$\Pi_{a,b}^{z,w} = \log(b - w) + \text{a locally analytic function of } z, w, a, b. \tag{4.24}$$

It follows that we can write

$$\Pi_{a,b}^{z,w} = \log p(z, w, a, b) + \tilde{\Pi}_{a,b}^{z,w}, \tag{4.25}$$

where

$$p(z, w, a, b) = \frac{(z - a)(w - b)}{(z - b)(w - a)} \tag{4.26}$$

is the cross-ratio introduced in §1.12 and where $\tilde{\Pi}_{a,b}^{z,w}$ is some function that is analytic in z, w, a, and b taking values in F. On exponentiation of (4.25), we find

$$e^{-\Pi_{a,b}^{z,w}} = \frac{e^{-\tilde{\Pi}_{a,b}^{z,w}}}{p(z, w, a, b)}. \tag{4.27}$$

This will be useful later.

The relation

$$\Pi_{a,b}^{w,z} = -\Pi_{a,b}^{z,w} \tag{4.28}$$

follows immediately by swapping z and w in the definition (4.1)–(4.2).

It is also easily established from the definition of $\Pi_{a,b}^{z,w}$ as an integral between a and b that

$$\Pi_{b,a}^{z,w} = -\Pi_{a,b}^{z,w}. \tag{4.29}$$

The following relations also follow from the basic properties of integration:

$$\Pi_{a,c}^{z,w} = \Pi_{a,b}^{z,w} + \Pi_{b,c}^{z,w} = \Pi_{a,b}^{z,w} - \Pi_{c,b}^{z,w}. \tag{4.30}$$

It is also easy to show that

$$\Pi_{a,b}^{z,w} = \Pi_{a,b}^{z,x} + \Pi_{a,b}^{x,w} = \Pi_{a,b}^{z,x} - \Pi_{a,b}^{w,x}. \tag{4.31}$$

Since $\mathcal{G}_0(z, a)$ is an analytic function of z, we can write

$$\frac{\partial \mathcal{G}_0(z, a)}{\partial a} - \frac{\partial \mathcal{G}_0(w, a)}{\partial a} = \int_w^z \frac{\partial^2 \mathcal{G}_0(z', a)}{\partial z' \partial a} dz', \tag{4.32}$$

for a suitable contour in F between w and z.

Now, using the fact that $G_0(z,a) = \mathrm{Im}[\mathcal{G}_0(z,a)]$, and using (4.32), we have

$$\Pi^{z,w}_{a,b} = -4\pi \int_b^a \int_w^z \frac{\partial^2 G_0(z',a')}{\partial z' \partial a'} dz' da', \tag{4.33}$$

from which the symmetry of the function, namely,

$$\Pi^{a,b}_{z,w} = \Pi^{z,w}_{a,b}, \tag{4.34}$$

can be seen on recalling the symmetry property (2.46) of $G_0(z,a)$.

It can be verified directly that the elementary function

$$\log p(z,w,a,b) \tag{4.35}$$

satisfies all the identities (4.28)–(4.31). Hence from (4.25) and the linearity of the identities (4.28)–(4.31) it also follows that

$$\tilde{\Pi}^{z,w}_{a,b} \tag{4.36}$$

satisfies the same identities. In particular, from (4.28), we have

$$\tilde{\Pi}^{w,z}_{a,b} = -\tilde{\Pi}^{z,w}_{a,b}, \tag{4.37}$$

which, together with the fact that $\tilde{\Pi}^{z,w}_{a,b}$ is known to be analytic in F as a function of all variables, implies that

$$\tilde{\Pi}^{a,a}_{a,a} = 0 \tag{4.38}$$

for any $a \in F$.

Finally, let us pick a point $z \in C'_j$ on the boundary of F for some choice of $j = 1, \ldots, M$. Then $\theta_j(z) \in C_j$ is also on the boundary of F. Now provided b and a are joined by a contour that does not pass through the barrier (L_j, L_j^+) introduced in Chapter 3,

$$\frac{1}{2\pi \mathrm{i}} \Pi^{\theta_j(z),z}_{a,b} = v_j(a) - v_j(b). \tag{4.39}$$

This follows because, by (4.1) and (4.2),

$$\frac{1}{2\pi \mathrm{i}} \Pi^{\theta_j(z),z}_{a,b} = \int_b^a \left[\frac{\partial \mathcal{G}_0(\theta_j(z),a')}{\partial a'} - \frac{\partial \mathcal{G}_0(z,a')}{\partial a'} \right] da' = \int_b^a \frac{dv_j}{da'} da', \tag{4.40}$$

where, in the second equality, we have used (3.35). Relation (4.39) now follows on performing the integration in (4.40) along a path that does not intersect the barrier (L_j, L_j^+).

Note too that if a is *also* a point on some C'_k for $k = 1, \ldots, M$ on the boundary of F, so that $\theta_k(a) \in C_k$ is also on the boundary of F, then (4.39) tells us that

$$\frac{1}{2\pi \mathrm{i}} \Pi^{\theta_j(z),z}_{\theta_k(a),a} = v_j(\theta_k(a)) - v_j(a) = \tau_{jk}, \tag{4.41}$$

where, in the second equality, we have used (3.42). Again, for this to hold we must ensure that the contour joining a and $\theta_k(a)$ does not pass across the barrier (L_j, L_j^+) defining a single-valued branch of v_j.

4.4 ▪ The Schwarz conjugate $\overline{\Pi}_{a,b}^{z,w}$

The Schwarz conjugate $\overline{\Pi}_{a,b}^{z,w}$ defined in the usual way by

$$\overline{\Pi}_{a,b}^{z,w} = \overline{\Pi_{\overline{a},\overline{b}}^{\overline{z},\overline{w}}} \tag{4.42}$$

satisfies the identity

$$\overline{\Pi}_{1/a,1/b}^{1/z,1/w} = \Pi_{a,b}^{z,w}. \tag{4.43}$$

To establish this fact we take a complex conjugate of (4.1):

$$\overline{\Pi_{a,b}^{1/\overline{z},1/\overline{w}}} = -2\pi\mathrm{i}\int_{\overline{b}}^{\overline{a}}\left[\overline{\left(\frac{\partial\mathcal{G}_0(1/\overline{z},a')}{\partial a'}\right)} - \overline{\left(\frac{\partial\mathcal{G}_0(1/\overline{w},a')}{\partial a'}\right)}\right]d\overline{a'}. \tag{4.44}$$

On use of (3.1) this becomes

$$\overline{\Pi}_{\overline{a},\overline{b}}^{1/z,1/w} = -2\pi\mathrm{i}\int_{\overline{b}}^{\overline{a}}\left[\frac{\partial\mathcal{G}_0(z,a')}{\partial\overline{a'}} - \frac{\partial\mathcal{G}_0(w,a')}{\partial\overline{a'}}\right]d\overline{a'}. \tag{4.45}$$

Now, treating $\mathcal{G}_0(z,a)$ as a function of a, the fundamental theorem of calculus implies

$$d\mathcal{G}_0(z,a') = \frac{\partial\mathcal{G}_0(z,a')}{\partial a'}da' + \frac{\partial\mathcal{G}_0(z,a')}{\partial\overline{a'}}d\overline{a'}; \tag{4.46}$$

hence

$$\mathcal{G}_0(z,a) - \mathcal{G}_0(z,b) = \int_b^a \frac{\partial\mathcal{G}_0(z,a')}{\partial a'}da' + \int_{\overline{b}}^{\overline{a}} \frac{\partial\mathcal{G}_0(z,a')}{\partial\overline{a'}}d\overline{a'}. \tag{4.47}$$

Hence, on rearrangement,

$$\int_{\overline{b}}^{\overline{a}} \frac{\partial\mathcal{G}_0(z,a')}{\partial\overline{a'}}d\overline{a'} = \mathcal{G}_0(z,a) - \mathcal{G}_0(z,b) - \int_b^a \frac{\partial\mathcal{G}_0(z,a')}{\partial a'}da'. \tag{4.48}$$

Use of (two instances) of (4.48) in (4.45) leads to

$$\overline{\Pi}_{\overline{a},\overline{b}}^{1/z,1/w} = 2\pi\mathrm{i}\int_b^a\left[\frac{\partial\mathcal{G}_0(z,a')}{\partial a'} - \frac{\partial\mathcal{G}_0(w,a')}{\partial a'}\right]da' \tag{4.49}$$

$$-2\pi\mathrm{i}\left[\mathcal{G}_0(z,a) - \mathcal{G}_0(z,b) - \mathcal{G}_0(w,a) + \mathcal{G}_0(w,b)\right]. \tag{4.50}$$

This can be rewritten as

$$\overline{\Pi}_{\overline{a},\overline{b}}^{1/z,1/w} - \Pi_{a,b}^{z,w} = -2\pi\mathrm{i}\left[\mathcal{G}_0(z,a) - \mathcal{G}_0(z,b) - \mathcal{G}_0(w,a) + \mathcal{G}_0(w,b)\right]. \tag{4.51}$$

Now, on use of the parametric identity (3.20), we deduce that

$$\underbrace{\overline{\Pi}_{\overline{a},\overline{b}}^{1/z,1/w} - \Pi_{a,b}^{z,w}}_{\text{analytic in } a,b} = -\left(\overline{\Pi}_{1/a,1/b}^{1/z,1/w} - \Pi_{1/\overline{a},1/\overline{b}}^{z,w}\right). \tag{4.52}$$

On equating the parts of this identity that are analytic in a and b we deduce that

$$\overline{\Pi}_{1/a,1/b}^{1/z,1/w} = \Pi_{a,b}^{z,w} + h(z,w), \tag{4.53}$$

where $h(z,w)$ is some function of z and w (but not of a or b). However on use of (4.34) it can be argued that we must have $h(z,w) = h(a,b)$ implying that this function h must be a constant. On substitution of (4.25) into (4.53) the terms involving the cross-ratio on both sides cancel; on evaluating the terms that remain after this cancellation when $a = b = z = w$, and on use of (4.38), it is found that this constant must be zero. On setting $h(z,w) = 0$ in (4.53) we arrive at the identity (4.43).

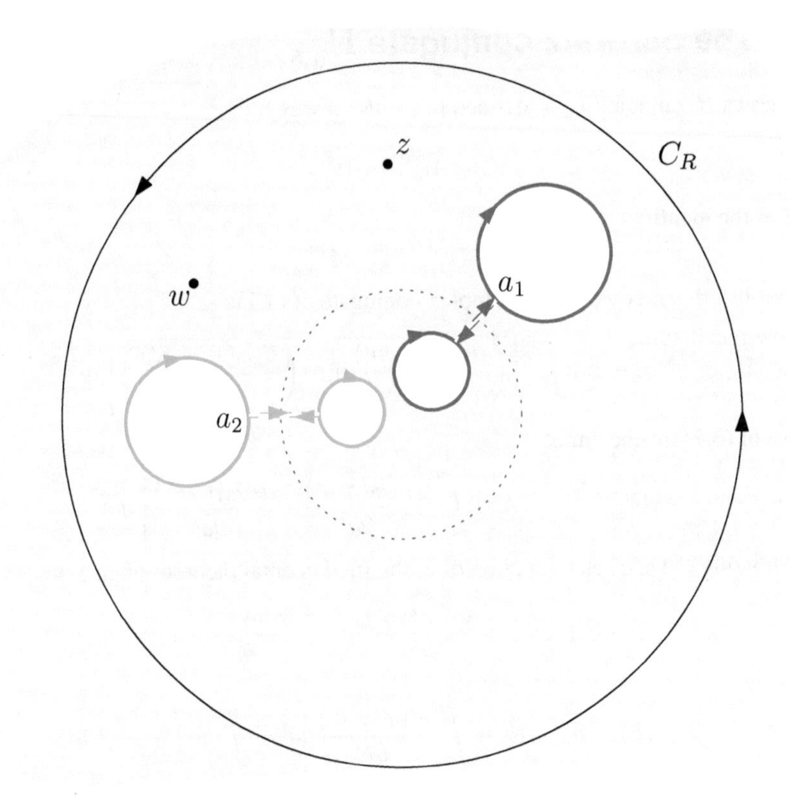

Figure 4.2. *The "dumbbell contour" ∂F_R. Integration around ∂F_R is around the boundary of the $(M+1)$ connected region interior to a large radius-R circle C_R and exterior to M closed dumbbell-shaped contours.*

4.5 ▪ Proof of the normalization (4.7)

To establish (4.20), and hence (4.7), it is necessary to consider, in the limit as $R \to \infty$, the integral of the differentials

$$v_j(a)d\Pi^{z,w}(a), \qquad j = 1, \ldots, M, \tag{4.54}$$

around what we will refer to as the "dumbbell contour" ∂F_R shown in Figure 4.2 for the case $M = 2$. It is the boundary of the $(M+1)$-connected region bounded by C_R, a large circular contour of radius R centered at the origin, together with M closed dumbbell contours, each of which is, for $k = 1, \ldots, M$, a loop comprising circles C_k and C'_k linked by the two sides (L_k, L_k^+) of the barriers introduced to define a single-valued branch of the function $v_k(z)$ traversed in opposite directions. In Figure 4.2 ∂F_c is the union of C_R and the closed red and green dumbbell contours. These are traversed in a clockwise sense, as indicated in the figure.

It follows from the residue theorem that

$$\lim_{R\to\infty} \left[\int_{\partial F_R} v_j(a)d\Pi^{z,w}(a) \right] = 2\pi \mathrm{i}(v_j(z) - v_j(w)), \qquad j = 1, \ldots, M, \tag{4.55}$$

where we have used the fact that the integrand is meromorphic in the region bounded by ∂F_R with a simple pole of residue $+1$ at $a = z$ and a simple pole of residue -1 at $a = w$ inherited from $d\Pi^{z,w}(a)$.

Let us focus on the left-hand side of (4.55) in the limit $R \to \infty$. Since the integrand can be shown to decay like $1/a^2$ as $a \to \infty$ (cf. Exercise 4.1), the contribution from C_R vanishes as $R \to \infty$, leaving

$$-\sum_{k=1}^{M}\Bigg\{\int_{C_k} v_j(a)d\Pi^{z,w}(a)+\int_{C'_k} v_j(a)d\Pi^{z,w}(a) \\ +\int_{L_k} v_j(a)d\Pi^{z,w}(a)+\int_{L_k^+} v_j(a)d\Pi^{z,w}(a)\Bigg\}. \quad (4.56)$$

We now recall some facts derived earlier. The first thing to note is that since integration is around the dumbbell contour which tracks along the barriers $\{(L_j, L_j^+)|j = 1,\ldots,M\}$ but does not cross any of them, the formulas derived in Chapter 3, many of which were derived by disallowing any crossing of these barriers, can be safely invoked. We also recall that $v_j(a)$ jumps by $+1$ as one travels continuously from L_j anticlockwise around C_j to meet L_j^+ and it is continuous across the other barriers (L_k, L_k^+) for $k \neq j$. Also, for each point $a \in C'_k$ the corresponding point on C_k is $\theta_k(a)$—see Figure 3.4—and, moreover, we recall the results (4.41) and (4.17), namely

$$v_j(\theta_k(a)) - v_j(a) = \tau_{jk}, \qquad d\Pi^{z,w}(\theta_k(a)) = d\Pi^{z,w}(a). \quad (4.57)$$

Therefore, if we denote by a_j the point where L_j and L_j^+ meet C'_j—the points a_1 and a_2 in the case $M = 2$ are shown in Figure 4.2—then (4.56) is

$$\begin{aligned}
&-\sum_{k=1}^{M}\left\{-\int_{C'_k} v_j(\theta_k(a))d\Pi^{z,w}(\theta_k(a))+\int_{C'_k} v_j(a)d\Pi^{z,w}(a)\right\}-\int_{L_j^+} d\Pi^{z,w}(a)\\
&=\sum_{k=1}^{M}\int_{C'_k}\tau_{jk}\, d\Pi^{z,w}(a)-\Pi^{z,w}_{a_j,\theta_j(a_j)}\\
&=\sum_{k=1}^{M}\tau_{jk}\int_{C'_k} d\Pi^{z,w}(a)+\Pi^{\theta_j(a_j),a_j}_{z,w},
\end{aligned} \quad (4.58)$$

where we have changed the integration variable $a \mapsto \theta_k(a)$ in the integral around C_k in (4.56). In the second equality in (4.58) we have made use of (4.29) followed by (4.34). But we also know from (4.39) that for $a_j \in C'_j$,

$$\Pi^{\theta_j(a_j),a_j}_{z,w} = 2\pi \mathrm{i}(v_j(z) - v_j(w)). \quad (4.59)$$

On use of (4.59) in (4.58), and on further substitution of that into the left-hand side of (4.55), we find a cancellation of the terms involving $v_j(z) - v_j(w)$ leading to the set of equations

$$\sum_{k=1}^{M}\tau_{jk}\int_{C'_k} d\Pi^{z,w}(a) = 0, \qquad j = 1,\ldots,M. \quad (4.60)$$

Now (2.30) reminds us that τ_{jk} is just a multiple of $\mathcal{P}^{-1}_{jk}$, which we know to be nonsingular and invertible, which leads us to the required result (4.20).

We have therefore established the key result (4.7) underlying the claim made in §4.1 that $\Pi^{z,w}_{a,b}$ is well defined up to integer multiples of $2\pi\mathrm{i}$ depending on the choice of contour.

This means that the exponentiated function $e^{-\Pi^{z,w}_{a,b}}$ is *uniquely* defined for any choice of z, w, a, and b in F and is independent of the choice of contour. It is this well-defined function $e^{-\Pi^{z,w}_{a,b}}$ that will now be used to define the prime function.

4.6 ▪ The square of the prime function $X(a, b)$

Since $e^{-\Pi_{a,b}^{z,w}}$ is well defined, so is the function $X(a,b)$ defined by the double limit

$$X(a,b) \equiv \lim_{\substack{z\to a\\ w\to b}} \left[-(z-a)(w-b)e^{-\Pi_{a,b}^{z,w}}\right]. \tag{4.61}$$

By collapsing an analytic function of four variables by a limiting process in this way we end up with an analytic function of just two variables a and b.

Several properties of this function $X(a,b)$ follow from its definition (4.61) and the properties of the function $\Pi_{a,b}^{z,w}$ we established earlier.

An immediate observation, following from (4.27), is that

$$X(a,b) \equiv \lim_{\substack{z\to a\\ w\to b}} \left[-(z-a)(w-b)e^{-\Pi_{a,b}^{z,w}}\right] \tag{4.62}$$

$$= \lim_{\substack{z\to a\\ w\to b}} \left[-(z-a)(w-b)\left\{\frac{(z-b)(w-a)}{(z-a)(w-b)}e^{-\tilde{\Pi}_{a,b}^{z,w}}\right\}\right] \tag{4.63}$$

$$= (a-b)^2 e^{-\tilde{\Pi}_{a,b}^{a,b}}. \tag{4.64}$$

Moreover,

$$\lim_{a\to b}\left[\frac{X(a,b)}{(a-b)^2}\right] = \lim_{a\to b}\left[e^{-\tilde{\Pi}_{a,b}^{a,b}}\right] = e^{-\tilde{\Pi}_{b,b}^{b,b}} = 1, \tag{4.65}$$

where we have used (4.64) and (4.38). It is clear that $X(a,b)$ has a second order zero at $a=b$.

Symmetry of $X(a,b)$: By the definition (4.61),

$$X(b,a) \equiv \lim_{\substack{z\to b\\ w\to a}} \left[-(z-b)(w-a)e^{-\Pi_{b,a}^{z,w}}\right]. \tag{4.66}$$

From (4.28) and (4.29),

$$\Pi_{b,a}^{z,w} = \Pi_{a,b}^{w,z}. \tag{4.67}$$

Hence

$$X(b,a) \equiv \lim_{\substack{z\to b\\ w\to a}} \left[-(z-b)(w-a)e^{-\Pi_{a,b}^{w,z}}\right] = X(a,b). \tag{4.68}$$

The Schwarz conjugate $\overline{X}(a,b)$: The Schwarz conjugate $\overline{X}(a,b) \equiv \overline{X(\overline{a},\overline{b})}$ is defined in the usual way, implying that

$$\overline{X}(1/a,1/b) = \overline{X(1/\overline{a},1/\overline{b})}, \tag{4.69}$$

where

$$X(1/\overline{a},1/\overline{b}) = \lim_{\substack{z\to 1/\overline{a}\\ w\to 1/\overline{b}}} \left[-\left(z-\frac{1}{\overline{a}}\right)\left(w-\frac{1}{\overline{b}}\right)e^{-\Pi_{1/\overline{a},1/\overline{b}}^{z,w}}\right]. \tag{4.70}$$

Hence

$$\overline{X}(1/a,1/b) = \lim_{\substack{\overline{z}\to 1/a\\ \overline{w}\to 1/b}} \left[-\left(\overline{z}-\frac{1}{a}\right)\left(\overline{w}-\frac{1}{b}\right)e^{-\overline{\Pi}_{1/a,1/b}^{\overline{z},\overline{w}}}\right]. \tag{4.71}$$

Redefining the variables $\overline{z} \mapsto 1/z, \overline{w} \mapsto 1/w$ does not affect the value of the limit and it then takes the form

$$\overline{X}(1/a,1/b) = \lim_{\substack{z\to a\\ w\to b}} \left[-\left(\frac{1}{z}-\frac{1}{a}\right)\left(\frac{1}{w}-\frac{1}{b}\right)e^{-\overline{\Pi}_{1/a,1/b}^{1/z,1/w}}\right]. \tag{4.72}$$

On use of the identity (4.43), we find

$$\overline{X}(1/a, 1/b) = \lim_{\substack{z\to a\\ w\to b}} \left[-\left(\frac{1}{z} - \frac{1}{a}\right)\left(\frac{1}{w} - \frac{1}{b}\right) e^{-\Pi_{a,b}^{z,w}} \right] \tag{4.73}$$

$$= \lim_{\substack{z\to a\\ w\to b}} \left[-\frac{1}{za}\frac{1}{wb}(z-a)(w-b)e^{-\Pi_{a,b}^{z,w}} \right] = \frac{1}{a^2b^2}X(a,b). \tag{4.74}$$

Identities of $X(a, b)$ associated with the holes: For any $j = 1, \ldots, M$ it follows from the definition (4.61) that

$$X(\theta_j(a), b) \equiv \lim_{\substack{z\to \theta_j(a)\\ w\to b}} \left[-(z - \theta_j(a))(w-b)e^{-\Pi_{\theta_j(a),b}^{z,w}} \right]. \tag{4.75}$$

Now exchange $z \mapsto \theta_j(z)$ in this limit:

$$X(\theta_j(a), b) \equiv \lim_{\substack{z\to a\\ w\to b}} \left[-(\theta_j(z) - \theta_j(a))(w-b)e^{-\Pi_{\theta_j(a),b}^{\theta_j(z),w}} \right]. \tag{4.76}$$

Hence

$$\frac{X(\theta_j(a), b)}{X(a,b)} = \lim_{\substack{z\to a\\ w\to b}} \left[\frac{(\theta_j(z) - \theta_j(a))(w-b)}{(z-a)(w-b)} e^{-\Pi_{\theta_j(a),b}^{\theta_j(z),w} + \Pi_{a,b}^{z,w}} \right] \tag{4.77}$$

$$= \lim_{\substack{z\to a\\ w\to b}} \left[\frac{(\theta_j(z) - \theta_j(a))}{(z-a)} e^{-\Pi_{\theta_j(a),b}^{\theta_j(z),w} + \Pi_{a,b}^{z,w}} \right]. \tag{4.78}$$

Now

$$\Pi_{a,b}^{z,w} - \Pi_{\theta_j(a),b}^{\theta_j(z),w} = \Pi_{a,b}^{z,w} - \Pi_{\theta_j(z),w}^{\theta_j(a),b}; \tag{4.79}$$

hence, on adding in a useful zero term,

$$\Pi_{a,b}^{z,w} - \Pi_{\theta_j(a),b}^{\theta_j(z),w} = \Pi_{a,b}^{z,w} \underbrace{-\Pi_{a,b}^{\theta_j(z),w} + \Pi_{\theta_j(z),w}^{a,b}}_{=0} - \Pi_{\theta_j(z),w}^{\theta_j(a),b}, \tag{4.80}$$

where we have used (4.34) in constructing this additional term. This can be written

$$\Pi_{a,b}^{z,w} - \Pi_{\theta_j(a),b}^{\theta_j(z),w} = -2\pi\mathrm{i}(v_j(a) - v_j(b)) - 2\pi\mathrm{i}(v_j(\theta_j(z)) - v_j(w)), \tag{4.81}$$

where we have used (4.39) twice in the right-hand side of (4.80). On use of (4.41) this relation becomes

$$\Pi_{a,b}^{z,w} - \Pi_{\theta_j(a),b}^{\theta_j(z),w} = -2\pi\mathrm{i}(v_j(a) - v_j(b)) - 2\pi\mathrm{i}(v_j(z) - v_j(w) + \tau_{jj}) \tag{4.82}$$

$$\to -2\pi\mathrm{i}\left[2(v_j(a) - v_j(b)) + \tau_{jj}\right] \quad \text{as } z \to a,\ w \to b. \tag{4.83}$$

Since, as $z \to a$,

$$\frac{\theta_j(z) - \theta_j(a)}{(z-a)} \to \theta_j'(a), \tag{4.84}$$

on substitution of (4.83) into (4.78) we deduce the functional relation

$$\frac{X(\theta_j(a), b)}{X(a,b)} = \theta_j'(a)e^{-2\pi\mathrm{i}(2v_j(a) - 2v_j(b) + \tau_{jj})}. \tag{4.85}$$

4.7 ▪ The prime function $\omega(a,b)$

The prime function $\omega(a,b)$ is now defined by

$$\omega(a,b) = X(a,b)^{1/2}, \tag{4.86}$$

where we choose the sign of the square root so that

$$\omega(a,b) \sim (a-b) \qquad \text{as} \quad a \to b. \tag{4.87}$$

The square root of the identities (4.68), (4.74), and (4.85) implies the following identities for $\omega(a,b)$:

$$\omega(b,a) = -\omega(a,b), \tag{4.88}$$

$$\overline{\omega}(1/a,1/b) = -\frac{1}{ab}\omega(a,b), \tag{4.89}$$

$$\omega(\theta_j(a),b) = -\frac{q_j}{1-\overline{\delta_j}a}e^{-2\pi\mathrm{i}(v_j(a)-v_j(b)+\tau_{jj}/2)}\omega(a,b), \qquad j=1,\ldots,M, \tag{4.90}$$

where, in (4.90), we have used the fact that

$$\theta_j(a) = \delta_j + \frac{q_j^2 a}{1-\overline{\delta_j}a}, \qquad \theta_j'(a) = \frac{q_j^2}{(1-\overline{\delta_j}a)^2} \tag{4.91}$$

and we have chosen

$$\sqrt{\theta_j'(a)} = -\frac{q_j}{(1-\overline{\delta_j}a)}. \tag{4.92}$$

The minus signs in the first two identities (4.88) and (4.89) are selected by examining the behavior of the square roots of both sides of the identities as $a \to b$. From (4.64),

$$X(a,b) = (a-b)^2\tilde{X}(a,b), \qquad \tilde{X}(a,b) \equiv e^{-\tilde{\Pi}_{a,b}^{a,b}}; \tag{4.93}$$

then (4.68) implies

$$(b-a)^2\tilde{X}(b,a) = (a-b)^2\tilde{X}(a,b) \tag{4.94}$$

and, hence, on taking the square root, that

$$(b-a)\tilde{X}(b,a)^{1/2} = \pm(a-b)\tilde{X}(a,b)^{1/2}. \tag{4.95}$$

This identity holds for all a and b including in the limit $b \to a$, where it is known from (4.38) that

$$\tilde{X}(b,a)^{1/2}, \tilde{X}(a,b)^{1/2} \to 1^{1/2}. \tag{4.96}$$

That is, both tend to the same quantity. Hence (4.95) is only consistent in this limit if the minus sign is chosen, leading to (4.88). Similar local arguments lead to the minus sign in (4.89).

The choice of sign in (4.90) requires more global consideration of the analytic continuation of the prime function into a neighboring copy of the fundamental region. Since this sign choice turns out to be largely inconsequential for all the applications to be considered in Part II, we do not include a detailed discussion of it here, although Exercise 4.9 gives an idea of the arguments needed to establish this sign choice.[9]

[9] This question of how to choose this sign is discussed by Baker [64]. However it is not broached in the papers by Burnside [72, 73] discussing the prime function. It was recently revisited by Bogatyrev [70].

The foundations of the framework are very nearly complete. It only remains, now that we have introduced the prime function, to write $\mathcal{G}_0(z,a)$ and the functions $\{v_j(z)|j = 1,\ldots,M\}$ in terms of it; this is done in the next two sections.

Remark: Our chosen notation is to denote all prime functions by $\omega(z,a)$, it being understood that this prime function is associated with some prespecified circular domain D_z. This has been done for convenience; otherwise we would have to write, say, $\omega(z,a;D_z)$ when writing down formulas, which can quickly become cumbersome. There are some applications, however, where the domain D_z changes, or evolves according to some specification, so that different prime functions are relevant to the same problem; in such cases it is necessary to include this extra functional dependence. In this monograph the thing to remember is that, except for the simple prime function for the unit disc, all prime functions $\omega(z,a)$ depend implicitly on additional geometrical data associated with some prespecified circular domain D_z.

4.8 ▪ A ratio of prime functions

Let us consider a ratio of prime functions defined by

$$R(z;a,b) = \frac{\omega(z,a)}{\omega(z,b)}, \tag{4.97}$$

where a and b are two distinct nonzero points. Since this new prime function satisfies the same identities (4.88) and (4.89) as used in Chapter 1 to deduce that when $z \in C_0$,

$$\overline{R(z;a,b)} = \frac{\overline{a}}{\overline{b}} R(z;1/\overline{a},1/\overline{b}), \qquad z \in C_0. \tag{4.98}$$

Now for points z on C_j, for $j = 1,\ldots,M$, we know that

$$z = \overline{S_j(z)}, \qquad j = 1,\ldots,M; \tag{4.99}$$

hence we can write

$$R(z;a,b) = \frac{\omega(\overline{S_j(z)},a)}{\omega(\overline{S_j(z)},b)} = \frac{\omega(\theta_j(1/\overline{z}),a)}{\omega(\theta_j(1/\overline{z}),b)}. \tag{4.100}$$

On use of identity (4.90) this becomes

$$R(z;a,b) = e^{2\pi \mathrm{i}(v_j(a)-v_j(b))}\frac{\omega(1/\overline{z},a)}{\omega(1/\overline{z},b)}. \tag{4.101}$$

On taking a complex conjugate,

$$\overline{R(z;a,b)} = e^{-2\pi \mathrm{i}(\overline{v_j(a)}-\overline{v_j(b)})}\frac{\overline{\omega}(1/z,\overline{a})}{\overline{\omega}(1/z,\overline{b})} = \frac{\overline{a}}{\overline{b}}e^{-2\pi \mathrm{i}(\overline{v_j(a)}-\overline{v_j(b)})}\frac{\omega(z,1/\overline{a})}{\omega(z,1/\overline{b})}, \tag{4.102}$$

where we have used (4.89).

In summary we have arrived at the useful result that

$$\overline{R(z;a,b)} = \begin{cases} \dfrac{\overline{a}}{\overline{b}} R(z;1/\overline{a},1/\overline{b}), & z \in C_0, \\ e^{-2\pi \mathrm{i}(\overline{v_j(a)}-\overline{v_j(b)})}\dfrac{\overline{a}}{\overline{b}} R(z;1/\overline{a},1/\overline{b}), & z \in C_j,\ j = 1,\ldots,M. \end{cases} \tag{4.103}$$

4.9 ▪ Representation of $\mathcal{G}_0(z,a)$ in terms of $\omega(.,.)$

Consider the function

$$\mathcal{H}(z,a) = \mathcal{G}_0(z,a) - \frac{1}{2\pi\mathrm{i}} \log\left[\frac{\omega(z,a)}{|a|\omega(z,1/\overline{a})}\right] \tag{4.104}$$

for values of z in the domain D_z. This function is analytic in D_z since the logarithmic singularities at $z = a$ of both functions on the right-hand side exactly cancel out. Now let

$$H(z,a) = \mathrm{Im}[\mathcal{H}(z,a)]. \tag{4.105}$$

It can be verified that $H(z,a)$, which is harmonic in D_z, vanishes on all the boundaries of D_z. To see this, one must use (2.50) together with the fact that if

$$w = \frac{1}{|a|}R(z;a,1/\overline{a}) = \frac{1}{|a|}\frac{\omega(z,a)}{\omega(z,1/\overline{a})} \tag{4.106}$$

then it follows from (4.103), with $b = 1/\overline{a}$, that

$$\overline{w} = \begin{cases} 1/w, & z \in C_0, \\ e^{2\pi\mathrm{i}(v_j(a)-\overline{v_j(a)})}/w, & z \in C_j,\ j = 1,\ldots,M. \end{cases} \tag{4.107}$$

$H(z,a)$ is therefore the zero function (cf. Exercise 2.4), and we deduce that to within a real function of a that is independent of z[10]

$$\mathcal{G}_0(z,a) = \frac{1}{2\pi\mathrm{i}} \log\left[\frac{\omega(z,a)}{|a|\omega(z,1/\overline{a})}\right]. \tag{4.108}$$

This is a representation of $\mathcal{G}_0(z,a)$ in terms of the prime function.

4.10 ▪ Representation of $\{v_j(z)|j = 1,\ldots,M\}$ in terms of $\omega(.,.)$

Let a be any nonzero point in D_z. The prime function identity (4.90) is

$$\omega(\theta_j(z),a) = -\frac{q_j}{(1-\overline{\delta_j}z)}e^{-2\pi\mathrm{i}(v_j(z)-v_j(a)+\tau_{jj}/2)}\omega(z,a), \qquad j = 1,\ldots,M. \tag{4.109}$$

By the antisymmetry of the prime function, this implies

$$\omega(a,\theta_j(z)) = -\frac{q_j}{(1-\overline{\delta_j}z)}e^{-2\pi\mathrm{i}(v_j(z)-v_j(a)+\tau_{jj}/2)}\omega(a,z), \qquad j = 1,\ldots,M, \tag{4.110}$$

and, on swapping the variables z and a, we find

$$\omega(z,\theta_j(a)) = -\frac{q_j}{(1-\overline{\delta_j}a)}e^{-2\pi\mathrm{i}(v_j(a)-v_j(z)+\tau_{jj}/2)}\omega(z,a), \qquad j = 1,\ldots,M. \tag{4.111}$$

On exchanging a for $1/\overline{a}$ in (4.111), we arrive at another expression:

$$\omega(z,\theta_j(1/\overline{a})) = -\frac{q_j}{(1-\overline{\delta_j}/\overline{a})}e^{-2\pi\mathrm{i}(v_j(1/\overline{a})-v_j(z)+\tau_{jj}/2)}\omega(z,1/\overline{a}), \qquad j = 1,\ldots,M. \tag{4.112}$$

[10] See also [46].

From (4.112) we can deduce

$$e^{2\pi i(v_j(z)-v_j(1/\overline{a})-\tau_{jj}/2)} = -\frac{(1-\overline{\delta_j}/\overline{a})}{q_j}\frac{\omega(z,\theta_j(1/\overline{a}))}{\omega(z,1/\overline{a})}, \qquad j=1,\ldots,M, \tag{4.113}$$

or, on taking a logarithm and dividing by $2\pi i$,

$$v_j(z) = \frac{1}{2\pi i}\log\left[\frac{\omega(z,\theta_j(1/\overline{a}))}{\omega(z,1/\overline{a})}\right] - \frac{1}{2\pi i}\log\left[-\frac{(1-\overline{\delta_j}/\overline{a})}{q_j}\right] + v_j(1/\overline{a}) + \frac{\tau_{jj}}{2} \tag{4.114}$$

for $j=1,\ldots,M$.

Equation (4.114) is an expression for the function $v_j(z)$ in terms of the prime function $\omega(z,.)$, at least up to the addition of a function that is independent of z (and which, from (4.114), we see depends on $v_j(1/\overline{a})$ and the diagonal elements of the period matrix). Despite appearances, the quantity on the right-hand side is independent of the choice of $1/\overline{a}$.

Substitution of the formulas (4.108) and (4.114) into the expression (2.53) for the functions $\{\mathcal{G}_j(z,a)|j=1,\ldots,M\}$ produces similar explicit expressions for the latter functions in terms of the prime function $\omega(.,.)$.

Exercises

4.1. **(Verify that C_R contribution vanishes)** Verify that

$$\lim_{R\to\infty}\int_{C_R} v_j(a)d\Pi^{z,w}(a) = 0, \qquad j=1,\ldots,M. \tag{4.115}$$

4.2. **(On the differential $d\Pi^{z,w}$)** Suppose D_z is an $(M+1)$-connected circular domain and pick n with $1\le n\le M$. Show that if we define the differential

$$d\hat{\Pi}^{z,w}(a) \equiv 2\pi i\left[\frac{\partial\mathcal{G}_n(z,a)}{\partial a} - \frac{\partial\mathcal{G}_n(w,a)}{\partial a}\right]da, \tag{4.116}$$

then

$$d\hat{\Pi}^{z,w}(a) = d\Pi^{z,w}(a). \tag{4.117}$$

4.3. **(Another differential similar to $d\Pi^{z,w}$)** Suppose D_z is an $(M+1)$-connected circular domain and pick n with $1\le n\le M$. Consider the differential defined by

$$d\hat{\Pi}^{z,w}(a') \equiv 2\pi i\left[\frac{\partial\mathcal{G}_0(z,a')}{\partial a'} - \frac{\partial\mathcal{G}_n(w,a')}{\partial a'}\right]da', \tag{4.118}$$

where $\mathcal{G}_n(z,a)$ is the analytic extension of one of the "other" modified Green's functions as defined in §2.7 of Chapter 2.

(a) Show that

$$d\hat{\Pi}^{z,w}(\theta_j(a)) = d\hat{\Pi}^{z,w}(a), \qquad j=1,\ldots,M. \tag{4.119}$$

Thus $d\hat{\Pi}^{z,w}(a)$ shares the property (4.8) of $d\Pi^{z,w}(a)$.

(b) Verify that, near $a = z$ and $a = w$,

$$d\hat{\Pi}^{z,w}(a') = \left[+\frac{1}{a'-z} + \text{a locally analytic function of } a'\right] da', \quad (4.120)$$

and near $a' = w$ it has the singular behavior

$$d\hat{\Pi}^{z,w}(a') = \left[-\frac{1}{a'-w} + \text{a locally analytic function of } a'\right] da', \quad (4.121)$$

which are analogous to the conditions (4.3) and (4.4) on $d\Pi^{z,w}(a)$.

(c) Verify, however, that $d\hat{\Pi}^{z,w}(a)$ does *not* satisfy the normalization condition (4.7).[11]

4.4. **($d\Pi^{z,w}(a)$ in terms of the prime function)** In §4.9 and §4.10 it was shown how, having defined the prime function, it is possible to write the analytic extensions $\{\mathcal{G}_j(z,a)|j = 0,\ldots,M\}$ of the modified Green's functions of D_z in terms of it. The same goes for the differential $d\Pi^{z,w}(a)$. Show that

$$d\Pi^{z,w}(a) = d\log\left[\frac{\omega(a,z)}{\omega(a,w)}\right]. \quad (4.122)$$

4.5. **($\Pi^{z,w}_{a,b}$ in terms of the prime function)** Show that

$$\Pi^{z,w}_{a,b} = \log\left[\frac{\omega(z,a)\omega(w,b)}{\omega(z,b)\omega(w,a)}\right]. \quad (4.123)$$

Remark: In the simply connected case when $\omega(z,a) = (z-a)$ the right-hand side is just the logarithm of the cross-ratio $p(z,w,a,b)$ defined in (1.87).

4.6. **(The differential $df_j(a)$ in (4.15))** Use the expression (4.108) for $\mathcal{G}_0(z,a)$ in terms of the prime function, together with the properties (4.88)–(4.90) of the prime function, to find an explicit expression for the differential $df_j(a)$ appearing in (4.15) in terms of the function $v_j(a)$ and the Möbius map $\theta_j(a)$.

4.7. **(Reflectionally symmetric domains)** Suppose all the circles $C_0, C_1, \ldots, C_M$ of a multiply connected circular domain D_z are centered on the real axis. This means that such a domain D_z is reflectionally symmetric about the real z axis. Use the result of Exercise 3.7 to show that, for this class of domains,

$$\overline{\Pi}^{z,w}_{a,b} = \Pi^{z,w}_{a,b} \quad (4.124)$$

and consequently that

$$\overline{X}(a,b) = X(a,b), \qquad \overline{\omega}(a,b) = \omega(a,b). \quad (4.125)$$

4.8. **(Use of the transformation formula)** Let γ be a point on circle C_j for $j = 1,\ldots,M$ of a multiply connected circular domain D_z and let $\omega(.,.)$ be its associated prime function. Show that

$$\frac{\omega(z,\gamma)}{\omega(z,1/\overline{\gamma})} = -\frac{q_j}{(1-\overline{\delta_j}z)} e^{-2\pi i(v_j(z)-v_j(1/\overline{\gamma})+\tau_{jj}/2)}. \quad (4.126)$$

[11]This exercise demonstrates that there are many choices of so-called third-kind (Abelian) differentials with poles of strength ± 1 but that choosing the correct normalization of such a differential is important in defining the prime function.

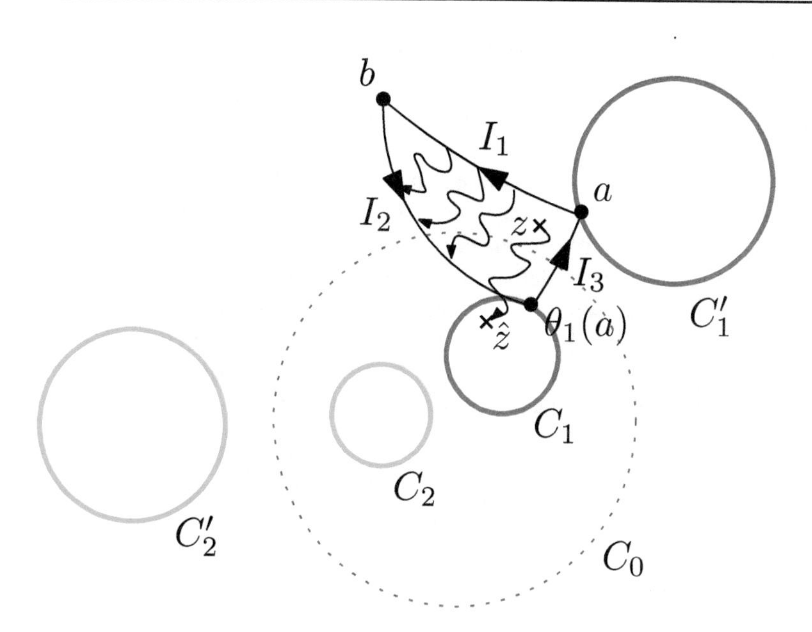

Figure 4.3. *The point z approaches a to the left of I_1 and I_3. In the analytic continuation the contour $-I_1$ and the point z are dragged continuously, as depicted by the squiggly lines, to contour I_2 and point $\hat{z}$, respectively.*

4.9. **(Alternative derivation of formula (4.90))** This exercise gives another derivation of the transformation formula (4.90); it also gives an indication of why we pick the minus sign in (4.90). Figure 4.3 shows a schematic of the particular situation to be considered. Pick some point a on C_j' for some choice of $j = 1, \ldots, M$—the case $j = 1$ is shown in Figure 4.3—and consider the limit as a point z in F tends to a. Pick a contour I_1 in F that directly joins a to b and let the point z tend to a on one side of the contour, for example, to the left of I_1 as shown in Figure 4.3.

(a) Use relations (4.86), (4.61), and (4.25) to execute the limit $w \to b$, but retain the limit as $z \to a$, to show that we can write

$$\omega(a, b) = (a - b) \lim_{z \to a} \left[e^{-\frac{1}{2} \int_b^a d\tilde{\Pi}^{z,b}(a')} \right], \tag{4.127}$$

where $d\tilde{\Pi}^{z,b}$ is defined by

$$d\Pi^{z,b}(a') = \frac{da'}{a' - z} - \frac{da'}{a' - b} + d\tilde{\Pi}^{z,b}(a'). \tag{4.128}$$

(b) Let I_3 be some contour joining $\theta_j(a)$ to a that will serve as a barrier that renders the function $v_j(z)$ single valued in the domain F provided it is not crossed. We also pick I_3 such that z is on the same side of I_3 as the point b, as shown in Figure 4.3. Verify that, with this choice of I_3,

$$-\int_{I_3} d\Pi^{z,b} = 2\pi\mathrm{i}(v_j(z) - v_j(b)). \tag{4.129}$$

(c) To define the analytic continuation of formula (4.127) we fix the end of the contour at b but continuously drag the end of the contour I_1 at a to the point $\theta_j(a)$ along I_3 and, simultaneously, continuously drag z to some other point $\hat{z}$ that remains to the left of the contour as it is being deformed (that is, we disallow any crossing of contours during the dragging process; the squiggly lines in Figure 4.3 are meant to indicate this continuous dragging process). Let the dragged contour now joining b to $\theta_j(a)$ be called I_2, as shown in Figure 4.3. Argue that the analytic continuation of formula (4.127) is given by

$$\omega(\theta_j(a), b) = (a - b) \lim_{\hat{z}\to\theta_j(a)} \left[e^{-\frac{1}{2}\int_{I_2} d\tilde{\Pi}^{\hat{z},b}(a')} \right], \tag{4.130}$$

where, from (4.128),

$$d\tilde{\Pi}^{\hat{z},b}(a') = d\Pi^{\hat{z},b}(a') + \frac{da'}{a' - b} - \frac{da'}{a' - \hat{z}}, \tag{4.131}$$

where $\hat{z}$ tends to $\theta_j(a)$ from the left of it, as shown in Figure 4.3.

(d) Show from the definitions (4.127) and (4.130) that

$$\frac{\omega(\theta_j(a), b)}{\omega(a, b)} = \lim_{\substack{z\to a\\ \hat{z}\to\theta_j(a)}} \left[\frac{(\theta_j(a) - b)}{(a - b)} e^{-\frac{1}{2}\left\{\int_{I_2} d\tilde{\Pi}^{\hat{z},b}(a') + \int_{I_1} d\tilde{\Pi}^{z,b}(a')\right\}} \right]. \tag{4.132}$$

(e) Show that the integrand of the first integral in the exponent of part (d) can be written as

$$d\tilde{\Pi}^{\hat{z},b}(a') = d\Pi^{z,b}(a') + d\Pi^{\hat{z},z}(a') + \underbrace{\left[\frac{da'}{a' - b} - \frac{da'}{a' - z}\right] + \left[\frac{da'}{a' - z}\right.}_{=0} \left. - \frac{da'}{a' - \hat{z}}\right]. \tag{4.133}$$

(f) Hence, show that the integral in the exponent in part (d) is

$$\int_{I_2} d\tilde{\Pi}^{\hat{z},b}(a') + \int_{I_1} d\tilde{\Pi}^{z,b}(a') = \int_{I_1\cup I_2} d\Pi^{z,b}(a') + \int_{I_2} d\Pi^{\hat{z},z} + \int_{I_1\cup I_2} \frac{da'}{a' - b} - \frac{da'}{a' - z} + \int_{I_2} \frac{da'}{a' - z} - \frac{da'}{a' - \hat{z}}. \tag{4.134}$$

(g) Use the residue theorem to confirm that

$$\int_C d\Pi^{z,b} = 2\pi\mathrm{i}, \qquad \int_C \frac{da'}{a' - b} - \frac{da'}{a' - z} = -2\pi\mathrm{i}, \tag{4.135}$$

$$\int_C \frac{da'}{a' - z} - \frac{da'}{a' - \hat{z}} = 2\pi\mathrm{i}, \tag{4.136}$$

where $C = I_1 \cup I_2 \cup I_3$ is the closed contour made up of the three contours I_1, I_2, and I_3.

(h) Use the results of part (g) to establish the following three results:

$$\int_{I_1\cup I_2} d\Pi^{z,b} = +2\pi\mathrm{i} - \int_{I_3} d\Pi^{z,b},$$

$$\int_{I_1\cup I_2} \frac{da'}{a'-b} - \frac{da'}{a'-z} = -2\pi\mathrm{i} - \int_{I_3} \frac{da'}{a'-b} - \frac{da'}{a'-z},$$

$$\int_{I_2} \frac{da'}{a'-z} - \frac{da'}{a'-\hat{z}} = +2\pi\mathrm{i} - \int_{I_1} \frac{da'}{a'-z} - \frac{da'}{a'-\hat{z}} - \int_{I_3} \frac{da'}{a'-z} - \frac{da'}{a'-\hat{z}}.$$

(i) Establish the two results

$$\int_{I_2} dv_j(a') = -\int_{I_1} dv_j(a') - \int_{I_3} dv_j(a') \tag{4.137}$$

and

$$\int_{I_3} dv_j(a') = -\tau_{jj}. \tag{4.138}$$

(j) Use the results of parts (b) and (h) in the expression in part (f) to deduce that

$$\begin{aligned}\int_{I_2} d\tilde{\Pi}^{\hat{z},b}(a') + \int_{I_1} d\tilde{\Pi}^{z,b}(a') &= 2\pi\mathrm{i} + 2\pi\mathrm{i}(v_j(z)-v_j(b)) + \int_{I_2} d\Pi^{\hat{z},z} \\ &\quad - \int_{I_1} \frac{da'}{a'-z} - \frac{da'}{a'-\hat{z}} + \int_{I_3} \frac{da'}{a'-\hat{z}} - \frac{da'}{a'-b}.\end{aligned}$$

(k) Now relate $\hat{z}$ to z by making the particular choice

$$\hat{z} = \theta_j(z) \tag{4.139}$$

and hence argue that

$$d\Pi^{\hat{z},z} = d\Pi^{\theta_j(z),z}(a') = 2\pi\mathrm{i}dv_j(a'). \tag{4.140}$$

(l) Now combine the results of parts (i) and (k) with that in part (j) to establish that

$$\begin{aligned}\int_{I_2} d\tilde{\Pi}^{\hat{z},b}(a') + \int_{I_1} d\tilde{\Pi}^{z,b}(a') &= 2\pi\mathrm{i} + 2\pi\mathrm{i}(v_j(z)-v_j(b)) + 2\pi\mathrm{i}(v_j(a)-v_j(b)) \\ &\quad +2\pi\mathrm{i}\tau_{jj} + 2\log\left[\frac{\theta_j(a)-b}{a-b}\right] \\ &\quad -\log\left[\frac{\theta_j(z)-\theta_j(a)}{z-a}\right].\end{aligned}$$

(m) Finally, by taking the limit $z \to a$, use the result in part (l) in the result in part (d) to deduce that

$$\frac{\omega(\theta_j(a),b)}{\omega(a,b)} = -\frac{q_j}{1-\overline{\delta_j}a} e^{-2\pi\mathrm{i}(v_j(a)-v_j(b)+\tau_{jj}/2)}. \tag{4.141}$$

Chapter 5

Doubly connected domains

5.1 ▪ The two-body problem

In applications it is often of interest to study the interaction of two entities in a two-dimensional medium. Understanding pairwise interactions gives valuable insights into the general N-body situation. The domain exterior to two entities, or objects, is an unbounded doubly connected domain. For this reason, the doubly connected case of our general framework for multiple connectivity is special, and so it is useful to study it in isolation.

For a doubly connected domain there is just a single interior circle C_1 in the preimage circular domain D_z. From §1.4 we know that an automorphism of a disc, which depends on three real parameters, one of which is a rotational degree of freedom, can be used to transform any given circular domain D_z to another circular domain. We can pick two of the real parameters in this automorphism to place the center δ_1 of circle C_1 at $\delta_1 = 0$, thereby reducing considerations to a concentric annulus. This choice puts restrictions on only two of the three parameters in the automorphism, leaving the rotational degree of freedom still to be fixed: it is clear that having centralized the circle C_1 we can still rotate the resulting circular domain about the origin through any angle without changing the geometry of the annulus.

It is therefore enough, when studying doubly connected domains, to develop the function theory in a concentric annulus

$$D_z = \{z \mid \rho < |z| < 1\}, \qquad 0 < \rho < 1. \tag{5.1}$$

5.2 ▪ The concentric annulus

The circular domain D_z in (5.1) corresponds to making the choices

$$\delta_1 = 0, \qquad q_1 = \rho \tag{5.2}$$

in the general formulation, as shown in Figure 5.1. Following the notational conventions introduced in Chapter 2, we use C_0 to denote $|z| = 1$ and C_1 to denote $|z| = \rho$. Circle C_0, and its images under the various conformal maps constructed herein, will be shown in blue; C_1 and its images will be shown in red.

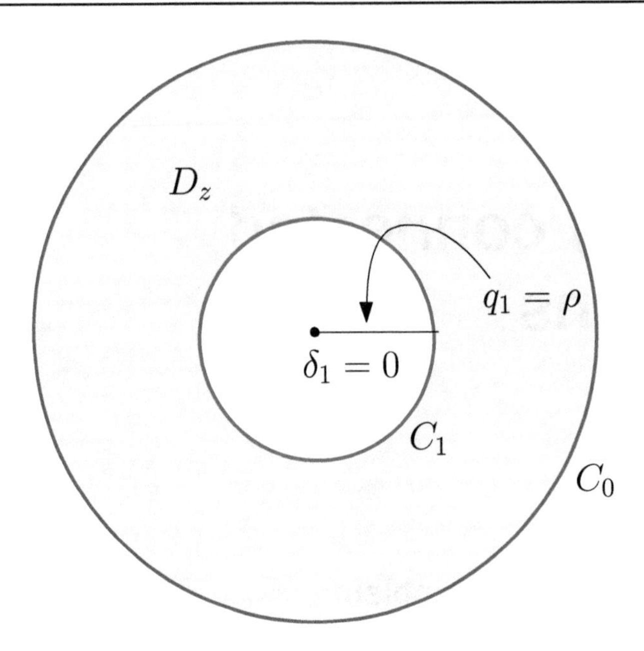

Figure 5.1. *Concentric annulus $\rho < |z| < 1$ with boundaries C_0 and C_1.*

We can list several reasons why this special case of a doubly connected circular domain is worth studying in isolation:

(a) The concentric annulus is the simplest nontrivial instance in which all the key elements of our general theory come into play.

(b) The functional identities satisfied by the prime function in the concentric annulus afford certain simplifications compared to the general case, making calculations slightly easier.

(c) We will see that the doubled domain F is bounded in this case, allowing the reader to see how the general theory of Chapters 2–4 carries over naturally to this case with only minor changes in detail.

(d) Existing analytical treatments of problems in doubly connected domains in the literature commonly use the theory of elliptic functions. Our different approach via the prime function obviates the need to introduce the reader to that theory. Some exercises have been included, however, that elucidate connections between the prime function approach and a more standard approach via elliptic functions. Examining these connections is useful in showing how the framework here extends beyond the limitations of elliptic function theory to capture the general multiply connected case.

This chapter closely follows the template set out in Chapter 1. It is more colorful, however, in that most of the schematics in this chapter are similar to those shown in Chapter 1, but now with additional red features. The idea is to highlight just how naturally our mathematical approach as applied to doubly connected domains generalizes the simply connected theory of Chapter 1.

5.3 ▪ The prime function identities

From the general theory of Chapters 2–4 it is known that associated with this concentric annulus D_z there exists a prime function, denoted by $\omega(z, a)$, with a single simple zero at $z = a$ in a fundamental region F. In this case the reflection of D_z in C_0, called D'_z, is the annulus $1 < |z| < 1/\rho$. This means that $F = D_z \cup D'_z$ is a larger "doubled-up" annulus,

$$\rho < |z| < 1/\rho, \tag{5.3}$$

obtained from the union of the annulus $\rho < |z| < 1$ with its reflection in C_0, i.e., $1 < |z| < 1/\rho$. Clearly, the doubled domain F is bounded in this case.

Remark: In accordance with the notational convention set out in §4.7, although we write simply $\omega(z, a)$ for the prime function, in this chapter it also depends on the parameter ρ.

The associated prime function $\omega(z, a)$ turns out to satisfy the following three functional identities:

$$\omega(a, z) = -\omega(z, a), \tag{5.4}$$

$$\overline{\omega}(1/z, 1/a) = -\frac{1}{za}\omega(z, a), \tag{5.5}$$

$$\omega(\rho^2 z, a) = -\frac{a}{z}\omega(z, a). \tag{5.6}$$

The first two identities, (5.4) and (5.5), are shared with the prime function associated with the unit disc considered in Chapter 1. It is the third identity, (5.6), that is new and which derives from the presence of the single hole in the unit disc.

Owing to the reflectional symmetry of the concentric annulus D_z about the real axis, the prime function $\omega(z, a)$ in this case enjoys the additional property that

$$\overline{\omega}(z, a) = \omega(z, a). \tag{5.7}$$

Recall that this is a feature also shared with the prime function associated with the unit disc considered in Chapter 1. As there, we will list the identity (5.7) separately from (5.4)–(5.6) since it arises only as a consequence of the reflectional symmetry of D_z about the real axis.

Let us examine the origin of the new identity (5.6) as it arises from the general considerations in Chapters 2–4.

The concentric annulus (5.1) corresponds to $M = 1$ with the choices (5.2). It follows from the well-known properties of the logarithm that

$$\sigma_1(z) = \frac{\log|z|}{\log\rho}, \qquad \hat{v}_1(z) = \mathrm{i}\frac{\log z}{\log\rho}. \tag{5.8}$$

Therefore we can immediately identify

$$v_1(z) = \frac{1}{2\pi\mathrm{i}}\log z, \tag{5.9}$$

where we have chosen a real additive degree of freedom associated with $v_1(z)$ to be zero, and

$$\mathcal{P}_{11} = \int_{C_1} d\hat{v}_1 = \int_{|z|=\rho} \frac{\mathrm{i}}{\log\rho}\frac{dz}{z} = -\frac{2\pi}{\log\rho}. \tag{5.10}$$

The period matrix in this case has just a single element:

$$\tau_{11} = 2\mathrm{i}\mathcal{P}_{11}^{-1} = -\frac{\mathrm{i}\log\rho}{\pi}. \tag{5.11}$$

Since $0 < \rho < 1$ this constant has strictly positive imaginary part.

Remark: The fact that we can identify $v_1(z)$ essentially by inspection is a special feature of the concentric annulus.

The third identity of the prime function can be written as

$$\omega(\theta_1(z), a) = \sqrt{\theta_1'(z)}e^{-2\pi\mathrm{i}(v_1(z)-v_1(a)+\tau_{11}/2)}\omega(z,a), \tag{5.12}$$

where $\theta_1(z)$ is the Möbius map

$$\theta_1(z) = \overline{S_1(1/\overline{z})} = \rho^2 z, \tag{5.13}$$

and where

$$S_1(z) = \frac{\rho^2}{z} \tag{5.14}$$

is the Schwarz function of the circle C_1, and, following (4.92), we choose

$$\sqrt{\theta_1'(z)} = -\rho. \tag{5.15}$$

On use of (5.9), (5.11), and (5.15), (5.12) can be written in the simpler form (5.6).

5.4 ▪ A ratio of prime functions

As before, we introduce a ratio of prime functions defined by

$$R(z; a, b) = \frac{\omega(z,a)}{\omega(z,b)}, \tag{5.16}$$

where a and b are arbitrary distinct points inside or on the boundary of D_z. This is a different function from that considered in Chapter 1 because the prime function has changed.

From (4.103), together with (5.9), and provided that a and b are distinct and nonzero, we deduce that this new function $R(z; a, b)$ has the useful properties

$$\overline{R(z;a,b)} = \begin{cases} \dfrac{\overline{a}}{\overline{b}} R(z; 1/\overline{a}, 1/\overline{b}), & z \in C_0, \\ \left(\dfrac{\overline{a}}{\overline{b}}\right)^2 R(z; 1/\overline{a}, 1/\overline{b}), & z \in C_1. \end{cases} \tag{5.17}$$

5.5 ▪ Modified Green's functions

It follows from the general theory of Chapter 4 that the function

$$\mathcal{G}_0(z,a) = \frac{1}{2\pi\mathrm{i}} \log\left(\frac{\omega(z,a)}{|a|\omega(z,1/\overline{a})}\right) \tag{5.18}$$

is the analytic extension of the modified Green's function $G_0(z,a)$ of the concentric annulus:

$$G_0(z,a) = \mathrm{Im}[\mathcal{G}_0(z,a)]. \tag{5.19}$$

From (2.50), we know that

$$G_0(z,a) = \begin{cases} 0, & z \in C_0, \\ \mathrm{Im}[v_1(a)], & z \in C_1, \end{cases} \tag{5.20}$$

and, from (2.52), we know that

$$\int_{C_0} d\mathcal{G}_0(z,a) = 1, \qquad \int_{C_1} d\mathcal{G}_0(z,a) = 0. \tag{5.21}$$

Unlike in the simply connected unit disc, there is a second modified Green's function associated with this doubly connected domain. Its analytic extension is

$$\mathcal{G}_1(z,a) \equiv \mathcal{G}_0(z,a) - v_1(z) - v_1(a) + \mathrm{i}[\mathcal{P}_{11}]^{-1}, \tag{5.22}$$

where we have used (2.53). It has the properties

$$\int_{C_0} d\mathcal{G}_1(z,a) = 0, \qquad \int_{C_1} d\mathcal{G}_1(z,a) = -1. \tag{5.23}$$

Just as we did in Chapter 1, we will give geometrical interpretations of all the functions to be introduced in what follows. Consider the conformal map defined by

$$w = \mathcal{G}_0(z,a) \tag{5.24}$$

for some real negative a inside the annulus, $-1 < a < -\rho$ say. We choose the branch cut joining the logarithmic branch points at $z = a$ and $z = 1/a$ to be along the negative real z axis. The image of the annulus D_z under this mapping is the region of a unit-width semistrip with $0 < \mathrm{Im}[w] < \infty$ exterior to a slit of finite length inside this strip and parallel to the real axis. The two sides of the branch cut correspond to the two sides of the semistrip parallel to the imaginary axis.

If we use the second modified Green's function $\mathcal{G}_1(z,a)$ to similarly define a conformal map to a complex w plane,

$$w = \mathcal{G}_1(z,a), \tag{5.25}$$

then this will effect a similar map to such a semistrip but with the role of the circles reversed: that is, circle C_1 will now be transplanted to the entire unit width of the strip with C_0 mapping to some interior slit of shorter width.

Figure 5.2 shows a schematic illustrating the two conformal mappings (5.24) and (5.25). Such mappings are useful when tackling problems associated with a periodic array of finite slots, plates, or intervals situated above an infinite wall or barrier.

5.6 ▪ Bounded circular slit maps

Now let

$$w = e^{2\pi \mathrm{i}\mathcal{G}_0(z,a)} = \frac{1}{|a|}\frac{\omega(z,a)}{\omega(z,1/\overline{a})}, \tag{5.26}$$

where $a \in \mathbb{C}$ is taken to be some nonzero interior point in D_z. This is the same formula as for the disc automorphism (1.31), but, since we have changed what we mean by the prime function, it is no longer a simple Möbius map as it was in Chapter 1.

From (5.20) and (5.9) we can work out that

$$|w| = \begin{cases} 1, & z \in C_0, \\ |a|, & z \in C_1. \end{cases} \tag{5.27}$$

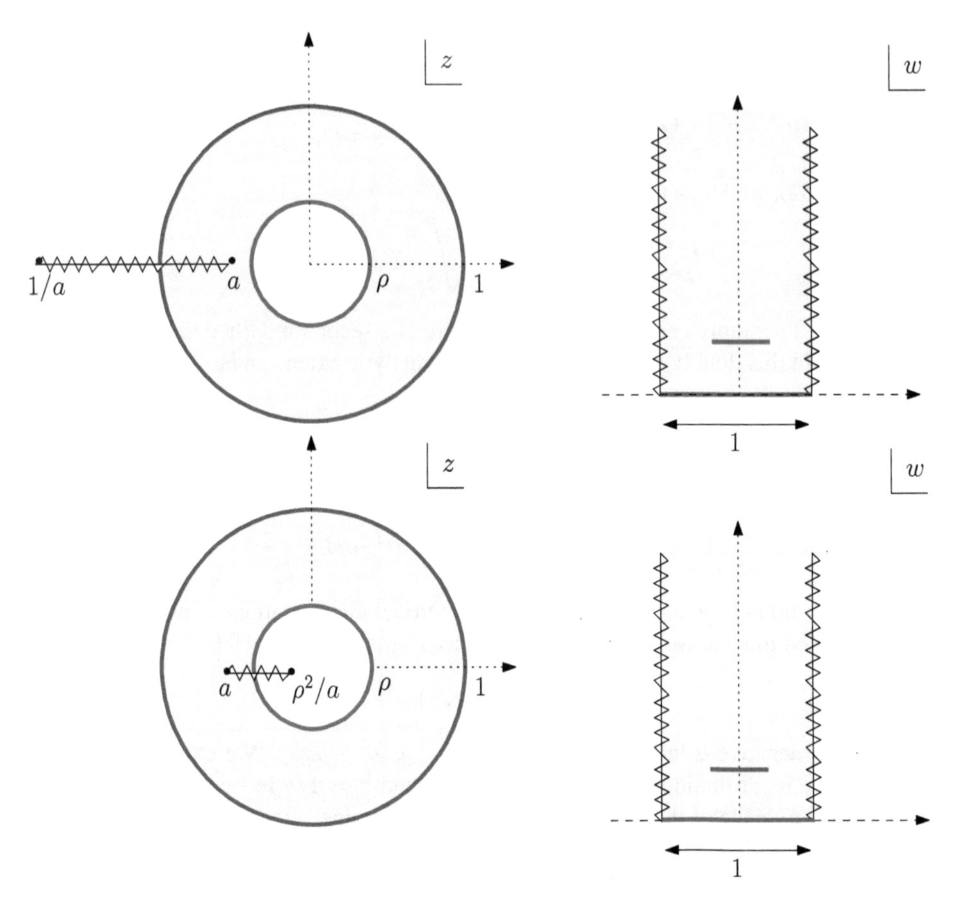

Figure 5.2. *The geometrical effect of (5.24) and (5.25) as conformal maps from the concentric annulus D_z. The two sides of the branch cut correspond to the two sides of the period window.*

Thus, the images of C_0 and C_1 lie on circles of radius 1 and $|a|$, respectively. It is not yet clear, however, if the images are the *whole* of these circles or just part of them. To determine this we need to study

$$[\arg w]_{C_0}, \qquad [\arg w]_{C_1}, \tag{5.28}$$

where we use square brackets to denote the change in the enclosed quantity on traversing the circles C_0 and C_1. Since $w = e^{2\pi i \mathcal{G}_0(z,a)}$ these quantities are precisely

$$2\pi \int_{C_0} d\mathcal{G}_0, \qquad 2\pi \int_{C_1} d\mathcal{G}_0, \tag{5.29}$$

which, from (5.21), we know to be 2π and 0, respectively. This means that the image of C_0 is the entire unit circle in the w plane while the image of C_1 is a circular arc slit, of finite length, inside the unit w circle. Such a conformal map is an example of a *bounded circular slit map*.

As in §1.4, we can use the argument principle to check that the mapping (5.26) is a univalent, or one-to-one, map from the interior of D_z to the interior of the image domain.

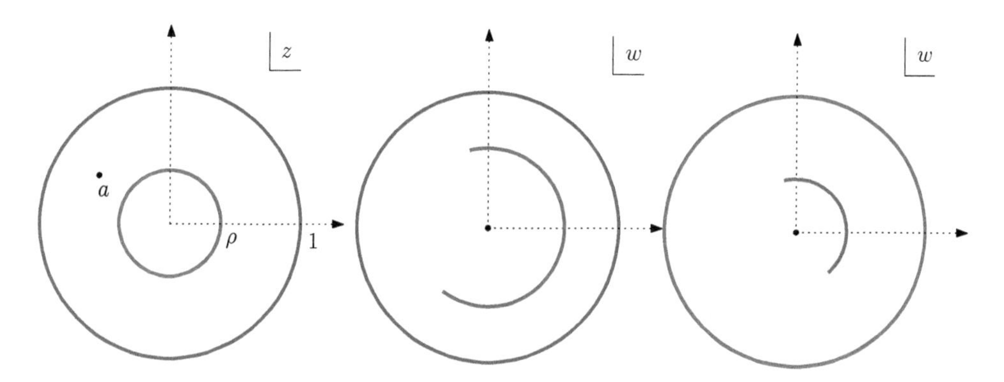

Figure 5.3. *Bounded circular slit maps: the two maps (5.26) and (5.31) are qualitatively similar in mapping D_z to the unit disc in the w plane with a circular arc slit. The difference is that in (5.26) it is C_0 that is transplanted to $|w| = 1$, while in (5.31) it is C_1 that is transplanted there.*

To do this, consider the function

$$f(z) = \frac{1}{|a|} \frac{\omega(z,a)}{\omega(z,1/\overline{a})} - \alpha, \tag{5.30}$$

where $\alpha \in \mathbb{C}$ is a constant assumed not to lie on the image circumferences. Since this function has no poles in D_z, the number of zeros of this function inside the image domain is given, according to the argument principle (1.34), by its change in argument as the boundary of ∂D_z is traversed, keeping the domain D_z on the left, divided by 2π. However, from the properties of the mapping (5.26) on C_0 and C_1 it is clear that the argument of $f(z)$ will not change around C_1 while it changes either by 2π if $|\alpha| < 1$ or by 0 if $|\alpha| > 1$. The number of zeros of $f(z)$ is, of course, the number of times the mapping attains the value α, and it is once for any point α inside the unit disc (and off the image of C_1). We have therefore established that the mapping to the bounded circular slit domain is one-to-one.

The argument principle can be used in a similar way to establish the univalency of all the conformal mappings to be introduced in the remainder of this chapter; see Exercise 5.5.

It is natural to consider a simple generalization of the map (5.26) defined by

$$w = e^{2\pi \mathrm{i} \mathcal{G}_1(z,a)}, \tag{5.31}$$

i.e., we have swapped $\mathcal{G}_1(z,a)$ for $\mathcal{G}_0(z,a)$ in the exponent. On use of the properties of the modified Green's function $\mathcal{G}_1(z,a)$ it is straightforward to show that this too is a bounded circular slit map, the difference being that it is now C_1 which is transplanted to the entire unit circle $|w| = 1$ with C_0 being transplanted to a finite-length circular slit inside the unit w disc.

Figure 5.3 shows a schematic of the geometrical effect of the two maps (5.26) and (5.31).

5.7 ▪ Unbounded circular slit maps

Consider the mapping given by

$$w = e^{2\pi \mathrm{i}(\mathcal{G}_0(z,a) - \mathcal{G}_0(z,b))}, \tag{5.32}$$

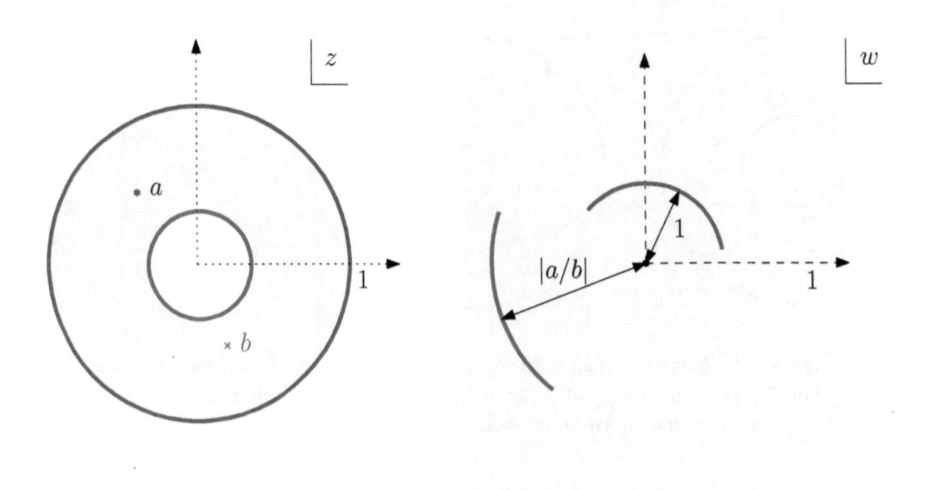

Figure 5.4. *Unbounded circular slit map: the map (5.33) transplants D_z to the plane exterior to two circular arc slits.*

where a and b are two distinct points in the annulus D_z. When rewritten in terms of the prime function this becomes

$$w = \frac{|b|}{|a|} \frac{\omega(z,a)\omega(z,1/\overline{b})}{\omega(z,b)\omega(z,1/\overline{a})}. \tag{5.33}$$

Clearly $z = a$ maps to $w = 0$ and $z = b$ maps to $w = \infty$, so the image domain is unbounded. From the properties (5.20),

$$|w| = \begin{cases} 1, & z \in C_0, \\ |a/b|, & z \in C_1, \end{cases} \tag{5.34}$$

so that the images of C_0 and C_1 lie on circles of radius 1 and $|a/b|$, respectively. To ascertain whether the images are closed circles or circular slits we need to find

$$[\arg w]_{C_0}, \qquad [\arg w]_{C_1}. \tag{5.35}$$

From (5.32) it can be verified that these quantities are

$$2\pi \int_{C_0} d\mathcal{G}_0(z,a) - d\mathcal{G}_0(z,b), \qquad 2\pi \int_{C_1} d\mathcal{G}_0(z,a) - d\mathcal{G}_0(z,b), \tag{5.36}$$

which, in view of (5.21), are both zero. The annulus D_z is transplanted to the entire w plane exterior to two circular slits, as shown in Figure 5.4.

(5.33) is referred to as an *unbounded circular slit map*.

If we choose $\mathcal{G}_1$ in place of $\mathcal{G}_0$ in (5.32), it is straightforward to use (2.53) to show that we would obtain a similar unbounded circular slit map (Exercise 5.9).

A useful special case is to make the choices

$$a = \mathrm{i}r, \qquad b = \overline{a}, \qquad \rho < r < 1 \tag{5.37}$$

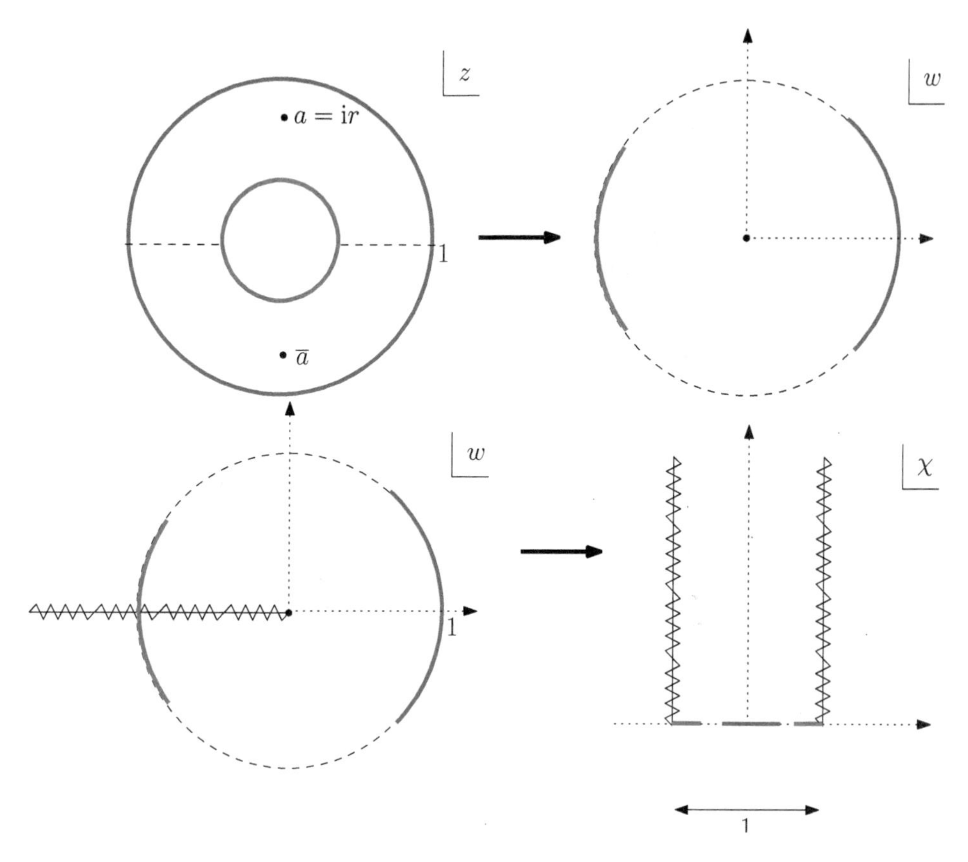

Figure 5.5. *The special case of an unbounded circular slit map (5.38) transplants the upper half annulus in the z plane to the unit w disc with C_0 and C_1 being mapped to two circular arcs on $|w| = 1$; the two segments of the real diameter inside the annulus map to the remainder of the unit w circle. On taking a logarithm of this map (5.41) takes the unit w disc to a unit-width semistrip with two slits along the real χ axis.*

and to premultiply by -1 so that (5.33) becomes

$$w = -\frac{\omega(z,a)\omega(z,1/a)}{\omega(z,\overline{a})\omega(z,1/\overline{a})}. \tag{5.38}$$

Now (5.34) implies

$$|w| = \begin{cases} 1, & z \in C_0, \\ 1, & z \in C_1, \end{cases} \tag{5.39}$$

meaning that both C_0 and C_1 map to circular slits on the unit circle $|w| = 1$. Moreover, for points z on the real axis inside the annulus, i.e., $-1 < z < -\rho$ and $\rho < z < 1$, it follows from the form of (5.38) and on use of (5.7) that

$$\overline{w} = \frac{1}{w}. \tag{5.40}$$

Hence these two portions of the real axis inside the annulus *also* map to the unit circle $|w| = 1$. Figure 5.5 shows the geometry of this mapping.[12]

[12] In Exercise 8.15 this unbounded circular slit map is related to the Jacobi cn function.

As in Chapter 1 we also consider a logarithm of (5.38) and study the map

$$\chi = \frac{1}{2\pi\mathrm{i}} \log w = \frac{1}{2\pi\mathrm{i}} \log \left[-\frac{\omega(z,a)\omega(z,1/a)}{\omega(z,\overline{a})\omega(z,1/\overline{a})} \right], \qquad a = \mathrm{i}r. \tag{5.41}$$

This transplants the annulus to an infinite strip region with C_0 and C_1 being transplanted to two finite-length slits along the real χ axis.

Figure 5.5 also shows the geometry of the map (5.41) in the χ plane.

5.8 ▪ Annular slit maps

Another mapping to consider is

$$w = e^{2\pi\mathrm{i}(\mathcal{G}_0(z,a) - \mathcal{G}_1(z,a))}, \tag{5.42}$$

where a is some point inside the annulus D_z. There was no analogue of this mapping in Chapter 1 because, when D_z is the unit disc, it is not possible to identify any other modified Green's functions beyond $G_0(z,a)$. If we consider

$$[\arg w]_{C_0}, \qquad [\arg w]_{C_1}, \tag{5.43}$$

these are precisely the quantities

$$2\pi \int_{C_0} d\mathcal{G}_0(z,a) - d\mathcal{G}_1(z,b), \qquad 2\pi \int_{C_1} d\mathcal{G}_0(z,a) - d\mathcal{G}_1(z,b), \tag{5.44}$$

which are both equal to 2π. This means that the images of *both* C_0 and C_1 completely encircle the origin $w = 0$. In this case (5.42) can be written as

$$w = ce^{2\pi\mathrm{i}v_1(z)} = cz, \qquad c = e^{2\pi\mathrm{i}(v_1(a) - \mathrm{i}[\mathcal{P}_{11}]^{-1})}, \tag{5.45}$$

where we have used (5.22) and (5.9). Therefore (5.42) turns out to be just a constant multiple of the identity map mapping D_z to another concentric annulus in a complex w plane.

While (5.42) is trivial for the concentric annulus, the analogues of the mapping (5.42) will be more interesting when we consider the triply and higher connected cases in Chapter 6. This is the class of *annular slit maps*, and the reason for this name will become clear in Chapter 6, when triply and higher connected cases are considered. In this doubly connected case the annulus is simply mapped to another annulus with no additional slits present.

We can also examine the effect of taking the logarithm of the right-hand side of (5.42). In this case we find, to within unimportant additive constants,

$$w = v_1(z) = \frac{1}{2\pi\mathrm{i}} \log z. \tag{5.46}$$

This mapping is not single valued in the annulus, but we can introduce a cross-cut, along the negative real axis say, and consider the image of this cut annulus under the mapping (5.46). Indeed, in this case, (5.46) is an elementary mapping that is well known to transplant the cut annulus to a rectangle in the w plane, as shown in Figure 5.6.

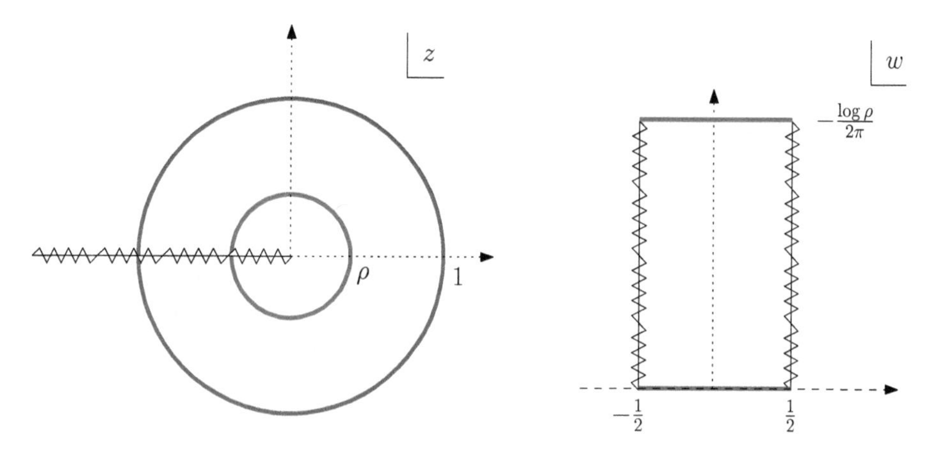

Figure 5.6. *Mapping of a cut annulus to a rectangle via (5.46). The two sides of the branch cut correspond to the two vertical sides of the rectangle.*

5.9 ▪ Parallel slit maps

Letting $a = a_x + \mathrm{i}a_y$ as we did in Chapter 1, we can consider the two maps

$$w_1 = -2\pi\mathrm{i}\frac{\partial \mathcal{G}_0(z,a)}{\partial(\mathrm{i}a_y)}, \qquad w_2 = -2\pi\mathrm{i}\frac{\partial \mathcal{G}_0(z,a)}{\partial a_x}. \tag{5.47}$$

Both parametric derivatives of $-2\pi\mathrm{i}\mathcal{G}_0(z,a)$ produce a simple pole, with unit residue at $z = a$. Thus the image of D_z will be an unbounded domain. Since $-2\pi\mathrm{i}\mathcal{G}_0(z,a)$ has constant real part on C_0 and C_1, its differentiation with respect to the purely imaginary parameter $\mathrm{i}a_y$ means that the imaginary part of w_1 is constant on both circles; thus the annulus D_z is transplanted to the entire w_1 plane exterior to two slits parallel to the real axis.

On the other hand, differentiation of $-2\pi\mathrm{i}\mathcal{G}_0(z,a)$ with respect to the real parameter a_x means that the real part of w_2 will be constant on both circles; thus D_z is transplanted to the entire w_2 plane exterior to two slits parallel to the imaginary axis.

Defining

$$\mathcal{K}(z,a) \equiv a\frac{\partial}{\partial a}\log\omega(z,a), \tag{5.48}$$

and dropping functions of a that are independent of z, we can introduce

$$\phi_0(z,a) = -\frac{1}{a}\mathcal{K}(z,a) + \frac{1}{\overline{a}}\mathcal{K}(z,1/\overline{a}), \tag{5.49}$$

$$\phi_{\pi/2}(z,a) = -\frac{1}{a}\mathcal{K}(z,a) - \frac{1}{\overline{a}}\mathcal{K}(z,1/\overline{a}) \tag{5.50}$$

and define

$$\phi_\theta(z,a) = e^{\mathrm{i}\theta}\left[\cos\theta\phi_0(z,a) - \mathrm{i}\sin\theta\phi_{\pi/2}(z,a)\right]. \tag{5.51}$$

It can be verified that $\phi_\theta(z,a)$ has a simple pole at $z = a$ with residue 1 and satisfies

$$\mathrm{Im}[e^{-\mathrm{i}\theta}\phi_\theta(z,a)] = \text{constant}, \qquad z \in C_0, C_1, \tag{5.52}$$

meaning that the images of C_0 and C_1 are both slits turned at angle θ to the positive real axis. By tuning the values of the parameters ρ, a, and b, these two slits with angle θ can be arranged to be in any desired position and of any desired lengths.

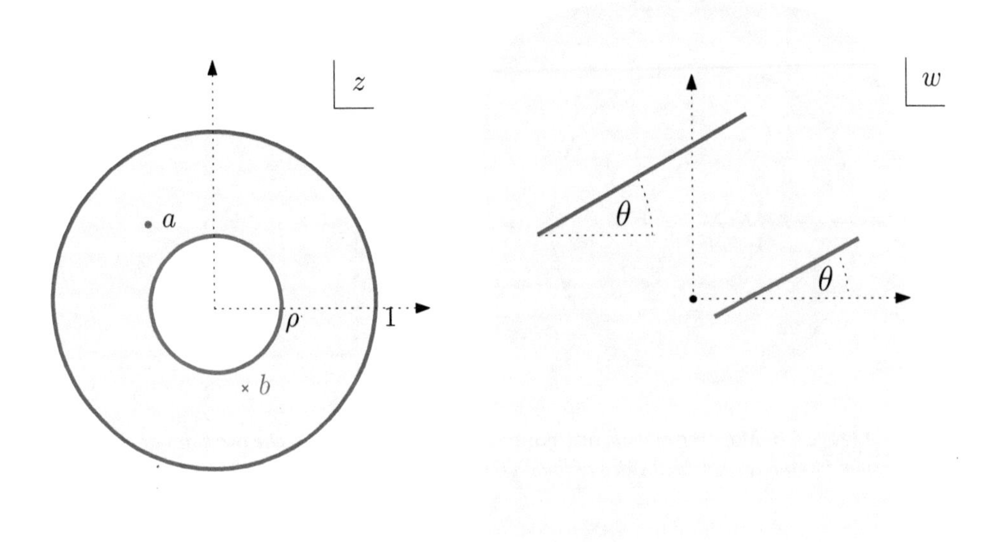

Figure 5.7. *The image of the annulus under the parallel slit mapping (5.51). Both C_0 and C_1 are mapped to finite-length slits that are parallel to each other.*

It now becomes clear why, in Chapter 1, we referred to the same formula (5.51) as a *parallel slit map*: for the concentric annulus this functional form of mapping in terms of the prime function has the effect of transplanting the two circular boundaries of D_z to two parallel slits.

5.10 ▪ Cayley-type maps

It is convenient to refer to a mapping of the form

$$w = R(z; a, b), \tag{5.53}$$

where $R(z; a, b)$ is defined in (5.16) and a and b are two distinct points taken *on the same boundary circle* of D_z as a *Cayley-type map*. We choose this name because in the simply connected case where D_z is the unit disc and the prime function is simply $\omega(z, a) = (z - a)$, such a mapping is precisely the well-known Cayley map discussed in Chapter 1. It is emphasized that the Cayley-type map (5.53) is no longer a simple Möbius map since the prime function has changed. Nevertheless it has important geometrical properties.

There are now two possibilities: either the points a and b are both on C_0, or they are both on C_1. Let us examine the function theoretical properties of such maps and then interpret them geometrically.

If both a and b are on C_0, then $1/\overline{a} = a$ and $1/\overline{b} = b$, so that (5.17) implies

$$\overline{R(z; a, b)} = \begin{cases} \dfrac{b}{a} R(z; a, b), & z \in C_0, \\ \left(\dfrac{b}{a}\right)^2 R(z; a, b), & z \in C_1. \end{cases} \tag{5.54}$$

This implies that

$$\arg[R(z;a,b))] = \text{constant} \tag{5.55}$$

on both C_0 and C_1.

On the other hand, if both a and b are on C_1, then $R(z;a,b)$ also has interesting properties. In this case, $1/\overline{a} = a/\rho^2$ and $1/\overline{b} = b/\rho^2$ and

$$\overline{R(z;a,b)} = \begin{cases} \dfrac{\overline{a}}{\overline{b}} R(z;a/\rho^2, b/\rho^2), & z \in C_0, \\ \left(\dfrac{\overline{a}}{\overline{b}}\right)^2 R(z;a/\rho^2, b/\rho^2), & z \in C_1. \end{cases} \tag{5.56}$$

But

$$R(z;a/\rho^2, b/\rho^2) = \frac{\omega(z, a/\rho^2)}{\omega(z, b/\rho^2)} = \frac{\omega(a/\rho^2, z)}{\omega(b/\rho^2, z)}, \tag{5.57}$$

where we have used (5.5). Now (5.6) implies that

$$\omega(z/\rho^2, a) = -\frac{z}{\rho^2 a}\omega(z, a), \tag{5.58}$$

which can be used to show that

$$R(z;a/\rho^2, b/\rho^2) = \frac{\omega(a/\rho^2, z)}{\omega(b/\rho^2, z)} = \frac{a}{b}\frac{\omega(z,a)}{\omega(z,b)}. \tag{5.59}$$

It follows from (5.56) that

$$\overline{R(z;a,b)} = \begin{cases} R(z;a,b), & z \in C_0, \\ \dfrac{b}{a} R(z;a,b), & z \in C_1, \end{cases} \tag{5.60}$$

implying again that

$$\arg[R(z;a,b))] = \text{constant} \tag{5.61}$$

on both C_0 and C_1.

Figure 5.8 shows the geometrical effects of these two Cayley-type maps.

As an example let us take $a = 1$ and $b = -1$ in (1.11), with a premultiplication by -1, leading to

$$w = -R(z;1,-1) = -\frac{\omega(z,+1)}{\omega(z,-1)}, \tag{5.62}$$

which, in Chapter 1, is precisely the formula (1.58) defining the Cayley map. From (5.17) we deduce that

$$\overline{w} = \begin{cases} -w, & z \in C_0, \\ +w, & z \in C_1. \end{cases} \tag{5.63}$$

At $z = 1$ we know that $w = 0$, while at $z = -1$, $w = \infty$, so we expect the image of $|z| = 1$ to be the whole of the imaginary w axis. However, w is bounded on the circle $|z| = \rho$, so the image of $|z| = \rho$ is a slit of finite length on the real axis. Figure 5.9 shows a schematic.

Figure 5.9 also shows the geometrical effect of a map resulting from taking a logarithm of (5.62):

$$\chi = \frac{1}{\pi \mathrm{i}} \log w = \frac{1}{\pi \mathrm{i}} \log[-R(z;1,-1)]. \tag{5.64}$$

It can be verified that such a map transplants the annulus to the region inside an infinite strip of unit length exterior to a finite-length slit parallel to the imaginary axis along the centerline of the strip.

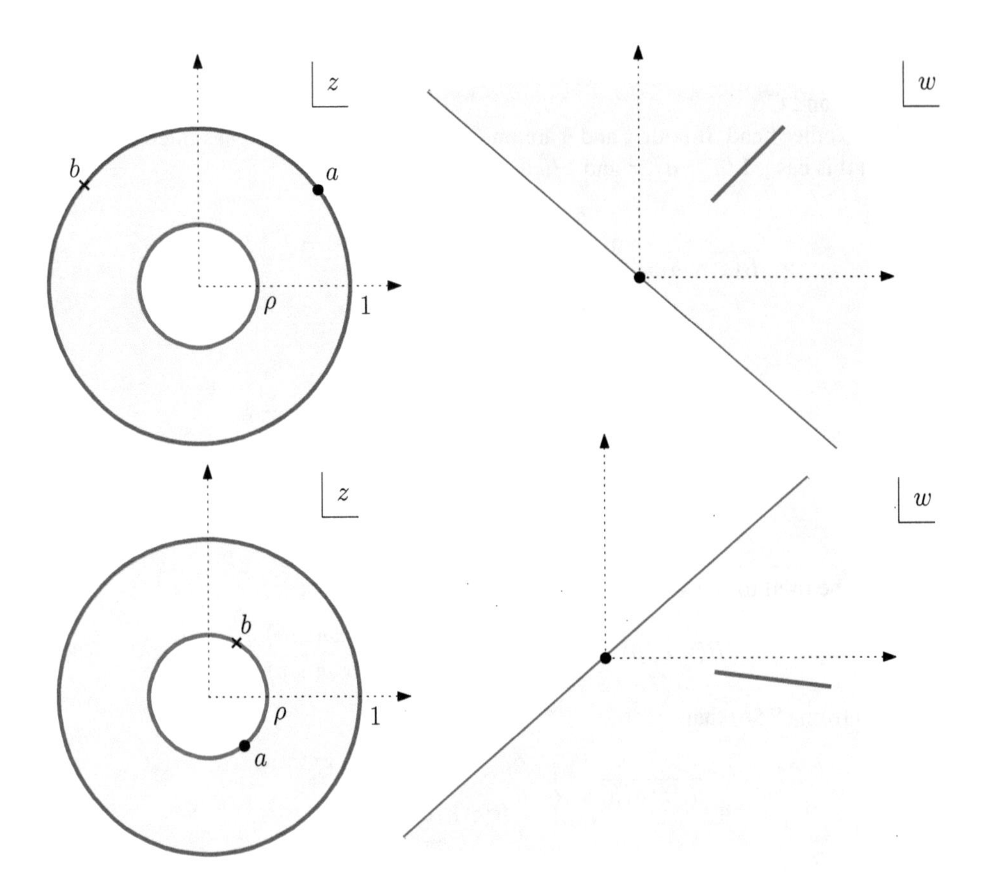

Figure 5.8. *The image of D_z under the Cayley-type mapping (5.53) when a and b are both on C_0 (upper) and when a and b are both on C_1 (lower).*

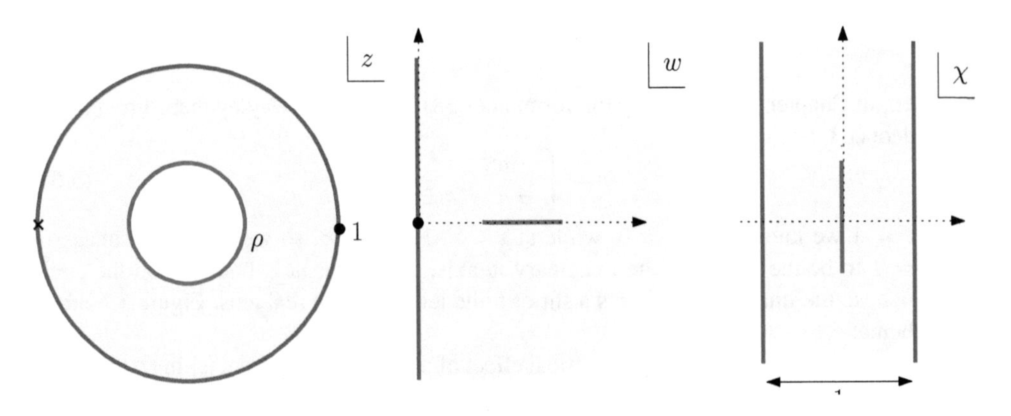

Figure 5.9. *The image of the special case (5.62) of a Cayley-type map in a w plane, with the image in a χ plane after a further logarithmic map (5.64).*

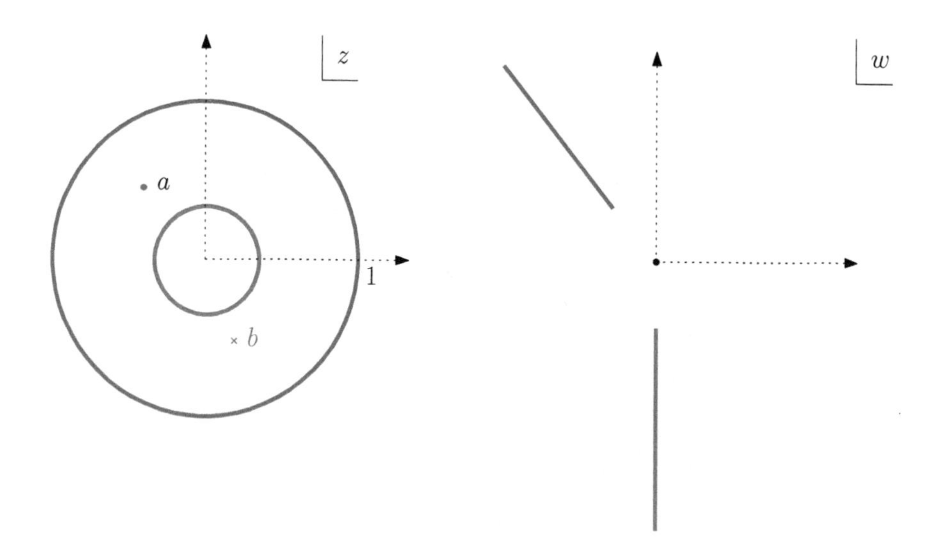

Figure 5.10. *Under the radial slit map (5.65) the circles C_0 and C_1 are transplanted to two finite-length slits on rays emanating from the origin $w = 0$.*

5.11 ▪ Radial slit maps

Consider

$$w = R(z; a, b)R(z; 1/\overline{a}, 1/\overline{b}) = \frac{\omega(z, a)\omega(z, 1/\overline{a})}{\omega(z, b)\omega(z, 1/\overline{b})}, \tag{5.65}$$

with a and b chosen to be two distinct points strictly inside the annulus $\rho < |z| < 1$. It is clear that $z = a$ maps to $w = 0$ and $z = b$ maps to $w = \infty$. (5.17) implies that

$$\overline{w} = \begin{cases} \dfrac{\overline{a}b}{a\overline{b}} w, & z \in C_0, \\ \left(\dfrac{\overline{a}b}{a\overline{b}}\right)^2 w, & z \in C_1, \end{cases} \tag{5.66}$$

or

$$\arg[w] = \begin{cases} \arg[a\overline{b}], & z \in C_0, \\ \arg[a^2\overline{b}^2], & z \in C_1. \end{cases} \tag{5.67}$$

This is a *radial slit mapping*. Both C_0 and C_1 are transplanted to finite-length slits along rays in the w plane emanating from the origin.

Figure 5.10 shows the image of D_z under a typical radial slit map.

A special case is to take a and $b \in \mathbb{R}$, so that (5.65) becomes

$$w = \frac{\omega(z, a)\omega(z, 1/a)}{\omega(z, b)\omega(z, 1/b)}, \qquad a, b \in \mathbb{R}; \tag{5.68}$$

then the images of both C_0 and of C_1 lie on the real axis. This map can be viewed as the two-slit generalization of the classical Joukowski mapping encountered in Exercise 1.15.

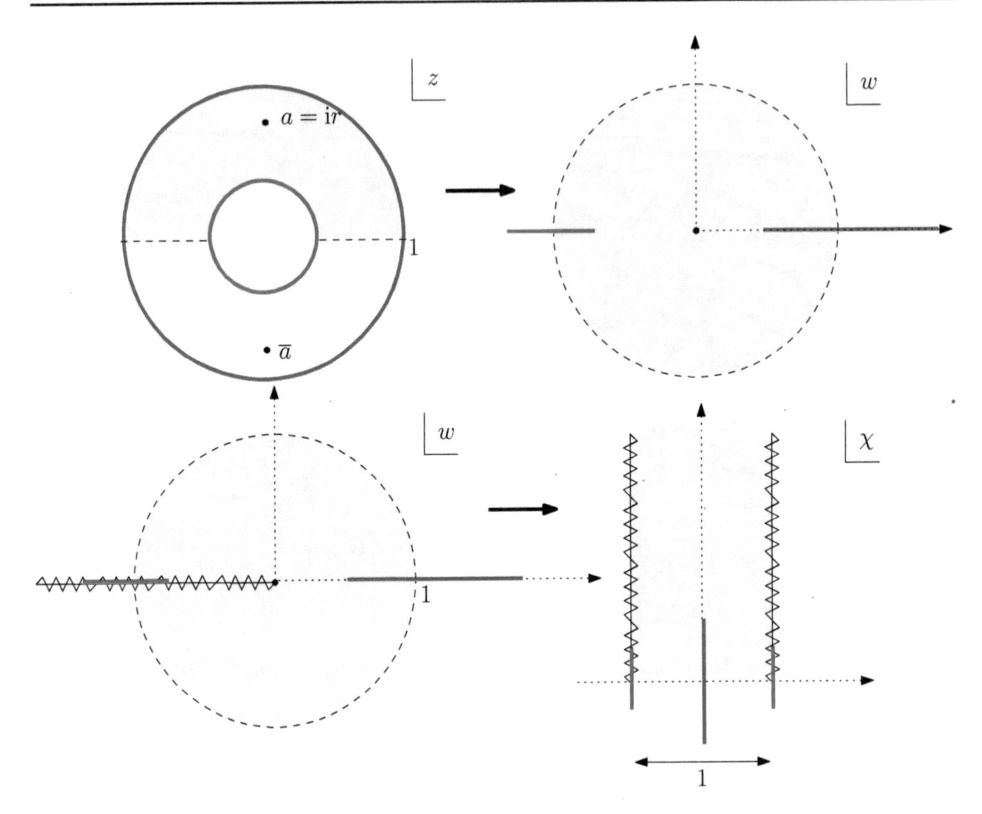

Figure 5.11. *The special case (5.70) of a radial slit map in which C_0 and C_1 are transplanted to radial slits passing through the unit circle with the real diameter $-1 < z < 1$ mapping to the unit circle $|w| = 1$. The upper half annulus is mapped to the unit w disc exterior to these slits. A further logarithmic transformation (5.73) takes the unit w disc, minus these two radial slits, to a unit-width semistrip in a χ plane with two slits parallel to the imaginary χ axis emanating from the real χ axis.*

Another special case involves the choice

$$a = ir, \qquad b = \overline{a}, \qquad 0 < r < 1. \tag{5.69}$$

We also include a premultiplication by -1 so that (5.65) becomes

$$w = -\frac{\omega(z, a)\omega(z, 1/\overline{a})}{\omega(z, \overline{a})\omega(z, 1/a)} \tag{5.70}$$

and (5.67) leads to

$$\arg[w] = \begin{cases} 0, & z \in C_0, \\ \pi, & z \in C_1. \end{cases} \tag{5.71}$$

It is easy to check that, for points z on the real axis inside the annulus, i.e., $-1 < z < -\rho$ and $\rho < z < 1$, we have

$$\overline{w} = \frac{1}{w}. \tag{5.72}$$

Hence these two portions of the real axis inside the annulus are also transplanted to the unit circle $|w| = 1$.

Figure 5.11 shows the correspondence between the z and w planes in this case.[13] Consider the following map obtained by taking a logarithm of (5.70):

$$\chi = \frac{1}{2\pi\mathrm{i}}\log w = \frac{1}{2\pi\mathrm{i}}\log\left[-\frac{\omega(z,a)\omega(z,1/\overline{a})}{\omega(z,\overline{a})\omega(z,1/a)}\right]. \tag{5.73}$$

This transplants the annulus to an infinite strip region with $-\infty < \mathrm{Re}[\chi] < \infty$ with C_0 and C_1 being transplanted to two finite-length slits parallel to the imaginary χ axis. For a suitable choice of the branch cut joining $\pm a$ in the annulus, the edges of the infinite strip can be made to be straight lines parallel to the imaginary χ axis.

Figure 5.11 shows a schematic of both maps (5.70) and (5.73).

5.12 ▪ The functions $\mathcal{K}(z,a)$ and $K(z,a)$

We have already introduced $\mathcal{K}(z,a)$ in (5.48)—it arose naturally when considering parallel slit maps. This function will turn out to be useful in the development so we record some of its properties.

From the identity (5.5) satisfied by the prime function it is straightforward to show that

$$\overline{\mathcal{K}}(1/z,1/a) = 1 - \mathcal{K}(z,a). \tag{5.74}$$

Using the extra identity (5.7), resulting from the reflectional symmetry of the circular annulus about the real axis, we can in fact write (5.74) as

$$\mathcal{K}(1/z,1/a) = 1 - \mathcal{K}(z,a). \tag{5.75}$$

The identity (5.6) leads to the additional functional identity

$$\mathcal{K}(\rho^2 z,a) = \mathcal{K}(z,a) + 1. \tag{5.76}$$

Just as we did in §1.11, we can also introduce

$$K(z,a) \equiv z\frac{\partial \log\omega(z,a)}{\partial z} \tag{5.77}$$

since, for the concentric annulus, this function also has useful properties. Like $\mathcal{K}(z,a)$ this function also has a simple pole at $z = a$. It follows immediately, from the antisymmetry (5.4) of the prime function, that these functions are related by

$$K(z,a) = \mathcal{K}(a,z). \tag{5.78}$$

It is straightforward to show, on use of the identity (5.5), that

$$\overline{K}(1/z,1/a) = 1 - K(z,a). \tag{5.79}$$

And, using the extra identity (5.7), we can write this as

$$K(1/z,1/a) = 1 - K(z,a). \tag{5.80}$$

Finally, the identity (5.6) leads to

$$K(\rho^2 z,a) = K(z,a) - 1. \tag{5.81}$$

[13] In Exercise 8.14 this radial slit map is related to the Jacobi sn function.

It should be clear from these functional relations that, for the concentric annulus, the two functions $\mathcal{K}(z,a)$ and $K(z,a)$ are closely related. Indeed, in Exercise 5.2, we demonstrate that

$$\mathcal{K}(z,a) + K(z,a) = 1. \tag{5.82}$$

This is the same identity as in (1.86) of Chapter 1, where it was a trivial matter to establish it. Remarkably it also holds for the analogous functions $\mathcal{K}(z,a)$ and $K(z,a)$ associated with the concentric annulus.[14]

Exercises

5.1. **(The differential $d\Pi^{z,w}(a)$ for the concentric annulus)** Use the properties (5.4) and (5.6) of the prime function to show that

$$\omega(z,\rho^2 a) = -\frac{z}{a}\omega(z,a). \tag{5.83}$$

Hence verify the identity

$$\mathcal{K}(z,\rho^2 a) = \mathcal{K}(z,a) - 1. \tag{5.84}$$

Show that

$$d\Pi^{z,w}(a) = (\mathcal{K}(z,a) - \mathcal{K}(w,a))\frac{da}{a}. \tag{5.85}$$

Verify that the differential $d\Pi^{z,w}(a)$ has a simple pole with residue $+1$ at $a = z$ and a simple pole with residue -1 at $a = w$ and that it is invariant when $a \mapsto \rho^2 a$; i.e., verify that

$$d\Pi^{z,w}(\rho^2 a) = d\Pi^{z,w}(a). \tag{5.86}$$

In this way, we have corroborated for this particular case some properties of $d\Pi^{z,w}(a)$ derived in the general treatment of Chapter 4.

5.2. **(Proof of the identity (5.82))** Let D_z be the annulus $\rho < |z| < 1$ and introduce the function

$$H(z,a) \equiv \mathcal{K}(z,a) + K(z,a), \tag{5.87}$$

where a is some point in the doubled domain $\rho < |z| < 1/\rho$.

(a) Show, on use of (5.76) and (5.81), that $H(z,a)$ satisfies the functional identity[15]

$$H(\rho^2 z, a) = H(z,a). \tag{5.88}$$

(b) Show that $H(z,a)$ is analytic and single-valued in the annulus $\rho < |z| < 1/\rho$.

(c) By writing the Laurent expansion

$$H(z,a) = \sum_{n=-\infty}^{\infty} c_n(a) z^n, \tag{5.89}$$

---

[14] This identity holds only for the disc and concentric annulus and does not generally hold for the prime functions of other circular domains.

[15] In Chapter 8 we will study a whole class of functions satisfying such a functional relation—they are known as loxodromic functions.

which is convergent for all $\rho < |z| < 1/\rho$, use the identity in part (a) for points z on the circle $|z| = 1/\rho$ to show that

$$H(z,a) = c_0(a), \tag{5.90}$$

where $c_0(a)$ is independent of z. *Hint:* Consider equating coefficients.

(d) Use (5.78) to show that

$$H(z,a) = H(a,z). \tag{5.91}$$

(e) Hence argue that

$$H(z,a) = c \tag{5.92}$$

for some constant c that is independent of both z and a (although it may depend on the parameter ρ).

(f) Now let

$$\omega(z,a) = (z-a)\tilde{\omega}(z,a). \tag{5.93}$$

Show that

$$H(z,a) = 1 + z\frac{\partial \log \tilde{\omega}(z,a)}{\partial z} + a\frac{\partial \log \tilde{\omega}(z,a)}{\partial a} = c. \tag{5.94}$$

Hence show that

$$H(z,a) = 1 + z\frac{\tilde{\omega}_1(z,a)}{\tilde{\omega}(z,a)} + a\frac{\tilde{\omega}_2(z,a)}{\tilde{\omega}(z,a)} = c, \tag{5.95}$$

where a subscript j, for $j = 1$ or 2, means partial differentiation with respect to the jth argument of the function.

(g) Use identity (5.4) to show that

$$\tilde{\omega}(z,a) = \tilde{\omega}(a,z) \tag{5.96}$$

and hence deduce that

$$\tilde{\omega}_1(z,a) = \tilde{\omega}_2(a,z). \tag{5.97}$$

(h) Use the two identities (5.5) and (5.7) to deduce that

$$\tilde{\omega}(1/z,1/a) = \tilde{\omega}(z,a) \tag{5.98}$$

and hence deduce that

$$-\frac{1}{z^2}\tilde{\omega}_1(1/z,1/a) = \tilde{\omega}_2(a,z). \tag{5.99}$$

(i) By setting $z = a = 1$ in (5.97) and (5.99) show that

$$\tilde{\omega}_1(1,1) = \tilde{\omega}_2(1,1) = 0. \tag{5.100}$$

(j) Hence, by evaluating (5.95) at $z = a = 1$, use the result in part (i) to show that $c = 1$ and therefore that

$$\mathcal{K}(z,a) + K(z,a) = 1. \tag{5.101}$$

5.3. **(Invariance to $z \mapsto cz, a \mapsto ca$)** In this exercise we will show that, for the concentric annulus, the functions

$$\frac{\omega(z,a)}{a}, \qquad K(z,a), \qquad \mathcal{K}(z,a) \tag{5.102}$$

are all invariant under the transformations[16]

$$z \mapsto cz, \qquad a \mapsto ca, \qquad c \in \mathbb{C}. \tag{5.103}$$

(a) All the functions in (5.102) are analytic functions of z and a. We can therefore consider introducing the (analytic) change of variables $(z,a) \mapsto (u,v)$ defined by

$$u = \frac{z}{a}, \qquad v = a. \tag{5.104}$$

Show that, for any differentiable function $\phi(z,a)$ of z and a,

$$\left.\frac{\partial \phi}{\partial u}\right|_v = a \left.\frac{\partial \phi}{\partial z}\right|_a, \tag{5.105}$$

$$\left.\frac{\partial \phi}{\partial v}\right|_u = -\frac{z}{a}\left.\frac{\partial \phi}{\partial z}\right|_a + \left.\frac{\partial \phi}{\partial a}\right|_z, \tag{5.106}$$

where $\partial\phi/\partial z|_a$ is used to denote the partial derivative with respect to z, keeping a fixed.

(b) Hence make the choice

$$\phi(z,a) = \frac{\omega(z,a)}{a} \tag{5.107}$$

to establish that

$$\left.\frac{\partial}{\partial v}\right|_u \left[\frac{\omega(z,a)}{a}\right] = \frac{\omega(z,a)}{a^2}\left[K(z,a) + \mathcal{K}(z,a) - 1\right]. \tag{5.108}$$

(c) Now use the result (5.101) of Exercise 5.2 to show that $\omega(z,a)/a$ depends only on $u = z/a$ and, as such, is invariant under the transformation (5.103).

(d) Use the result of part (c) to show that the function $K(z,a)$ defined by

$$K(z,a) = z\frac{\partial}{\partial z}\log\omega(z,a) \tag{5.109}$$

is also invariant under the transformation (5.103).

(e) Deduce from part (d) and the identity (5.101) that $\mathcal{K}(z,a)$ is similarly invariant under the transformation (5.103).

5.4. **(The value of $K(\sqrt{\rho}, 1/\sqrt{\rho})$)** Show that, when D_z is the concentric annulus $\rho < |z| < 1$,

$$K(\sqrt{\rho}, 1/\sqrt{\rho}) = 0. \tag{5.110}$$

Hint: Combine the identities (5.80) and (5.81) and evaluate at $z = 1/\rho, a = 1$ to show that $K(\rho, 1) = 0$, and then use the result of Exercise 5.3.

[16]The result of this exercise explains why, in many of the author's published papers involving a concentric annulus, the function $P(z/a) \propto -\omega(z,a)/a$ is introduced and the function $K(z,a)$ is written as $K(z/a)$.

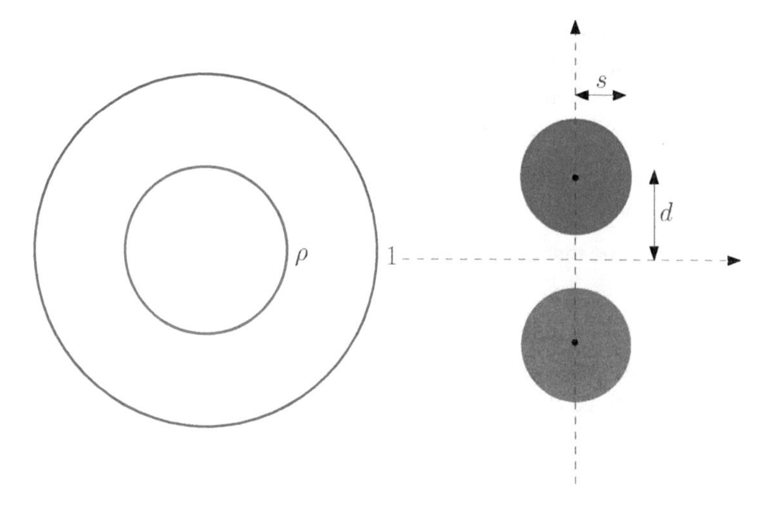

Figure 5.12. *Conformal mapping from a concentric annulus $\rho < |z| < 1$ to the unbounded region exterior to two circular discs of equal size (Exercise 5.7).*

5.5. **(Univalency and the argument principle)** Use the argument principle to establish that the interior of the concentric annulus D_z is transplanted in a one-to-one fashion to the various target domains of the unbounded circular slit map, the parallel slit maps, the Cayley-type map, and the radial slit map. *Hint:* Note that the Cayley-type map can be viewed as having "half a pole" inside D_z.

5.6. **(Blaschke-type products)** It is required to find a function $f(z)$ with the following properties:

   (i) $f(z)$ is analytic, and single-valued, everywhere in the concentric annulus D_z.

   (ii) $f(z)$ has precisely N zeros inside D_z at given locations $\{a_j | j = 1, \ldots, N\}$.

   (iii) The modulus of $f(z)$ is unity on the boundary C_0 of D_z and has constant modulus on the boundary C_1.

   By thinking about the bounded circular slit mappings (5.26) construct an example of a function $f(z)$ having the properties (i)–(iii). What is the modulus on C_1 of the function $f(z)$ you have constructed? Can you find another function with the same properties (i)–(iii) but with a different modulus on C_1?

5.7. **(Mapping to the exterior of two circular discs)** Show that the conformal mapping shown in Figure 5.12 from a concentric annulus $\rho < |z| < 1$ to the unbounded region in a complex w-plane exterior to two circular discs of radius s centered at $\pm id$, where $s < d$, is the Möbius map

$$w = f(z) = \mathrm{i}R \left[\frac{\sqrt{\rho} - z}{\sqrt{\rho} + z} \right], \tag{5.111}$$

   where

$$R = \sqrt{d^2 - s^2}, \qquad \rho = \left[\frac{\sqrt{d+s} + \sqrt{d-s}}{\sqrt{d+s} - \sqrt{d-s}} \right]^2 = \left[\frac{d + \sqrt{d^2 - s^2}}{s} \right]^2. \tag{5.112}$$

5.8. **(Mapping to the exterior of a circular disc above a wall)** Suppose we wish to find a conformal mapping from a concentric annulus in a complex z plane to the unbounded region in the upper half w plane and exterior to a circular disc of radius s and centered at $\mathrm{i}d$. We will derive this mapping from the result of Exercise 5.7.

(a) Show that the mapping (5.111) of Exercise 5.7 maps the circle $|z| = \sqrt{\rho}$ to the real axis between the two discs centered at $\pm \mathrm{i}d$.

(b) Hence argue that the required mapping from a concentric annulus

$$\hat{\rho} < |z| < 1 \tag{5.113}$$

to the region in the upper half w plane and exterior to a circular disc of radius s and centered at $\mathrm{i}d$ is

$$w = f(z) = \mathrm{i}R\left[\frac{\hat{\rho} - z}{\hat{\rho} + z}\right], \tag{5.114}$$

where

$$R = \sqrt{d^2 - s^2}, \qquad \hat{\rho} = \left[\frac{d + \sqrt{d^2 - s^2}}{s}\right]. \tag{5.115}$$

5.9. **(On unbounded circular slit maps)** Verify that if we define a map

$$w = e^{2\pi \mathrm{i}(\mathcal{G}_1(z,a) - \mathcal{G}_1(z,b))}, \tag{5.116}$$

where a and b are distinct points in D_z—this map is similar to (5.32) but has $\mathcal{G}_1$ replacing the $\mathcal{G}_0$ that appears there—then the result is an unbounded circular slit map qualitatively similar to that given by (5.32).

5.10. **(Two views of a two-slit map)** Consider the conformal mapping to the unbounded domain in a w-plane exterior to two slits $[t, s]$ and $[r, 1]$ on the real axis where $t < s < r < 1$. Two of the three real degrees of freedom of the Riemann mapping theorem have been used to insist that the preimage circle C_1 of the second slit is concentric with C_0. We use up the final degree of freedom to insist that $z = 1$ is the preimage of $w = 1$. By the reflectional symmetry of the domain with respect to the real axis we expect the preimage of $w = \infty$ to be at some point $z = b \in \mathbb{R}$.

(a) We can construct this map as an instance of a radial slit mapping. Show that the relevant function is

$$w = f_1(z) = \frac{\omega(1,b)\omega(1,1/b)\omega(z,a)\omega(z,1/a)}{\omega(1,a)\omega(1,1/a)\omega(z,b)\omega(z,1/b)}, \tag{5.117}$$

which depends on three real parameters: ρ, b, and a. These parameters determine the values of r, s, and t, the endpoints of the slits.

(b) The same mapping can be derived as an instance of a parallel slit mapping. Show that the required map is

$$w = f_2(z) = \frac{\mathcal{K}(z,b) - \mathcal{K}(z,1/b) - (\mathcal{K}(a,b) - \mathcal{K}(a,1/b))}{\mathcal{K}(1,b) - \mathcal{K}(1,1/b) - (\mathcal{K}(a,b) - \mathcal{K}(a,1/b))}. \tag{5.118}$$

The function $f_2(z)$ must be identical to $f_1(z)$. Indeed, $f_2(z)$ can be thought of as the "partial fraction" decomposition of the function $f_1(z)$ (although remember that we are no longer dealing with rational functions).

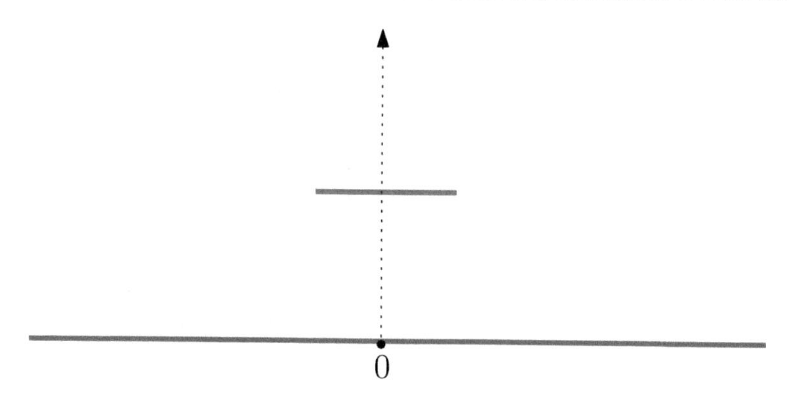

Figure 5.13. *A "microstrip" (a flat plate, or slit) above an infinite wall (Exercise 5.11).*

5.11. **("Microstrip" geometry)** Consider the map

$$w = \mathrm{i}(\mathcal{K}(z,1) - \mathcal{K}(-1,1)). \tag{5.119}$$

(a) Show that it transplants the concentric annulus $\rho < |z| < 1$ to the unbounded region in the upper half w plane exterior to a slit of finite length in the upper half plane and parallel to the real axis.[17] *Hint:* Think about the parallel slit maps of §5.9 in the limit $a \to 1$.

(b) Using the Schwarz reflection principle, show that this same conformal mapping transplants the annulus

$$\rho < |z| < 1/\rho \tag{5.120}$$

to the unbounded region exterior to two slits of equal length that are reflections of each other about the real w axis.

5.12. **(Inclined flat plate above an infinite wall)** The Cayley-type map

$$w = -\mathrm{i}\,\frac{\omega(z,+1)}{\omega(z,-1)} \tag{5.121}$$

is a rotation of the image of the mapping (5.62) through angle $\pi/2$ and it therefore transplants the annulus $\rho < |z| < 1$ to the upper half w plane exterior to a finite-length slit along the imaginary w axis. A simple generalization of this map is

$$w = e^{-\mathrm{i}\theta}\frac{\omega(z,-e^{2\mathrm{i}\theta})}{\omega(z,-1)}, \tag{5.122}$$

which reduces to (5.121) when $\theta = \pi/2$. Show that the map (5.122) transplants the concentric annulus $\rho < |z| < 1$ to the unbounded region in the upper half w plane exterior to a slit of finite length inclined at angle θ to the positive real axis, where $0 < \theta < \pi$.[18]

5.13. **(On Cayley-type maps)** Consider the Cayley-type map of the concentric annulus D_z given by

$$w = -\frac{1}{\tau}\frac{\omega(z,\tau)}{\omega(z,\overline{\tau})}, \tag{5.123}$$

where $\tau \in C_0^+$ is a point on the upper half unit circle $|\tau| = 1, \mathrm{Im}[\tau] > 0$.

[17] This map is essentially the one used in [117] using a related function theory.

[18] This mapping was first used in [50].

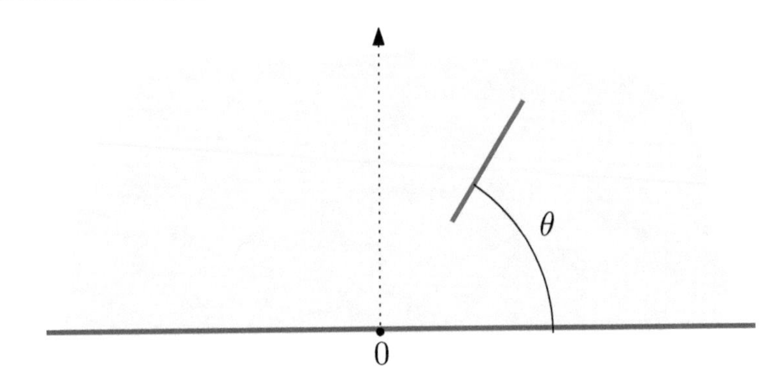

Figure 5.14. *A flat plate, or slit, inclined at angle θ above an infinite flat wall (Exercise 5.12).*

(a) Show that this mapping transplants D_z to the lower half w plane with a radial slit excised with C_0 being transplanted to the real w axis.

(b) On the other hand, consider the Cayley-type map of the concentric annulus D_z given by

$$w = +\frac{1}{\tau}\frac{\omega(z,\tau)}{\omega(z,\overline{\tau})}, \tag{5.124}$$

where $\tau \in C_1^+$ is a point on the upper half circle $|\tau| = \rho, \mathrm{Im}[\tau] > 0$. Show that this mapping also transplants D_z to the lower half w plane with a radial slit excised but now with C_1 being transplanted to the real w axis.

5.14. **(Two flat plates at an angle)** Consider the map from the annulus $\rho < |z| < 1$ given by

$$w = e^{-\mathrm{i}\theta} A \frac{\omega(z,1)\omega(z,e^{2\mathrm{i}\theta})}{\omega(z,\sqrt{\rho})\omega(z,1/\sqrt{\rho})}, \tag{5.125}$$

where $A > 0$ is some real parameter. Show that this maps to the unbounded region exterior to a finite slit along the real w axis and covering the origin $w = 0$ and another slit oriented at angle θ, as shown in Figure 5.15.

5.15. **(Weis–Fogh mechanism)** In a two-dimensional model of the Weis–Fogh mechanism of insect wings in flight, two detached flat plates of equal length inclined at a fixed angle 2θ with $0 < \theta < \pi$ to each other move in opposite directions with their midpoints at a fixed $y = Y$ position but moving apart at equal speeds so that the midpoints are at $\pm X(t) + \mathrm{i}Y$. By considering a radial slit map, show that the conformal mapping from a time-dependent concentric annulus

$$\rho(t) < |z| < 1 \tag{5.126}$$

to the unbounded region exterior to *both* flat plates is of the functional form

$$w = \mathrm{i}e^{-3\mathrm{i}\theta} A(t) \frac{\omega(z,\sqrt{\rho(t)}e^{2\mathrm{i}\theta})\omega(z,e^{2\mathrm{i}\theta}/\sqrt{\rho(t)})}{\omega(z,\sqrt{\rho(t)})\omega(z,1/\sqrt{\rho(t)})} - \mathrm{i}d(t), \tag{5.127}$$

where $A(t)$ and $d(t)$ are real time-evolving parameters. Give a prescription for the determination of the three real parameters $A(t), d(t)$, and $\rho(t)$ to ensure the plates have unit length and have midpoints at the required locations $(\pm X(t), Y)$, as shown in Figure 5.16.[19]

[19]This mapping was first used in [15].

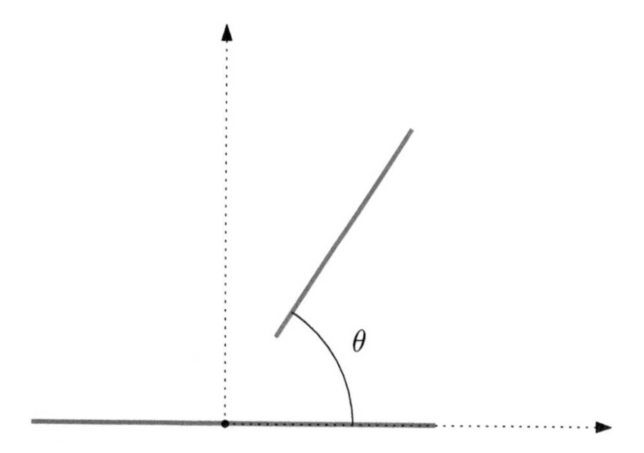

Figure 5.15. *A flat plate, or slit, inclined at angle θ relative to another flat plate (Exercise 5.14).*

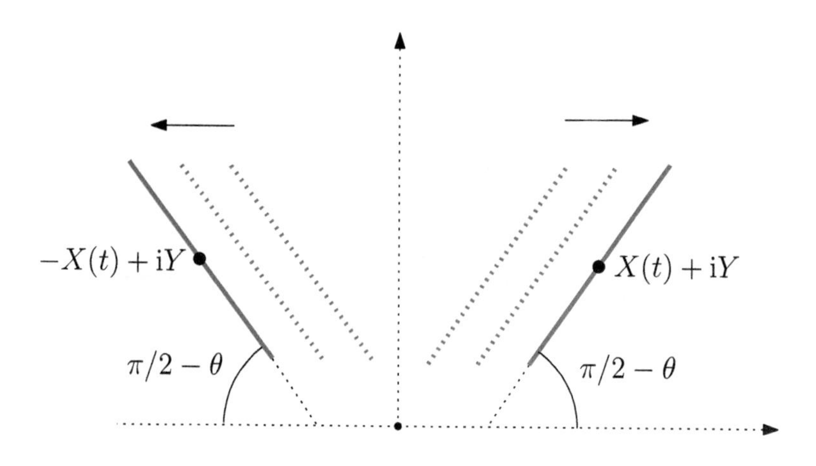

Figure 5.16. *Two-plate model of the Weis–Fogh mechanism (Exercise 5.15). Two plates, or slits, move apart at equal speed but with fixed inclination to the real axis.*

5.16. **(Disc with a slit)** Show that the map

$$w = \frac{\omega(z,-1) + \omega(z,+1)}{\omega(z,-1) - \omega(z,+1)} \tag{5.128}$$

transplants the annulus $\rho < |z| < 1$ to a unit disc in the w plane exterior to a slit $[-r, r]$ along the real w axis, as shown in Figure 5.17. Find how to generalize this

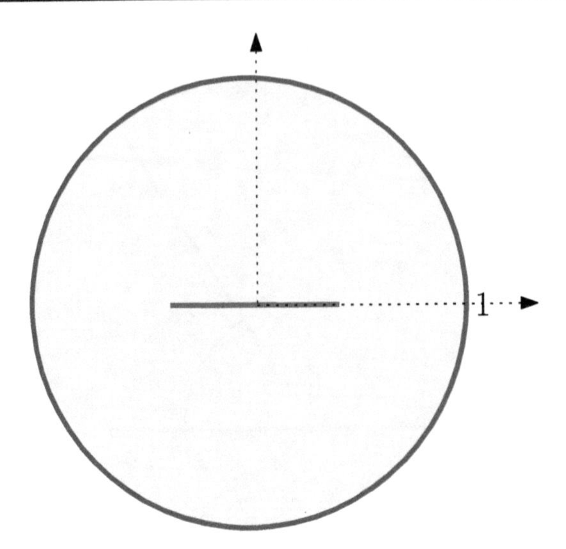

Figure 5.17. *The unit disc with an interior slit along its real diameter (Exercise 5.16).*

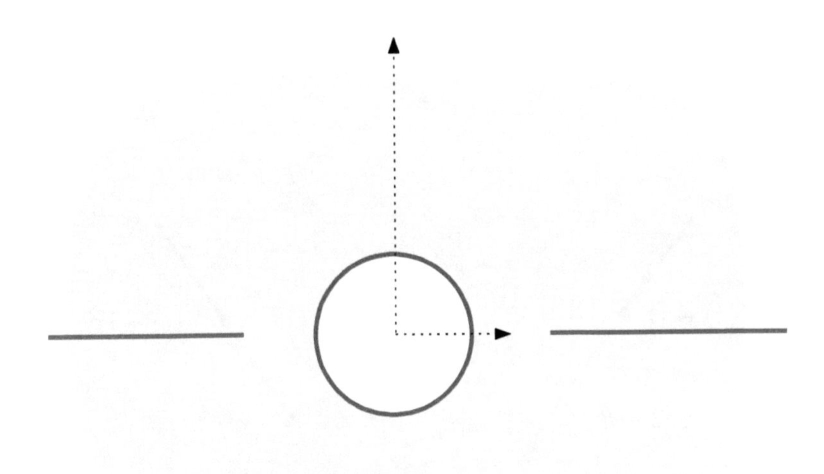

Figure 5.18. *The region exterior to a circular disc blocking a gap in an infinite wall (Exercise 5.17).*

mapping so that the slit occupies some given nonsymmetric interval $[s, r]$ along the real axis of the disc.

5.17. **(Cylinder blocking a gap in a wall)** Use the result in Exercise 5.16 to find the conformal map from the annulus $\rho < |z| < 1$ to the unbounded region in a complex w plane exterior to an infinite wall $-\infty < w < -1, 1 < w < \infty$ and exterior to a circular cylinder of radius s sitting in the middle of the gap between $-1 < w < 1$, as depicted in Figure 5.18.

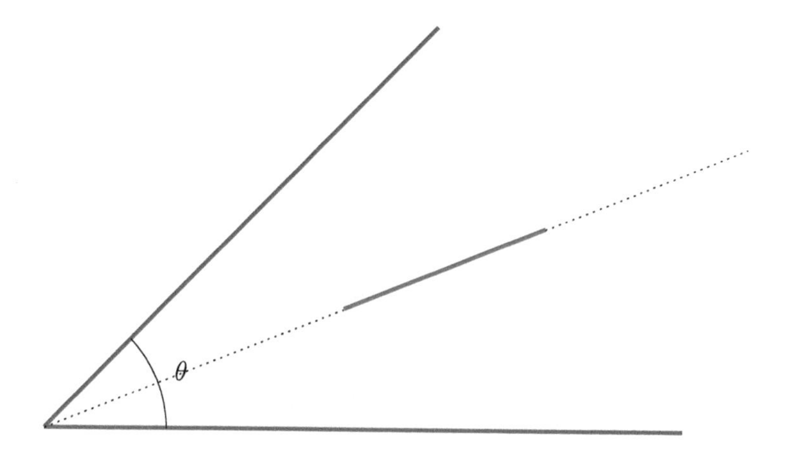

Figure 5.19. *A flat plate, or slit, lying along the bisector of a wedge region of opening angle θ (Exercise 5.19).*

5.18. **(Another view of a two-slit map)** Combine the classical Joukowski map

$$\frac{1}{2}\left(\frac{1}{z}+z\right) \tag{5.129}$$

with the result in Exercise 5.16 to show that a map from the annulus $\rho < |z| < 1$ to the unbounded region exterior to two slits on the real axis is

$$w = \frac{\omega(z,-1)^2 + \omega(z,+1)^2}{\omega(z,-1)^2 - \omega(z,+1)^2}. \tag{5.130}$$

Verify that the image of C_0 is the interval $[-1,1]$.

5.19. **(Slit in a wedge)** Consider the domain shown in Figure 5.19.

(a) By considering the use of Cayley-type maps, find the functional form of the conformal mapping from a concentric annulus to the unbounded region in a wedge of opening angle θ exterior to a finite-length radial slit along the perpendicular bisector of the wedge, as shown in Figure 5.19.

(b) Generalize this result to find the mapping when the radial slit is located on a ray at any given angle ϕ.

5.20. **(Connection with bipolar coordinates)** This exercise introduces the notion of *bipolar coordinates* and shows that they are really just a disguised form of the Möbius map

$$z = a\left[\frac{\sqrt{\rho}+\zeta}{\sqrt{\rho}-\zeta}\right], \qquad a \in \mathbb{R}^+, \tag{5.131}$$

from a concentric annulus to the region exterior to two discs.

(a) Show that the map (5.131) transplants the annulus

$$\rho < |\zeta| < 1 \tag{5.132}$$

to the exterior of two equal-size circular discs centered on the real axis at $\pm\Delta$ and each of radius Q. Find Δ and Q in terms of a and ρ.

(b) Show that

$$\frac{\zeta}{\sqrt{\rho}} = \frac{z/a - 1}{z/a + 1}. \tag{5.133}$$

(c) Now consider the conformal map

$$u = -\mathrm{i} \log \left(\frac{\zeta}{\sqrt{\rho}} \right). \tag{5.134}$$

Show that the annulus $\rho < |\zeta| < 1$ is transplanted by this map to a rectangle in the complex u plane.

(d) Hence show that

$$z = \mathrm{i} a \cot \left(\frac{u}{2} \right). \tag{5.135}$$

(e) If we set $u = \sigma + \mathrm{i}\tau$, derive the relations

$$x = \frac{a \sinh \tau}{\cosh \tau - \cos \sigma}, \qquad y = \frac{a \sin \sigma}{\cosh \tau - \cos \sigma}. \tag{5.136}$$

The change of variable from (x, y) to (σ, τ) in (5.136) is known as the bipolar coordinate transformation. It is clear from this exercise that this change of variables is essentially equivalent to performing the analysis in a concentric annulus where analytic functions can easily be described by Laurent series.

Chapter 6

Triply and higher connected domains

6.1 ▪ The N-body problem: $N \geq 3$

When three (or more) two-dimensional objects interact in some ambient medium, that medium is triply (or higher) connected. It is important in applications to be able to study these multibody interactions. The general theory of Chapters 2–4 can be useful here.

Chapter 5 showed how the general theory of Chapters 2–4 plays out in the context of the doubly connected concentric annulus; that theory is relevant to the two-body problem. We now turn to the higher connected case, which pertains to the N-body problem with $N \geq 3$. In fact it will only be necessary to study the triply connected case in detail because, as will be outlined in §6.13, the framework for quadruply connected, and all higher connected, cases is entirely analogous. This is one of the advantages of our mathematical approach.

6.2 ▪ A triply connected circular domain

When triply connected domains are of interest we are encouraged to consider a triply connected circular domain D_z comprising the unit disc, with boundary C_0, and two excised interior circular discs with boundaries C_1 and C_2, as shown in Figure 6.1. Just as we did in Chapter 5, we emulate the template set out in Chapter 1 to underline the naturalness of the approach. This chapter is more colorful than Chapter 5: in Figure 6.1 C_0 is blue and C_1 is red, just as they were in Chapter 5, and C_2 is green. All the new features consequent to adding this additional hole in D_z will be shown in green. Images of these circles under any conformal mappings introduced in the development will be shown in the same color as their preimage circles.

6.3 ▪ The prime function identities

The theory of Chapters 2–4 guarantees the existence of a prime function associated with the triply connected circular domain in Figure 6.1, which we will again denote by $\omega(z, a)$. Its dependence on the conformal moduli δ_1, δ_2, q_1, and q_2 will be implicitly understood. It is useful to collect what we know to be the properties of this prime function. It has a

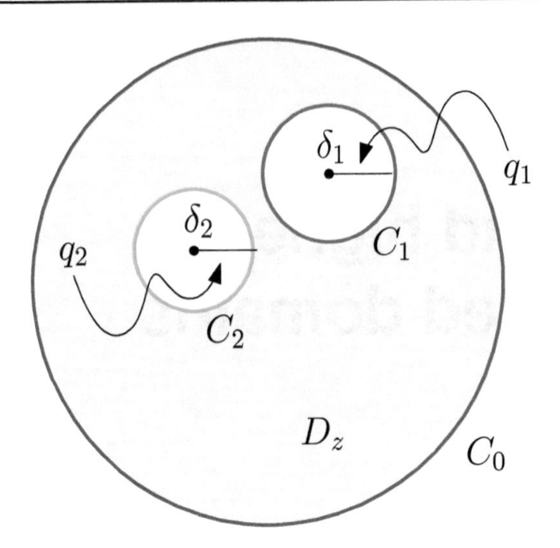

Figure 6.1. *A triply connected circular domain D_z with boundary circles C_0, C_1, and C_2. Circle C_1 has center at δ_1 and radius q_1; C_2 has center at δ_2 and radius q_2.*

simple zero at $z = a$ and satisfies the identities

$$\omega(a, z) = -\omega(z, a), \tag{6.1}$$

$$\overline{\omega}(1/z, 1/a) = -\frac{1}{za}\omega(z, a), \tag{6.2}$$

$$\omega(\theta_1(z), a) = -\frac{q_1}{1 - \overline{\delta_1} z} e^{-2\pi \mathrm{i}(v_1(z) - v_1(a) + \tau_{11}/2)} \omega(z, a), \tag{6.3}$$

$$\omega(\theta_2(z), a) = -\frac{q_2}{1 - \overline{\delta_2} z} e^{-2\pi \mathrm{i}(v_2(z) - v_2(a) + \tau_{22}/2)} \omega(z, a), \tag{6.4}$$

where $\theta_1(z)$ and $\theta_2(z)$ are the Möbius maps

$$\theta_1(z) = \overline{S_1(1/\overline{z})}, \qquad \theta_2(z) = \overline{S_2(1/\overline{z})}, \tag{6.5}$$

with

$$S_1(z) = \overline{\delta_1} + \frac{q_1^2}{z - \delta_1}, \qquad S_2(z) = \overline{\delta_2} + \frac{q_2^2}{z - \delta_2}, \tag{6.6}$$

which are the Schwarz functions of the circles C_0, C_1, and C_2, respectively. The functions $\{v_1(z), v_2(z)\}$, as defined in §2.5, are both analytic in D_z. They are not single-valued in D_z, however, and they satisfy

$$\int_{C_j} dv_i = \delta_{ij}. \tag{6.7}$$

These functions have constant imaginary parts on the three boundaries of D_z. For $i = 1, 2$, (2.32) tells us that

$$\mathrm{Im}[v_i(z)] = \begin{cases} 0, & z \in C_0, \\ \dfrac{\tau_{ij}}{2\mathrm{i}}, & z \in C_j,\ j = 1, 2, \end{cases} \tag{6.8}$$

where τ_{ij} is the period matrix of D_z defined in (2.30).

Unlike the prime functions for the unit disc and the concentric annulus, which are circular domains that are reflectionally symmetric about the real axis, we will only have the additional identity $\overline{\omega}(z,a) = \omega(z,a)$ if the two circles C_1 and C_2 are both centered on the real diameter of the unit circle so that this triply connected D_z would then enjoy reflectional symmetry about the real axis. Of course this is a special case, and we want henceforth to consider the circles C_1 and C_2 to be in general position inside the unit disc.

6.4 ▪ A ratio of prime functions

If we introduce the ratio of prime functions

$$R(z;a,b) = \frac{\omega(z,a)}{\omega(z,b)}, \tag{6.9}$$

then, for our triply connected circular domain D_z, we know from (4.103) that

$$\overline{R(z;a,b)} = \begin{cases} \frac{\overline{a}}{\overline{b}} R(z;1/\overline{a},1/\overline{b}), & z \in C_0, \\ e^{-2\pi \mathrm{i}(\overline{v_1(a)}-\overline{v_1(b)})} \frac{\overline{a}}{\overline{b}} R(z;1/\overline{a},1/\overline{b}), & z \in C_1, \\ e^{-2\pi \mathrm{i}(\overline{v_2(a)}-\overline{v_2(b)})} \frac{\overline{a}}{\overline{b}} R(z;1/\overline{a},1/\overline{b}), & z \in C_2. \end{cases} \tag{6.10}$$

In this expression we assume only that a and b are two nonzero points with $a \neq b$.

6.5 ▪ Modified Green's function

From the analysis of Chapter 4, the function

$$\mathcal{G}_0(z,a) = \frac{1}{2\pi \mathrm{i}} \log\left(\frac{\omega(z,a)}{|a|\omega(z,1/\overline{a})}\right), \tag{6.11}$$

where $\omega(z,a)$ is the prime function associated with D_z, is the analytic extension of the modified Green's function $G_0(z,a)$ in the same domain:

$$G_0(z,a) = \mathrm{Im}[\mathcal{G}_0(z,a)]. \tag{6.12}$$

From (2.50),

$$G_0(z,a) = \begin{cases} 0, & z \in C_0, \\ \mathrm{Im}[v_1(a)], & z \in C_1, \\ \mathrm{Im}[v_2(a)], & z \in C_2, \end{cases} \tag{6.13}$$

and from (2.52),

$$\int_{C_0} d\mathcal{G}_0(z,a) = 1, \qquad \int_{C_1} d\mathcal{G}_0(z,a) = 0, \qquad \int_{C_2} d\mathcal{G}_0(z,a) = 0. \tag{6.14}$$

Consider the conformal map defined by

$$w = \mathcal{G}_0(z,a). \tag{6.15}$$

We select the parameter a to be real, with $-1 < a < 0$ inside D_z, and choose the branch cut of $\mathcal{G}_0(z,a)$ joining the logarithmic branch points at $z = a$ and $z = 1/a$ to lie along the negative real z axis. Then the image of D_z under this mapping is a semistrip exterior to

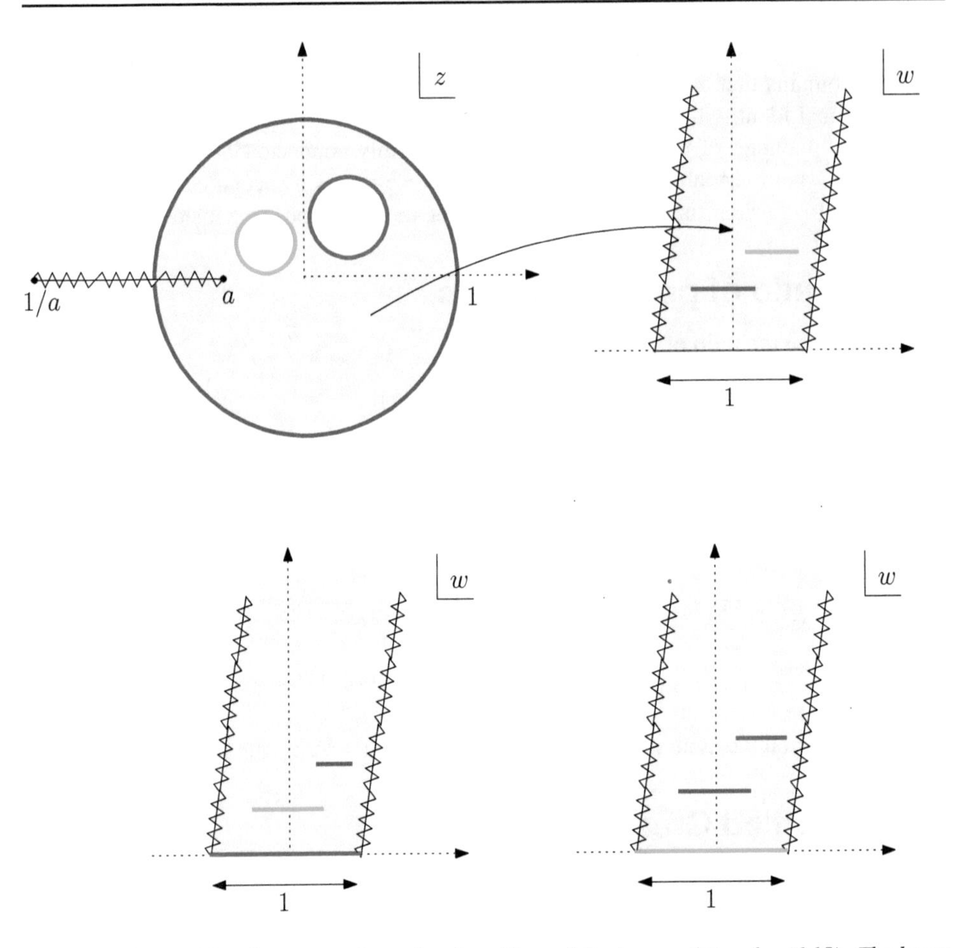

Figure 6.2. *The upper schematics show D_z and the image of it under (6.15). The lower schematics show the image of D_z under the mappings (6.16) and (6.18).*

two slits of finite length parallel to the real axis inside this strip, as shown schematically in Figure 6.2.

For this triply connected D_z we know there to be two other possible modified Green's functions with analytic extensions given by $\mathcal{G}_j(z, a)$ for $j = 1, 2$, as defined in (2.53). Thus we might also consider, for example, the mapping defined by

$$w = \mathcal{G}_1(z, a). \tag{6.16}$$

Since the image of this map also has constant imaginary part on all the boundary circles of D_z with

$$\int_{C_0} d\mathcal{G}_1(z, a) = 0, \qquad \int_{C_1} d\mathcal{G}_1(z, a) = -1, \qquad \int_{C_2} d\mathcal{G}_1(z, a) = 0, \tag{6.17}$$

the mapping (6.16) produces the same qualitative effect as the mapping (6.15), the difference being that the image of circle C_1 is the open unit-length interval, while C_0 and C_2 now map to finite-length slits parallel to the real axis.

The other possible choice,

$$w = \mathcal{G}_2(z, a), \tag{6.18}$$

leads to a similar map with C_2 mapping to the open unit-length interval and C_0 and C_1 being transplanted to finite-length slits parallel to the real axis.

Figure 6.2 shows the image of the circular domain D_z under all three conformal mappings (6.15), (6.16), and (6.18).

6.6 ▪ Bounded circular slit map

Now let

$$w = e^{2\pi i \mathcal{G}_0(z,a)} = \frac{1}{|a|}\frac{\omega(z,a)}{\omega(z,1/\overline{a})}, \tag{6.19}$$

where $a \in \mathbb{C}$ is taken to be some point in D_z. From (6.13) we deduce that

$$|w| = \begin{cases} 1, & z \in C_0, \\ e^{-2\pi \mathrm{Im}[v_1(a)]}, & z \in C_1, \\ e^{-2\pi \mathrm{Im}[v_2(a)]}, & z \in C_2. \end{cases} \tag{6.20}$$

To study whether the image of each circle is a full circle or a circular slit, we need to know the values of

$$[\arg w]_{C_0}, \qquad [\arg w]_{C_1}, \qquad [\arg w]_{C_2}. \tag{6.21}$$

But since $w = e^{2\pi i \mathcal{G}_0(z,a)}$, these quantities are precisely

$$2\pi \int_{C_0} d\mathcal{G}_0(z,a), \qquad 2\pi \int_{C_1} d\mathcal{G}_0(z,a), \qquad 2\pi \int_{C_2} d\mathcal{G}_0(z,a), \tag{6.22}$$

which we know to be 2π, 0, and 0, respectively. The image of C_0 is therefore the entire unit circle in the w-plane, while the images of C_1 and C_2 are circular slits, of finite length, inside the unit w circle.

The argument principle can be used, just as it was in §5.6, to establish the univalency of this mapping, and of all the conformal mappings to be introduced in the remainder of this chapter; see Exercise 6.2.

There are three different modified Green's functions in this triply connected setting, so we could alternatively have introduced the mapping

$$w = e^{2\pi i \mathcal{G}_1(z,a)}, \tag{6.23}$$

where $\mathcal{G}_1(z,a)$ is defined in (2.53). This produces the same effect as the mapping (6.19), the difference being that it is the image of circle C_1 that is the unit w-circle with C_0 and C_2 being transplanted to finite-length circular slits inside the unit disc in the w plane.

A third possible mapping,

$$w = e^{2\pi i \mathcal{G}_2(z,a)}, \tag{6.24}$$

gives rise to the final remaining case where the image of circle C_2 is the unit w-circle with C_0 and C_1 being transplanted to finite-length circular slits inside the unit disc in the w plane.

All three cases of these bounded circular slit maps are illustrated in Figure 6.3.

6.7 ▪ Unbounded circular slit maps

Consider the mapping given by

$$w = e^{2\pi i (\mathcal{G}_0(z,a) - \mathcal{G}_0(z,b))} = \frac{|b|}{|a|}\frac{\omega(z,a)\omega(z,1/\overline{b})}{\omega(z,b)\omega(z,1/\overline{a})}, \tag{6.25}$$

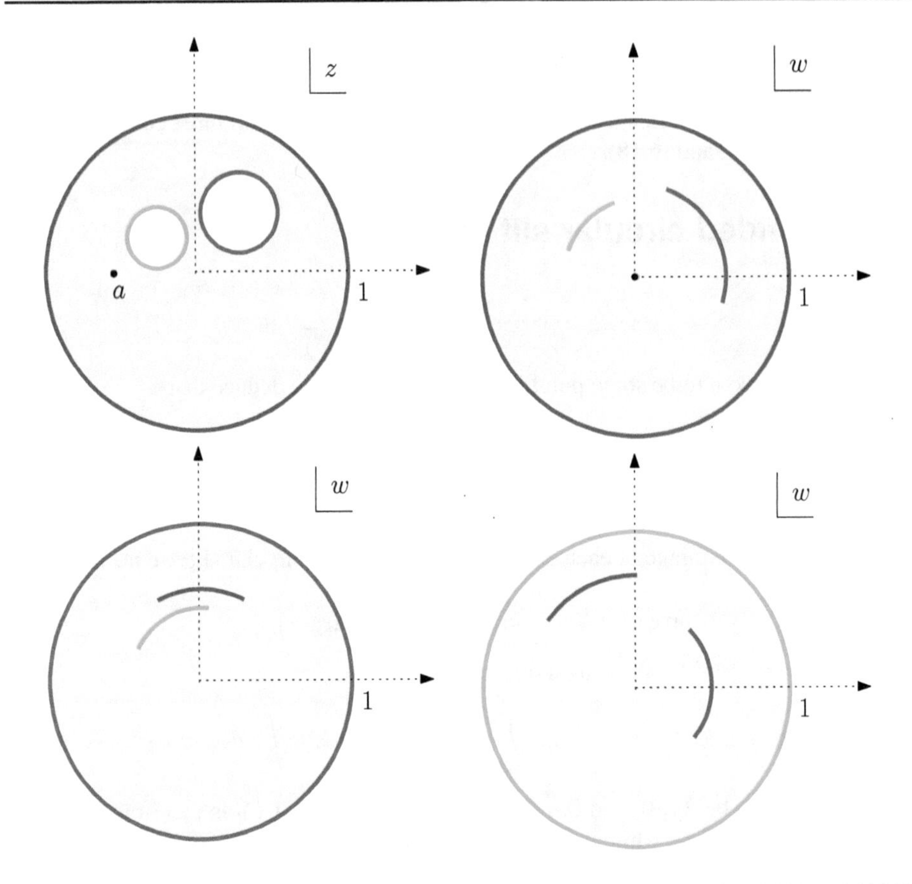

Figure 6.3. *Triply connected circular domain D_z and its image under (6.19), (6.23), and (6.24).*

where a and b are two distinct nonzero points in D_z. Clearly $z = a$ maps to $w = 0$ and $z = b$ maps to $w = \infty$, so the image domain is unbounded. From (6.13) it follows that

$$|w| = \begin{cases} 1, & z \in C_0, \\ e^{-2\pi(\mathrm{Im}[v_1(a)] - \mathrm{Im}[v_1(b)])}, & z \in C_1, \\ e^{-2\pi(\mathrm{Im}[v_2(a)] - \mathrm{Im}[v_2(b)])}, & z \in C_2, \end{cases} \tag{6.26}$$

so that the images of all boundary circles of D_z lie on arcs of circles centered at the origin $w = 0$. Whether these are full circles or finite-length circular slits will be determined by the values of

$$[\arg w]_{C_0}, \qquad [\arg w]_{C_1}, \qquad [\arg w]_{C_2}. \tag{6.27}$$

But since $w = e^{2\pi \mathrm{i}(\mathcal{G}_0(z,a) - \mathcal{G}_0(z,b))}$, these quantities are precisely

$$2\pi \int_{C_j} d(\mathcal{G}_0(z,a) - \mathcal{G}_0(z,b)), \qquad j = 0, 1, 2, \tag{6.28}$$

which, by (6.14), are all zero. Thus D_z is transplanted to the entire w plane exterior to three finite-length arcs of circles centered at $w = 0$.

If we choose either $\mathcal{G}_1$ or $\mathcal{G}_2$ in place of $\mathcal{G}_0$ in (6.25), it is straightforward to show that we will obtain an unbounded circular slit map of the kind just described (Exercise 6.5).

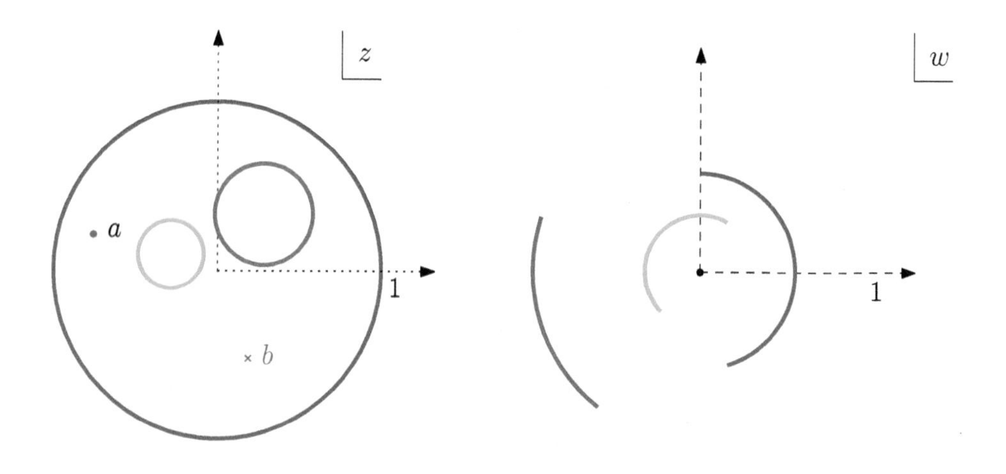

Figure 6.4. *The image of the triply connected circular domain D_z under the unbounded circular slit map (6.25).*

A special case is to take

$$a = \mathrm{i}r, \qquad b = \overline{a}, \qquad 0 < r < 1, \tag{6.29}$$

and to consider the situation in which both C_1 and C_2 are centered on the real z axis so that $\delta_1, \delta_2 \in \mathbb{R}$. We also premultiply by -1 so that (6.25) becomes

$$w = -\frac{\omega(z,a)\omega(z,1/a)}{\omega(z,\overline{a})\omega(z,1/\overline{a})}. \tag{6.30}$$

Now (6.26) implies

$$|w| = \begin{cases} 1, & z \in C_0, \\ e^{-2\pi(\mathrm{Im}[v_1(a)]-\mathrm{Im}[v_1(\overline{a})])}, & z \in C_1, \\ e^{-2\pi(\mathrm{Im}[v_2(a)]-\mathrm{Im}[v_2(\overline{a})])}, & z \in C_2. \end{cases} \tag{6.31}$$

However it was shown in Exercise 3.7 that, for this class of reflectionally symmetric domains D_z, from (3.57) we have

$$\overline{v_j}(z) = -v_j(z) + d_j, \qquad j = 1, 2, \tag{6.32}$$

for some real constants d_1 and d_2 (which can be set equal to zero without loss of generality). Hence, for $j = 1, 2$,

$$\mathrm{Im}[v_j(a)] - \mathrm{Im}[v_j(\overline{a})] = \mathrm{Im}[v_j(a)] + \mathrm{Im}[\overline{v_j}(a)] = \mathrm{Im}[v_j(a)] - \mathrm{Im}[v_j(a)] = 0. \tag{6.33}$$

Thus all the boundary circles C_0, C_1, and C_2 map to circular slits on the unit circle $|w| = 1$. Moreover, it can also be shown for such domains that

$$\overline{\omega}(z,a) = \omega(z,a), \tag{6.34}$$

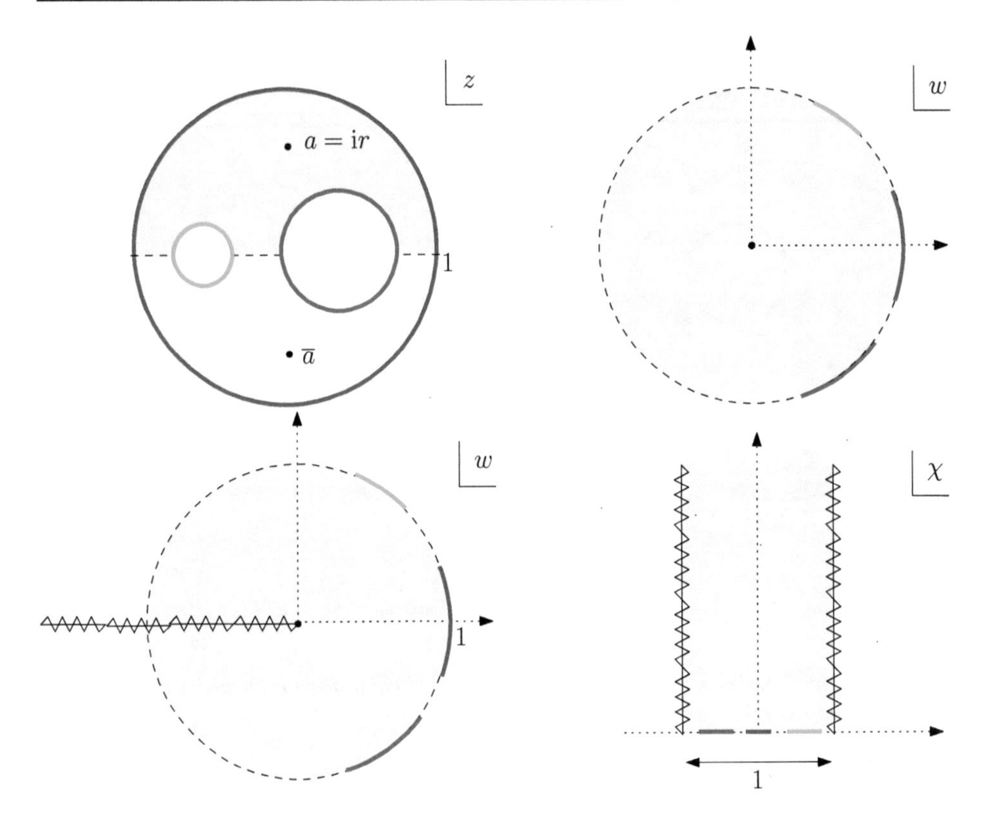

Figure 6.5. *The special case (6.30) of an unbounded circular slit map and, in the lower schematics, the effect of composing it with a logarithmic map as in (6.36). All circles C_0, C_1, and C_2 are centered on the real diameter.*

implying that for points z on the real axis $-1 < z < 1$,

$$\overline{w} = \frac{1}{w}. \tag{6.35}$$

Hence, in this case, the real diameter of the unit disc *also* maps to the unit circle $|w| = 1$.

The logarithm of (6.30) leads to consideration of the conformal mapping

$$w = \frac{1}{2\pi \mathrm{i}} \log \left[-\frac{\omega(z,a)\omega(z,1/a)}{\omega(z,\overline{a})\omega(z,1/\overline{a})} \right]. \tag{6.36}$$

This transplants D_z to a semi-infinite strip of unit width in the upper half plane with each of the boundary circles of D_z being transplanted to a finite-length slit on the real axis as shown in Figure 6.5.

6.8 ▪ Annular slit maps

Two other mappings that closely resemble (6.25) are

$$w = e^{2\pi \mathrm{i}(\mathcal{G}_0(z,a) - \mathcal{G}_j(z,a))}, \qquad j = 1, 2, \tag{6.37}$$

where a is some point inside D_z and we have exchanged one of the two $\mathcal{G}_0$ functions appearing in (6.25) with a $\mathcal{G}_j$ function for $j = 1$ or 2. Inspection of (2.53) shows that

(6.37) can be written as

$$w = c e^{2\pi i v_j(z)}, \qquad j = 1, 2, \tag{6.38}$$

where c is some constant. Since $\mathrm{Im}[v_j(z)]$ is constant on all boundary circles of ∂D_z, the image of each lies on an arc of a circle in the w plane centered at $w = 0$. To study whether these are full circles or finite-length circular slits, we need to know

$$[\arg w]_{C_0}, \qquad [\arg w]_{C_1}, \qquad [\arg w]_{C_2}. \tag{6.39}$$

But these are the quantities

$$2\pi \int_{C_k} dv_j, \qquad k = 0, 1, 2. \tag{6.40}$$

From this we see, on use of (2.38), that if $j = 1$ in (6.38), then the circles C_0 and C_1 map to full circles, while C_2 maps to a circular slit; if $j = 2$ in (6.38), the circles C_0 and C_2 map to full circles, while C_1 maps to a circular slit. These are examples of *annular slit maps*.

There is clearly a third case of such an annular slit map,

$$w = e^{2\pi i(\mathcal{G}_1(z,a) - \mathcal{G}_2(z,a))}, \tag{6.41}$$

and this has the effect of taking C_1 and C_2 to closed circles with C_0 being transplanted to a finite-length circular arc slit in the annular region enclosed by these two circles. Illustrative schematics of all three such annular slit maps are shown in Figure 6.6.

We can take a logarithm of the annular slit maps (6.38). Modulo unimportant scalings and shifts this results in consideration of the maps

$$w = v_j(z), \qquad j = 1, 2. \tag{6.42}$$

Suppose we choose δ_1 and δ_2 to be on the real axis inside the unit disc; then $v_j(z)$ satisfies property (6.32). The map (6.42), with $j = 1$ say, will not be single-valued around C_1 or C_0, but we can pick a branch cut along the real axis joining δ_1 to δ_1', and it can be seen that the interior of D_z maps to a rectangle in the w plane with C_0 and C_1 mapping to the two sides of unit length parallel to the real axis, the two sides of the branch cut (that are inside D_z) mapping to the two sides of the rectangle parallel to the imaginary axis, and C_2 being transplanted to a finite-length slit inside the rectangle that is parallel to the real axis.

6.9 ▪ Parallel slit maps

On isolating the real and imaginary parts of the parameter a, i.e., $a = a_x + i a_y$, we can consider the two maps

$$w_1 = -2\pi i \frac{\partial \mathcal{G}_0(z,a)}{\partial (i a_y)}, \qquad w_2 = -2\pi i \frac{\partial \mathcal{G}_0(z,a)}{\partial a_x}. \tag{6.43}$$

Both parametric derivatives of $-2\pi i \mathcal{G}_0(z,a)$ produce a simple pole, with unit residue at $z = a$. Since $-2\pi i \mathcal{G}_0(z,a)$ has constant real part on $C_j, j = 0, 1, 2$, its differentiation with respect to the purely imaginary parameter $i a_y$ means that the imaginary part of w_1 is constant on all three circles; thus the unit z disc is transplanted to the entire w_1 plane exterior

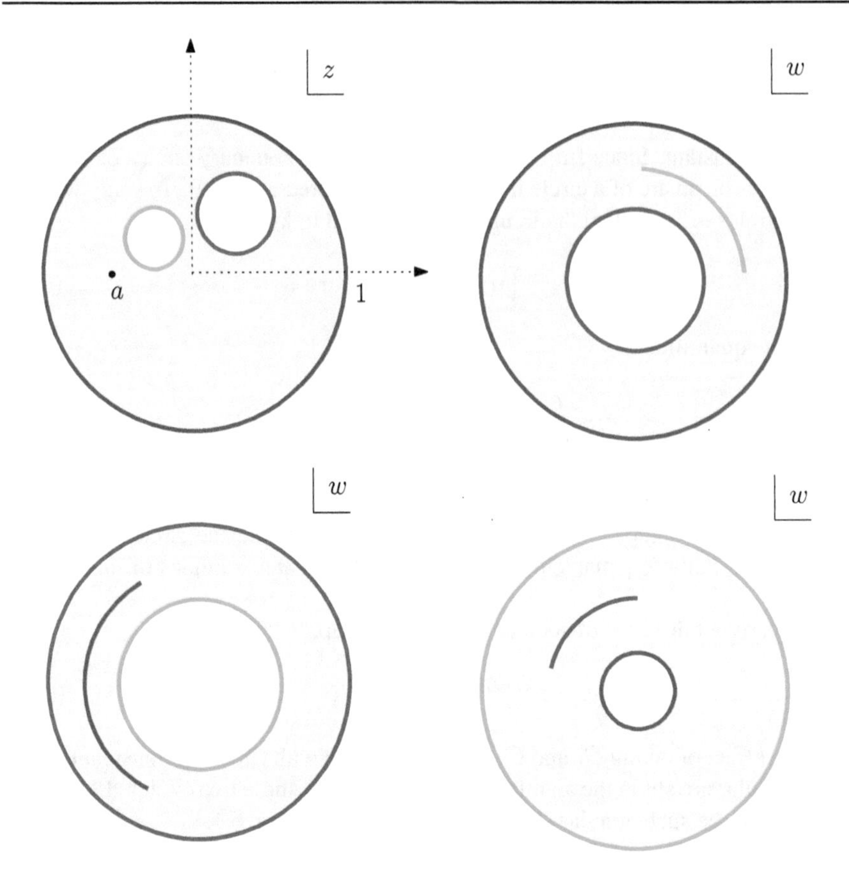

Figure 6.6. *The triply connected circular domain D_z and its image under the three possible annular slit maps (6.37) and (6.41).*

to three slits parallel to the real axis. On the other hand, differentiation of $-2\pi \mathrm{i}\mathcal{G}_0(z,a)$ with respect to the real parameter a_x means that the real part of w_2 will be constant on all boundary circles; thus the interior of the unit z disc to the entire w_2 plane exterior to three slits parallel to the imaginary axis.

Defining

$$\mathcal{K}(z,a) \equiv a\frac{\partial}{\partial a}\log\omega(z,a), \tag{6.44}$$

we can introduce

$$\phi_0(z,a) = -\frac{1}{a}\mathcal{K}(z,a) + \frac{1}{\overline{a}}\mathcal{K}(z,1/\overline{a}), \tag{6.45}$$

$$\phi_{\pi/2}(z,a) = -\frac{1}{a}\mathcal{K}(z,a) - \frac{1}{\overline{a}}\mathcal{K}(z,1/\overline{a}), \tag{6.46}$$

and define

$$\phi_\theta(z,a) = e^{\mathrm{i}\theta}\left[\cos\theta\phi_0(z,a) - \mathrm{i}\sin\theta\phi_{\pi/2}(z,a)\right]. \tag{6.47}$$

It can be verified that $\phi_\theta(z,a)$ has a simple pole at $z=a$ with residue 1 and satisfies

$$\mathrm{Im}[e^{-\mathrm{i}\theta}\phi_\theta(z,a)] = \text{constant}, \qquad z \in C_0, C_1, C_2, \tag{6.48}$$

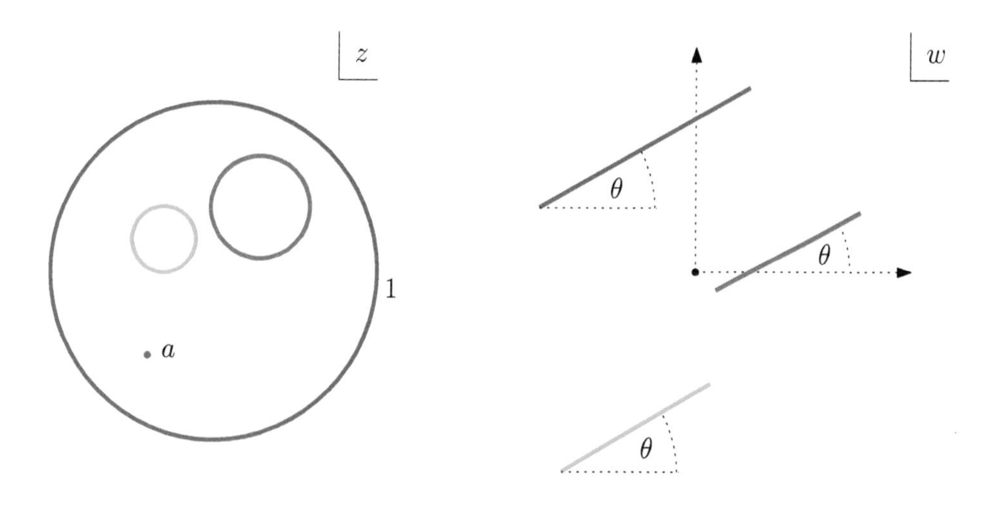

Figure 6.7. *The image of the triply connected circular domain D_z under the parallel slit map (6.49). Each circle is transplanted to a finite-length slit inclined at angle θ to the positive real axis.*

meaning that the images of C_j for $j = 0, 1, 2$ are all slits turned at angle θ to the positive real axis.

A parallel slit map, parametrized by some angle θ with $0 \leq \theta < \pi$ to a complex w plane, can now be defined by

$$w = \phi_\theta(z, a). \tag{6.49}$$

Under such a mapping D_z is transplanted to the unbounded w plane exterior to three parallel finite-length slits all oriented at angle θ to the real w axis. These slits are the images of C_0, C_1, and C_2 as shown in Figure 6.7.

What happens if we exchange $\mathcal{G}_0(z, a)$ in (5.47) for $\mathcal{G}_j(z, a)$ with $j = 1$ or 2? From (2.53) it is easy to show that nothing new is obtained and that such a construction results in the same mappings as those just constructed (see Exercise 6.6).

6.10 ▪ Cayley-type maps

As in §5.10 of Chapter 5 we refer to a mapping of the form

$$w = R(z; a, b), \tag{6.50}$$

where $R(z; a, b)$ is defined in (6.9) and a and b are two distinct points taken on the *same* boundary circle of D_z as a Cayley-type mapping. There are now *three* possibilities: a and b both on C_0, a and b both on C_1, or a and b both on C_2.

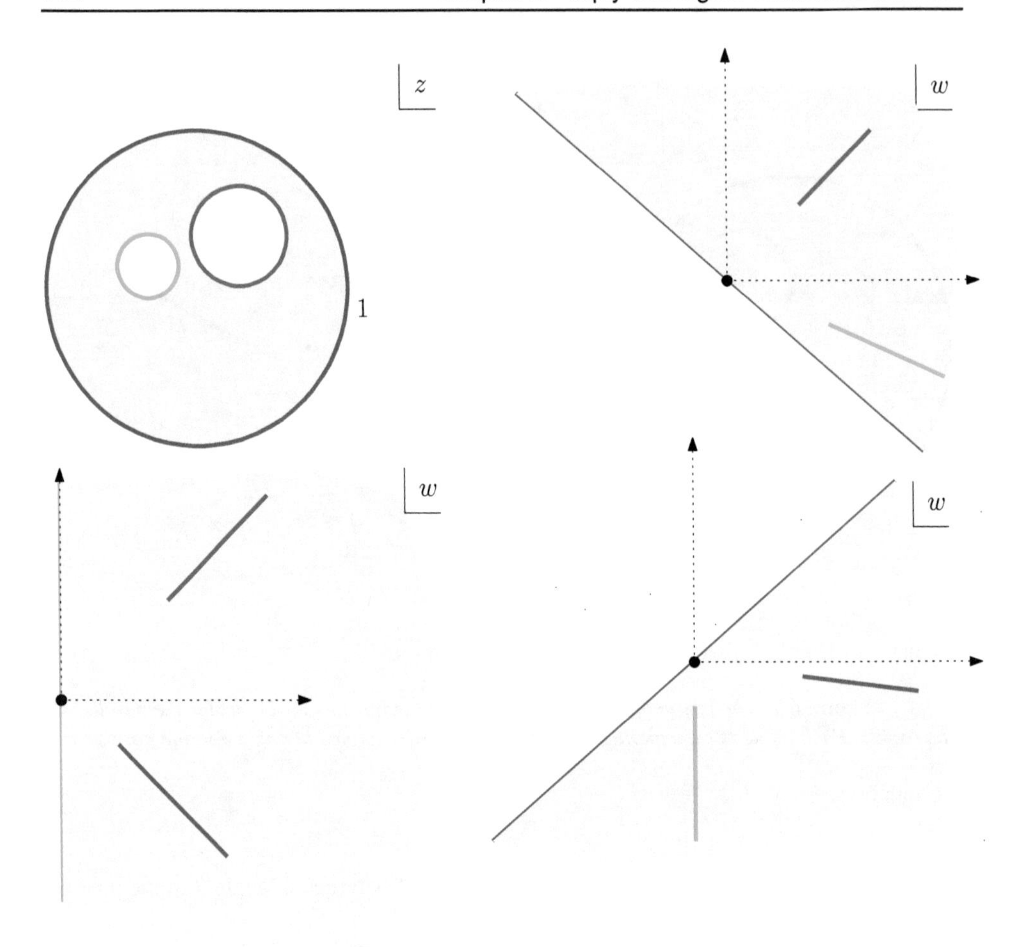

Figure 6.8. *Triply connected circular domain D_z and the effect of the three Cayley-type slit maps (6.50): a and b both on C_0 (top right), a and b both on C_1 (lower right), and a and b both on C_2 (lower left).*

Consider first the case where both a and b are distinct points on C_0. Then $1/\overline{a} = a$ and $1/\overline{b} = b$ so that (6.10) implies

$$\overline{R(z;a,b)} = \begin{cases} \dfrac{b}{a} R(z;a,b), & z \in C_0, \\ \\ e^{-2\pi \mathrm{i}(\overline{v_1(a)} - \overline{v_1(b)})} \dfrac{b}{a} R(z;a,b), & z \in C_1, \\ \\ e^{-2\pi \mathrm{i}(\overline{v_2(a)} - \overline{v_2(b)})} \dfrac{b}{a} R(z;a,b), & z \in C_2. \end{cases} \tag{6.51}$$

Since it follows from this that the quantity

$$\frac{R(z;a,b)}{\overline{R(z;a,b)}} \tag{6.52}$$

is constant on all the boundary circles of D_z, we have shown that

$$\arg[R(z;a,b)] = \text{constant}, \qquad z \in \partial D_z. \tag{6.53}$$

On the other hand, if both points a and b lie on circle C_k for $k = 1$ or $k = 2$, then $R(z; a, b)$ also has interesting properties. Indeed it can be shown that

$$\overline{R(z;a,b)} = \begin{cases} \dfrac{b-\delta_k}{a-\delta_k} e^{2\pi \mathrm{i}(v_k(a)-v_k(b))} R(z;a,b), & z \in C_0, \\ \dfrac{b-\delta_k}{a-\delta_k} e^{-2\pi \mathrm{i}(v_j(a)-v_j(b))} e^{2\pi \mathrm{i}(v_k(a)-v_k(b))} R(z;a,b), & z \in C_j,\ j=1,2. \end{cases} \tag{6.54}$$

In order to establish this result one option is, again, to make use of the key result (6.10); Exercise 6.16 asks the reader to derive (6.54) in this way. However, since it is an instructive exercise in the use of the prime function identities, we will proceed slightly differently.

First consider the case where $z \in C_0$. On use of the antisymmetry of the prime function,

$$R(z;a,b) = \frac{\omega(1/\overline{z}, \theta_k(1/\overline{a}))}{\omega(1/\overline{z}, \theta_k(1/\overline{b}))} = \frac{\omega(\theta_k(1/\overline{a}), 1/\overline{z})}{\omega(\theta_k(1/\overline{b}), 1/\overline{z})}, \tag{6.55}$$

where, for the first equation, we have used the facts that

$$z = \frac{1}{\overline{z}}, \qquad a = \theta_k(1/\overline{a}), \qquad b = \theta_k(1/\overline{b}). \tag{6.56}$$

An application of the identities (6.3)–(6.4) implies

$$R(z;a,b) = \frac{1-\overline{\delta_k/b}}{1-\overline{\delta_k/a}} e^{-2\pi \mathrm{i}(v_k(1/\overline{a})-v_k(1/\overline{b}))} \frac{\omega(1/\overline{a}, 1/\overline{z})}{\omega(1/\overline{b}, 1/\overline{z})}. \tag{6.57}$$

This can be written

$$R(z;a,b) = \frac{1-\overline{\delta_k/b}}{1-\overline{\delta_k/a}} e^{-2\pi \mathrm{i}(\overline{v_k(a)}-\overline{v_k(b)})} \frac{\omega(1/\overline{z}, 1/\overline{a})}{\omega(1/\overline{z}, 1/\overline{b})}, \tag{6.58}$$

where we have used (3.8), i.e., $\overline{v_k}(1/a) = v_k(a)$. On taking a complex conjugate we find

$$\overline{R(z;a,b)} = \frac{1-\delta_k/b}{1-\delta_k/a} e^{2\pi \mathrm{i}(v_k(a)-v_k(b))} \frac{\overline{\omega}(1/z, 1/a)}{\overline{\omega}(1/z, 1/b)} \tag{6.59}$$

$$= \frac{b-\delta_k}{a-\delta_k} e^{2\pi \mathrm{i}(v_k(a)-v_k(b))} R(z;a,b), \tag{6.60}$$

where, in the second equation, we have used (6.2).

On the other hand, if $z \in C_j$ for $j = 1, 2$, then we can write

$$R(z;a,b) = \frac{\omega(z,a)}{\omega(z,b)} = \frac{\omega(\theta_j(1/\overline{z}), \theta_k(1/\overline{a}))}{\omega(\theta_j(1/\overline{z}), \theta_k(1/\overline{b}))}, \tag{6.61}$$

where we have used the facts that

$$z = \theta_j(1/\overline{z}), \qquad a = \theta_k(1/\overline{a}), \qquad b = \theta_k(1/\overline{b}). \tag{6.62}$$

On use of the identities (6.3)–(6.4) this becomes

$$R(z;a,b) = e^{2\pi \mathrm{i}(v_j(\theta_k(1/\overline{a}))-v_j(\theta_k(1/\overline{b})))} \frac{\omega(1/\overline{z}, \theta_k(1/\overline{a}))}{\omega(1/\overline{z}, \theta_k(1/\overline{b}))}. \tag{6.63}$$

It is now useful to invoke the transformation property (3.43) of the function $v_j(z)$ to write

$$R(z;a,b) = e^{2\pi \mathrm{i}(v_j(1/\overline{a})-v_j(1/\overline{b}))} \frac{\omega(1/\overline{z}, \theta_k(1/\overline{a}))}{\omega(1/\overline{z}, \theta_k(1/\overline{b}))} \tag{6.64}$$

or, on use of (3.8) and the antisymmetry of ω,

$$R(z;a,b) = e^{2\pi i(\overline{v_j(a)} - \overline{v_j(b)})} \frac{\omega(\theta_k(1/\overline{a}), 1/\overline{z})}{\omega(\theta_k(1/\overline{b}), 1/\overline{z})}. \tag{6.65}$$

We can now make use of the identities (6.3)–(6.4) for a second time to find

$$R(z;a,b) = \frac{1 - \overline{\delta_k/b}}{1 - \overline{\delta_k/a}} e^{2\pi i(\overline{v_j(a)} - \overline{v_j(b)})} e^{-2\pi i(v_k(1/\overline{a}) - v_k(1/\overline{b}))} \frac{\omega(1/\overline{a}, 1/\overline{z})}{\omega(1/\overline{b}, 1/\overline{z})}. \tag{6.66}$$

This can be written as

$$R(z;a,b) = \frac{1 - \overline{\delta_k/b}}{1 - \overline{\delta_k/a}} e^{2\pi i(\overline{v_j(a)} - \overline{v_j(b)})} e^{-2\pi i(\overline{v_k(a)} - \overline{v_k(b)})} \frac{\omega(1/\overline{z}, 1/\overline{a})}{\omega(1/\overline{z}, 1/\overline{b})}, \tag{6.67}$$

where we have again used (3.8) and the antisymmetry of $\omega(.,.)$. Now we can take the complex conjugate of this equation to find

$$\overline{R(z;a,b)} = \frac{1 - \delta_k/b}{1 - \delta_k/a} e^{-2\pi i(v_j(a) - v_j(b))} e^{2\pi i(v_k(a) - v_k(b))} \frac{\overline{\omega}(1/z, 1/a)}{\omega(1/z, 1/b)} \tag{6.68}$$

$$= \frac{b - \delta_k}{a - \delta_k} e^{-2\pi i(v_j(a) - v_j(b))} e^{2\pi i(v_k(a) - v_k(b))} R(z;a,b), \tag{6.69}$$

where we have used (6.2). Thus we arrive at (6.54).

Putting all this together means that when a and b are both located on *any* of the boundary circles of D_z, then

$$\arg[R(z;a,b)] = \text{constant}, \qquad z \in \partial D_z. \tag{6.70}$$

The Cayley-type mapping $R(z;a,b)$ therefore has constant argument on all the boundary circles of D_z.

A special case is to take $a = +1$ and $b = -1$ and to premultiply by -1 so that the mapping is

$$w = -R(z;1,-1) = -\frac{\omega(z,+1)}{\omega(z,-1)}. \tag{6.71}$$

This Cayley-type mapping is such that, from (6.51),

$$\overline{w} = \begin{cases} -w, & z \in C_0, \\ -e^{-2\pi i(\overline{v_1(1)} - \overline{v_1(-1)})} w, & z \in C_1, \\ -e^{-2\pi i(\overline{v_2(1)} - \overline{v_2(-1)})} w, & z \in C_2. \end{cases} \tag{6.72}$$

Hence C_0 maps to the imaginary axis while C_1 and C_2 map to slits of finite length lying on a ray whose argument is determined by the real quantities $v_1(1) - v_1(-1)$ and $v_2(1) - v_2(-1)$.

Figure 6.9 shows an example where all three circles C_0, C_1, and C_2 are centered on the real axis so that D_z is reflectionally symmetric about that axis. In this case, it is possible to use this reflectional symmetry, together with the property (2.38), to argue that $v_1(1) - v_1(-1) = v_2(1) - v_2(-1) = 1/2$ and thus that in this case (6.72) becomes

$$\overline{w} = \begin{cases} -w, & z \in C_0, \\ +w, & z \in C_1, \\ +w, & z \in C_2. \end{cases} \tag{6.73}$$

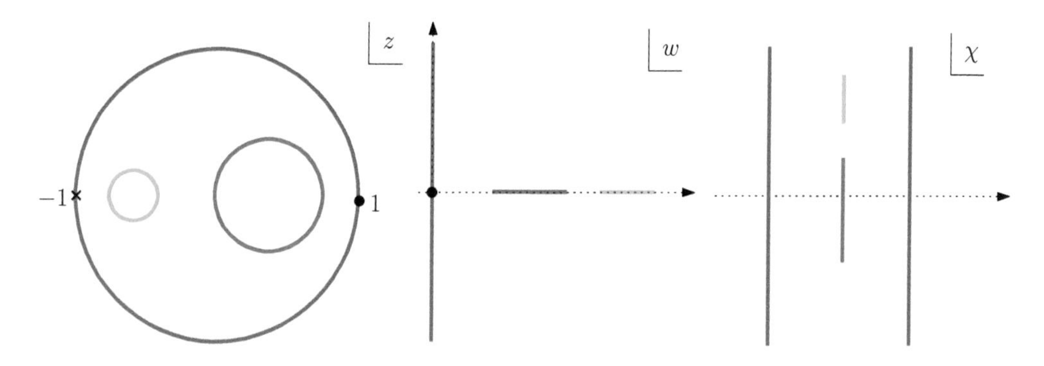

Figure 6.9. *Cayley-type map from the z to the w plane, as in (6.71), and the logarithm of this mapping to a complex χ plane, as in (6.74). All three circles C_0, C_1, and C_2 are centered on the real axis.*

Maps produced by taking logarithms of Cayley-type maps turn out to be particularly useful in applications. Figure 6.9 also shows the effect on D_z of the mapping obtained by taking a logarithm of (6.71), that is,

$$\chi = \frac{1}{\pi \mathrm{i}} \log w = \frac{1}{\pi \mathrm{i}} \log[-R(z; 1, -1)]. \tag{6.74}$$

If we use the principal branch of the logarithm, such a map transplants D_z to the unbounded region in a complex χ plane comprising the infinite strip of unit width exterior to two finite-length slits along the imaginary χ axis.

6.11 ▪ Radial slit maps

Consider the special choice

$$w = R(z; a, b)R(z; 1/\overline{a}, 1/\overline{b}) = \frac{\omega(z, a)\omega(z, 1/\overline{a})}{\omega(z, b)\omega(z, 1/\overline{b})} \tag{6.75}$$

with a and b chosen to be two distinct nonzero points inside D_z. It is clear that $z = a$ maps to $w = 0$ and $z = b$ maps to $w = \infty$. Relation (6.10) implies that

$$\overline{w} = \begin{cases} \dfrac{\overline{a}b}{a\overline{b}} w, & z \in C_0, \\[2ex] \dfrac{\overline{a}b}{a\overline{b}} e^{-2\pi \mathrm{i}(\overline{v_1(a)} - \overline{v_1(b)} + \overline{v_1(1/\overline{a})} - \overline{v_1(1/\overline{b})})} w, & z \in C_1, \\[2ex] \dfrac{\overline{a}b}{a\overline{b}} e^{-2\pi \mathrm{i}(\overline{v_2(a)} - \overline{v_2(b)} + \overline{v_2(1/\overline{a})} - \overline{v_2(1/\overline{b})})} w, & z \in C_2. \end{cases} \tag{6.76}$$

We conclude that

$$\arg[w] = \text{constant} \qquad \text{on } C_j, \;\; j = 0, 1, 2. \tag{6.77}$$

This is a *radial slit mapping* in which all three circles C_0, C_1, and C_2 are transplanted to rays in the w plane.

A special case is to take a and b to be real and all circles $C_j, j = 0, 1, 2$, to be centered on the real z axis. The images of all the circles $C_j, j = 0, 1, 2$, then lie on the real axis.

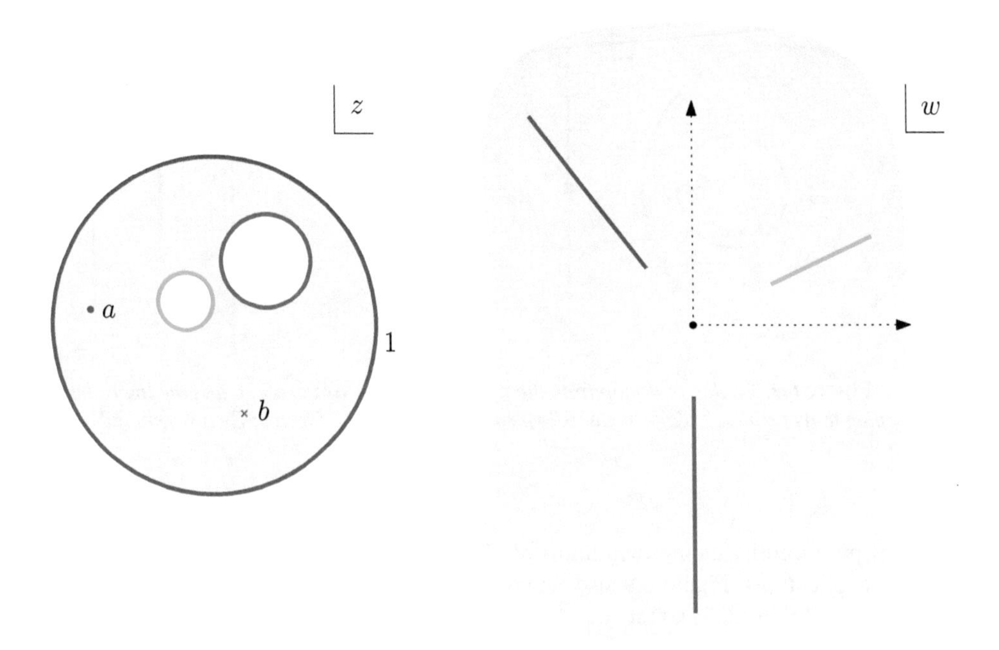

Figure 6.10. *Triply connected circular domain D_z and its image under a radial slit map of the form (6.75). The point $z = a$ maps to $w = 0$ and $z = b$ is the preimage of $w = \infty$.*

This map can be viewed as the three-slit generalization of the classical Joukowski mapping of Exercise 1.15.

Another special case is to take

$$a = \mathrm{i}r, \qquad b = \overline{a}, \qquad 0 < r < 1, \tag{6.78}$$

and to consider the situation in which both C_1 and C_2 are centered on the real z axis so that $\delta_1, \delta_2 \in \mathbb{R}$. We also premultiply by -1 so that (6.75) becomes

$$w = -\frac{\omega(z,a)\omega(z,1/\overline{a})}{\omega(z,\overline{a})\omega(z,1/a)}. \tag{6.79}$$

Then (6.76) is equivalent to the conditions

$$\overline{w} = \begin{cases} w, & z \in C_0, \\ e^{-2\pi\mathrm{i}(\overline{v_1(a)} - \overline{v_1(\overline{a})} + \overline{v_1(1/\overline{a})} - \overline{v_1(1/a)})} w, & z \in C_1, \\ e^{-2\pi\mathrm{i}(\overline{v_2(a)} - \overline{v_2(\overline{a})} + \overline{v_2(1/\overline{a})} - \overline{v_2(1/a)})} w, & z \in C_2. \end{cases} \tag{6.80}$$

On use of the fact that $\overline{v_j}(1/z) = v_j(z)$ for $j = 1, 2$ and (6.32),

$$\overline{v_j(a)} - \overline{v_j(\overline{a})} + \overline{v_j(1/\overline{a})} - \overline{v_j(1/a)} = \overline{v_j(a)} - \overline{v_j(\overline{a})} + v_j(a) - v_j(\overline{a}) \tag{6.81}$$

$$= 2(\mathrm{Re}[v_j(a)] - \mathrm{Re}[v_j(\overline{a})]). \tag{6.82}$$

Hence

$$\arg[w] = \begin{cases} 0, & z \in C_0, \\ 2\pi\mathrm{Re}[v_j(a) - v_j(\overline{a})], & z \in C_j, j = 1, 2. \end{cases} \tag{6.83}$$

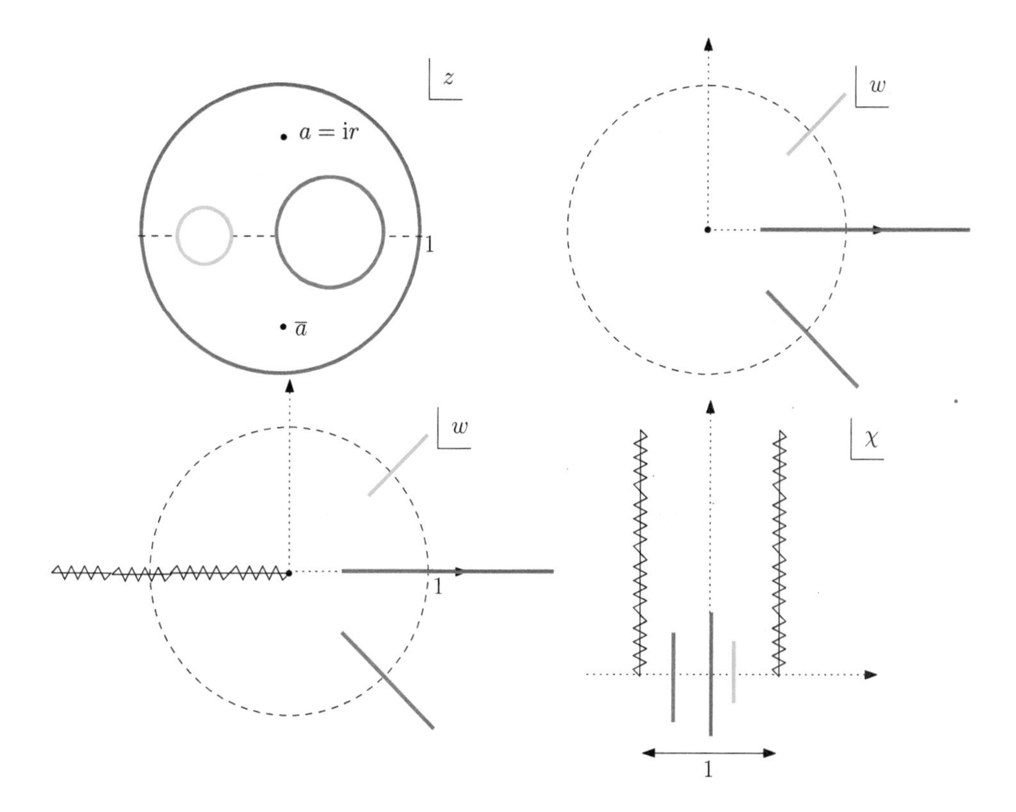

Figure 6.11. *The special case of the radial slit map given in (6.79). The logarithm of this mapping (6.85) to the complex χ plane is also shown.*

For a domain D_z that is reflectionally symmetric about the real z axis, $\overline{\omega}(z,a) = \omega(z,a)$, and it is easy to check that, for points z on the real axis $-1 < z < 1$,

$$\overline{w} = \frac{1}{w}. \tag{6.84}$$

Hence this real diameter of the unit disc maps to the unit circle $|w| = 1$.

It is of interest to consider the logarithm of (6.79):

$$\chi = \frac{1}{2\pi \mathrm{i}} \log \left[-\frac{\omega(z,a)\omega(z,1/\overline{a})}{\omega(z,\overline{a})\omega(z,1/a)} \right]. \tag{6.85}$$

Under such a mapping D_z is transplanted to an infinite strip parallel to the imaginary axis in a complex χ plane exterior to three finite-length slits parallel to the imaginary χ axis and symmetric about the real χ axis, as shown in Figure 6.11.

6.12 ▪ The functions $\mathcal{K}(z,a)$ and $K(z,a)$

The function $\mathcal{K}(z,a)$ introduced in (6.44) has a simple pole at $z = a$. It can be shown from the identity (6.2) satisfied by the prime function that

$$\overline{\mathcal{K}}(1/z, 1/a) = 1 - \mathcal{K}(z,a). \tag{6.86}$$

From the transformation formulas (6.3)–(6.4) we know that

$$\omega(\theta_j(z), a) = -\frac{q_j}{1-\overline{\delta_j} z} e^{-2\pi i(v_j(z)-v_j(a)+\tau_{jj}/2)}\omega(z,a), \qquad j=1,2, \tag{6.87}$$

which, on taking a partial logarithmic derivative with respect to a and multiplying by a, gives

$$\mathcal{K}(\theta_j(z), a) = \mathcal{K}(z,a) + 2\pi i a v_j'(a), \qquad j=1,2. \tag{6.88}$$

Since $\mathcal{K}(z,a)$ acquires an additive z-independent term as $z \mapsto \theta_j(z)$, it is called a *quasi-automorphic function.* We will study such functions in more detail in Chapter 9.

As in Chapter 5 and earlier in this chapter, we can also introduce the logarithmic derivative of $\omega(z,a)$ in the z variable:

$$K(z,a) \equiv z\frac{\partial \log \omega(z,a)}{\partial z}. \tag{6.89}$$

It is immediately clear from the antisymmetry of the prime function (6.1) that $\mathcal{K}(z,a)$ and $K(z,a)$ are related by

$$K(z,a) = \mathcal{K}(a,z). \tag{6.90}$$

It can also be shown that $K(z,a)$ satisfies the identity

$$\overline{K}(1/z, 1/a) = 1 - K(z,a). \tag{6.91}$$

In principle we can write down an identity for $K(\theta_j(z), a)$ for $j=1,2$ deriving from the transformation formulas (6.3)–(6.4), but it turns out that such identities have limited use in applications, so we leave this as an exercise for the reader.

6.13 ▪ Quadruply and higher connected domains

For quadruply and higher connected domains the framework described in this chapter for triply connected domains carries over directly. The only difference, for a quadruply connected domain, say, is that the relevant circular domain D_z now has three holes with boundaries C_1, C_2, and C_3, as shown in Figure 6.12, and the prime function relevant to this domain must now be used in all the same formulas just presented. Once again we denote the relevant prime function by $\omega(.,.)$.

As an example of how naturally everything extends to any $M > 2$ we can again introduce the ratio of these prime functions

$$R(z;a,b) = \frac{\omega(z,a)}{\omega(z,b)} \tag{6.92}$$

and establish, from (4.103), that for two nonzero points a and b with $a \neq b$,

$$\overline{R(z;a,b)} = \begin{cases} \dfrac{\overline{a}}{\overline{b}} R(z; 1/\overline{a}, 1/\overline{b}), & z \in C_0, \\ e^{-2\pi i(\overline{v_k(a)} - \overline{v_k(b)})} \dfrac{\overline{a}}{\overline{b}} R(z; 1/\overline{a}, 1/\overline{b}), & z \in C_k, \ k=1,\ldots,M. \end{cases} \tag{6.93}$$

The result (6.10) is just the $M=2$ case of (6.93).

We end by recording several formulas giving useful functions defined in a quadruply connected domain D_z and illustrating their geometrical interpretation as conformal maps in Figure 6.13. The new circle C_3, and all its images under various conformal mappings, will be shown in magenta.

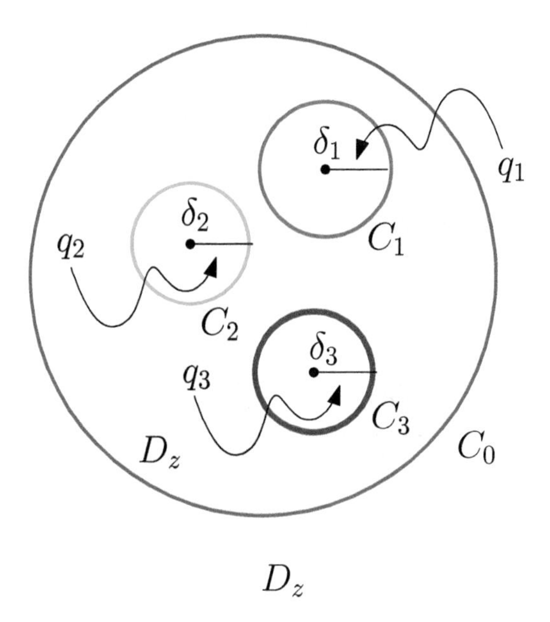

Figure 6.12. *A quadruply connected circular domain D_z with boundary circles C_0, C_1, C_2, and C_3. C_0 is the unit circle; the circles C_1, C_2, and C_3 have centers at δ_1, δ_2, and δ_3 and radii q_1, q_2, and q_3.*

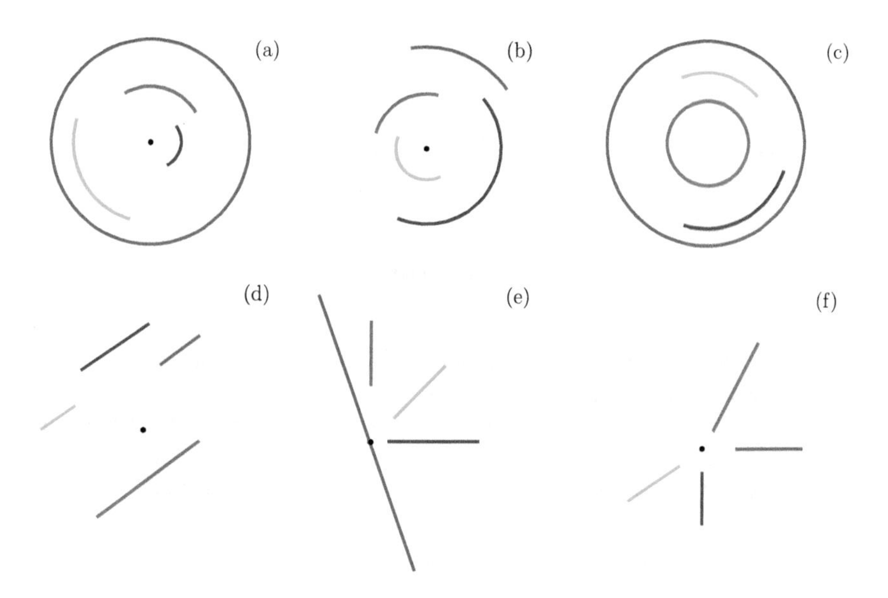

Figure 6.13. *The basic conformal slit mappings from a quadruply connected circular domain such as that in Figure 6.12. (a) A bounded circular slit map (6.94); (b) an unbounded circular slit map (6.95); (c) an annular slit map (6.96); (d) a parallel slit map (6.97); (e) a Cayley-type map (6.98); and (f) a radial slit map (6.99).*

The mapping given by (6.19), that is,

$$w = \frac{1}{|a|} \frac{\omega(z,a)}{\omega(z,1/\overline{a})}, \tag{6.94}$$

is one of the *bounded circular slit maps* whose image of D_z is shown in subplot (a) of Figure 6.13.

The mapping given by (6.25), that is,

$$w = \frac{|b|}{|a|} \frac{\omega(z,a)\omega(z,1/\overline{b})}{\omega(z,b)\omega(z,1/\overline{a})}, \tag{6.95}$$

where a and b are two nonzero points inside D_z, is the *unbounded slit map* whose image of D_z is shown in subplot (b) of Figure 6.13.

The mapping given by (6.38) with $j = 1$, say, that is,

$$w = e^{2\pi \mathrm{i}(\mathcal{G}_0(z,a) - \mathcal{G}_1(z,a))} = c e^{2\pi \mathrm{i} v_1(z)}, \qquad c = \text{constant}, \tag{6.96}$$

is one of the *annular slit maps* whose image of D_z is shown in subplot (c) of Figure 6.13.

The mapping given by (6.47), that is,

$$w = \phi_\theta(z,a) = e^{\mathrm{i}\theta} \left[\cos\theta \phi_0(z,a) - \mathrm{i}\sin\theta \phi_{\pi/2}(z,a)\right], \tag{6.97}$$

where $\phi_0(z,a)$ and $\phi_{\pi/2}(z,a)$ are defined in (6.45) and (6.46), with $\mathcal{K}(z,a)$ defined in (6.44), is one of the *parallel slit maps* whose image of D_z is shown in subplot (d) of Figure 6.13. The images of all the boundary circles of D_z are parallel slits oriented at angle θ to the positive real axis in the image domain.

The mapping given by (6.50), that is,

$$w = \frac{\omega(z,a)}{\omega(z,b)}, \tag{6.98}$$

where a and b are two distinct points taken on C_0, say, is one of the *Cayley-type maps* whose image of D_z is shown in subplot (e) of Figure 6.13.

The mapping given by (6.75), that is,

$$w = \frac{\omega(z,a)\omega(z,1/\overline{a})}{\omega(z,b)\omega(z,1/\overline{b})}, \tag{6.99}$$

where a and b are two distinct nonzero points inside D_z, is a *radial slit map* whose image of D_z is shown in subplot (f) of Figure 6.13.

This array of formulas giving the conformal mappings, in terms of the prime function, from a multiply connected circular domain to conformally equivalent slit domains, as shown in Figure 6.13, was originally discussed by Crowdy and Marshall [47].

Exercises

6.1. **(An observation on triply connected domains)** For the doubly connected domains of Chapter 5 we used a disc automorphism to argue that it is enough to consider C_1 to be concentric with C_0, that is, $\delta_1 = 0$. Once this is done, we still have a rotational degree of freedom left in choosing this automorphism, and this can be used to offer a special advantage to the case of triply connected circular domains.

(a) By thinking about disc automorphisms, argue that any triply connected circular domain is conformally equivalent to a triply connected circular domain D_ζ, where both C_1 and C_2 are centered on the real diameter $(-1, 1)$, i.e.,

$$\delta_1, \delta_2 \in \mathbb{R}. \tag{6.100}$$

(b) Convince yourself that this is no longer possible for quadruply and higher connected domains.

(c) How many real degrees of freedom characterize a triply connected circular domain once all the freedoms of a disc automorphism are factored out?

6.2. **(Univalency and the argument principle)** Use the argument principle to establish that the interior of the triply connected circular domain D_z is transplanted in a one-to-one fashion to the various target domains in the unbounded circular slit map, the parallel slit map, the Cayley-type map, and the radial slit map. *Hint:* Note that the Cayley-type map can be viewed as having "half a pole" inside D_z.

6.3. **(Blaschke-type products)** It is required to find a function $f(z)$ with the following properties:

(i) $f(z)$ is analytic, and single-valued, everywhere in a triply connected circular domain D_z.

(ii) $f(z)$ has precisely N zeros inside D_z at given locations $\{a_j | j = 1, \ldots, N\}$.

(iii) The modulus of $f(z)$ is unity on the boundary C_0 of D_z and also has constant modulus on the boundaries C_1 and C_2.

By thinking about the bounded circular slit mappings (6.19) construct an example of a function $f(z)$ having the properties (i)–(iii). What is the modulus on C_1 and C_2 of the function $f(z)$ you have constructed? Can you find another function with the same properties (i)–(iii) but with a different modulus on C_1 and C_2?

6.4. **(Two views of a three-slit map)** Consider the conformal mapping from the triply connected circular domain D_z to the unbounded domain in a w-plane exterior to three slits $[v, u]$, $[t, s]$, and $[r, 1]$ on the real axis where $v < u < t < s < r < 1$. Two of the three real degrees of freedom of the Riemann mapping theorem have been used to insist that the preimage circle C_1 of the second slit is concentric with C_0, i.e., $\delta_1 = 0$. We use up the final degree of freedom to insist that $z = 1$ is the preimage of $w = 1$. By the reflectional symmetry of the domain with respect to the real axis, we expect the preimage of $w = \infty$ to be at some point $z = b \in \mathbb{R}$; we also expect the center of C_2 to be on the real axis, i.e., $\delta_2 \in \mathbb{R}$.

(a) We can construct this map as an instance of a radial slit mapping. Show that the relevant function is

$$w = f_1(z) = \frac{\omega(1, b)\omega(1, 1/b)\omega(z, a)\omega(z, 1/a)}{\omega(1, a)\omega(1, 1/a)\omega(z, b)\omega(z, 1/b)}, \tag{6.101}$$

which depends on five real parameters: q_1, δ_2, q_2, b, and a. These parameters determine the values of r, s, t, u, and v, the endpoints of the slits.

(b) The same mapping can be constructed as an instance of a parallel slit mapping. Show that the required map is

$$w = f_2(z) = \frac{\mathcal{K}(z, b) - \mathcal{K}(z, 1/b) - (\mathcal{K}(a, b) - \mathcal{K}(a, 1/b))}{\mathcal{K}(1, b) - \mathcal{K}(1, 1/b) - (\mathcal{K}(a, b) - \mathcal{K}(a, 1/b))}. \tag{6.102}$$

The function $f_2(z)$ must be identical to $f_1(z)$. Indeed, $f_2(z)$ is simply the "partial fraction" decomposition of the function $f_1(z)$ although we are no longer dealing with rational functions.

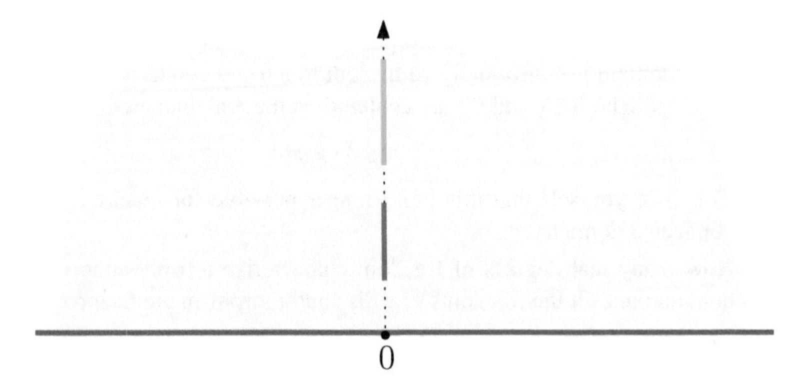

Figure 6.14. *Two flat plates, or slits, vertically aligned above an infinite-length wall (Exercise 6.7).*

6.5. **(Observation on unbounded circular slit maps)** Verify that if we define a map

$$w = e^{2\pi \mathrm{i}(\mathcal{G}_1(z,a)-\mathcal{G}_1(z,b))}, \tag{6.103}$$

where a and b are distinct points in D_z—this map is similar to (6.25) but has $\mathcal{G}_1$ replacing the $\mathcal{G}_0$ that appears there—then the result is an unbounded circular slit map qualitatively similar to that given by (6.25).

6.6. **(Observation on parallel slit maps)** Make use of (2.53) to verify that if the construction of the parallel slit mappings in §6.9 is repeated but with either $\mathcal{G}_1(z,a)$ or $\mathcal{G}_2(z,a)$ replacing $\mathcal{G}_0(z,a)$ in §6.9, then the same class of parallel slit mappings appearing in (6.47) is obtained.

6.7. **(Two flat plates perpendicular to a wall)** Suppose that both C_1 and C_2 are centered on the real z axis so that $\delta_1, \delta_2 \in \mathbb{R}$. Use the result of Exercise 3.7 and (6.72) to show that a rotation by $\pi/2$ of the Cayley-type mapping (6.71) given by

$$w = -\mathrm{i}\,\frac{\omega(z,+1)}{\omega(z,-1)} \tag{6.104}$$

transplants D_z to the upper half w-plane exterior to two slits of finite length sitting on the positive imaginary w axis (see Figure 6.14).

6.8. **(On Cayley-type maps)** Let D_z be a triply connected circular domain with the two interior circles C_1 and C_2 centered at δ_1 and δ_2 (respectively) on the real z axis so that D_z is reflectionally symmetric with respect to that axis.

(a) Consider the Cayley-type map from D_z given by

$$w = -\frac{1}{\tau}\frac{\omega(z,\tau)}{\omega(z,\overline{\tau})}, \tag{6.105}$$

where $\tau \in C_0^+$ is a point on the upper half unit circle $|\tau| = 1, \mathrm{Im}[\tau] > 0$. Show that this mapping transplants D_z to the lower half w plane with two radial slits excised and C_0 being transplanted to the real w axis.

(b) Now consider the two Cayley-type maps, for $j = 1, 2$, from D_z given by

$$w = +\frac{1}{\tau-\delta_j}\frac{\omega(z,\tau)}{\omega(z,\overline{\tau})}, \tag{6.106}$$

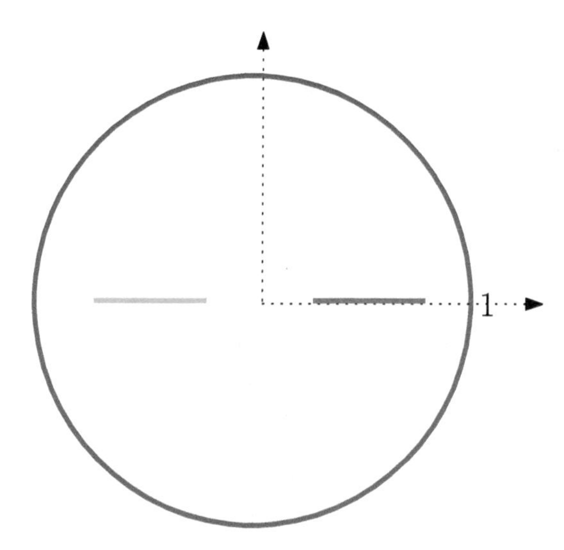

Figure 6.15. *The unit disc with two interior flat plates, or slits, along the real diameter (Exercise 6.9).*

where $\tau \in C_j^+$ is a point on the upper half circle $|\tau - \delta_j| = q_j, \text{Im}[\tau - \delta_j] > 0$. Show that each mapping also transplants D_z to the lower half w plane with two radial slits excised but now with C_j being transplanted to the real w axis.

6.9. **(Disc with two slits)** Suppose again that both C_1 and C_2 are centered on the real z axis with

$$\delta_1 = -\delta_2 = \delta \in \mathbb{R}, \qquad q_1 = q_2 = q, \qquad 0 < q < d < 1. \tag{6.107}$$

Show that the map

$$w = \frac{\omega(z, -1) + \omega(z, +1)}{\omega(z, -1) - \omega(z, +1)} \tag{6.108}$$

transplants D_z to the unit disc in the w plane exterior to two slits $[-s, -r]$ and $[r, s]$ along the real w axis (see Figure 6.15). Find expressions for r and s in terms of q and d.

6.10. **(Cylinder between two slits)** Use the result in Exercise 6.9 to find the conformal map from D_z to the unbounded region exterior to a circular cylinder with two equal-length slits $[-s, -r]$ and $[r, s]$ at either side of it.

6.11. **(Yet another view of a three-slit map)** Combine the classical Joukowski map

$$\frac{1}{2}\left(\frac{1}{z} + z\right) \tag{6.109}$$

with the result in Exercise 6.9 to show that a map from the annulus $\rho < |z| < 1$ to the unbounded region exterior to three slits on the real axis is

$$w = \frac{\omega(z, -1)^2 + \omega(z, +1)^2}{\omega(z, -1)^2 - \omega(z, +1)^2}. \tag{6.110}$$

Verify that the image of C_0 is the interval $[-1, 1]$.

6.12. **(Three gaps in an infinite wall)** Show that the conformal mapping from a circular preimage domain to the unbounded region exterior to an infinite wall with three gaps, say,

$$[-1,-1/2-r], \qquad [-1/2+r,1/2-r], \qquad [1/2+r,1], \tag{6.111}$$

is given by

$$w = \frac{\omega(z,-1)^2 - \omega(z,+1)^2}{\omega(z,-1)^2 + \omega(z,+1)^2}. \tag{6.112}$$

Suppose the preimage domain D_z assumes the same symmetries with respect to reflection in both the real and imaginary axes as the target domain. Then the centers and radii of circles C_1 and C_2 can be taken to be

$$\delta_1 = -\delta_2 = -\delta, \qquad q_1 = q_2 = q. \tag{6.113}$$

Write down two equations that must be solved to find the two real parameters δ and q as functions of r. Show that the limit $r \to 0$ of a simply connected domain in which the two finite-wall regions vanish corresponds to

$$q \to 0, \qquad \delta \to 2-\sqrt{3}. \tag{6.114}$$

6.13. **(Swapping the correspondences)** Consider the conformal mapping to the unbounded region exterior to an infinite wall with three gaps as just discussed in Exercise 6.12. Let us call this mapping

$$w = \mathcal{Z}_1(z). \tag{6.115}$$

This mapping will have the form (6.112). We can equivalently consider the conformal mapping from a different circular preimage region, $\tilde{D}_z$ say, to the unbounded region exterior to the three finite-width gaps (6.111). Let us call this mapping

$$w = \mathcal{Z}_2(z). \tag{6.116}$$

This mapping will have the form (6.110). We can again suppose that the preimage domain $\tilde{D}_z$ assumes the same symmetries with respect to reflection in both the real and imaginary axes as the target domain. Then the centers and radii of circles $\tilde{C}_1$ and $\tilde{C}_2$ can be taken to be

$$\tilde{\delta}_1 = -\tilde{\delta}_2 = -\tilde{\delta}, \qquad \tilde{q}_1 = \tilde{q}_2 = \tilde{q}. \tag{6.117}$$

Write down two equations that must be solved to find the two real parameters $\tilde{\delta}$ and $\tilde{q}$ as functions of r.

Remark: Figure 6.16 shows graphs of δ, q and $\tilde{\delta}, \tilde{q}$ as functions of r obtained by solving the equations derived in this exercise and in Exercise 6.12. Notice that the value of $q \to 0$ as $r \to 0$ and $\tilde{q} \to 0$ as $r \to 1/2$ so that when the radius of the interior circles gets large in one parametrization of the domain the radius in the alternative parametrization of the same domain gets small. This feature can often be usefully exploited in practice (see Chapter 22).

6.14. **(Two equal flat plates above and parallel to an infinite wall)** Suppose again that both C_1 and C_2 are centered on the real z axis with

$$\delta_1 = -\delta_2 = \delta \in \mathbb{R}, \qquad q_1 = q_2 = q, \qquad 0 < q < d < 1. \tag{6.118}$$

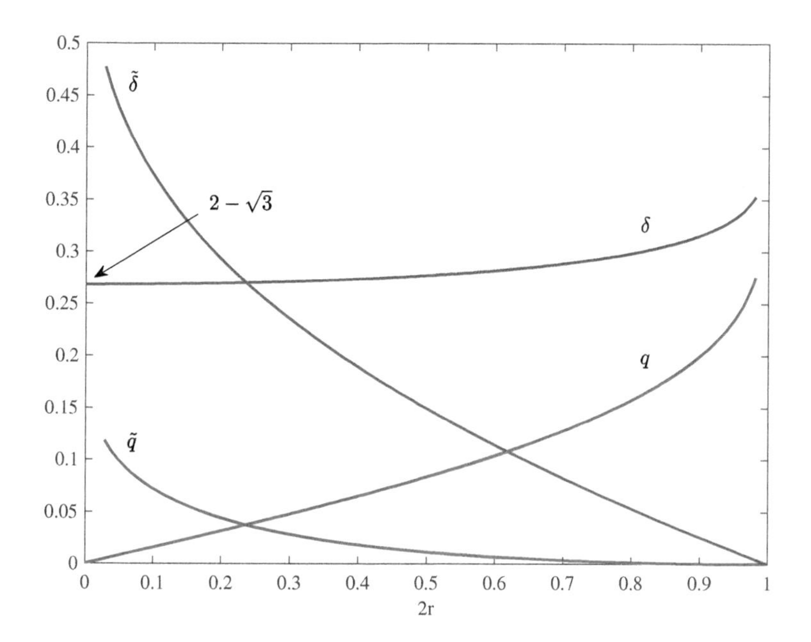

Figure 6.16. *Graphs of the parameters δ, q and $\tilde{\delta}, \tilde{q}$ as functions of r (Exercises 6.12–6.13).*

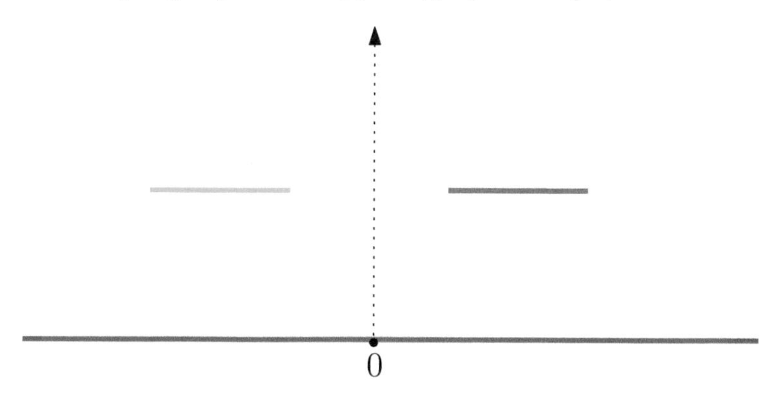

Figure 6.17. *Two flat plates, or slits, above an infinite-length wall and parallel to it (Exercise 6.14).*

Show that the map

$$w = \mathrm{i}\,(\mathcal{K}(z, \mathrm{i}) - \mathcal{K}(-\mathrm{i}, \mathrm{i})) \tag{6.119}$$

transplants D_z to the upper half plane exterior to two equal-length slits parallel to the real w axis (see Figure 6.17).

6.15. **(Vertical stack of three parallel plates)** Suppose (see Figure 6.18) again that D_z is such that both C_1 and C_2 are centered on the real z axis with

$$\delta_1 = -\delta_2 = \delta \in \mathbb{R}, \qquad q_1 = q_2 = q, \qquad 0 < q < d < 1. \tag{6.120}$$

Use the result of Exercise 1.14 to map D_z to the circular domain $\tilde{D}_z$ for which the two interior circles $\tilde{C}_1$ and $\tilde{C}_2$ are

$$|z| = \rho, \qquad |z - D| = Q, \tag{6.121}$$

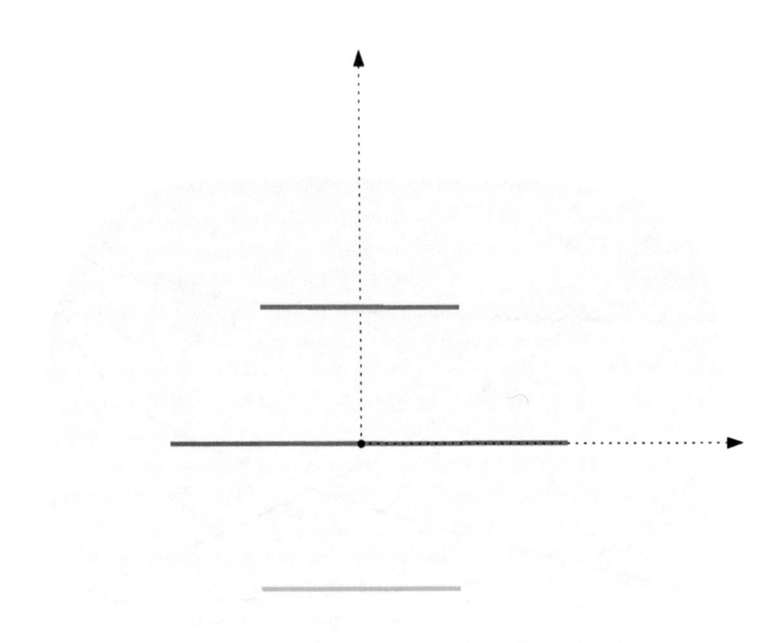

Figure 6.18. *Three vertically stacked flat plates, or slits, with reflectional symmetry about the middle plate (Exercise 6.15).*

where ρ, D, and Q are reported in Exercise 1.14. Show that

$$w = -\frac{\mathrm{i}}{a}\mathcal{K}(z,-a) - \frac{\mathrm{i}}{a}\mathcal{K}(z,-1/a), \tag{6.122}$$

where

$$a = \frac{(1+\delta^2-q^2) - [(1+\delta^2-q^2)^2 - 4\delta^2]^{1/2}}{2\delta}, \tag{6.123}$$

maps $\tilde{D}_z$ to the unbounded region exterior to three flat plates parallel to the real w axis and stacked parallel to the imaginary w axis with the upper and lower slit being of identical length. Write down a condition on q and d such that all plates have the same length.

6.16. **(Another proof of the properties of Cayley-type maps)** Make use of the result (6.93), which is valid for arbitrary nonzero choices of the parameters a and b, and any $M \geq 1$, to show that when both a and b lie on C_k for $k = 1, \ldots, M$, then

$$\overline{R(z;a,b)} = \begin{cases} \dfrac{b-\delta_k}{a-\delta_k} e^{2\pi \mathrm{i}(v_k(a)-v_k(b))} R(z;a,b), & z \in C_0, \\[2ex] \dfrac{b-\delta_k}{a-\delta_k} e^{-2\pi \mathrm{i}(v_j(a)-v_j(b))} e^{2\pi \mathrm{i}(v_k(a)-v_k(b))} R(z;a,b), & z \in C_j, \\ & j = 1, \ldots, M. \end{cases} \tag{6.124}$$

Chapter 7

Schwarz–Christoffel mappings

7.1 ▪ Beyond slit domains

In Chapter 6, we derived explicit formulas, in terms of the relevant prime function, for the conformal mappings from a triply connected circular preimage domain to the unbounded domains exterior to three straight slits, shown in blue, red, and green, in various configurations. Among them are configurations where three slits are in parallel and where all three slits lie on rays emanating from the origin. Of course, these are special configurations. We are left wondering about the case shown in Figure 7.1, where the three slits are in an arbitrary configuration. What can be done in this case to construct the relevant conformal mapping function from a circular preimage domain D_z? Can these mappings similarly be expressed in terms of the prime function?

Since the image curves in Figure 7.1 are all straight lines, it is reasonable to describe the domains exterior to them as polygonal. And the required conformal function is an example of what is known as a *Schwarz–Christoffel (S–C) mapping* to an unbounded triply connected polygonal domain. Formulas for such mappings can indeed be expressed in terms of the prime function. Such formulas were written down for the first time in [25, 21]. The parallel slit and radial slit mappings already encountered are special cases of these more general multiply connected S–C mappings.

7.2 ▪ What is a Schwarz–Christoffel mapping?

A simply connected polygon is a bounded or unbounded domain in the plane, without holes, whose boundaries are a union of straight line segments or, what amounts to the same thing, circular arcs having zero curvature. A classical S–C mapping refers to a conformal mapping from a canonical simply connected domain, such as the unit disc D_z in a parametric z plane, to such a simply connected polygon. The Riemann mapping theorem guarantees the existence of such a mapping, but it does not show how to construct it. Historically, the S–C mapping formula has been an important tool in applications since it provides one of the few constructive algorithms for mapping one region to another. The classical formula dates back to the mid-19th century, but only recently has that formula been generalized to account for polygonal domains with arbitrary connectivity.[20]

[20] The existence of a conformal mapping from a circular preimage to a multiply connected polygon follows from multiply connected generalizations of the Riemann mapping theorem [82].

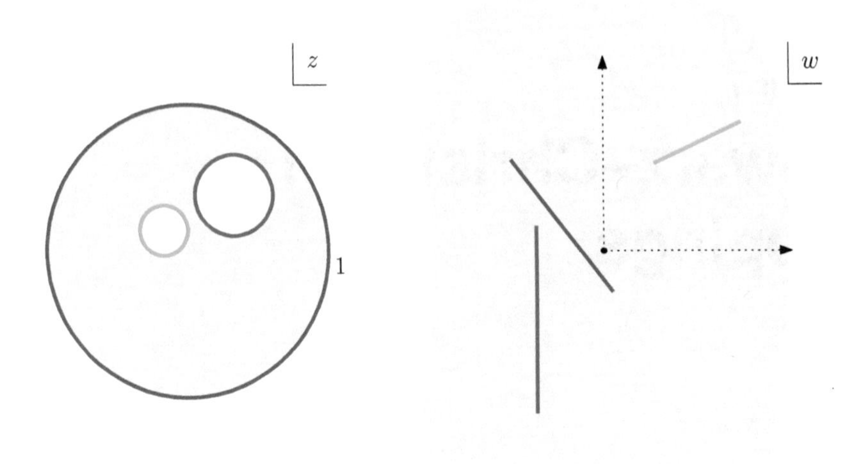

Figure 7.1. *Three slits in an arbitrary configuration. The required mapping is an example of a Schwarz–Christoffel mapping to an unbounded triply connected polygonal domain.*

Suppose the target polygon of interest lies in a complex w plane. If the domain is unbounded, then the point at infinity in the w plane must have a preimage in D_z which we can take—at the price of using up two of three available degrees of freedom—to be at the origin $z = 0$. The classical formula for an S–C map to an arbitrary unbounded n_0-sided polygonal is then given by

$$w = f(z) = A \int^z \prod_{k=1}^{n_0} (z' - a_k^{(0)})^{\beta_k^{(0)}} \frac{dz'}{z'^2} + B, \qquad A, B \in \mathbb{C}, \tag{7.1}$$

where A and B are complex-valued constants, $\{\pi\beta_k^{(0)} | k = 1, \dots, n_0\}$ are the *turning angles* at each vertex, and $\{a_k^{(0)} | k = 1, \dots, n_0\}$ are the prevertices, that is, the preimages of the vertices of the polygon. To use up a single remaining rotational degree of freedom in this formula we can arbitrarily specify the preimage of one of the vertices.

We encountered some examples of S–C mappings in Chapter 1: the parallel slit maps of §1.6 and the radial slit maps of §1.8 are both mappings from the unit disc D_z to unbounded polygonal domains. These mappings were constructed using alternative arguments in Chapter 1, but it is instructive to ask how the general S–C formula (7.1) can be used to retrieve them.

Let us note first that, in Exercise 1.15, it was shown how the parallel slit map of §1.6 and the radial slit map given in §1.8 are just different functional representations of the same mapping: indeed, the parallel slit map was just the "partial fraction" decomposition of the rational function given by the radial slit map. In that same exercise, it was shown that as the preimage b of infinity in D_z is taken to the origin, i.e., as $b \to 0$, both functional forms reduce to the mapping

$$w = f(z) = \frac{1}{z} + z. \tag{7.2}$$

This is the more familiar form of the classical Joukowski map.

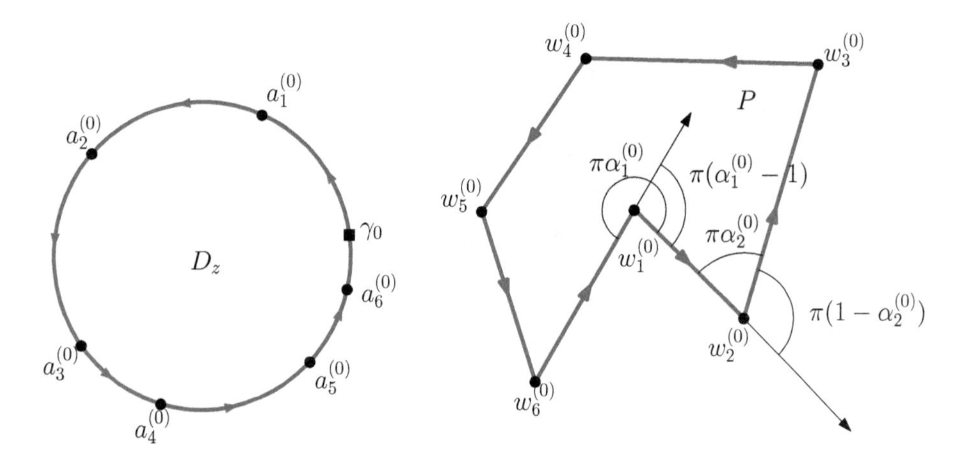

Figure 7.2. *Conformal mapping of the unit disc D_z to a bounded simply connected polygon P. Following the path through the prevertex $a_1^{(0)}$, where the domain is on the left, the trajectory of the image turns to the right through angle $\pi(\alpha_1^{(0)}-1)$.*

Consider now using (7.1) to construct the S–C mapping to the unbounded polygonal domain exterior to a single slit along the real axis and with $z=0$ being the preimage of the point at infinity in the target w plane. Such an unbounded polygon has two vertices, and two sides: the "top side" and the "bottom side" of the slit. Therefore, $n_0=2$. The single remaining degree of freedom in formula (7.1) can be used to set the preimage of one of the edges to be at $z=1$, or

$$a_1^{(0)}=1. \tag{7.3}$$

Since the polygon in the w plane is reflectionally symmetric about the real w axis, and having chosen the preimage of the point at infinity to be at $z=0$, it can be argued on the basis of symmetry that the preimage of the second vertex must be at $z=-1$. Hence,

$$a_2^{(0)}=-1. \tag{7.4}$$

The parameters $\beta_1^{(0)}$ and $\beta_2^{(0)}$ are related to the so-called turning angles at each vertex; this is explained in the next section. For this example these two turning angles are identical and equal to π. This means that

$$\beta_1^{(0)}=\beta_2^{(0)}=1. \tag{7.5}$$

Therefore, (7.1) tells us that the required mapping must be

$$w=f(z)=A\int_{z_0}^{z}(z'^2-1)\frac{dz'}{z'^2}=A\int_{z_0}^{z}\left(1-\frac{1}{z'^2}\right)dz'=A\left(z+\frac{1}{z}\right)+B, \qquad A\in\mathbb{C}, \tag{7.6}$$

where, by choosing $A=1$ and $B=0$, we retrieve (7.2).

We have therefore confirmed that all of our different approaches to the same problem of mapping a disc to the region exterior to a single slit are consistent.

7.3 ▪ Turning angles

Figure 7.2 shows the set-up for the classical S–C mapping problem: how to find a function mapping the unit disc D_z, say, to some simply connected polygon P. Let this S–C map be

$$w=f(z). \tag{7.7}$$

We define the internal angle of a polygon P to be the angle between two sides of a polygon as the sides of the polygon are traversed while keeping the interior of the polygon to the left. Let $w_j^{(0)}$ be a typical vertex of P where j varies between 1 and N, the total number of vertices. Let the preimage of this vertex in a complex z plane be at $z = a_j^{(0)}$. Angles are not preserved by $f(z)$ as z passes through $a_j^{(0)}$. As shown in Figure 7.2, if the internal angle of the vertex at $w_j^{(0)}$ is $\pi\alpha_j^{(0)}$, then, near $a_j^{(0)}$, the derivative function $f'(z)$ must have the local form

$$f'(z) \sim (z - a_j^{(0)})^{\beta_j^{(0)}} g_j^{(0)}(z), \tag{7.8}$$

where $\beta_j^{(0)} = \alpha_j^{(0)} - 1$ and the function $g_j^{(0)}(z)$ is locally analytic at $z = a_j^{(0)}$.

The quantity

$$\pi\beta_j^{(0)} = \pi(\alpha_j^{(0)} - 1) \tag{7.9}$$

is called the *turning angle* at the vertex where $\pi\alpha_j^{(0)}$ is the internal angle of the polygon at that vertex. If the domain is to the left as the boundary of the polygon is traversed, then the turning angle is the angle *to the right* that one must turn at a vertex to keep following the boundary. In Figure 7.2, for example, at vertex $w_1^{(0)}$ the internal angle is greater than π and the deflection angle is $\pi(\alpha_1^{(0)} - 1)$ to the right. This is a positive turning angle. At vertex $w_2^{(0)}$ the internal angle is less than π and the deflection angle is $\pi(1 - \alpha_2^{(0)})$ to the left. However a deflection to the left is equivalent to a deflection through a *negative* angle to the right, and the turning angle at $w_2^{(0)}$ is therefore $\pi(\alpha_2^{(0)} - 1)$. This is a negative turning angle.

Since the internal angles of a bounded polygon sum to 2π, and since making a single anticlockwise traversal of the boundary of a simply connected polygon keeping the interior to the left requires turning to the left through a total angle of 2π, then we must have

$$\sum_{j=1}^{N} \beta_j = -2. \tag{7.10}$$

If there is no corner at the point $z = a$ on C_0, then $f'(z)$ is analytic at a and the corresponding value is $\beta = 0$.

If the point $z = a$ in C_0 is the preimage of the edge of a slit, then the corresponding value is $\beta = 1$ and $f'(z)$ has a simple zero at $z = a$.

7.4 ▪ The classical S–C mapping formula

There are a several different derivations of the classical S–C formula in the literature. Here we give a novel derivation of it based on use of the suite of conformal slit maps introduced in Chapter 1. This approach provides a natural route to generalization to find the corresponding formulas for S–C mappings to multiply connected polygons.

Bounded case: Suppose a bounded polygon P in a complex w plane has polygonal boundary P_0 with n_0 sides, as shown in Figure 7.2, and let $\{w_k^{(0)} | k = 1, \dots, n_0\}$ be its vertices. The kth straight side of polygon P_0 is given by a linear relation of the form

$$\overline{w} = \epsilon_k^{(0)} w + \kappa_k^{(0)}, \tag{7.11}$$

where $\epsilon_k^{(0)}$ and $\kappa_k^{(0)}$ are a set of complex constants and $|\epsilon_k^{(0)}| = 1$. Equation (7.11) holds for $k = 1, \dots, n_0$ corresponding to the n_0 sides. Let the conformal mapping we seek be denoted by

$$w = f(z). \tag{7.12}$$

Let the preimages of the vertices $\{w_k^{(0)} | k = 1, \ldots, n_0\}$ on C_0 be at the prevertices $\{a_k^{(0)} | k = 1, \ldots, n_0\}$ with

$$w_k^{(0)} = f(a_k^{(0)}). \tag{7.13}$$

Let the turning angle parameters at these vertices be denoted by $\{\beta_k^{(0)} | k = 1, \ldots, n_0\}$. As just discussed, a necessary condition on these parameters for a simply connected, bounded polygon is (7.10).

The mapping $f(z)$ has the following properties:

(1) $f(z)$ is an analytic function for $z \in D_z$ and is a one-to-one map between D_z and P. A necessary condition is that $f'(z) \neq 0$ inside D_z.

(2) $f'(z)$ has the following local form at the prevertices:

$$f'(z) = (z - a_k^{(0)})^{\beta_k^{(0)}} g_k^{(0)}(z), \tag{7.14}$$

where $g_k^{(0)}(z)$ is analytic at $a_k^{(0)}$.

(3) The quantity

$$zf'(z) \tag{7.15}$$

must have piecewise-constant argument on C_0. Since $f'(z)$ cannot vanish inside D_z, then $zf'(z)$ has just one simple zero inside D_z at $z = 0$.

Condition (1) is a general property of a one-to-one conformal mapping, while condition (2) derives from the local conditions at the vertices. Condition (3) comes from substitution of (7.12) into (7.11),

$$\overline{w} = \overline{f(z)} = \overline{f}(1/z) = \epsilon_k^{(0)} f(z) + \kappa_k^{(0)}, \tag{7.16}$$

where we have used the fact that $\overline{z} = 1/z$ on C_0. Differentiation with respect to z leads to

$$-\frac{1}{z^2}\overline{f'}(1/z) = \epsilon_k^{(0)} f'(z), \tag{7.17}$$

implying that

$$\frac{zf'(z)}{\overline{zf'(z)}} = -\frac{1}{\epsilon_k^{(0)}}. \tag{7.18}$$

The left-hand side of (7.18) can be related to the argument of the quantity $zf'(z)$; indeed,

$$\frac{zf'(z)}{\overline{zf'(z)}} = e^{2i\arg[zf'(z)]}, \tag{7.19}$$

and the right-hand side of (7.18) is constant. In general it is a different constant for each k. The conclusion is that $zf'(z)$ must have piecewise constant argument on C_0, as stated in condition (3).

The construction of the mapping formula proceeds by using Cayley maps and radial slit maps, both of which have constant argument for z on C_0, as building blocks to construct the quantity $zf'(z)$.

The construction proceeds as follows. Pick an arbitrary point γ_0 on C_0. It is natural to propose that

$$zf'(z) = \left\{ \prod_{k=1}^{n_0} \left(R(z; a_k^{(0)}, \gamma_0) \right)^{\beta_k^{(0)}} \right\} C(z), \tag{7.20}$$

where the functional form of a "correction function" $C(z)$ remains to be found. Since each pair $(a_k^{(0)}, \gamma_0)$ lies on C_0, for each k, $R(z; a_k^{(0)}, \gamma_0)$ is an example of a Cayley map. The function in curly brackets in (7.20) has the correct local behavior at the prevertices $\{a_k^{(0)}\}$, and, being a product of powers of Cayley maps, it has piecewise constant argument on C_0. Unfortunately, it also has an unwanted second order zero at γ_0; this is because of condition (7.10). But we can choose $C(z)$ to get rid of this unwanted zero while retaining the property that $zf'(z)$ has piecewise constant argument on C_0.

To achieve this, let $C(z)$ be the radial slit map

$$C(z) = \frac{z}{\omega(z, \gamma_0)\omega(z, 1/\overline{\gamma_0})} = \frac{z}{\omega(z, \gamma_0)^2}, \tag{7.21}$$

which has a second order pole at γ_0 and a simple zero at $z = 0$. This was a special case of the general class of radial slit mappings introduced in (1.67) of Chapter 1. As a radial slit map, this choice of $C(z)$ has constant argument on C_0. This will ensure that $zf'(z)$ as given in (7.20) will have piecewise constant argument on C_0. This choice of $C(z)$ also ensures that a possible second order zero of $zf'(z)$ at γ_0 has been removed and replaced by a simple zero at $z = 0$. Indeed, with this choice of $C(z)$ the function $zf'(z)$ in (7.20) now has all the desired properties.

It remains to argue that we have in fact produced the required form of $zf'(z)$. To do so consider the ratio

$$\mathcal{R}(z) \equiv zf'(z) \Bigg/ \left\{ \frac{z}{\omega(z, \gamma_0)^2} \prod_{k=1}^{n_0} \left(R(z; a_k^{(0)}, \gamma_0) \right)^{\beta_k^{(0)}} \right\}. \tag{7.22}$$

It is clearly analytic inside D_z and has constant argument on C_0 since any branch points, zeros, or poles of the numerator and denominator cancel out. Since $f'(z)$ must not vanish in D_z, neither does $\mathcal{R}(z)$; we also note that the simple zero of the numerator at $z = 0$ is removed by the same simple zero of the denominator. The function $\log \mathcal{R}(z)$ is therefore analytic in D_z with constant imaginary part on C_0; it must therefore be a constant. It follows that $\mathcal{R}(z)$ itself is constant and hence that

$$zf'(z) = A \frac{z}{\omega(z, \gamma_0)^2} \prod_{k=1}^{n_0} \left(R(z; a_k^{(0)}, \gamma_0) \right)^{\beta_k^{(0)}}, \tag{7.23}$$

or, after simplification,

$$f'(z) = A \prod_{k=1}^{n_0} \omega(z, a_k^{(0)})^{\beta_k^{(0)}}, \tag{7.24}$$

where we have used (7.10). An integration yields

$$f(z) = A \int^z \prod_{k=1}^{n_0} \omega(z', a_k^{(0)})^{\beta_k^{(0)}} dz' + B, \qquad A, B \in \mathbb{C}, \tag{7.25}$$

which is the classical *S–C formula*.

An alternative derivation of this result is explored in Exercise 7.2.

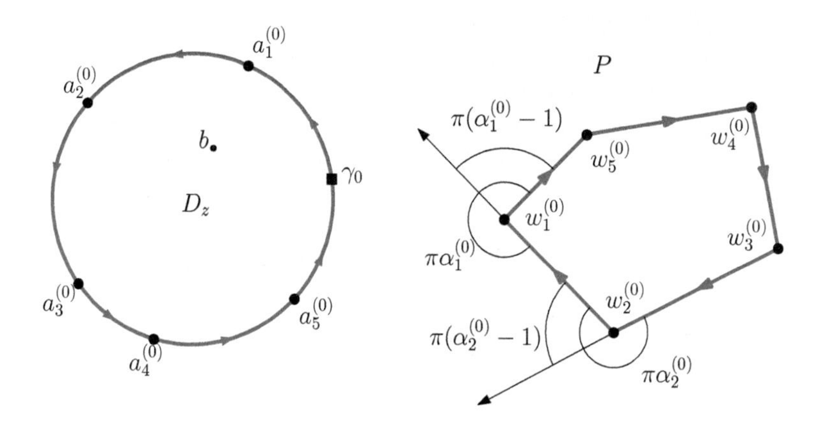

Figure 7.3. *S–C map from D_z to the unbounded simply connected polygon P. The point b inside D_z maps to infinity in the w plane.*

Unbounded case: The construction of the formula for the S–C map to an unbounded polygonal region P requires only minor modifications. Figure 7.3 shows a schematic of this situation. The principal difference in this case is that such a map is not analytic inside D_z but must have the behavior

$$f(z) = \frac{a_\infty}{z-b} + \text{a locally analytic function} \tag{7.26}$$

at some point b strictly inside D_ζ chosen to map to infinity, where a_∞ is some constant. Another difference is that the sum of the turning angles is now $+2$:

$$\sum_{k=1}^{n_0} \beta_k^{(0)} = +2. \tag{7.27}$$

The S–C mapping

$$w = f(z) \tag{7.28}$$

to such an unbounded polygonal region therefore has the following three properties:

(1) It must be an analytic function everywhere inside D_z except for a simple pole at $z = b$:

$$f(z) = \frac{a_\infty}{z-b} + \text{a locally analytic function,} \tag{7.29}$$

where a_∞ is some constant. It is a one-to-one map between D_z and P. A necessary condition is that $f'(z) \neq 0$ in D_z.

(2) At the prevertices $f(z)$ must have the local behavior

$$f'(z) = (z - a_k^{(0)})^{\beta_k^{(0)}} g_k^{(0)}(z), \tag{7.30}$$

where $g_k^{(0)}(z)$ is analytic at $a_k^{(0)}$.

(3) The quantity

$$zf'(z) \tag{7.31}$$

must have piecewise-constant argument on C_0. Since $f'(z)$ cannot vanish in D_z, then $zf'(z)$ has just one simple zero at $z = 0$.

We mimic the construction in the bounded case. Again we pick an arbitrary point γ_0 on C_0 and propose that $zf'(z)$ has the form given in (7.20).

The only change from the analysis of the bounded case is that the form of the required correction function $C(z)$ is now different. The quantity in curly brackets in (7.20) still has the required branch point behavior at the prevertices $\{a_k^{(0)}\}$ and has piecewise constant argument on C_0. But, owing to condition (7.27), it also has an unwanted second order pole at γ_0. More than that, it also fails to have the required double pole at $z = b$ and the required simple zero at $z = 0$.

To fix these deficiencies, we propose the correction function given by

$$C(z) = \left[\frac{\omega(z, \gamma_0)\omega(z, 1/\overline{\gamma_0})}{\omega(z, b)\omega(z, 1/\overline{b})}\right] \frac{z}{\omega(z, b)\omega(z, 1/\overline{b})}, \tag{7.32}$$

which is a product of two radial slit mappings. As such, it has constant argument for $z \in C_0$. It is easy to check that this choice obliterates the unwanted second order pole at γ_0 in (7.20) and adds in the required second order pole at $z = b$ as well as a simple zero at $z = 0$. Indeed, with the choice (7.32) for $C(z)$, the mapping (7.20) has all the properties required of $zf'(z)$.

It follows, by consideration of a ratio akin to (7.22) and following similar arguments as in the bounded case, that, for some constant $A \in \mathbb{C}$,

$$zf'(z) = A \frac{z\omega(z, \gamma_0)^2}{\omega^2(z, b)\omega^2(z, 1/\overline{b})} \prod_{k=1}^{n_0} \left(R(z; a_k^{(0)}, \gamma_0)\right)^{\beta_k^{(0)}}, \qquad A \in \mathbb{C}, \tag{7.33}$$

or, after simplification,

$$f'(z) = \frac{A}{\omega^2(z, b)\omega^2(z, 1/\overline{b})} \prod_{k=1}^{n_0} \omega(z, a_k^{(0)})^{\beta_k^{(0)}}. \tag{7.34}$$

A final integration yields

$$w = f(z) = A \int^z \prod_{k=1}^{n_0} \omega(z', a_k^{(0)})^{\beta_k^{(0)}} \frac{dz'}{\omega^2(z', b)\omega^2(z', 1/\overline{b})} + B, \qquad A, B \in \mathbb{C}. \tag{7.35}$$

This is the classical S–C formula mapping to the unbounded region exterior to a single bounded polygon.

In the special case where we choose $b = 0$, we must take a limit of (7.35) as $b \to 0$ with, in order to give a meaningful limit, the factor A rescaled as $\tilde{A}/\overline{b}^2$, leading to

$$f(z) = \tilde{A} \int^z \prod_{k=1}^{n_0} \omega(z', a_k^{(0)})^{\beta_k^{(0)}} \frac{dz'}{z'^2} + B, \qquad \tilde{A}, B \in \mathbb{C}. \tag{7.36}$$

This is precisely the classical result (7.1) once the substitution $\omega(z, a) = (z - a)$ for the prime function is made.

7.5 ▪ S–C maps to bounded multiply connected polygons

The classical authors did not show how to extend their formulas, just rederived here, to account for multiply connected polygons. However the relevant formula can be found by

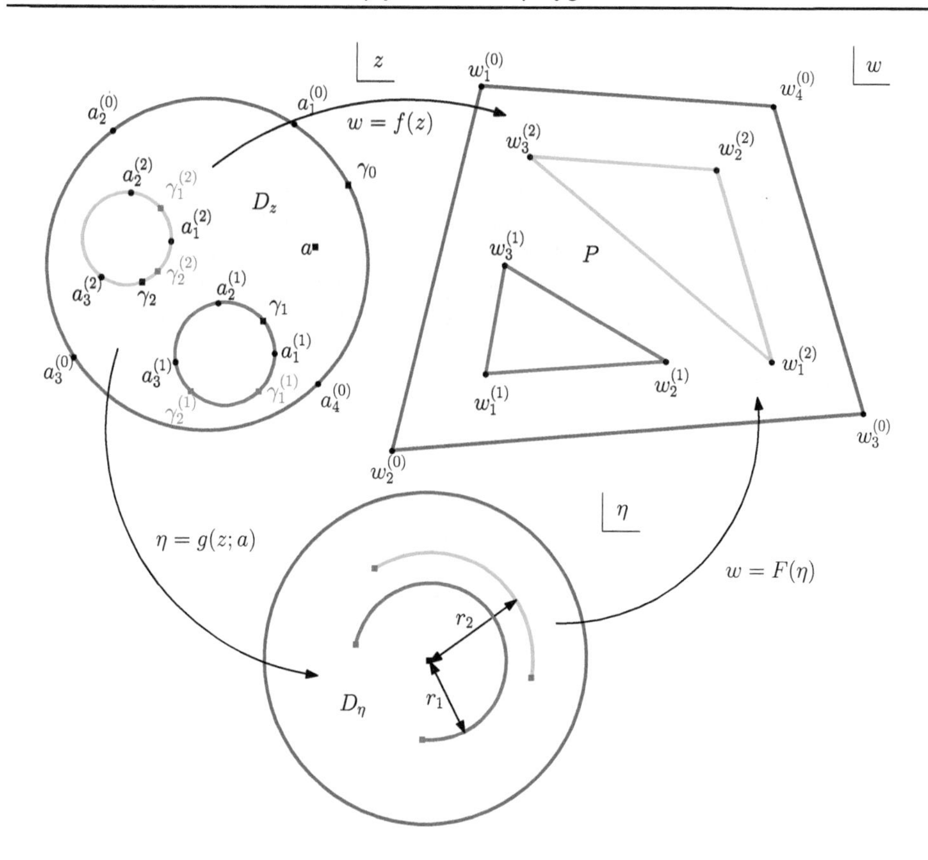

Figure 7.4. *The multiply connected S–C formula can be derived with the aid of an intermediate circular slit mapping that makes all preimage curves concentric circular slits.*

natural extension of the derivation just presented, and by using the generalized formulas for the Cayley-type mappings and radial slit mappings of Chapter 6 in terms of the prime function as basic building blocks. This strategy was first implemented by the author in [25] for the case of bounded multiply connected polygons.

For ease of presentation, we will focus here on a triply connected case. The extension to polygons of any finite connectivity will be obvious.

Suppose we seek the S–C map

$$w = f(z) \tag{7.37}$$

from a triply connected circular domain D_z to the interior of a bounded triply connected polygon P. Let the outer polygonal boundary of P be P_0, and let the two interior polygonal boundaries be P_1 and P_2. The preimages of these boundaries will be denoted by C_0, C_1, and C_2 as usual. Figure 7.4 shows a schematic of this triply connected scenario.

Suppose there are $n_j \geq 0$ vertices on the boundary P_j. Let the corresponding n_j prevertices on each circle C_j be denoted by

$$\{a_k^{(j)} | k = 1, \ldots, n_j \text{ and } j = 0, 1, 2\}, \tag{7.38}$$

and let the turning angle parameters at these vertices be denoted by

$$\{\beta_k^{(j)} | k = 1, \ldots, n_j \text{ and } j = 0, 1, 2\}, \tag{7.39}$$

where

$$\sum_{k=1}^{n_0} \beta_k^{(0)} = -2, \qquad \sum_{k=1}^{n_1} \beta_k^{(1)} = +2, \qquad \sum_{k=1}^{n_2} \beta_k^{(2)} = +2. \tag{7.40}$$

The signs in (7.40) are different for the outer boundary P_0 and the two interior boundaries P_1 and P_2. They are easy to remember, however, since the turning angles for the outer boundary P_0 must behave just as they did for the outer boundary of a bounded simply connected polygon, while the turning angles for the interior boundary of each hole must behave just as they did when mapping to the unbounded region exterior to a single polygonal hole.

On the portion of circle C_j for $j = 0, 1, 2$ which is the preimage of the side of the polygon described by

$$\overline{w} = \epsilon_k^{(j)} w + \kappa_k^{(j)}, \qquad j = 0, 1, 2, \tag{7.41}$$

we have

$$\overline{f}(S_j(z)) = \epsilon_k^{(j)} f(z) + \kappa_k^{(j)}, \qquad j = 0, 1, 2, \tag{7.42}$$

where $S_j(z)$ is the Schwarz function of circle C_j. On differentiation with respect to z we find

$$S_j'(z)\overline{f}'(S_j(z)) = \epsilon_k^{(j)} f'(z). \tag{7.43}$$

Now since

$$S_0(z) = \frac{1}{z}, \qquad S_0'(z) = -\frac{1}{z^2}, \tag{7.44}$$

while, for $j = 1, 2$,

$$S_j(z) = \overline{\delta_j} + \frac{q_j^2}{z - \delta_j}, \qquad S_j'(z) = -\frac{q_j^2}{(z - \delta_j)^2}, \tag{7.45}$$

(7.43) is clearly no longer equivalent to $zf'(z)$ having piecewise constant argument on C_j. For this reason we can no longer directly emulate the construction of the classical S–C formula just presented.

The way around this apparent obstruction is to introduce an intermediate conformal mapping to a circular slit domain [25].

This intermediate mapping is shown in Figure 7.4. Consider mapping the circular region D_z to a bounded circular slit domain D_η by the bounded circular slit mapping

$$\eta = g(z; a) \equiv \frac{\omega(z, a)}{|a|\omega(z, 1/\overline{a})} \tag{7.46}$$

for some arbitrary choice of a inside D_z. Since a is assumed fixed, we consider $g(z; a)$ to be an analytic function of z and use the notation $g'(z; a)$ to denote its derivative with respect to z. Since C_1 and C_2 are each transplanted under this map to bounded circular slits, $g'(z; a)$ must have two simple zeros for each of these two circles. Let $\gamma_1^{(1)}$ and $\gamma_2^{(1)}$ be the preimages on C_1 of the mapping (7.46). Hence,

$$g'(\gamma_1^{(1)}; a) = g'(\gamma_2^{(1)}; a) = 0. \tag{7.47}$$

Similarly, let $\gamma_1^{(2)}$ and $\gamma_2^{(2)}$ be the preimages on C_2 of the mapping (7.46). Hence,

$$g'(\gamma_1^{(2)}; a) = g'(\gamma_2^{(2)}; a) = 0. \tag{7.48}$$

Clearly the values of $\gamma_1^{(j)}$ and $\gamma_2^{(j)}$ for $j = 1, 2$ will depend on the choice of a, and on the geometry of the circular preimage domain D_z. We also suppose that C_1 and C_2 map to circular arcs $\tilde{C}_1$ and $\tilde{C}_2$ of radii r_1 and r_2 in the η plane.

Now consider the composition of functions given by

$$w = f(z) = F(g(z; a)) = F(\eta). \tag{7.49}$$

In this way, $F(\eta)$ is the conformal mapping that transplants D_η to the polygon P. Equation (7.41) now says that

$$\overline{w} = \overline{F(\eta)} = \epsilon_k^{(j)} F(\eta) + \kappa_k^{(j)}, \qquad j = 0, 1, 2, \tag{7.50}$$

or, equivalently,

$$\overline{F}(r_j^2/\eta) = \epsilon_k^{(j)} F(\eta) + \kappa_k^{(j)}, \qquad j = 0, 1, 2, \tag{7.51}$$

where we also define $r_0 = 1$. On differentiation with respect to η we find

$$-\frac{r_j^2}{\eta^2}\overline{F}'(r_j^2/\eta) = \epsilon_k^{(j)} F'(\eta), \qquad j = 0, 1, 2, \tag{7.52}$$

or

$$\frac{\eta F'(\eta)}{\overline{\eta F'(z)}} = -\frac{1}{\epsilon_k^{(j)}}, \qquad j = 0, 1, 2. \tag{7.53}$$

We are using $F'(\eta)$ to denote the derivative of $F(\eta)$ with respect to its argument. This set of conditions is similar to those encountered earlier in the derivation of the classical simply connected S–C formula, and it tells us that the quantity $\eta F'(\eta)$ must have piecewise constant argument on all boundaries of D_η and, hence, also on all boundaries of the original circular domain D_z.

By the chain rule we have

$$\eta F'(\eta) = \frac{g(z; a)}{g'(z; a)} \frac{df}{dz}. \tag{7.54}$$

From this it is clear that, when viewed as a function of z, $\eta F'(\eta)$ will inherit all the same singularities at the prevertices as df/dz. However, it will also have a simple zero at $z = a$, from the appearance of $\eta = g(z; a)$ in formula (7.54), and four simple poles, one at $\gamma_1^{(j)}$ and another at $\gamma_2^{(j)}$ on C_j for each $j = 1$ and $j = 2$. These poles are associated with the two zeros of $g'(z; a)$ on each C_j corresponding to the edges of the circular slit images in D_η.

The function $F(\eta)$ therefore has the following properties:

(1) $F(\eta)$ must be an analytic function everywhere inside D_z. In order that the mapping is one-to-one, $F'(\eta)$ must not vanish inside D_z.

(2) Near $z = a_k^{(j)}$ the function $\eta(z)F'(\eta)$ has the local behavior

$$\eta F'(\eta) = (z - a_k^{(j)})^{\beta_k^{(j)}} g_k^{(j)}(z), \tag{7.55}$$

where $g_k^{(j)}(z)$ is locally analytic at $a_k^{(j)}$ and the parameters $\{\beta_k^{(j)}\}$ satisfy (7.40).

(3) $\eta F'(\eta)$ must have a simple zero at $z = a$ and simple poles at $\gamma_1^{(j)}$ and $\gamma_2^{(j)}$ for $j = 1, 2$.

(4) $\eta F'(\eta)$ must have piecewise-constant argument on $\{C_j | j = 0, 1, 2\}$.

Now we are in a position to emulate the construction of the classical S–C formula introduced earlier. In what follows it is the function $\eta F'(\eta)$ that will be constructed using ratios of prime functions as building blocks.

First, pick arbitrary points γ_0 on C_0, γ_1 on C_1, and γ_2 on C_2. It is natural to propose that $\eta F'(\eta)$ has the form

$$\eta F'(\eta) = \left\{ \prod_{k=1}^{n_0} \left(R(z;a_k^{(0)},\gamma_0)\right)^{\beta_k^{(0)}} \prod_{k=1}^{n_1} \left(R(z;a_k^{(1)},\gamma_1)\right)^{\beta_k^{(1)}} \prod_{k=1}^{n_2} \left(R(z;a_k^{(2)},\gamma_2)\right)^{\beta_k^{(2)}} \right\} C(z), \tag{7.56}$$

where, as before, the correction function $C(z)$ remains to be determined. Since any pair $(a_k^{(0)},\gamma_0)$ lies on C_0, any pair $(a_k^{(1)},\gamma_1)$ lies on C_1, and any pair $(a_k^{(2)},\gamma_2)$ lies on C_2, the term in curly brackets is a product of powers of Cayley-type mappings. These are known from Chapter 6 to have constant argument on all the boundaries of D_z. It follows that the product of powers of such Cayley-type mappings as given in curly brackets in (7.56) will therefore have piecewise constant argument on all the boundaries of D_z. Moreover the choice of term in curly brackets ensures that $\eta F'(\eta)$ has the correct local form (7.55) at the prevertices $\{a_k^{(j)}|k=1,\ldots,n_j, j=0,1,2\}$. Unfortunately the curly-bracketed term also has an unwanted second order zero at γ_0 and unwanted second order poles at both γ_1 and γ_2.

As before we find the correction function $C(z)$ that gets rid of these unwanted features, and also adds in needed features, all the while retaining the property that $\eta F'(\eta)$ has piecewise constant argument on the boundaries of D_z.

Consider the choice

$$\begin{aligned} C(z) = &\left[R(z;\gamma_1,\gamma_1^{(1)})R(z;\gamma_1,\gamma_2^{(1)})R(z;\gamma_2,\gamma_1^{(2)})R(z;\gamma_1,\gamma_2^{(2)})\right] \\ &\times R(z;a,\gamma_0)R(z;1/\overline{a},1/\overline{\gamma_0}). \end{aligned} \tag{7.57}$$

The square brackets contain a product of four Cayley-type mappings; the term outside the brackets is a radial slit mapping. Thus this choice of $C(z)$ has constant argument on all the boundaries of D_z. By virtue of the term in square brackets this choice removes the unwanted second order pole at γ_1 and puts simple poles at $\gamma_1^{(1)}$ and $\gamma_2^{(1)}$; similarly, the unwanted second order pole at γ_2 is removed in favor of simple poles at $\gamma_1^{(2)}$ and $\gamma_2^{(2)}$. The radial slit mapping sitting outside the square brackets in (7.57) removes the unwanted second order zero at γ_0 and adds in simple zeros at a and $1/\overline{a}$.

The resulting function, which has all the properties required of $\eta F'(\eta)$, simplifies to

$$\begin{aligned} &\frac{\omega(z,a)\omega(z,1/\overline{a})}{\omega(z,\gamma_1^{(1)})\omega(z,\gamma_2^{(1)})\omega(z,\gamma_1^{(2)})\omega(z,\gamma_2^{(2)})} \\ &\times \prod_{k=1}^{n_0}[\omega(z,a_k^{(0)})]^{\beta_k^{(0)}} \prod_{k=1}^{n_1}[\omega(z,a_k^{(1)})]^{\beta_k^{(1)}} \prod_{k=1}^{n_2}[\omega(z,a_k^{(2)})]^{\beta_k^{(2)}}. \end{aligned} \tag{7.58}$$

By construction this function has piecewise constant argument on C_0, C_1, and C_2, and it has the required simple zero at $z=a$ and simple poles at $\gamma_1^{(j)}$ and $\gamma_2^{(j)}$ for $j=1,2$. It can be argued that the function must be proportional to $\eta F'(\eta)$. Introducing the ratio of $\eta F'(\eta)$ and the function (7.58) we see that it is analytic, and single-valued, everywhere in D_z and has no zeros there. It also has constant argument on all boundaries of D_z. The logarithm of this ratio is therefore analytic and single-valued in D_z with constant imaginary part on

C_0, C_1, and C_2. It must therefore be constant. Now since

$$\frac{g(z,a)}{g'(z,a)} = \frac{z}{K(z,a) - K(z,1/\overline{a})}, \tag{7.59}$$

where $K(.,.)$ was introduced in §6.12, setting (7.54) proportional to (7.58) leads to

$$\frac{df}{dz} = A\, S(z,a) \prod_{j=0}^{2} \prod_{k=1}^{n_j} [\omega(z, a_k^{(j)})]^{\beta_k^{(j)}}, \tag{7.60}$$

where

$$S(z,a) \equiv \frac{\omega(z,a)\omega(z,1/\overline{a})}{z\omega(z,\gamma_1^{(1)})\omega(z,\gamma_2^{(1)})\omega(z,\gamma_1^{(2)})\omega(z,\gamma_2^{(2)})} [K(z,a) - K(z,1/\overline{a})] \tag{7.61}$$

for some constant $A \in \mathbb{C}$. A final integration leads to

$$w = f(z) = A \int^{z} S(z',a) \prod_{j=0}^{2} \prod_{k=1}^{n_j} [\omega(z', a_k^{(j)})]^{\beta_k^{(j)}} dz' + B, \tag{7.62}$$

where $A, B \in \mathbb{C}$.

Formula (7.62) is the general formula for an S–C mapping to a bounded triply connected polygon.

Remark: It is remarkable just how many of the basic functions introduced in Chapter 6 come into play here in the construction of the S–C formula (7.62). We used the Cayley-type mappings of §6.10, the radial slit mappings of §6.11, and the bounded circular slit mappings of §6.6 to arrive at the final result (7.62). This underscores the central lesson of Chapters 1 and 5–6: having introduced the prime function, it is useful to equip oneself with a suite of other useful basic functions built from it that can then be strategically deployed to solve more complicated problems.

The derivation above has been given for the triply connected case $M = 2$. It should be clear that all of the steps can be repeated, with an identical line of reasoning, for any $M \geq 1$. The general formula for the S–C mapping to an $M+1$ connected polygon P from a conformally equivalent circular preimage domain D_z is given by

$$w = f(z) = A \int^{z} S(z',a) \prod_{j=0}^{M} \prod_{k=1}^{n_j} [\omega(z', a_k^{(j)})]^{\beta_k^{(j)}} dz' + B, \qquad A, B \in \mathbb{C}, \tag{7.63}$$

where $\omega(.,.)$ is the prime function associated with D_z and

$$S(z,a) \equiv \frac{\omega(z,a)\omega(z,1/\overline{a})}{z \prod_{j=1}^{M} \omega(z,\gamma_1^{(j)})\omega(z,\gamma_2^{(j)})} [K(z,a) - K(z,1/\overline{a})]. \tag{7.64}$$

In this formula there are $M+1$ polygonal sides. The polygon that is the image of circle C_j, for $j = 0, 1, \ldots, M$, has n_j vertices at locations

$$w_k^{(j)} = f(a_k^{(j)}), \qquad k = 1, \ldots, n_j, \tag{7.65}$$

with corresponding turning angles $\pi\beta_k^{(j)}$ for $k = 1, \ldots, n_j$.

Expressions (7.63)–(7.64) for the multiply connected S–C formula in terms of the prime function were first derived by the author in [25].

In the special case of simply connected domains where $M = 0$, formula (7.63) holds with

$$S(z, a) = 1. \tag{7.66}$$

This retrieves the classical S–C formula (7.25).

In the special case of doubly connected domains from a concentric annulus $\rho < |z| < 1$ where $M = 1$, formula (7.63) holds, and the function $S(z, a)$ simplifies to

$$S(z, a) = \frac{C}{z^2}, \qquad C \in \mathbb{C}. \tag{7.67}$$

This simplification of $S(z, a)$ in this case is the topic of Exercise 8.7.[21]

7.6 ▪ S–C maps to unbounded multiply connected domains

The analogous formula to an unbounded $(M + 1)$-connected polygonal region can be constructed in a similar way with the final result

$$w = f(z) = A \int^z S_\infty(z', a, b) \prod_{j=0}^{M} \prod_{k=1}^{n_j} [\omega(z', a_k^{(j)})]^{\beta_k^{(j)}} dz' + B, \tag{7.68}$$

where $A, B \in \mathbb{C}$ and

$$S_\infty(z, a, b) \equiv \frac{S(z, a)}{\omega(z, b)^2 \omega(z, 1/\overline{b})^2}, \tag{7.69}$$

where $S(z, a)$ is defined in (7.64) and where the point b is taken to be the preimage of the point at infinity in the w plane.

The derivation of (7.68) is the topic of Exercise 7.7. The formula was first derived by the author in [21].[22]

7.7 ▪ S–C maps to periodic domains

The theory of S–C mappings can be generalized in several other directions. Consider the four conformal mappings from the concentric annulus

$$\rho < |z| < 1 \tag{7.70}$$

given by

$$w = f_1(z) = \frac{1}{2\pi \mathrm{i}} \log z, \tag{7.71}$$

$$w = f_2(z) = \frac{1}{2\pi \mathrm{i}} \log \left[\frac{\omega(z, a_\infty)}{|a_\infty| \omega(z, 1/\overline{a_\infty})} \right], \tag{7.72}$$

$$w = f_3(z) = \frac{1}{2\pi \mathrm{i}} \log \left[\frac{\omega(z, a_\infty) \omega(z, 1/\overline{a_\infty})}{\omega(z, \overline{a_\infty}) \omega(z, 1/a_\infty)} \right], \tag{7.73}$$

$$w = f_4(z) = \frac{1}{2\pi \mathrm{i}} \log \left[\frac{\omega(z, a_\infty) \omega(z, 1/a_\infty)}{\omega(z, \overline{a_\infty}) \omega(z, 1/\overline{a_\infty})} \right]. \tag{7.74}$$

[21] See [25] for a discussion of other derivations of the doubly connected S–C formula by other authors.

[22] Since this original work, other authors have explored using alternative choices of the intermediate slit mapping [123]. Before [21] was published other authors had tackled the same problem using different ideas, and without using the prime function [77]. DeLillo [78] has since explored the connection between these two different approaches.

All of these were encountered in Chapter 5. The mapping $f_1(z)$ is elementary; it is precisely the function $v_1(z)$. It transplants the annulus to a rectangle as shown in Figure 5.6. The mapping $f_2(z)$ in (7.72) is the function $\mathcal{G}_0(z, a_\infty)$ studied in §5.5 and whose image is shown in Figure 5.2. The function $f_3(z)$ is the logarithm of the unbounded circular slit mapping as given in (5.41) and shown in Figure 5.5; similarly, $f_4(z)$ is the logarithm of the unbounded radial slit mapping as given in (5.73) and shown in Figure 5.11. All four are explicit examples of what we refer to as S–C maps to singly periodic domains. Under these mappings a circular preimage domain D_z is transplanted to a period window of a singly periodic target domain having boundaries that are straight line segments.

There are three different cases of such mappings, which we will call cases 1, 2, and 3, as depicted, in the case $M = 2$, in Figure 7.5. All three cases refer to domains with straight line boundaries that are periodic in the direction of the real axis. The turning angle parameters in each case satisfy the following conditions:

$$\textbf{Case 1}: \sum_{k=1}^{n_j} \beta_k^{(j)} = +2, \qquad j = 0, 1, \ldots, M. \tag{7.75}$$

$$\textbf{Case 2}: \sum_{k=1}^{n_0} \beta_k^{(0)} = 0, \sum_{k=1}^{n_j} \beta_k^{(j)} = +2, \qquad j = 1, \ldots, M. \tag{7.76}$$

$$\textbf{Case 3}: \sum_{k=1}^{n_0} \beta_k^{(0)} = 0, \sum_{k=1}^{n_1} \beta_k^{(1)} = 0, \sum_{k=1}^{n_j} \beta_k^{(j)} = +2, \qquad j = 2, \ldots, M. \tag{7.77}$$

In case 1, the period window is infinite as $y \to \pm\infty$; in case 2, it is only infinite in one direction, either $y \to +\infty$ or $y \to -\infty$; in case 3, the period window is bounded. There is an associated form of the S–C mapping for each case. The following results are drawn from the original work in [2].

Case 1: In case 1, the relevant S–C formula to the unbounded period window exterior to a finite collection of $M + 1$ polygonal objects is

$$w = f(z) = A \int^z S_{\text{per}}^{(1)}(z', a, a_{\infty^+}, a_{\infty^-}) \prod_{j=0}^{M} \prod_{k=1}^{n_j} [\omega(z', a_k^{(j)})]^{\beta_k^{(j)}} dz' + B, \tag{7.78}$$

where $A, B \in \mathbb{C}$ and

$$S_{\text{per}}^{(1)}(z, a, a_{\infty^+}, a_{\infty^-}) \equiv \frac{S(z,a)}{\omega(z, a_{\infty^+})\omega(z, 1/\overline{a_{\infty^+}})\omega(z, a_{\infty^-})\omega(z, 1/\overline{a_{\infty^-}})}, \tag{7.79}$$

and where, for a period window extending to $y \to \pm\infty$, the point a_{∞^+} is the preimage inside D_ζ of the part of the period window where $y \to +\infty$ and the point a_{∞^-} is the preimage inside D_ζ of the part of the period window where $y \to -\infty$. The derivation of (7.78)–(7.79) is the topic of Exercise 7.8. It is a natural extension of the derivation detailed in §7.5.

Case 2: In case 2, the relevant S–C formula to the unbounded period window extending to $y \to +\infty$ (or $y \to -\infty$) exterior to a finite collection of M polygonal objects is

$$w = f(z) = A \int^z S_{\text{per}}^{(2)}(z', a, a_{\infty^+}) \prod_{j=0}^{M} \prod_{k=1}^{n_j} [\omega(z', a_k^{(j)})]^{\beta_k^{(j)}} dz' + B, \tag{7.80}$$

where $A, B \in \mathbb{C}$,

$$S_{\text{per}}^{(2)}(z, a, a_{\infty^+}) \equiv \frac{S(z,a)}{\omega(z, a_{\infty^+})\omega(z, 1/\overline{a_{\infty^+}})}, \tag{7.81}$$

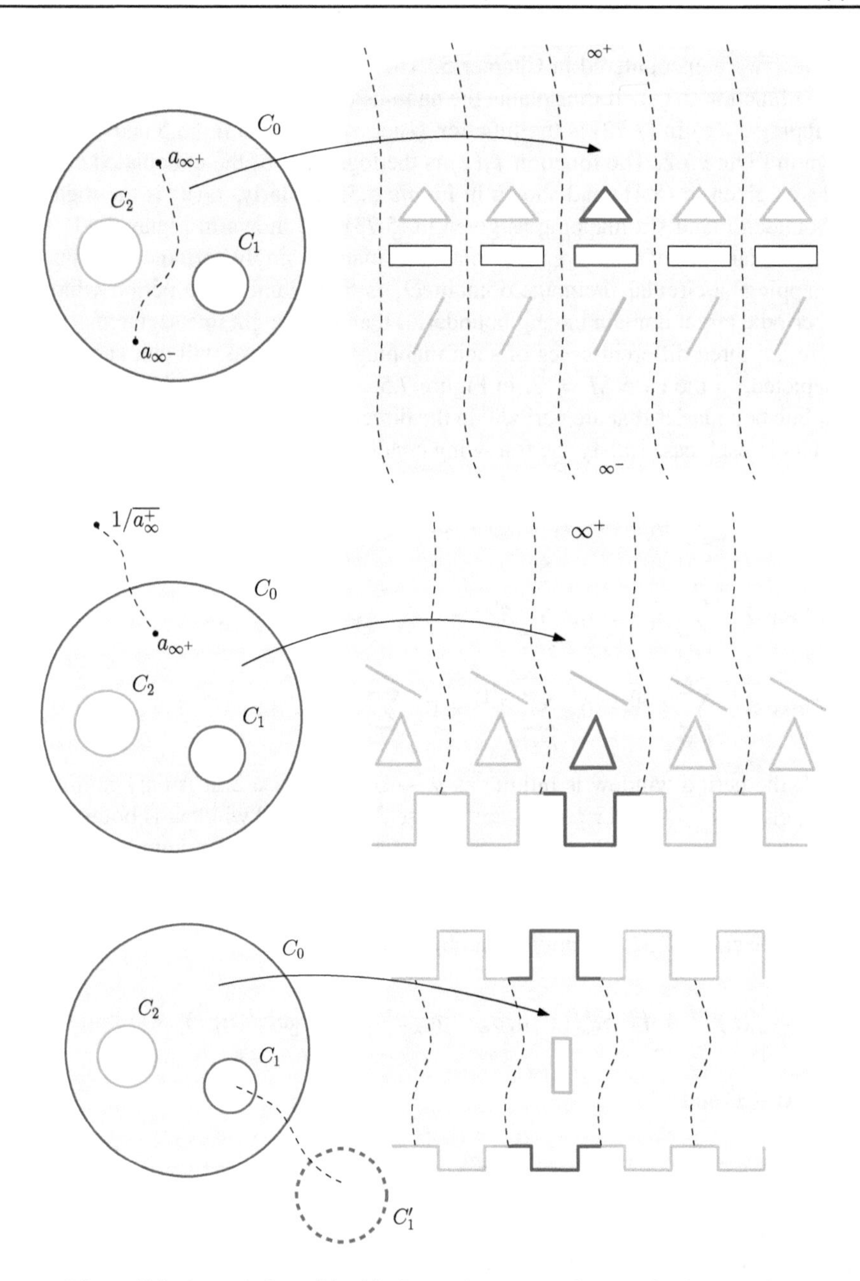

Figure 7.5. *Cases* 1, 2, *and* 3 *of S–C mappings to singly periodic domains from a circular preimage* D_z *to a single period window having boundaries that are straight line segments. The period window can be unbounded as* $y \to \pm\infty$ *(case* 1*), unbounded in one direction, say* $y \to \infty$ *(case* 2*), or bounded (case* 3*).*

and, for a period window extending to $y \to \pm\infty$, the point a_{∞^+} is the preimage inside D_ζ of the part of the period window where $y \to +\infty$. The derivation of (7.80)–(7.81) is the topic of Exercise 7.9.

Case 3: In case 3, the relevant S–C formula to the bounded period window exterior to a finite collection of $M-1$ polygonal objects, and assuming that the images of both C_0

and C_1 span the period window, is

$$w = f(z) = A \int^z S_{\rm per}^{(3)}(z', a) \prod_{j=0}^{M} \prod_{k=1}^{n_j} [\omega(z', a_k^{(j)})]^{\beta_k^{(j)}} dz' + B, \tag{7.82}$$

where $A, B \in \mathbb{C}$ and

$$S_{\rm per}^{(3)}(z, a) \equiv S(z, a) \frac{\omega(z, \gamma_1^{(1)})}{\overline{\omega(z, 1/\overline{\gamma_1^{(1)}})}}. \tag{7.83}$$

The derivation of (7.82)–(7.83) is the topic of Exercise 7.10. There is an alternative way to write $S_{\rm per}^{(3)}(z, a)$ which is explored in Exercise 7.11.

7.8 ▪ The parameter problem

A feature of all of the S–C mapping formulas derived in this chapter is that their functional forms can be written down explicitly in terms of the prime function with just a finite set of parameters left to be determined. These are often called *accessory parameters.* It is worth checking that the number of accessory parameters in the mapping function equals the number of degrees of freedom needed to specify the target domain.

Let us check the parameter count for the S–C formulas (7.63)–(7.64) from a circular preimage D_z to a conformally equivalent bounded $(M + 1)$-connected polygon P.

First note that the parameter a is arbitrary and, for a given set of conformal moduli $\{q_j, \delta_j | j = 1, \ldots, M\}$, the parameters $\{\gamma_1^{(j)}, \gamma_2^{(j)} | j = 1, \ldots, M\}$ can be directly computed from the chosen value of a; the latter $2M$ values are slaved to the choice of a and the conformal moduli and should not therefore be considered to be independent free parameters. Furthermore, the final mapping formulas can be shown to be independent of the choice of a. We therefore discount both a and the set $\{\gamma_1^{(j)}, \gamma_2^{(j)} | j = 1, \ldots, M\}$ when tallying the number of degrees of freedom in the mapping. It is still necessary in practice, however, to solve for the set $\{\gamma_1^{(j)}, \gamma_2^{(j)} | j = 1, \ldots, M\}$ as functions of a and $\{q_j, \delta_j | j = 1, \ldots, M\}$ in any given problem, as will be seen in §7.9.

The complex-valued parameters A and B represent a total of 4 real degrees of freedom; the conformal moduli $\{q_j, \delta_j | j = 1, \ldots, M\}$ provide a further $3M$ real degrees of freedom; on any circle C_j for $j = 0, 1, \ldots, M$, there are n_j real degrees of freedom associated with the angular location of the n_j prevertices on that circle C_j, giving a total of $N \equiv \sum_{j=0}^{M} n_j$ real degrees of freedom. At the same time, we must remove three real degrees of freedom associated with an automorphism of the unit disc; these are the usual 3 real degrees of freedom associated with the Riemann mapping theorem. This produces a total count of $4 + 3M + N - 3 = 3M + N + 1$ real parameters.

On the other hand, the constraints on the mapping dictated by the geometry of the target are as follows. For each of the $M + 1$ polygonal boundaries we must specify the location of one of the vertices and the direction from that vertex of one of the straight line edges, making a total of $3(M + 1)$ real conditions. Once the direction of an edge is given, the mapping formula automatically ensures that the mapping "turns" through the correct angle at each vertex but it does not guarantee that the length of each side is as required; the n_j real parameters associated with the angular locations of prevertices $a_k^{(j)}$ on C_j must be chosen to enforce that the side lengths are correct. This provides a total of N real constraints. Not all of these conditions are independent because, by Cauchy's theorem, the mapping

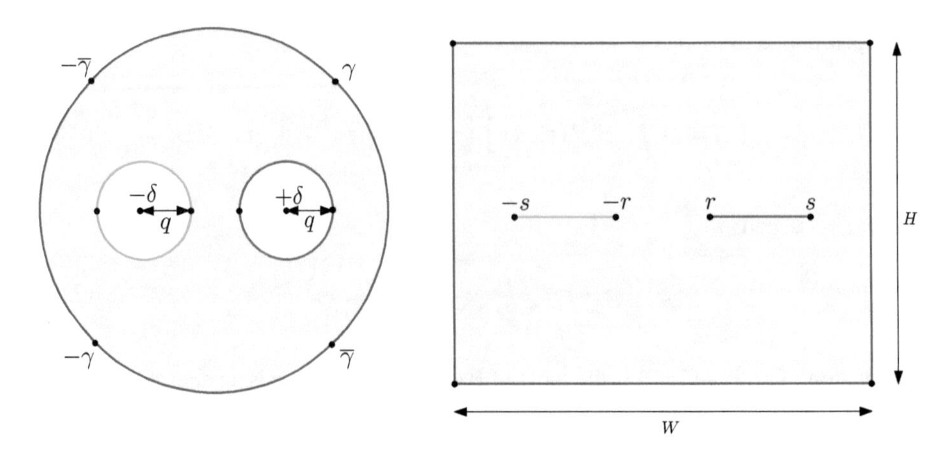

Figure 7.6. *S–C map from a triply connected circular domain to a rectangle of width W and height H with two excised slits $[-s, -r]$ and $[r, s]$ along its centerline.*

automatically satisfies the single complex equation

$$\int_{\partial P} df = 0 \tag{7.84}$$

where integration is around all the boundary components of the domain, keeping P to the left. This removes the need to impose two of the real conditions just discussed. In total, we therefore have $3(M + 1) + N - 2 = 3M + N + 1$ real conditions, which is consistent with the number of free parameters.

7.9 ▪ Illustrative examples

Geometrical symmetries can be exploited to reduce the number of accessory parameters to be determined. By way of example, and to illustrate how the finite set of parameters in the mapping function is determined, let us construct the map from a circular preimage domain D_z to the triply connected polygon comprising the interior of a rectangle centered at the origin and with two symmetrically disposed slits, $[-s, -r]$ and $[r, s]$, along its centerline, as shown in Figure 7.6. It is natural to choose the preimage domain D_z to have the same symmetries as the target domain; we therefore pick the circles C_1 and C_2 such that

$$\delta_1 = -\delta_2 = \delta, \qquad q_2 = q_1 = q \tag{7.85}$$

for some real parameters $\delta, q > 0$. On grounds of symmetry we can therefore argue that the derivative of the map has the form

$$f'(z) = AS(z, a)\frac{\omega(z, \delta + q)\omega(z, \delta - q)\omega(z, -\delta + q)\omega(z, -\delta - q)}{\omega(z, \gamma)^{1/2}\omega(z, \overline{\gamma})^{1/2}\omega(z, -\gamma)^{1/2}\omega(z, -\overline{\gamma})^{1/2}}, \tag{7.86}$$

where γ is the preimage in the first quadrant of the z plane of the top right-hand vertex; the three other prevertices have been placed at $\overline{\gamma}$, $-\overline{\gamma}$, and $-\gamma$. The preimages of the two edges of the internal slits, which are simple zeros of $f'(z)$, are taken to be at the four points

$$\pm\delta \pm q. \tag{7.87}$$

In this construction it is natural to make the choice $a = 0$ in the intermediate slit map (7.46):

$$\eta = g(z,0) = \frac{\omega(z,0)}{\tilde{\omega}(z,\infty)}, \tag{7.88}$$

where the function $\tilde{\omega}(z,a)$ is defined via the relation

$$\omega(z,a) = (z-a)\tilde{\omega}(z,a). \tag{7.89}$$

The function $S(z,a) = S(z,0)$ therefore takes the form

$$S(z,0) = \frac{\tilde{\omega}(z,\infty)\frac{\partial \omega}{\partial z}(z,0) - \omega(z,0)\frac{\partial \tilde{\omega}}{\partial z}(z,\infty)}{\omega(z,\lambda)\omega(z,\overline{\lambda})\omega(z,-\overline{\lambda})\omega(z,-\lambda)}, \tag{7.90}$$

where λ is a point on C_1 satisfying

$$\frac{\partial g(\lambda,0)}{\partial z} = 0. \tag{7.91}$$

We know that $\partial g/\partial z$ has two simple zeros on C_1, and these will be symmetrically disposed with respect to the real axis. Let us denote these points by

$$\lambda = \delta + qe^{\pm i\theta_\lambda}. \tag{7.92}$$

(7.91) is an equation to determine the real parameter θ_λ. On integration, the mapping is given by

$$f(z) = \int_0^z AS(z',0)\frac{\omega(z',\delta+q)\omega(z',\delta-q)\omega(z',-\delta+q)\omega(z',-\delta-q)}{\omega(z',\gamma)^{1/2}\omega(z',\overline{\gamma})^{1/2}\omega(z',-\gamma)^{1/2}\omega(z',-\overline{\gamma})^{1/2}}dz', \tag{7.93}$$

where we have picked a constant of integration to ensure that $f(0) = 0$.

The derivative of the mapping (7.93) depends on the following five real parameters:

$$A,\ \delta,\ q,\ \theta_\lambda,\ \arg[\gamma], \tag{7.94}$$

which have yet to be determined. The parameter count here is clear: these five parameters must be found to ensure that (7.91) is satisfied, that the height and width of the outer rectangle are H and W, respectively, and finally that the edges of the two slits are at $[-s,-r]$ and $[r,s]$. By the symmetry of the configuration, this is a total of five real conditions to determine the five parameters (7.94).

Figure 7.7 shows an orthogonal grid constructed using such an S–C mapping. Contour lines of $\mathrm{Re}[g(z,0)]$ and $\mathrm{Im}[g(z,0)]$ are shown.

In a second case study we replace the two slits in the previous example by two holes taking the shape of equilateral triangles with centroids symmetrically disposed on the centerline of the rectangle and such that there is reflectional symmetry about perpendicular axes parallel to the sides of the rectangle. Figure 7.8 shows a schematic.[23]

This target polygon has the same geometrical symmetries as the first example, so, as before, we seek a preimage domain that shares these symmetries. It can be argued that the derivative of the map has the form

$$f'(z) = AS(z,0)\frac{[\omega(z,\delta-q)\omega(z,-\delta+q)\omega(z,\chi)\omega(z,\overline{\chi})\omega(z,-\chi)\omega(z,-\overline{\chi})]^{2/3}}{\omega(z,\gamma)^{1/2}\omega(z,\overline{\gamma})^{1/2}\omega(z,-\gamma)^{1/2}\omega(z,-\overline{\gamma})^{1/2}}, \tag{7.95}$$

[23]This geometry is Example 7 of [68] on the computation of the capacity of planar condensers.

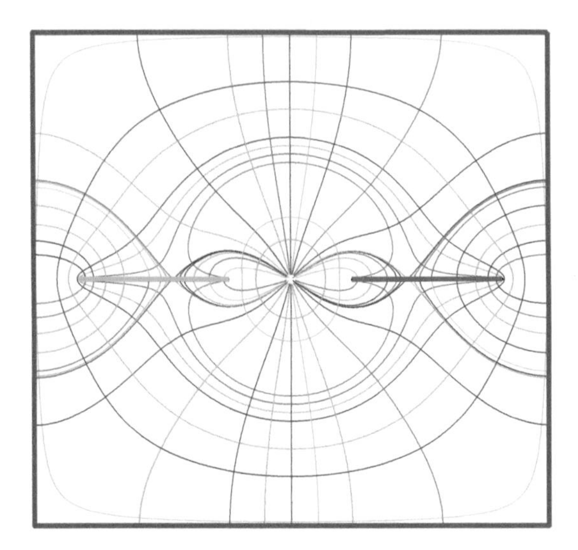

Figure 7.7. *Orthogonal grid in a rectangular box with two plates along the centerline.*

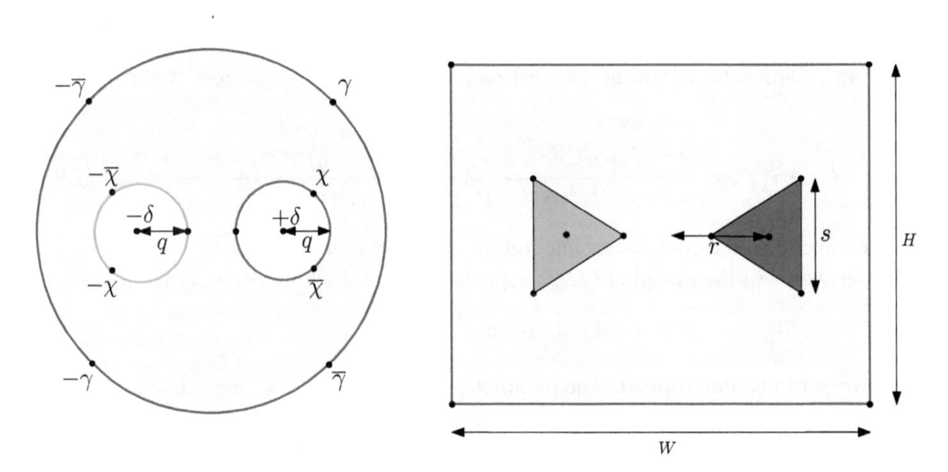

Figure 7.8. *S–C map from a triply connected circular domain to a rectangle of width W and height H with two equilateral triangles along its centerline.*

where A is a real scaling parameter; γ, which is on C_0, is the preimage in the first quadrant of the z plane of the top right-hand vertex; and χ, which is on C_1, is the preimage in the first quadrant of the z plane of the top right-hand corner of one of the triangular holes. We can write

$$\chi = \delta + qe^{\pm i\theta_\chi} \tag{7.96}$$

for some θ_χ. The function $S(z, 0)$ is again given by (7.90). The exponent $2/3$ in the product of prime functions in the numerator ensures that the interior angles at each vertex of the two triangular holes are all $\pi/3$.

A count of the real parameters in the derivative of the mapping (7.95) shows a total of six, namely,

$$A, \ \ \delta, \ \ q, \ \ \theta_\lambda, \ \ \arg[\gamma], \ \ \theta_\chi. \tag{7.97}$$

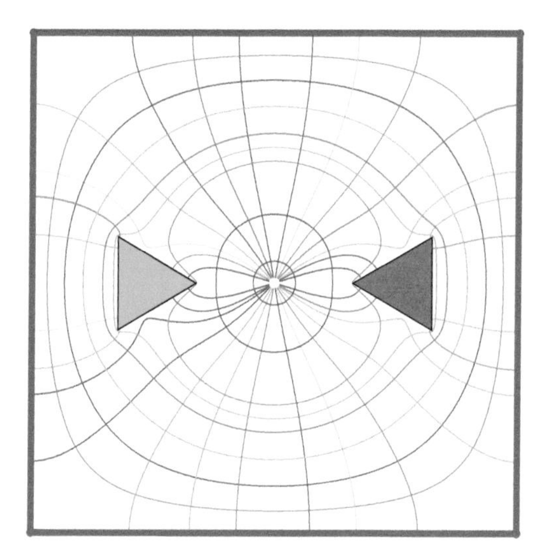

Figure 7.9. *Orthogonal grid in a rectangular box with triangular holes along the centerline.*

This is one more than in the previous example, even though the number of geometrical parameters, i.e., the width W and height H of the rectangle and the centroid position r and side length s of the triangular holes, is the same. We must ensure that condition (7.91) is satisfied, which provides one real constraint on the six parameters; in this case, however, it is found that for given values of the four geometrical parameters W, H, r, and s, only for a particular choice of the six parameters (7.97) will the centroids of the two triangular holes be located on the centerline of the rectangle as required.

7.10 ▪ Other uses of Cayley-type maps

The basic ideas underlying our construction of the multiply connected S–C formulas are versatile. In particular, much use can be made of the fact that Cayley-type maps have constant argument on $\{C_j | j = 0, 1, \ldots, M\}$. They come in handy when one wants to construct conformal mappings between polygons. The final two examples show these considerations in action.

A slit in a channel: In Part I we consider conformal maps of the form

$$z = f(\zeta) = \frac{1}{\pi \mathrm{i}} \log \left[-\frac{\omega(\zeta, +1)}{\omega(\zeta, -1)} \right], \tag{7.98}$$

where $\omega(\zeta, .)$ is the prime function associated with an $(M+1)$-connected circular domain D_ζ in a parametric ζ plane. These are logarithms of Cayley-type maps, and they transplant D_ζ into a channel region in the complex z plane,

$$-\frac{1}{2} < \mathrm{Re}[z] < \frac{1}{2}, \tag{7.99}$$

exterior to M finite-length slits all parallel to the channel walls at $\mathrm{Re}[z] = \pm 1/2$. It is natural to ask about the corresponding maps from D_ζ to channel regions with slits that are

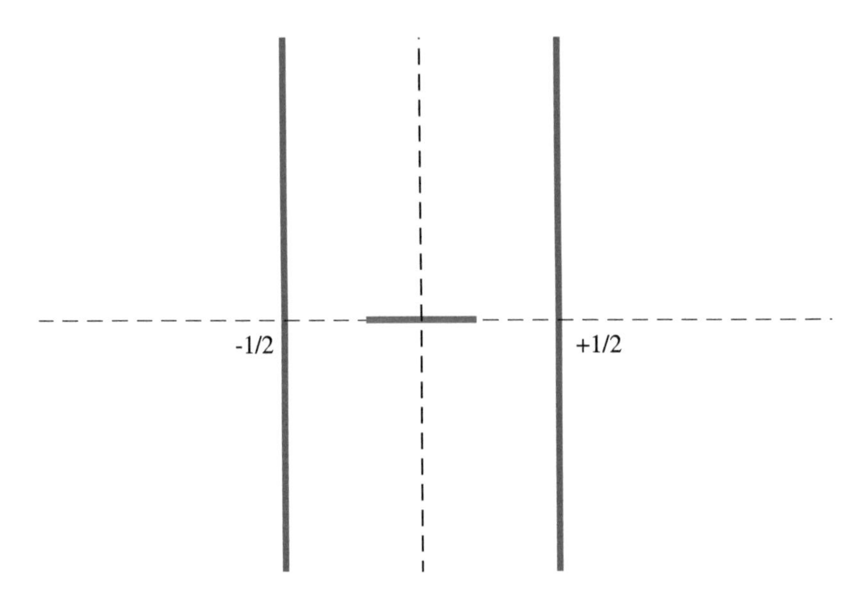

Figure 7.10. *Channel region with a symmetrically disposed slit perpendicular to the walls.*

perpendicular to the walls, or inclined at any other desired angle. For example, what is the mapping

$$w = g(\zeta), \tag{7.100}$$

to a complex w plane, say, from the annulus

$$\rho < |\zeta| < 1 \tag{7.101}$$

to a channel region (7.99) exterior to a finite-length, symmetrically disposed slit perpendicular to the two walls, as shown in Figure 7.10? This can be constructed from formula (7.62) with $S(z, a)$ given by (7.67); indeed, it is easy to argue, on grounds of symmetry, that the preimages of the ends of the slit are at $\pm i\rho$. Hence

$$g(\zeta) = A \int_{\rho}^{\zeta} \left[\frac{\omega(\zeta', i\rho)\omega(\zeta', -i\rho)}{\omega(\zeta', +1)\omega(\zeta', -1)} \right] \frac{d\zeta'}{\zeta'^2}, \tag{7.102}$$

where A is chosen to ensure the width of the channel is unity.

However there is another approach that can often be useful. It involves considering the quantity dw/dz where we exploit the fact that we *know* the mapping (7.98) from the annulus (7.101) to a channel region in the complex z plane with a slit parallel to the channel walls. The idea is to consider mapping this slit channel in the z plane to the different slit channel in the w plane. Since all the boundaries of both slit channels are made up of straight lines, the quantity dw/dz is piecewise constant, which reminds us of the Cayley-type maps considered in Chapters 5 and 6. Indeed on tracing around the boundary of the slit channel in the z plane it is clear that the argument of dw/dz is zero on the channel walls—this is because the infinitesimal elements dz and dw both have the same argument—but swaps between the values $\pm\pi/2$ as you pass through the edges of the slits in the z and w planes. Indeed a little thought on where dw/dz must vanish and where it is singular reveals that it can be written, as a function of ζ, using Cayley-type maps:

$$\frac{dw}{dz} = A\left[\frac{\omega(\zeta, \mathrm{i}\rho)\omega(\zeta, -\mathrm{i}\rho)}{\omega(\zeta, \rho)\omega(\zeta, -\rho)}\right], \tag{7.103}$$

where A is a constant to be determined. By the chain rule,

$$\frac{dw}{dz} = \frac{g'(\zeta)}{f'(\zeta)} = A\left[\frac{\omega(\zeta, \mathrm{i}\rho)\omega(\zeta, -\mathrm{i}\rho)}{\omega(\zeta, \rho)\omega(\zeta, -\rho)}\right]. \tag{7.104}$$

But we know $f(\zeta)$ from (7.98), and hence we know that

$$f'(\zeta) = \frac{1}{\pi \mathrm{i}\zeta}\left[K(\zeta, +1) - K(\zeta, -1)\right]. \tag{7.105}$$

Therefore we deduce

$$g(\zeta) = \frac{B}{\pi \mathrm{i}}\int_{\rho}^{\zeta}\left[K(\zeta', +1) - K(\zeta', -1)\right]\frac{\omega(\zeta', +\mathrm{i}\rho)\omega(\zeta', -\mathrm{i}\rho)}{\omega(\zeta', +\rho)\omega(\zeta', -\rho)}\frac{d\zeta'}{\zeta'}, \tag{7.106}$$

where

$$B = \frac{\omega(1, +\rho)\omega(1, -\rho)}{\omega(1, +\mathrm{i}\rho)\omega(1, -\mathrm{i}\rho)}, \tag{7.107}$$

in order that the logarithmic singularities of $f(\zeta)$ and $g(\zeta)$ at ± 1 have the same strength. The mapping (7.106) depends on the single real parameter ρ, the choice of which determines the length of the slit or flat plate.

Remark: In Chapter 8, we introduce the notion of a *loxodromic function*. On use of the theory of such functions we will be able to show that (7.102) and (7.106) are equivalent (see Exercise 8.8).

Periodic fins in a channel: We can use the same idea in a different example. Consider the mapping

$$z = \mathcal{Z}(\zeta) = \mathrm{i}A\log\zeta, \qquad A \in \mathbb{R}, \tag{7.108}$$

from the concentric annulus

$$\rho < |\zeta| < 1 \tag{7.109}$$

to a rectangle. The two vertical sides of the rectangle are the images of the two sides of a logarithmic branch cut from 0 to ∞ in the ζ plane.

Now suppose we want to map the annulus to a single period of a periodic polygon such as a channel region with a staggered periodic array of fins on the top and bottom wall, as shown in Figure 7.11. Let this mapping be

$$w = \mathcal{W}(\zeta). \tag{7.110}$$

One way to construct the form of this mapping is to use the theory of §7.7. It is an example of the case 3 mapping discussed there.

However, there is another way to arrive at the same result. Consider the mapping directly from the rectangle in the z plane to the periodic polygon in the w plane, where the vertical sides of the rectangle in the z plane map to the two vertical sides of a period window in the w plane. It is clear that

$$\frac{dw}{dz} \tag{7.111}$$

is piecewise constant as z traverses the top and bottom of the z rectangle and, hence, as ζ traverses C_0 and C_1.

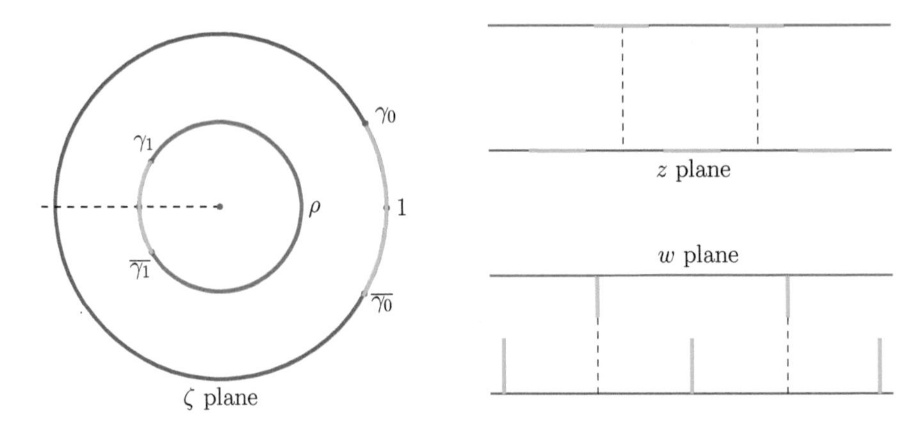

Figure 7.11. *Channel with staggered periodic fins on each wall in a w plane. One route to constructing the mapping from an annulus in a ζ plane is via a mapping to a rectangle in a complex z plane.*

We can therefore construct dw/dz, as a function of ζ, using Cayley-type maps in the annulus. Indeed, a little thought reveals that we must have

$$\frac{dw}{dz} = \frac{B\omega(\zeta, 1)\omega(\zeta, -\rho)}{[\omega(\zeta, \gamma_0)\omega(\zeta, \overline{\gamma_0})\omega(\zeta, \gamma_1)\omega(\zeta, \overline{\gamma_1})]^{1/2}}, \tag{7.112}$$

where $\gamma_0 \in C_0$ and $\gamma_1 \in C_1$, $\omega(.,.)$ is the prime function for the annulus and $B \in \mathbb{R}$ is a constant. Notice that we have put a zero of this function on the positive real axis on C_0 and on the negative real axis on C_1. If we take the branch cut of the logarithm to be on the negative real axis, then this will mean that the two fins on the upper wall are on the edges of the period window. Since

$$\mathcal{Z}'(\zeta) = \frac{\mathrm{i}A}{\zeta}, \tag{7.113}$$

by the chain rule,

$$W'(\zeta) = \frac{dw}{dz}\mathcal{Z}'(\zeta) = \mathrm{i}R\left[\frac{\omega(\zeta, 1)\omega(\zeta, -\rho)}{[\omega(\zeta, \gamma_0)\omega(\zeta, \overline{\gamma_0})\omega(\zeta, \gamma_1)\omega(\zeta, \overline{\gamma_1})]^{1/2}}\right]\frac{1}{\zeta}, \tag{7.114}$$

where R is a constant. This yields the expression

$$W(\zeta) = \mathrm{i}R\int_{\gamma_0}^{\zeta} \frac{\omega(\zeta', 1)\omega(\zeta', -\rho)}{[\omega(\zeta', \gamma_0)\omega(\zeta', \overline{\gamma_0})\omega(\zeta', \gamma_1)\omega(\zeta', \overline{\gamma_1})]^{1/2}}\frac{d\zeta'}{\zeta'}, \tag{7.115}$$

which depends on four real parameters: $R, \rho, \arg[\gamma_0]$, and $\arg[\gamma_1]$. These can be adjusted to ensure that the height and width of the period window as well as the lengths of the protruding barriers on the top and bottom walls of the channel are as required.

Exercises

7.1. **(Study's formula for curvature)** Derive Study's formula for the curvature κ of the image of a circle under an analytic transformation $w = f(z)$:

$$\kappa = \pm\frac{1}{|zf'(z)|}\mathrm{Re}\left[1 + \frac{zf''(z)}{f'(z)}\right]. \tag{7.116}$$

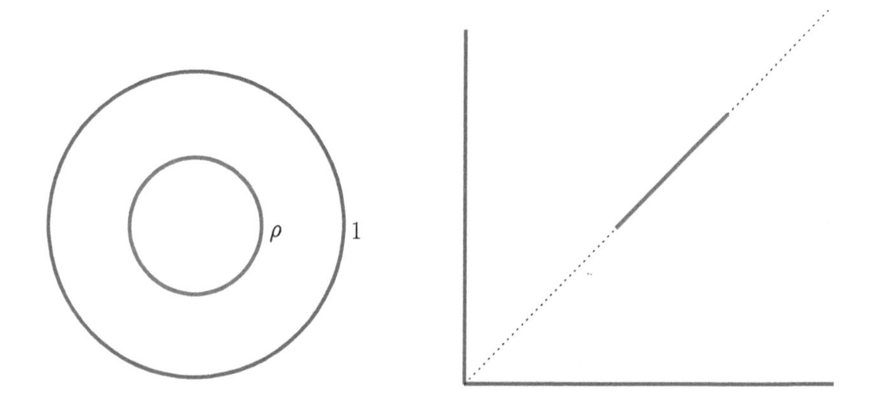

Figure 7.12. *S–C mapping from an annulus to a quarter plane with a slit on the bisector (Exercise 7.3).*

7.2. **(Pre-Schwarzian approach to classical S–C maps)** Use Study's formula from Exercise 7.1 to argue that if $f(z)$ is an S–C map from the unit circle $|z| = 1$, then if

$$T(z) \equiv \frac{zf''(z)}{f'(z)}, \tag{7.117}$$

then

$$\mathrm{Re}[T(z)] = -1. \tag{7.118}$$

Furthermore, argue that if the prevertices of an N-sided polygon are at $\{z_n | n = 1, \ldots, N\}$ with turning angles $\{\pi\beta_n | n = 1, \ldots, N\}$, then $T(z)$ has simple poles at $\{z_n | n = 1, \ldots, N\}$ with residue $z_n\beta_n$. Combine these facts to find an alternative derivation of the classical S–C formula (7.25).

7.3. **(S–C mapping to first quadrant with a slit)** By considering a composition of simpler conformal mappings, find an analytical formula, in terms of the prime function of the concentric annulus from the annulus

$$\rho < |\zeta| < 1, \tag{7.119}$$

for the conformal mapping from this annulus to the region in the first quadrant exterior to a finite-length slit along the bisecting ray with argument $\pi/4$, as shown in Figure 7.12. *Hint:* Make use of the Cayley-type mappings of §5.10.

7.4. **(Classical S–C formula from a half plane)** It is known that the Cayley map

$$\zeta = -\mathrm{i}\frac{\omega(z,+1)}{\omega(z,-1)} = \mathrm{i}\left(\frac{1-z}{1+z}\right), \tag{7.120}$$

where $\omega(z, a) = z - a$, transplants the unit z disc to the upper half ζ plane. By considering how formula (7.25) transforms under this mapping, show that the S–C formula for the mapping from this upper half ζ plane to a bounded polygon in a complex w plane is

$$w = F(\zeta) = A \int^{\zeta} \prod_{k=1}^{n_0} \omega(\zeta', \zeta_k^{(0)})^{\beta_k^{(0)}} d\zeta', \qquad A \in \mathbb{C}. \tag{7.121}$$

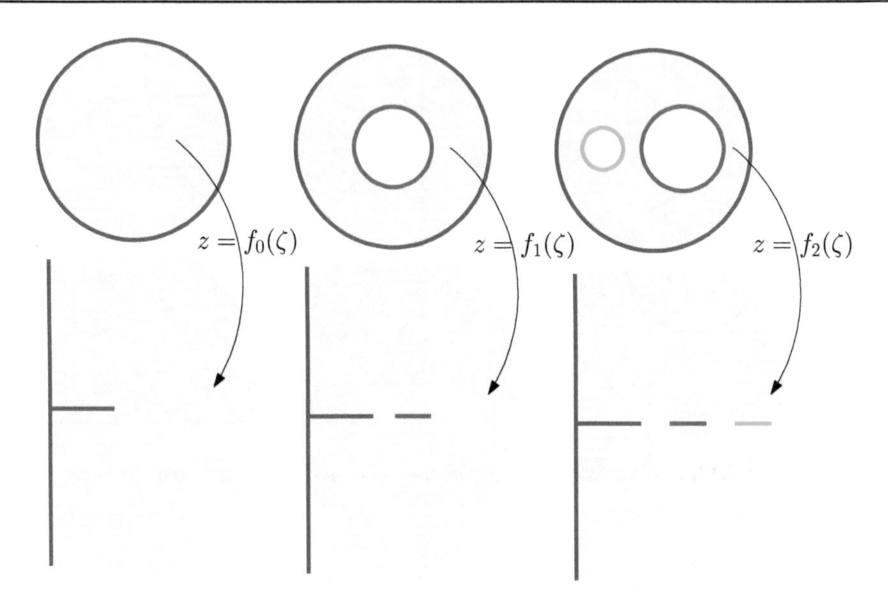

Figure 7.13. *S–C mappings $f_j(\zeta)$ for $j = 0,\ 1,$ and 2 to simply, doubly, and triply connected domains (Exercise 7.5).*

This is the classical S–C formula when the preimage domain is the upper half plane instead of the unit disc.

7.5. **(Analytical examples of multiply connected S–C mappings)** Figure 7.13 shows three S–C mappings $z = f_j(\zeta)$ for $j = 0$, 1, and 2. The domain involved in each case is $(j + 1)$ connected.

(a) Find an analytical expression for the simply connected mapping $f_0(\zeta)$ from the unit ζ disc shown in Figure 7.13 by composing a Cayley map to a right half plane with the relevant simply connected S–C mapping from Exercise 7.4 that transplants the right half plane to the right half plane with a reentrant slit along the positive real axis. (*Note*: The final formula should not involve any indefinite integrals.)

(b) By exchanging the Cayley map in part (a) for a *Cayley-type* map, as discussed in Chapters 5 and 6, find analytical expressions for the doubly and triply connected mappings $f_1(\zeta)$ and $f_2(\zeta)$.

(c) Show that the same functional form of the mapping works for any number of finite-length slits along the positive real axis.

7.6. **(S–C formula to doubly connected domains)** The construction of the S–C formula to a doubly connected polygon from a concentric annulus $\rho < |z| < 1$ has some special properties which we will explore in this exercise. Consider the conformal map $w = f(z)$ from $\rho < |z| < 1$ to a bounded doubly connected polygon D_w. Since both boundary circles of the preimage annulus are concentric, there is no need for the intermediate slit map introduced in §7.5. The S–C map $w = f(z)$ can therefore be constructed *directly* by using the Cayley-type maps and radial slit maps associated with D_z to construct the function

$$z\frac{df}{dz}, \tag{7.122}$$

which must have piecewise constant argument on both boundaries of the preimage annulus. Let the prevertices on C_0 be at $\{a_k^{(0)}|k=1,\dots,n_0\}$, and let the prevertices on C_1 be at $\{a_k^{(1)}|k=1,\dots,n_1\}$. The turning angle parameters at these prevertices are denoted $\{\beta_k^{(0)}|k=1,\dots,n_0\}$ and $\{\beta_k^{(1)}|k=1,\dots,n_1\}$, respectively. As in Chapter 7, in order that the polygonal boundaries are closed it is necessary that

$$\sum_{k=1}^{n_0}\beta_k^{(0)}=-2,\qquad \sum_{k=1}^{n_1}\beta_k^{(1)}=+2. \tag{7.123}$$

By emulating the constructive steps in the derivation of the multiply connected S–C formula, establish that the mapping from D_z to D_w has a derivative given by

$$\frac{df}{dz}=\frac{A}{z^2}\prod_{k=1}^{n_0}[\omega(z,a_k^{(0)})]^{\beta_k^{(0)}}\prod_{k=1}^{n_1}[\omega(z,a_k^{(1)})]^{\beta_k^{(1)}},\qquad A\in\mathbb{C}, \tag{7.124}$$

where $\omega(.,.)$ is the prime function associated with D_z.

Remark: Yet another derivation of this same result is the subject of Exercise 8.6.

7.7. **(S–C formula to unbounded polygonal domains)** This exercise modifies the construction of the S–C mapping formula to a bounded polygonal domain to the case of an unbounded region exterior to a finite collection of polygonal objects.[24] We focus on the triply connected case for simplicity, but all the steps will be generalizable to any number of polygonal objects. The aim is to find the function $f(z)$ where

$$w=f(z) \tag{7.125}$$

is the conformal mapping taking a triply connected circular domain D_z with boundary circles $C_j, j=0,1,2$, to the exterior of three polygons $P_j, j=0,1,2$. Suppose there are $n_j\geq 0$ vertices on the boundary $P_j, j=0,1,2$. Let the corresponding n_j prevertices on each circle C_j be denoted by

$$\{a_k^{(j)}|k=1,\dots,n_j \text{ and } j=0,1,2\}, \tag{7.126}$$

and let the turning angle parameters at these vertices be denoted by

$$\{\beta_k^{(j)}|k=1,\dots,n_j \text{ and } j=0,1,2\}. \tag{7.127}$$

We suppose some point b in D_z is the preimage of the point at infinity so that, near b,

$$f(z)=\frac{A}{z-b}+\text{a locally analytic function}, \tag{7.128}$$

where A is some constant.

(a) Verify that

$$\sum_{k=1}^{n_0}\beta_k^{(0)}=+2,\qquad \sum_{k=1}^{n_1}\beta_k^{(1)}=+2,\qquad \sum_{k=1}^{n_2}\beta_k^{(2)}=+2. \tag{7.129}$$

[24]The original derivation was given in [21].

(b) Introduce an intermediate mapping from a triply connected circular region D_z to a bounded circular slit domain D_η in a complex η plane:

$$\eta = g(z; a), \tag{7.130}$$

where a is some arbitrarily chosen point in D_z. The derivative of this function will have simple zeros at two points on each of C_1 and C_2, denoted by $\gamma_1^{(j)}$ and $\gamma_2^{(j)}$ for $j = 1, 2$. Now consider the function $F(\eta)$ defined by

$$w = f(z) = F(g(z; a)) = F(\eta). \tag{7.131}$$

Verify that

$$\eta \frac{dF}{d\eta} \tag{7.132}$$

must have piecewise constant argument on C_0, C_1, and C_2.

(c) Now suppose that $\eta F'(\eta)$ has the form (7.56), where the correction function $C(z)$ is to be determined. Show that the choice

$$C(z) = \frac{\omega(z,\gamma_1)}{\omega(z,\gamma_1^{(1)})} \frac{\omega(z,\gamma_1)}{\omega(z,\gamma_1^{(2)})} \times \frac{\omega(z,\gamma_2)}{\omega(z,\gamma_2^{(1)})} \frac{\omega(z,\gamma_2)}{\omega(z,\gamma_2^{(2)})} \tag{7.133}$$

$$\times \frac{\omega(z,\gamma_0)\omega(z,1/\overline{\gamma_0})}{\omega(z,b)\omega(z,1/\overline{b})} \times \frac{\omega(z,a)\omega(z,1/\overline{a})}{\omega(z,b)\omega(z,1/\overline{b})} \tag{7.134}$$

leads to a form of $\eta F'(\eta)$ that satisfies all the requirements above.

(d) Hence derive formulas (7.68)–(7.69).

7.8. **(Periodic S–C formula: case 1)** This exercise derives a general S–C mapping formula to a single period window exterior to an L-periodic array, in the direction of the real w axis, of polygonal objects sitting in each period window. We study the case of three objects per period, but all the steps are generalizable to any number of polygonal objects. The aim is to find the function $f(z)$ where

$$w = f(z) \tag{7.135}$$

is the conformal mapping taking a triply connected circular domain D_z with boundary circles C_j for $j = 0, 1, 2$ to the exterior of three polygons P_j for $j = 0, 1, 2$ in a given period window. We suppose some point a_{∞^+} in D_z is the preimage of the point where $\mathrm{Im}[w] \to +\infty$ so that

$$f(z) = -\frac{\mathrm{i}L}{2\pi} \log(z - a_{\infty^+}) + \text{a locally analytic function}, \tag{7.136}$$

where L is the period in the x direction; there is another point a_{∞^-} in D_z that is the preimage of the point where $\mathrm{Im}[w] \to -\infty$, so that

$$f(z) = +\frac{\mathrm{i}L}{2\pi} \log(z - a_{\infty^-}) + \text{a locally analytic function}. \tag{7.137}$$

The two sides of a branch cut joining a_{∞^+} and a_{∞^-} will correspond to the two edges of the period window. Suppose there are $n_j \geq 0$ vertices on the boundary P_j for $j = 0, 1, 2$. Let the corresponding n_j prevertices on each circle C_j be denoted by

$$\{a_k^{(j)} | k = 1, \ldots, n_j \text{ and } j = 0, 1, 2\}, \tag{7.138}$$

and let the turning angle parameters at these vertices be denoted by

$$\{\beta_k^{(j)} | k = 1, \ldots, n_j \text{ and } j = 0, 1, 2\}. \tag{7.139}$$

(a) Verify that

$$\sum_{k=1}^{n_0} \beta_k^{(0)} = +2, \qquad \sum_{k=1}^{n_1} \beta_k^{(1)} = +2, \qquad \sum_{k=1}^{n_2} \beta_k^{(2)} = +2. \tag{7.140}$$

(b) Introduce an intermediate mapping from a triply connected circular region D_z to a bounded circular slit domain D_η in a complex η plane:

$$\eta = g(z; a), \tag{7.141}$$

where a is some arbitrarily chosen point in D_z that is distinct from a_{∞^+} and a_{∞^-}. The derivative of this function will have simple zeros at two points on each of C_1 and C_2, denoted by $\gamma_1^{(j)}$ and $\gamma_2^{(j)}$ for $j = 1, 2$. Now consider the function $F(\eta)$ defined by

$$w = f(z) = F(g(z; a)) = F(\eta). \tag{7.142}$$

Verify that

$$\eta \frac{dF}{d\eta} \tag{7.143}$$

must have piecewise constant argument on C_0, C_1, and C_2.

(c) On use of the chain rule, verify that the function $\eta dF/d\eta$, in addition to branch points at the prevertices $\{a_k^{(j)} | k = 1, \ldots, n_j \text{ and } j = 0, 1, 2\}$, must have simple poles at $z = a_{\infty^+}$ and a_{∞^-}, a simple zero at $z = a$, and simple poles at $z = \gamma_1^{(j)}$ and $z = \gamma_2^{(j)}$ for $j = 1, 2$.

(d) Pick arbitrary points γ_0, γ_1, and γ_2 on C_0, C_1, and C_2, respectively, and suppose that $\eta F'(\eta)$ has the form (7.56), where the correction function $C(z)$ is to be determined. Show that the choice

$$C(z) = \frac{\omega(z,\gamma_1)}{\omega(z,\gamma_1^{(1)})} \frac{\omega(z,\gamma_1)}{\omega(z,\gamma_1^{(2)})} \times \frac{\omega(z,\gamma_2)}{\omega(z,\gamma_2^{(1)})} \frac{\omega(z,\gamma_2)}{\omega(z,\gamma_2^{(2)})} \tag{7.144}$$

$$\times \frac{\omega(z,\gamma_0)\omega(z,1/\overline{\gamma_0})}{\omega(z,a_{\infty^+})\omega(z,1/\overline{a_{\infty^+}})} \times \frac{\omega(z,a)\omega(z,1/\overline{a})}{\omega(z,a_{\infty^-})\omega(z,1/\overline{a_{\infty^-}})} \tag{7.145}$$

leads to a form of $\eta F'(\eta)$ that satisfies all the requirements in (b) and (c).

(e) Hence derive formulas (7.78)–(7.79).

(f) Check that the counting of parameters in (7.78)–(7.79) is consistent with the constraints on the mapping to a target period window for case 1.

7.9. **(Periodic S–C formula: case 2)** This exercise concerns formula (7.80) for the periodic S–C mapping to domains of type 2.

(a) Adapt the arguments of Exercise 7.8 to derive (7.80)–(7.81).

(b) Check that the counting of parameters in (7.80) is consistent with the constraints on the mapping to a target period window for case 2.

7.10. **(Periodic S–C formula: case 3)** This exercise concerns formula (7.82) for the periodic S–C mapping to domains of type 3.

(a) Adapt the arguments of Exercise 7.8 to derive (7.82)–(7.83).

(b) Check that the counting of parameters in (7.82) is consistent with the constraints on the mapping to a target period window for case 3.

7.11. **(Alternative form for $S^{(3)}_{\rm per}(z,a)$)** Use the general properties of the prime function to establish that the function $S^{(3)}_{\rm per}(z,a)$ given in (7.83) and appearing in the formula (7.82) for case 3 of the periodic S–C mapping can also be written in the alternative form

$$S^{(3)}_{\rm per}(z,a) = -\frac{q_1}{1-\overline{\delta_1/\gamma_1}} e^{-2\pi{\rm i}(v_1(1/\overline{\gamma_1})-v_1(z)+\tau_{11}/2)} S(z,a). \tag{7.146}$$

7.12. **(Dependence on the choice of a)** By considering a ratio

$$\frac{S(z,a_1)}{S(z,a_2)}, \tag{7.147}$$

where $a_1 \neq a_2$ are two distinct interior points of a multiply connected circular domain D_z, show that it does not matter which choice of $a \in D_z$ is made in the S–C formula (7.62). *Hint: You may find Exercise 2.4 useful here.*[25]

[25] See also the appendix to [16], where this matter is discussed.

Chapter 8

Loxodromic functions

8.1 ▪ Extending the function theory: Loxodromic functions

In Chapters 10 and 11 we plan to expand the class of planar regions amenable to analysis within our mathematical framework based on the prime function beyond slit domains and polygons. Before we can do this, however, we need to add a few more tools to our function theoretic toolbox. That is the aim of this chapter, and Chapter 9 to follow.

This chapter introduces the notion of a loxodromic function. A loxodromic function can be thought of as a special case, for $M = 1$, of a more general automorphic function which can be defined for any $M \geq 1$. We might therefore have condensed Chapters 8 and 9 into a single chapter. However, following in the spirit of Chapters 5 and 6, where the theory for the concentric annulus was presented separately from that for a general circular domain when $M \geq 2$, we have split our treatment over two chapters. The topic of automorphic functions for $M \geq 2$ is reserved for Chapter 9, where, it is hoped, the development will be seen to be a natural analogue of what we do here for the concentric annulus.

As in Chapter 5, the details for the concentric annulus afford various simplifications that make the treatment easier to follow.

8.2 ▪ Radial and circular slit maps as loxodromic functions

It turns out that we have already come across examples of loxodromic functions in Chapter 4, where the focus was on function theory associated with the concentric annulus

$$\rho < |z| < 1, \qquad 0 < \rho < 1. \tag{8.1}$$

Suppose the image of this annulus under the mapping

$$w = \mathcal{A}(z) \tag{8.2}$$

is the unbounded region exterior to two slits on the real axis in the complex w plane; such a mapping was considered in (5.68) of Chapter 5 as a special case of a radial slit mapping. Since C_0 maps to a slit on the real axis, we have, for $z \in C_0$,

$$\overline{\mathcal{A}(z)} = \mathcal{A}(z) \qquad \text{or} \qquad \overline{\mathcal{A}}(1/z) = \mathcal{A}(z), \tag{8.3}$$

where we have used the fact that $\overline{z} = 1/z$ on C_0. But C_1 also maps to a slit on the real axis, so we have, for $z \in C_1$,

$$\overline{\mathcal{A}(z)} = \mathcal{A}(z) \quad \text{or} \quad \overline{\mathcal{A}}(\rho^2/z) = \mathcal{A}(z), \tag{8.4}$$

where we have used the fact that $\overline{z} = \rho^2/z$ on C_1. The two relations

$$\overline{\mathcal{A}}(1/z) = \mathcal{A}(z), \qquad \overline{\mathcal{A}}(\rho^2/z) = \mathcal{A}(z) \tag{8.5}$$

can be analytically continued off the two circles on which they were originally derived. Indeed, from them we infer that

$$\overline{\mathcal{A}}(1/z) = \overline{\mathcal{A}}(\rho^2/z), \tag{8.6}$$

or, on taking a complex conjugate and renaming the variable $1/\overline{z} \mapsto z$, we find the functional relation

$$\mathcal{A}(\rho^2 z) = \mathcal{A}(z). \tag{8.7}$$

This functional relation has been deduced based on general considerations. However, in Chapter 5 we derived the functional form (5.68) of such a radial slit mapping, namely,

$$\mathcal{A}(z) \equiv \frac{\omega(z,a)\omega(z,1/a)}{\omega(z,b)\omega(z,1/b)}, \tag{8.8}$$

where $a, b \in D_z$ are real and $\omega(.,.)$ is the prime function associated with the concentric annulus and satisfying the functional identities (5.4)–(5.6). It is instructive to check property (8.7) directly. Identity (5.6) provides that

$$\mathcal{A}(\rho^2 z) = \frac{\omega(\rho^2 z,a)\omega(\rho^2 z,1/a)}{\omega(\rho^2 z,b)\omega(\rho^2 z,1/b)} = \frac{(-a/z)\omega(z,a)(-1/(za))\omega(z,1/a)}{(-b/z)\omega(z,b)(-1/(zb))\omega(z,1/b)} = \mathcal{A}(z), \tag{8.9}$$

which confirms that the functional form (8.8) of $\mathcal{A}(z)$ does indeed satisfy (8.7).

In this case the doubled domain F is the annular region $\rho < |z| < 1/\rho$. The function $\mathcal{A}(z)$ in (8.8) is analytic in the annulus $\rho < |z| < 1$ except for a simple pole singularity at $b \in D_z$; it has another simple pole at $1/b$ which lies in the "other half" of F. $\mathcal{A}(z)$ is therefore a meromorphic function in F, possessing only simple pole singularities.

Functions that are meromorphic in the annulus $\rho < |z| < 1/\rho$ and satisfy the functional identity (8.7) are known as *loxodromic functions*.

The functional relation (8.7) can be used to provide the analytic continuation of $\mathcal{A}(z)$ outside F, indeed, to all the images of F under iteration of the transformation $\theta_1(z) = \rho^2 z$ and its inverse. On setting $z = z/\rho^2$ in (8.7) we find

$$\mathcal{A}(z) = \mathcal{A}(z/\rho^2). \tag{8.10}$$

Relations (8.7) and (8.10) can be used to deduce that $\mathcal{A}(z)$ has an infinite number of simple poles at the points

$$b\rho^{2n}, \qquad \frac{\rho^{2n}}{b}, \qquad n \in \mathbb{Z}. \tag{8.11}$$

Since knowledge of the singularity structure in F determines it, through the identity (8.7), we have that in every other annulus $\rho^{2n+1} < |z| < \rho^{2n-1}$ for $n \in \mathbb{Z}$ we call F a *fundamental annulus*. In view of this observation the "other" annuli, namely $\rho^{2n+1} < |z| < \rho^{2n-1}$ for $n \neq 0$, are referred to as *equivalent annuli* to the fundamental annulus F.

The slit mapping encountered in (5.38) of Chapter 5, the special case of an unbounded circular slit map to two finite-length slits on the unit circle as shown in Figure 5.5, is another example of a loxodromic function. We will reuse notation and let this different mapping be denoted by the same function name:

$$w = \mathcal{A}(z). \tag{8.12}$$

Since the image of C_0 lies on the unit w circle,

$$\mathcal{A}(z)\overline{\mathcal{A}(z)} = 1, \tag{8.13}$$

implying that

$$\mathcal{A}(z) = \frac{1}{\overline{\mathcal{A}}(1/z)}. \tag{8.14}$$

Since the image of C_1 also lies on the unit w circle, we have

$$\mathcal{A}(z)\overline{\mathcal{A}(z)} = 1, \tag{8.15}$$

implying that

$$\mathcal{A}(z) = \frac{1}{\overline{\mathcal{A}}(\rho^2/z)}. \tag{8.16}$$

Both (8.14) and (8.16) can be analytically continued off the circles where they were derived and together imply that

$$\frac{1}{\overline{\mathcal{A}}(1/z)} = \frac{1}{\overline{\mathcal{A}}(\rho^2/z)}, \tag{8.17}$$

which leads again to the functional relation (8.7).

We can also verify this functional relation directly from the known formula for the mapping as found in (5.38) of Chapter 5:

$$\mathcal{A}(z) = -\frac{\omega(z,a)\omega(z,1/a)}{\omega(z,\overline{a})\omega(z,1/\overline{a})}, \qquad a = ir, \qquad \rho < r < 1. \tag{8.18}$$

It is clear that this function has simple poles in $\rho < |z| < 1/\rho$ at $\overline{a}$ and $1/\overline{a}$, and identity (5.6) can be used to reconfirm that it satisfies (8.7). The function $\mathcal{A}(z)$ in (8.18) is therefore another example of a loxodromic function.

Loxodromic functions are intimately related to elliptic functions, which are defined as doubly periodic meromorphic functions. We point this out because many problems in doubly connected domains solved in the extant literature often make use of the theory of elliptic functions, of Jacobi or Weierstrass, say, to solve them. This monograph advocates a different approach to the same class of problems using the theory of loxodromic functions to be developed in this chapter, following naturally from the prime function theory developed earlier.

It turns out that there is a direct connection between the loxodromic radial slit mapping (8.8) just considered and the classical Jacobi sn function; that connection is explored in Exercise 8.14 using a geometrical argument based on conformal geometry. An analogous connection exists between the loxodromic unbounded circular slit mapping (8.18) and the classical Jacobi cn function; that connection is explored in Exercise 8.15.

Further connections of the function theory in the annulus $\rho < |z| < 1/\rho$ developed in this chapter with the classical function theories of Jacobi and Weierstrass are elucidated in Exercises 14.6–14.8 in Chapter 14.

8.3 ▪ Schottky groups

Loxodromic functions are meromorphic functions satisfying the functional relation

$$\mathcal{A}(\rho^2 z) = \mathcal{A}(z). \tag{8.19}$$

If the structure of the pole singularities of $\mathcal{A}(z)$ in the annulus

$$\rho < |z| < 1/\rho \tag{8.20}$$

is known, then (8.19) can be used to analytically (or meromorphically) continue $\mathcal{A}(z)$ into other regions of the complex z plane. Indeed, on repeated use of (8.19), it follows that

$$\mathcal{A}(\rho^{2n} z) = \mathcal{A}(z), \qquad n \in \mathbb{Z}. \tag{8.21}$$

A loxodromic function is therefore invariant under the action on its argument z of the set of Möbius maps denoted by

$$\Theta = \{\rho^{2n} z | n \in \mathbb{Z}\}. \tag{8.22}$$

It is easy to check that the set Θ is a *group* of Möbius maps where the group action is composition of Möbius maps. The entire group can be generated by repeated composition of the two generating maps

$$\theta_1(z) = \rho^2 z, \qquad \theta_1^{-1}(z) = \rho^{-2} z. \tag{8.23}$$

The identity map corresponds to $n = 0$ and is clearly an element of Θ; for any $n > 0$ the inverse mapping to $\rho^{2n} z$ is the mapping $\rho^{-2n} z$, which is also in Θ, and the composition of $\rho^{2m} z$ and $\rho^{2n} z$ for any integers m and n is $\rho^{2(m+n)} z$, also contained in Θ.

The group Θ is a simple example of an infinite, free *Schottky group*; this is a group of Möbius maps generated by a given finite set of generating Möbius maps, in this case, the maps (8.23). Schottky groups of this kind, but with more generators, were introduced in Exercise 3.3.

For a loxodromic function the domain F is the annulus $\rho < |z| < 1/\rho$. In the theory of automorphic functions it is known as the *domain of discontinuity* of this group action [81]. Practically speaking, once the poles of a loxodromic function in F have been specified, this dictates the structure of its poles in all regions of the complex plane that are images of this domain of discontinuity under the action of the Schottky group.

8.4 ▪ Properties of loxodromic functions

The two slit maps (8.8) and (8.18) are examples of loxodromic functions. There are lots of other such functions and they turn out to be useful in a wide variety of applications. In §8.6 we will explore two methods of constructing them. Before doing so it is useful to survey some of their general properties.[26]

The first property concerns the number of zeros and poles of a loxodromic function in F. Let $\mathcal{A}(z)$ be a loxodromic function satisfying

$$\mathcal{A}(\theta_1(z)) = \mathcal{A}(\rho^2 z) = \mathcal{A}(z) \tag{8.24}$$

and having Z zeros and N poles in F counted by multiplicity. By the argument principle,

$$Z - N = \frac{1}{2\pi \mathrm{i}} \int_{\partial F_c} \frac{d\mathcal{A}(z)}{\mathcal{A}(z)}, \tag{8.25}$$

[26]We present only the essential elements of the theory of loxodromic functions needed to solve problems to follow later in this book. Other aspects can be found in [122].

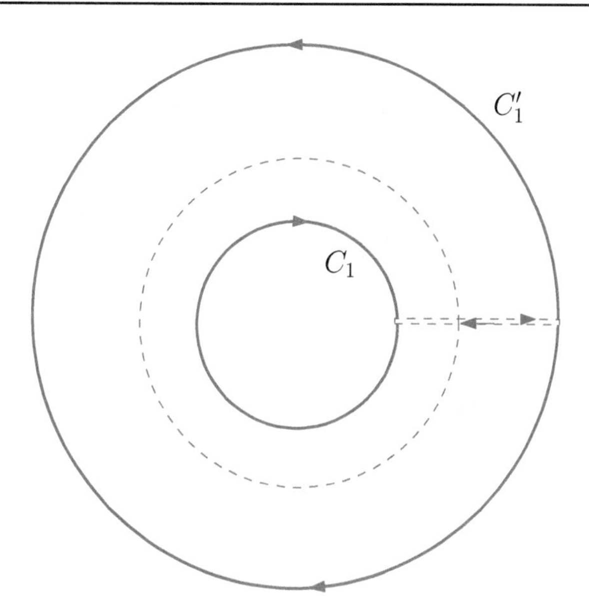

Figure 8.1. *The closed contour ∂F_c. This is the analogue of the contour ∂F_R of Figure 3.2 in the case of a concentric annulus with $M = 1$ and where F is now bounded.*

where the integral is taken around the contour ∂F_c shown in Figure 8.1. The contour ∂F_c is the analogue, when F is bounded, of the contour ∂F_R introduced in Figure 3.2 when F was unbounded. The right-hand side of (8.25) can be decomposed as

$$\int_{\partial F_c} \frac{d\mathcal{A}(z)}{\mathcal{A}(z)} = \int_{C_1'} \frac{d\mathcal{A}(z)}{\mathcal{A}(z)} - \int_{C_1} \frac{d\mathcal{A}(z)}{\mathcal{A}(z)} = \int_{C_1'} \left[\frac{d\mathcal{A}(z)}{\mathcal{A}(z)} - \frac{d\mathcal{A}(\theta_1(z))}{\mathcal{A}(\theta_1(z))} \right], \quad (8.26)$$

where we have effected the change of variable $z \mapsto \theta_1(z)$ in the integral around C_1. But by (8.24)

$$\frac{d\mathcal{A}(\theta_1(z))}{\mathcal{A}(\theta_1(z))} = \frac{d\mathcal{A}(z)}{\mathcal{A}(z)}, \quad (8.27)$$

implying that

$$Z - N = \frac{1}{2\pi \mathrm{i}} \int_{\partial F_c} \frac{d\mathcal{A}(z)}{\mathcal{A}(z)} = 0. \quad (8.28)$$

A loxodromic function therefore has the same number of zeros as poles in the fundamental annulus F.

A second property is to show that the N zeros and N poles of a loxodromic function cannot be arbitrarily chosen—there is a constraint between them. There are two forms of this constraint, and, as we will see later, each form is appropriate to a particular choice of representation of the loxodromic function.

Suppose that $\mathcal{A}(z)$ has N simple poles in F located at positions $\{b_n | n = 1, \dots, N\}$; the assumption that the poles are simple can be relaxed but is made for ease of exposition. Therefore, near b_n,

$$\mathcal{A}(z) = \frac{r_n}{z - b_n} + \text{a locally analytic function}, \qquad n = 1, \dots, N, \quad (8.29)$$

where r_n is the residue of $\mathcal{A}(z)$ at b_n. We know there will be N zeros, so let these be at positions $\{a_n | n = 1, \dots, N\}$.

Consider the contour integral

$$\int_{\partial F_c} v_1(z)\frac{d\mathcal{A}(z)}{\mathcal{A}(z)}. \tag{8.30}$$

By the residue theorem, and the fact that the logarithmic derivative of $\mathcal{A}(z)$ has a simple pole of residue $+1$ at each zero a_n and a simple pole of residue -1 at each pole b_n, this integral equals

$$2\pi\mathrm{i}\sum_{n=1}^{N}(v_1(a_n)-v_1(b_n)). \tag{8.31}$$

On the other hand, on adding up all the directed boundary contributions, the integral (8.30) also equals

$$\int_{C_1'} v_1(z)\frac{d\mathcal{A}(z)}{\mathcal{A}(z)} - \int_{C_1} v_1(z)\frac{d\mathcal{A}(z)}{\mathcal{A}(z)} - \int_{L_1} v_1(z)\frac{d\mathcal{A}(z)}{\mathcal{A}(z)} - \int_{L_1^+} v_1(z)\frac{d\mathcal{A}(z)}{\mathcal{A}(z)}. \tag{8.32}$$

On changing the integration variable $z \mapsto \theta_1(z)$ in the integral around C_1 this becomes

$$\int_{C_1'} v_1(z)\frac{d\mathcal{A}(z)}{\mathcal{A}(z)} - \int_{C_1'} v_1(\theta_1(z))\frac{d\mathcal{A}(\theta_1(z))}{\mathcal{A}(\theta_1(z))} - \int_{L_1} v_1(z)\frac{d\mathcal{A}(z)}{\mathcal{A}(z)} - \int_{L_1^+} v_1(z)\frac{d\mathcal{A}(z)}{\mathcal{A}(z)}. \tag{8.33}$$

Now $v_1(a)$ jumps by $+1$ as one travels continuously from any point on L_1 anticlockwise around C_1 to meet the corresponding point on L_1^+. We also know that

$$v_1(\theta_1(z)) - v_j(z) = \tau_{11}. \tag{8.34}$$

Hence, on use of the loxodromic property of $\mathcal{A}(z)$, (8.33) is

$$-\tau_{11}\int_{C_1'}\frac{d\mathcal{A}(z)}{\mathcal{A}(z)}dz - \int_{L_1^+}\frac{d\mathcal{A}(z)}{\mathcal{A}(z)}dz \tag{8.35}$$

$$= -\tau_{11}\int_{C_1'} d\log\mathcal{A}(z) + \int_{L_1} d\log\mathcal{A}(z). \tag{8.36}$$

But the quantities

$$\frac{1}{2\pi\mathrm{i}}\int_{L_1} d\log\mathcal{A}(z), \qquad \frac{1}{2\pi\mathrm{i}}\int_{C_1'} d\log\mathcal{A}(z) \tag{8.37}$$

must be integers because $\mathcal{A}(z)$ returns to its same value at each end of L_1 (by its invariance under the group action), while C_1' is a closed curve. Hence, on equating (8.31) with (8.36) and using (8.37) we find

$$\sum_{n=1}^{N}(v_1(a_n)-v_1(b_n)) = m_1 + \tau_{11}n_1 \tag{8.38}$$

for some integers m_1, n_1. Now (5.11) implies that (8.38) is

$$\frac{1}{2\pi\mathrm{i}}\log\prod_{n=1}^{N}\left(\frac{a_n}{b_n}\right) = m_1 - \frac{\mathrm{i}n_1\log\rho}{\pi}. \tag{8.39}$$

Equivalently, on exponentiation,

$$\prod_{n=1}^{N} \left(\frac{a_n}{b_n} \right) = \rho^{2n_1}. \tag{8.40}$$

Condition (8.40) is one form of the constraint relating the zeros $\{a_n | n = 1, \ldots, N\}$ and poles $\{b_n | n = 1, \ldots, N\}$ of a loxodromic function.

There is a different form of this constraint. Consider the integral

$$\int_{\partial F_c} \mathcal{A}(z) dv_1(z). \tag{8.41}$$

The integrand is meromorphic in F, so, by the residue theorem, we must have

$$\int_{\partial F_c} \mathcal{A}(z) dv_1(z) = 2\pi \mathrm{i} \sum_{n=1}^{N} r_n v_1'(b_n) = \sum_{n=1}^{N} \frac{r_n}{b_n}, \tag{8.42}$$

where we have used (8.29) and the known functional form (5.9) of $v_1(z)$. On the other hand, adding up the contributions from the separate components of ∂F_c gives

$$\int_{\partial F_c} \mathcal{A}(z) dv_1(z) = \int_{C_1'} \mathcal{A}(z) dv_1(z) - \int_{C_1} \mathcal{A}(z) dv_1(z) \tag{8.43}$$

$$= \int_{C_1'} \{ \mathcal{A}(z) dv_1(z) - \mathcal{A}(\theta_1(z)) dv_1(\theta_1(z)) \}, \tag{8.44}$$

where we have made the substitution $z \mapsto \theta_1(z)$ in the integral around C_1 and used the fact that the integrand is continuous across the cut (L_1, L_1^+). Since $\mathcal{A}(z)$ is loxodromic, and since

$$dv_1(\theta_1(z)) = dv_1(z), \tag{8.45}$$

then

$$\int_{\partial F_c} \mathcal{A}(z) dv_1(z) = 0. \tag{8.46}$$

Consequently, we conclude that

$$\sum_{n=1}^{N} \frac{r_n}{b_n} = 0. \tag{8.47}$$

This is another form of the constraint on the poles and zeros of the loxodromic function. In this case, the dependence of these conditions on the zeros of the loxodromic function is hidden in the residues.

An important observation is that if $N = 1$, then to satisfy (8.47) we must have $r_1 = 0$. This removes the simple pole at b_1. We see from this that it is not possible to find a loxodromic function with just a single simple pole in F.

8.5 ▪ The quasi-loxodromic function $\mathcal{K}(z,a)$

The function $\mathcal{K}(z,a)$ introduced in §5.12, while not a loxodromic function itself, turns out to be useful in *constructing* loxodromic functions. To illustrate this we now derive an alternative representation of the loxodromic radial slit map (8.8). Since, from (5.76), we see that setting $z \mapsto \rho^2 z$ means that we *almost* retrieve $\mathcal{K}(z,a)$ except for an additive term that is independent of z, we say that $\mathcal{K}(z,a)$ is a *quasi-loxodromic* function.

Pick points $a, b \in \mathbb{R}$ in the annulus $\rho < |z| < 1/\rho$ and define

$$\mathcal{H}(z,b) \equiv \mathcal{K}(z,b) - \mathcal{K}(z,1/b) - [\mathcal{K}(a,b) - \mathcal{K}(a,1/b)]. \tag{8.48}$$

This function has simple poles at b and $1/b$, and it is clear by the choice of constant term in square brackets in (8.48) that it vanishes at $z = a$. Moreover, on use of (5.76), it is easily checked that $\mathcal{H}(z,b)$ satisfies

$$\mathcal{H}(\rho^2 z, b) = \mathcal{H}(z,b). \tag{8.49}$$

$\mathcal{H}(z,b)$ is therefore a loxodromic function with simple poles at b and $1/b$. Since it vanishes at $z = a$, the identity (5.75) can be used to verify that

$$\mathcal{H}(1/a, b) = 0. \tag{8.50}$$

So $\mathcal{H}(z,b)$ also vanishes at $1/a$.

Let $\mathcal{A}(z)$ be the loxodromic radial slit map defined in (8.8). The ratio

$$\mathcal{R}(z) \equiv \frac{\mathcal{H}(z,b)}{\mathcal{A}(z)} \tag{8.51}$$

satisfies the identity

$$\mathcal{R}(\rho^2 z) = \mathcal{R}(z). \tag{8.52}$$

It is analytic and single-valued everywhere in the annulus $\rho < |z| < 1/\rho$ since the simple poles in the numerator and denominator at b and $1/b$ cancel out, and any poles at a and $1/a$ due to the zeros of the denominator are removed by the zeros of the numerator at those points. It follows that $\mathcal{R}(z)$ must be a constant function (see Exercise 8.1) and, hence, that $\mathcal{H}(z,b)$ and $\mathcal{A}(z)$ are proportional:

$$\mathcal{H}(z,b) = c\mathcal{A}(z), \qquad c \in \mathbb{C}. \tag{8.53}$$

The value of the constant c can be determined by evaluating both sides at any nonsingular point, for example, at $z = 1$:

$$c = \frac{\mathcal{H}(1,b)}{\mathcal{A}(1)}. \tag{8.54}$$

As a result, the following identity has been derived:

$$\begin{aligned} &\frac{\omega(1,b)\omega(1,1/b)}{\omega(1,a)\omega(1,1/a)} \left[\frac{\omega(z,a)\omega(z,1/a)}{\omega(z,b)\omega(z,1/b)}\right] \\ &= \frac{\mathcal{K}(z,b) - \mathcal{K}(z,1/b) - [\mathcal{K}(a,b) - \mathcal{K}(a,1/b)]}{\mathcal{K}(1,b) - \mathcal{K}(1,1/b) - [\mathcal{K}(a,b) - \mathcal{K}(a,1/b)]}. \end{aligned} \tag{8.55}$$

This is exactly the result of Exercise 5.10, where "two views of a two-slit map" were considered. In that exercise formula (8.55) was derived by identifying the conformal map from the annulus $\rho < |z| < 1$ to two slits on the real axis as an instance of a radial slit map on the one hand, and as a parallel slit map on the other. We see now that the same formula *also* arises by finding two different representations of a loxodromic function: one as a ratio of products of prime functions and the other as a sum of multiples of the quasi-loxodromic function $\mathcal{K}(z,a)$.

8.6 ▪ Two ways to construct loxodromic functions

The identity (8.55) between two distinct representations of a conformal slit map, where both sides are loxodromic functions, provides the clue that there are at least two different ways to construct more general loxodromic functions. One approach, reflected in the form of the left-hand side of (8.55), is to consider ratios of products of prime functions. An advantage to this method is that a representation of the poles and zeros of the function appears explicitly in the expression for the function.

For some $N \geq 2$ consider the function given by

$$\mathcal{A}(z) = A \prod_{n=1}^{N} \frac{\omega(z, a_n)}{\omega(z, b_n)}, \tag{8.56}$$

where $\{a_n, b_n | n = 1, \ldots, N\}$ are a fixed set of distinct complex numbers. We know that $\omega(z, a)$ has a simple zero at $z = a$ and, by virtue of the identity (5.6), at all equivalent points $z = a\rho^{2n}, n \in \mathbb{Z}$. Hence $\mathcal{A}(z)$ clearly has simple zeros at $\{a_n | n = 1, \ldots, N\}$ and simple poles at $\{b_n | n = 1, \ldots, N\}$, as well as at all equivalent points. Notice that the representation (8.56) has picked out a single representative of each of these zeros and poles to appear in the formula. Now

$$\mathcal{A}(\rho^2 z) = A \prod_{n=1}^{N} \frac{\omega(\rho^2 z, a_n)}{\omega(\rho^2 z, b_n)}, \tag{8.57}$$

which, on use of the identity (5.6), becomes

$$\mathcal{A}(\rho^2 z) = A \prod_{n=1}^{N} \left(\frac{a_n}{b_n}\right) \prod_{n=1}^{N} \frac{\omega(z, a_n)}{\omega(z, b_n)} = \left\{\prod_{n=1}^{N} \left(\frac{a_n}{b_n}\right)\right\} \mathcal{A}(z). \tag{8.58}$$

$\mathcal{A}(z)$ will be a loxodromic function, satisfying (8.7), provided the constants $\{a_n, b_n\}$ satisfy the single constraint

$$\prod_{n=1}^{N} \left(\frac{a_n}{b_n}\right) = 1. \tag{8.59}$$

This is an instance of (8.40) with $n_1 = 0$.

Another function with the same zeros and poles is

$$\tilde{\mathcal{A}}(z) = Az \prod_{n=1}^{N} \frac{\omega(z, a_n)}{\omega(z, b_n)} = z\mathcal{A}(z). \tag{8.60}$$

In this case,

$$\tilde{\mathcal{A}}(\rho^2 z) = (\rho^2 z)\mathcal{A}(\rho^2 z) = \rho^2 z \left\{\prod_{n=1}^{N} \left(\frac{a_n}{b_n}\right)\right\} \mathcal{A}(z) = \rho^2 \left\{\prod_{n=1}^{N} \left(\frac{a_n}{b_n}\right)\right\} \tilde{\mathcal{A}}(z), \tag{8.61}$$

which is a loxodromic function provided the constants $\{a_n, b_n | n = 1, \ldots, N\}$ satisfy the single constraint

$$\rho^2 \prod_{n=1}^{N} \left(\frac{a_n}{b_n}\right) = 1. \tag{8.62}$$

Condition (8.62) is different from (8.59); it is an instance of (8.40) with $n_1 = -1$.

However if we replace, say, the zero at a_1 by an equivalent zero at $\rho^2 a_1$ so that the set of representative zeros in the formula is now

$$\tilde{a}_1 = \rho^2 a_1, \quad \tilde{a}_2 = a_2, \quad \ldots, \quad \tilde{a}_N = a_N, \tag{8.63}$$

then condition (8.62) becomes

$$\prod_{n=1}^{N} \left(\frac{\tilde{a}_n}{\tilde{b}_n} \right) = 1, \tag{8.64}$$

which is identical in form to (8.59). This is not surprising because if we write down the function (8.56) with the new set of representative zeros (8.64), and the same representative poles $\{b_n | n = 1, \ldots, N\}$, we get

$$A \prod_{n=1}^{N} \frac{\omega(z, \tilde{a}_n)}{\omega(z, \tilde{b}_n)} = A \frac{\omega(z, \rho^2 a_1)}{\omega(z, b_1)} \prod_{n=2}^{N} \frac{\omega(z, a_n)}{\omega(z, b_n)} = \frac{A}{a_1} z \prod_{n=1}^{N} \frac{\omega(z, a_n)}{\omega(z, b_n)}, \tag{8.65}$$

where we have used the identity (5.6). This is just a multiple of $\tilde{\mathcal{A}}(z)$.

We can generalize this construction further and consider the function

$$A z^M \prod_{n=1}^{N} \frac{\omega(z, a_n)}{\omega(z, b_n)}, \qquad M \in \mathbb{Z}, \tag{8.66}$$

which is also single-valued in the annulus $\rho < |z| < 1/\rho$ with zeros at $\{a_n | n = 1, \ldots, N\}$ and simple poles at $\{b_n | n = 1, \ldots, N\}$, as well as at all equivalent points. It is easy to check, from (5.6), that the condition on the parameters appearing in this formula is

$$\rho^{2M} \prod_{n=1}^{N} \left(\frac{a_n}{b_n} \right) = 1. \tag{8.67}$$

This is an instance, corresponding to $n_1 = -M$, of the constraint (8.40) derived earlier from general considerations.

An alternative construction of a loxodromic function with simple poles at $\{b_n | n = 1, \ldots, N\}$ is to consider the function

$$\mathcal{A}(z) = \sum_{n=1}^{N} A_n \mathcal{K}(z, b_n) + A_0, \tag{8.68}$$

where $\{A_n | 0, 1, \ldots, N\}$ are complex constants. Now,

$$\mathcal{A}(\rho^2 z) = \sum_{n=1}^{N} A_n \mathcal{K}(\rho^2 z, b_n) + A_0 = \sum_{n=1}^{N} A_n (\mathcal{K}(z, b_n) + 1) + A_0 = \mathcal{A}(z) + \sum_{n=1}^{N} A_n, \tag{8.69}$$

where we have used (5.76). Hence $\mathcal{A}(z)$ will be a loxodromic function provided the constants $\{A_n\}$ satisfy

$$\sum_{n=1}^{N} A_n = 0. \tag{8.70}$$

From (8.68), and the properties of $\mathcal{K}(.,.)$, it follows that near $z = b_n$,

$$\mathcal{A}(z) = -\frac{A_n b_n}{z - b_n} + \text{locally analytic function}, \tag{8.71}$$

and then condition (8.70) is precisely equivalent to the constraint (8.47) derived earlier from general considerations.

These constructions can be extended in a natural way to find representations of loxodromic functions having higher order poles.

Exercises

8.1. **(A Liouville-type theorem for loxodromic functions)** Let $\mathcal{A}(z)$ be a loxodromic function with no poles in $\rho < |z| < 1/\rho$. By considering a Laurent expansion representation of $\mathcal{A}(z)$ show that it must be a constant. *Hint:* This same result was used in Exercise 5.2.

8.2. **(The function $L(z,a)$)** Take a point a in the annulus $\rho < |z| < 1/\rho$ and consider the function

$$L(z,a) \equiv z\frac{\partial}{\partial z}K(z,a), \tag{8.72}$$

where $K(z,a)$ was introduced in §5.12.

(a) Show that, near $z = a$,

$$K(z,a) = \frac{a}{z-a} + \text{a locally analytic function} \tag{8.73}$$

and hence that

$$L(z,a) = -\frac{a^2}{(z-a)^2} - \frac{a}{z-a} + \text{a locally analytic function.} \tag{8.74}$$

It is clear that $L(z,a)$ has a second order pole at $z = a$.

(b) Verify that

$$L(\rho^2 z, a) = L(z,a), \tag{8.75}$$

and deduce that $L(z,a)$ is a loxodromic function.

(c) Show that $L(z,a)$ satisfies the functional identities

$$\overline{L}(1/z, 1/a) = L(1/z, 1/a) = L(z,a). \tag{8.76}$$

8.3. **(The function $M(z,a)$)** Define the function

$$M(z,a) \equiv z\frac{\partial}{\partial z}L(z,a). \tag{8.77}$$

(a) Verify that $M(z,a)$ has a third order pole at $z = a$.

(b) Verify that $M(z,a)$ is a loxodromic function.

(c) Show that $M(z,a)$ satisfies the functional identities

$$\overline{M}(1/z, 1/a) = M(1/z, 1/a) = -M(z,a). \tag{8.78}$$

8.4. **(Loxodromic function with higher order poles)** This exercise concerns the construction of loxodromic functions with higher order poles.

(a) Use the results of Exercise 8.2 to find a representation of a loxodromic function with a general second order pole at some point $z = a$ in the annulus $\rho < |z| < 1/\rho$.

(b) Use the results of Exercises 8.2 and 8.3 to find a representation of a loxodromic function with a general third order pole at some point $z = a$ in the annulus $\rho < |z| < 1/\rho$.

8.5. **(Special radial slit map as a loxodromic function)** Verify that the special instance of the radial slit map of the concentric annulus given in (5.70), i.e., the function

$$f(z) = -\frac{\omega(z,a)\omega(z,1/\overline{a})}{\omega(z,\overline{a})\omega(z,1/a)}, \qquad a = ir,\ 0 < r < 1, \tag{8.79}$$

is a loxodromic function.

8.6. **(New derivation of doubly connected S–C)** This exercise is an extension of Exercise 7.2; it gives a different derivation of the doubly connected S–C formula from that given in Chapter 7. Instead of using conformal slit maps, we make use of the theory of loxodromic functions. Consider the conformal map $w = f(z)$ from the annulus $\rho < |z| < 1$ to a bounded doubly connected polygon. Let the prevertices on C_0 be at $\{a_k^{(0)} | k = 1, \ldots, n_0\}$, and let the prevertices on C_1 be at $\{a_k^{(1)} | k = 1, \ldots, n_1\}$. The turning angle parameters at these prevertices are denoted $\{\beta_k^{(0)} | k = 1, \ldots, n_0\}$ and $\{\beta_k^{(1)} | k = 1, \ldots, n_1\}$, respectively. As in Chapter 7, in order that the polygonal boundaries are closed it is necessary that

$$\sum_{k=1}^{n_0} \beta_k^{(0)} = -2, \qquad \sum_{k=1}^{n_1} \beta_k^{(1)} = +2. \tag{8.80}$$

(a) Define

$$T(z) \equiv \frac{zf''(z)}{f'(z)}. \tag{8.81}$$

Show that, on both C_0 and C_1,

$$\text{Re}[T(z)] = -1. \tag{8.82}$$

(b) Show that, at any prevertex $a_k^{(j)}$ for $j = 0, 1$, $T(z)$ has a simple pole of strength $\beta_k^{(j)} a_k^{(j)}$.

(c) Show that $T(z)$ is a loxodromic function.

(d) Hence show, using the result of Exercise 8.2, that

$$T(z) = \sum_{k=1}^{n_0} \beta_k^{(0)} K(z, a_k^{(0)}) + \sum_{k=1}^{n_1} \beta_k^{(1)} K(z, a_k^{(1)}) - 2. \tag{8.83}$$

(e) Hence show that

$$\frac{df}{dz} = \frac{A}{z^2} \prod_{k=1}^{n_0} [\omega(z, a_k^{(0)})]^{\beta_k^{(0)}} \prod_{k=1}^{n_1} [\omega(z, a_k^{(1)})]^{\beta_k^{(1)}}, \tag{8.84}$$

where A is some complex constant.[27]

[27] This derivation was presented in [36].

8.7. **(The S–C function $S(z,a)$ for a concentric annulus)** Let $w = \eta(z,a)$ be the bounded circular slit mapping from a concentric annulus $\rho < |z| < 1$ to the interior of the unit w disc $|w| < 1$ exterior to a circular arc slit, as given in (5.26). The point a inside the annulus maps to $w = 0$. This circular slit mapping was used in Chapter 7 in the construction of the formula for an S–C mapping to a doubly connected polygon.

(a) Show that, for the concentric annulus, the function $S(z,a)$ in (7.64) is

$$S(z,a) = \frac{\omega(z,a)\omega(z,1/\overline{a})}{z\omega(z,\gamma_1^{(1)})\omega(z,\gamma_2^{(1)})}\left[K(z,a) - K(z,1/\overline{a})\right], \tag{8.85}$$

where $\gamma_1^{(1)}$ and $\gamma_2^{(1)}$ are the two zeros of the derivative $\partial\eta/\partial z$ on C_1.

(b) Use the properties of $K(z,a)$ established in Exercise 5.1 to show that

$$K(z,a) - K(z,1/\overline{a}) \tag{8.86}$$

is a loxodromic function with simple poles in the fundamental annulus F given by $\rho < |z| < 1/\rho$ at $z = a$ and $z = 1/\overline{a}$.

(c) Show that

$$\frac{\partial\eta}{\partial z} = \frac{\eta(z,a)}{z}\left[K(z,a) - K(z,1/\overline{a})\right]. \tag{8.87}$$

Hence argue that the loxodromic function (8.86) has simple zeros at $\gamma_1^{(1)}$ and $\gamma_2^{(1)}$.

(d) Show, on use of the identities (5.4)–(5.6) satisfied by the prime function of the concentric annulus, that the function

$$\mathcal{A}(z,a) \equiv \frac{\omega(z,\gamma_1^{(1)})\omega(z,\gamma_2^{(1)})}{z\omega(z,a)\omega(z,1/\overline{a})} \tag{8.88}$$

is also a loxodromic function with the same poles and zeros in F as the function (8.86).

(e) Hence show that

$$K(z,a) - K(z,1/\overline{a}) = B\mathcal{A}(z,a) \tag{8.89}$$

for some constant B.

(f) Finally, combine the results in (a), (d), and (e) to show that

$$S(z,a) = \frac{B}{z^2}. \tag{8.90}$$

8.8. Use the two different representations of a loxodromic function introduced in this chapter to demonstrate that formulas (7.102) and (7.106), for the doubly connected S–C mapping to a channel with a single slit perpendicular to the channel walls, are equivalent.

8.9. **(Trisecant Fay identity)** Define the cross-ratio-like quantity

$$\tilde{p}(z,w,a,b) \equiv \frac{\omega(z,a)\omega(w,b)}{\omega(z,b)\omega(w,a)}. \tag{8.91}$$

By showing that the left-hand side is a loxodromic function as a function both of z and w, establish the identity

$$\frac{\omega(z/w,k)\omega(b/a,k)}{\omega(zb/(wa),k)}\tilde{p}(z,w,a,b) + \frac{\omega(z/a,k)\omega(b/w,k)}{\omega(zb/(wa),k)}\tilde{p}(z,a,w,b) = \omega(1,k). \tag{8.92}$$

8.10. **("Partial fraction"–type decomposition for integration)**[28] This exercise is concerned with the indefinite integral

$$I(z) = \int^z \left[\frac{\omega(z',1/\gamma_1)^2\omega(z',1/\gamma_2)^2}{\omega(z',\alpha)\omega(z',1/\alpha)\omega(z',\beta)\omega(z',1/\beta)}\right] dz', \tag{8.93}$$

where $\omega(.,.)$ is the prime function for the concentric annulus

$$\rho < |z| < 1 \tag{8.94}$$

and the parameters γ_1, γ_2, α, and β are real and satisfy

$$\alpha\beta = \gamma_1\gamma_2 = -\rho. \tag{8.95}$$

(a) Show that the function defined by

$$Q(z) = z\left[\frac{\omega(z,1/\gamma_1)^2\omega(z,1/\gamma_2)^2}{\omega(z,\alpha)\omega(z,1/\alpha)\omega(z,\beta)\omega(z,1/\beta)}\right] \tag{8.96}$$

is a loxodromic function. Identify the poles of this function.

(b) Hence show that we can write

$$Q(z) = aK(z,\alpha) + bK(z,1/\alpha) + cK(z,\beta) + dK(z,1/\beta) + e \tag{8.97}$$

for some constants a, b, c, d, and e satisfying

$$a+b+c+d=0. \tag{8.98}$$

Here $K(.,.)$ is the function defined in (5.77).

(c) Show, on use of the relation (8.95), that

$$a+c=0, \qquad b+d=0, \qquad b=-\chi a, \tag{8.99}$$

where

$$\chi = \frac{1}{\rho^2\alpha^2}\left[\frac{\omega(\alpha,\gamma_1)\omega(\alpha,\gamma_2)}{\omega(\alpha,1/\gamma_1)\omega(\alpha,1/\gamma_2)}\right]^2. \tag{8.100}$$

(d) Show that an expression for e is

$$e = -[aK(1/\gamma_1,\alpha) + bK(1/\gamma_1,1/\alpha) + cK(1/\gamma_1,\beta) + dK(1/\gamma_1,1/\beta)]. \tag{8.101}$$

[28] The calculation in this exercise was used in a study of compressible vortex streets [39].

(e) Show that

$$I(z) = a\left[\log\left(\frac{\omega(z,\alpha)}{\omega(z,\beta)}\right) - \chi\log\left(\frac{\omega(z,1/\alpha)}{\omega(z,1/\beta)}\right)\right] + e\log z + f, \quad (8.102)$$

where f is a constant.

8.11. **(Evaluation of an integral)** By using the ideas of Exercise 8.10, find an alternative expression for the indefinite integral

$$\int^z L(z',1)^2 dz'. \quad (8.103)$$

8.12. **(Map of upper half annulus to upper half plane)** Show that the conformal mapping

$$w = L(z,1) \quad (8.104)$$

transplants the upper half annulus $\rho < |z| < 1, \text{Im}[z] > 0$ to the upper half w plane.

8.13. **(Nonmeromorphic functions with the loxodromic property)** Consider the function

$$f(z) = -\frac{\mathrm{i}}{2\pi}\log\left(\frac{z}{\sqrt{\rho}}\right) - \frac{\mathrm{i}\log\rho}{\pi}K\left(z,\frac{1}{\sqrt{\rho}}\right). \quad (8.105)$$

Show that it satisfies the functional relation

$$f(\rho^2 z) = f(z). \quad (8.106)$$

Is $f(z)$ a loxodromic function?

8.14. **(Radial slit maps and the Jacobi sn function)** This exercise elucidates a functional connection between the loxodromic radial slit maps from a concentric annulus constructed in Chapter 5 with a classical elliptic function known as the Jacobi sn function.

(a) Show that the mapping from the upper half ζ annulus $\rho < |\zeta| < 1, \text{Im}[\zeta] > 0$ to the square $-1 < \text{Re}[w] < 1, 0 < \text{Im}[w] < 2$ is

$$w = \frac{\log(\zeta/\sqrt{\rho})}{\log(1/\sqrt{\rho})}, \qquad \rho = e^{-\pi}. \quad (8.107)$$

(b) Verify that this function maps $\zeta = \sqrt{\rho}$ to $w = 0$, the point $\zeta = 1$ to $w = 1$, and the point $\zeta = -1$ to $w = 1 + 2\mathrm{i}$.

(c) Show that the inverse function is

$$\zeta = e^{\pi(w-1)/2}. \quad (8.108)$$

(d) Use the result of Exercise 7.4 to argue that the S–C map from an upper half χ plane to the same square $-1 < \text{Re}[w] < 1, 0 < \text{Im}[w] < 2$ satisfies the differential equation

$$\frac{dw}{d\chi} = \frac{C}{\sqrt{(1-\chi^2)(1-k'^2\chi^2)}}, \qquad C = \left(\int_0^{\pi/2}\frac{dt}{\sqrt{1-k'^2\sin^2 t}}\right)^{-1}, \quad (8.109)$$

where the points $\chi = \pm 1$ map to $w = \pm 1$, the points $\chi = \pm 1/k'$ map to $w = \pm 1 + 2\mathrm{i}$, and the point $\chi = 0$ maps to $w = 0$.

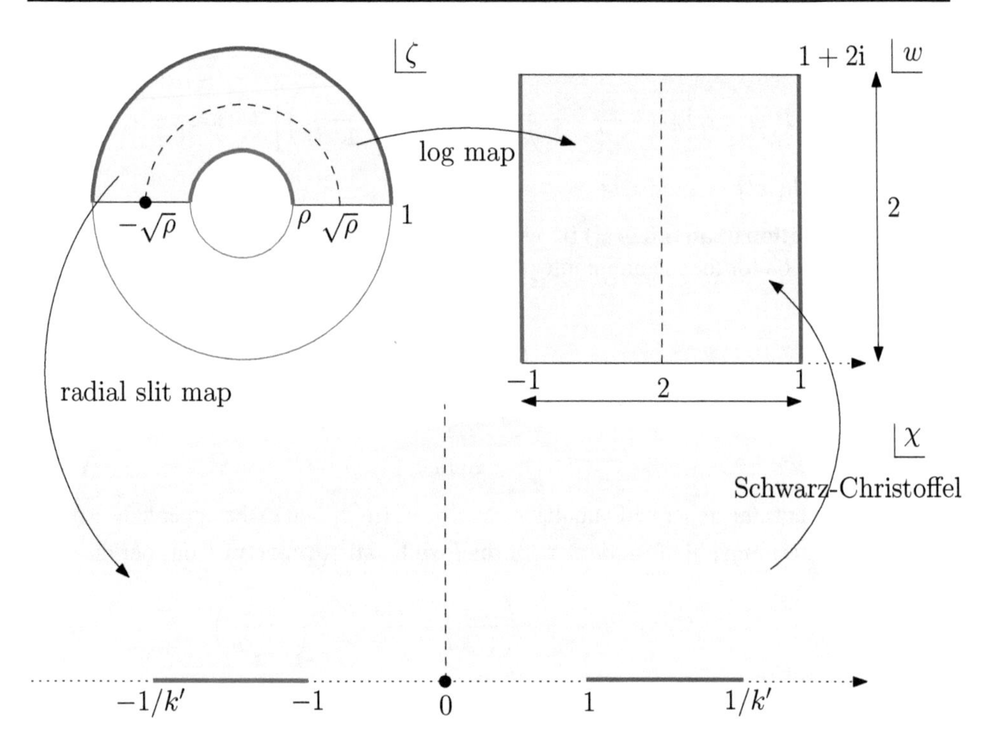

Figure 8.2. *Sequence of conformal maps used to derive the identity (8.116) connecting the Jacobi sn function with a radial slit map.*

(e) We now introduce the fact that the solution of the first order differential equation

$$\left(\frac{d\chi}{dw}\right)^2 = (1-\chi^2)(1-k'^2\chi^2) \tag{8.110}$$

having a simple zero at $w = 0$ is known to be given by the Jacobi sn function:

$$\chi = \mathrm{sn}(w, k'). \tag{8.111}$$

Given this fact, show using (8.109) that we can write

$$\chi = \mathrm{sn}(w/C, k'). \tag{8.112}$$

(f) Show, by considering a radial slit mapping of the form (8.8), that the upper half annulus in the ζ plane can be mapped directly to the χ plane by means of

$$\chi = \left[\frac{\omega(1,-\sqrt{\rho})\omega(1,-1/\sqrt{\rho})}{\omega(1,\sqrt{\rho})\omega(1,1/\sqrt{\rho})}\right]\frac{\omega(\zeta,-\sqrt{\rho})\omega(\zeta,-1/\sqrt{\rho})}{\omega(\zeta,\sqrt{\rho})\omega(\zeta,1/\sqrt{\rho})}, \tag{8.113}$$

where

$$\frac{1}{k'} = \left[\frac{\omega(1,-\sqrt{\rho})\omega(1,-1/\sqrt{\rho})}{\omega(1,\sqrt{\rho})\omega(1,1/\sqrt{\rho})}\right]\frac{\omega(-1,-\sqrt{\rho})\omega(-1,-1/\sqrt{\rho})}{\omega(-1,\sqrt{\rho})\omega(-1,1/\sqrt{\rho})}. \tag{8.114}$$

Recall that we established earlier in this chapter that such a radial slit map is a loxodromic function.

(g) Finally, combine (8.108), (8.112), and (8.113) to establish the following functional identity between this loxodromic function and the Jacobi sn function:

$$\mathrm{sn}(w/C, k') = \left[\frac{\omega(1, -\sqrt{\rho})\omega(1, -1/\sqrt{\rho})}{\omega(1, \sqrt{\rho})\omega(1, 1/\sqrt{\rho})}\right] \tag{8.115}$$

$$\times \frac{\omega(e^{\pi(w-1)/2}, -\sqrt{\rho})\omega(e^{\pi(w-1)/2}, -1/\sqrt{\rho})}{\omega(e^{\pi(w-1)/2}, \sqrt{\rho})\omega(e^{\pi(w-1)/2}, 1/\sqrt{\rho})}, \tag{8.116}$$

where ρ, C, and k' are given by (8.107), (8.109), and (8.114).

8.15. **(Circular slit maps and the Jacobi cn function)** This exercise builds on Exercise 8.14 to elucidate a similar connection between the loxodromic unbounded circular slit map (8.18) and the classical elliptic function known as the Jacobi cn function.

(a) Show that the shift and rotation of the square domain in the w plane of Exercise 8.14 given by

$$\tilde{w} = e^{\mathrm{i}\pi/4}(w - \mathrm{i}) \tag{8.117}$$

is a square now centered at the origin in the $\tilde{w}$ plane with vertices at $\tilde{w} = \pm 1, \pm \mathrm{i}$.

(b) With the variable ζ associated with the same annulus $\rho < |\zeta| < 1$ as in Exercise 8.14 verify that ζ and $\tilde{w}$ are related via

$$\zeta = \mathrm{i}\sqrt{\rho}e^{\pi/2(\tilde{w}e^{-\mathrm{i}\pi/4})}. \tag{8.118}$$

(c) Show that the S–C map from a unit $\tilde{\chi}$ disc to this square domain in the $\tilde{w}$ plane satisfies the differential equation

$$\frac{d\tilde{w}}{d\tilde{\chi}} = \frac{B}{\sqrt{(1-\tilde{\chi}^2)(1+\tilde{\chi}^2)}}, \tag{8.119}$$

where, on the grounds of symmetry, we take the prevertices to be at $\tilde{\chi} = \pm 1, \pm \mathrm{i}$ and we insist that

$$\int_0^1 \frac{B d\tilde{\chi}}{\sqrt{(1-\tilde{\chi}^2)(1+\tilde{\chi}^2)}} = \sqrt{2}, \tag{8.120}$$

which is a statement of the condition that the difference between the image of $\tilde{\chi} = 1$ and $\tilde{\chi} = 0$ is $\sqrt{2}$.

(d) It is known that $y = \mathrm{cn}(x, k)$ is the solution of the differential equation

$$\left(\frac{dy}{dx}\right)^2 = (1 - y^2)(1 - k^2 + k^2 y^2) \tag{8.121}$$

having a simple zero at $x = -K$, where

$$K \equiv \int_0^{\pi/2} \frac{dt}{\sqrt{1 - \sin^2 t/2}}. \tag{8.122}$$

By picking $k = 1/\sqrt{2}$, show from (8.119) that we can write

$$\tilde{\chi} = \mathrm{cn}\left(\frac{\sqrt{2}\tilde{w}}{B} + A, \frac{1}{\sqrt{2}}\right), \tag{8.123}$$

where A is some constant.

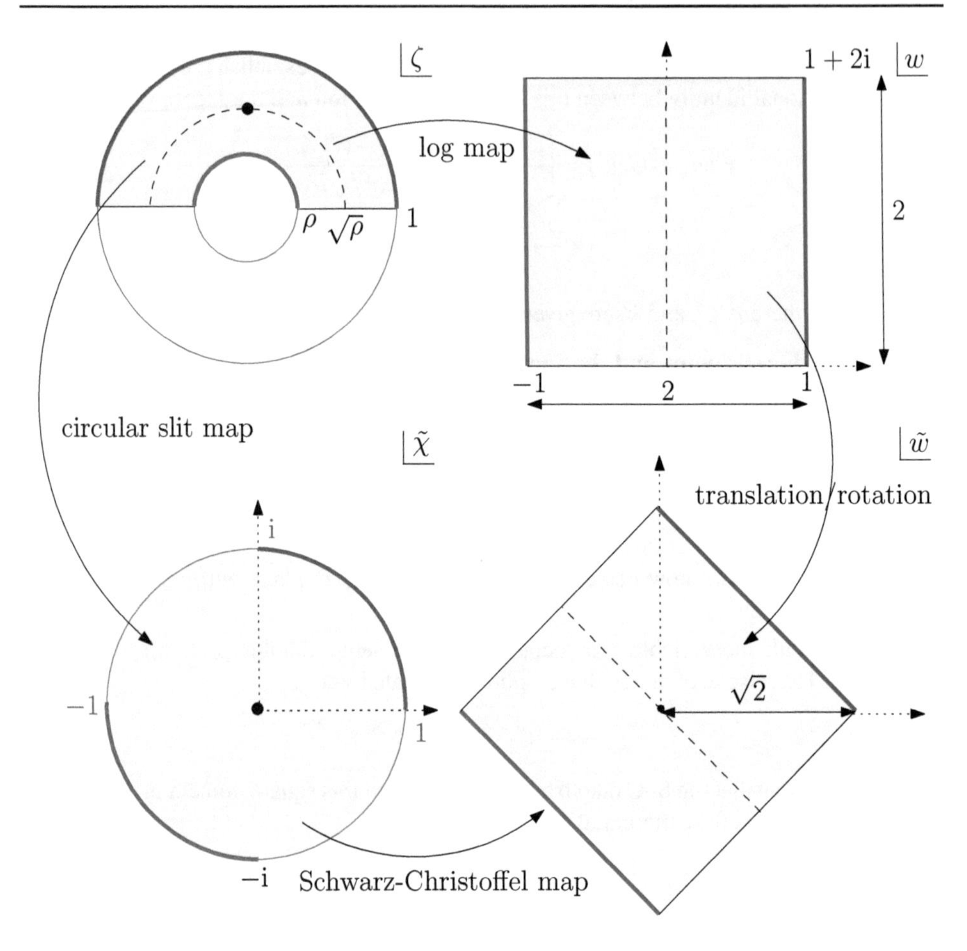

Figure 8.3. *Sequence of conformal maps used to derive the identity (8.132) connecting the Jacobi cn function with a circular slit map.*

(e) Verify that

$$\frac{2}{B} = K \tag{8.124}$$

and

$$A = -K, \tag{8.125}$$

where the latter condition is a result of insisting that $\chi = 0$ corresponds to $\tilde{w} = 0$. Hence deduce that

$$\tilde{\chi} = \mathrm{cn}\left(\frac{K\tilde{w}}{\sqrt{2}} - K, \frac{1}{\sqrt{2}}\right). \tag{8.126}$$

(f) Show, from the theory presented in Chapter 5, that the mapping from the upper half annulus in the ζ plane directly to the interior of the unit disc in a $\tilde{\chi}$ plane is the unbounded circular slit map

$$\tilde{\chi} = -e^{\mathrm{i}\pi/4}\left[\frac{\omega(\zeta,\alpha)\omega(\zeta,1/\alpha)}{\omega(\zeta,\overline{\alpha})\omega(\zeta,1/\overline{\alpha})}\right], \qquad \alpha = \mathrm{i}\sqrt{\rho} = \mathrm{i}e^{-\pi/2}. \tag{8.127}$$

Verify the correspondence of points given by

$$\zeta = 1 \mapsto \tilde{\chi} = 1, \tag{8.128}$$

$$\zeta = -1 \mapsto \tilde{\chi} = \mathrm{i}, \tag{8.129}$$

$$\zeta = -\rho \mapsto \tilde{\chi} = -1, \tag{8.130}$$

$$\zeta = \rho \mapsto \tilde{\chi} = -\mathrm{i}. \tag{8.131}$$

(g) Hence deduce that functional identity

$$\mathrm{cn}\left(\frac{Kwe^{\mathrm{i}\pi/4}}{\sqrt{2}} - K, \frac{1}{\sqrt{2}}\right) = -e^{\mathrm{i}\pi/4}\left[\frac{\omega(\mathrm{i}\sqrt{\rho}e^{\pi w/2}, \alpha)\omega(\mathrm{i}\sqrt{\rho}e^{\pi w/2}, 1/\alpha)}{\omega(\mathrm{i}\sqrt{\rho}e^{\pi w/2}, \overline{\alpha})\omega(\mathrm{i}\sqrt{\rho}e^{\pi w/2}, 1/\overline{\alpha})}\right]. \tag{8.132}$$

This identity shows that the Jacobi cn function has an interpretation in terms of the unbounded circular slit maps of Chapter 5.

(h) By finding a Cayley map directly from the $\tilde{\chi}$ plane of Figure 8.3 to the upper half χ plane of Figure 8.2 show that

$$\frac{1}{k'} = \frac{\sqrt{2}+1}{\sqrt{2}-1}. \tag{8.133}$$

This gives an alternative, and simpler, expression for $1/k'$ than that given in (8.114).

Chapter 9

Automorphic functions

9.1 ▪ Extending the function theory: Automorphic functions

Loxodromic functions lie within the wider class of *automorphic functions*. Automorphic functions are the natural analogues of loxodromic functions when $M \geq 2$ and the underlying circular domain D_z has more than one hole. In an effort to underline this we emulate the presentation of Chapter 8 as closely as possible.

Automorphic functions will be an important tool in subsequent chapters; we will need to use them in Chapters 10 and 11. We aim here to give the applied scientist the essential elements of the theory of automorphic functions to solve some of the problems to follow later in this book. A much more comprehensive treatment is given in the monograph by Ford [81].[29]

9.2 ▪ Radial and circular slit maps as automorphic functions

Just as certain doubly connected conformal slit maps from Chapter 5 provided examples of loxodromic functions, their triply connected analogues from Chapter 6 provide examples of more general automorphic functions.

To see this, consider a triply connected circular domain D_z in which the two circles C_1 and C_2 are centered on the real axis inside the unit disc $|z| < 1$ so that D_z is reflectionally symmetric about the real z axis. Let

$$w = \mathcal{A}(z) \tag{9.1}$$

be the radial slit mapping transplanting D_z to three slits on the real axis in a complex w plane. Since C_0 is transplanted to a slit on the real axis, for $z \in C_0$,

$$\overline{\mathcal{A}(z)} = \mathcal{A}(z) \qquad \text{or} \qquad \overline{\mathcal{A}}(1/z) = \mathcal{A}(z), \tag{9.2}$$

where we have used the fact that $\overline{z} = 1/z$ on C_0. The circles C_j, for $j = 1, 2$, are also transplanted to slits on the real w axis. This means that, for $z \in C_j$,

$$\overline{\mathcal{A}(z)} = \mathcal{A}(z) \qquad \text{or} \qquad \overline{\mathcal{A}}(S_j(z)) = \mathcal{A}(z), \qquad j = 1, 2, \tag{9.3}$$

[29]Readers should be advised, however, that Ford's book contains no mention of the prime function.

where $S_j(z)$ is the Schwarz function of circle C_j introduced in §3.6 of Chapter 3. The relations

$$\overline{\mathcal{A}}(1/z) = \mathcal{A}(z), \qquad \overline{\mathcal{A}}(S_j(z)) = \mathcal{A}(z), \qquad j = 1, 2, \tag{9.4}$$

can be analytically continued off the circles on which they were originally derived. From them we infer that

$$\overline{\mathcal{A}}(S_j(z)) = \overline{\mathcal{A}}(1/z), \qquad j = 1, 2, \tag{9.5}$$

or, on taking a complex conjugate,

$$\mathcal{A}(\overline{S_j}(1/\overline{z})) = \mathcal{A}(1/\overline{z}). \tag{9.6}$$

Now $1/\overline{z}$ is an arbitrary complex variable, so replacing $1/\overline{z}$ by z in (9.6) implies that $\mathcal{A}(z)$ satisfies the functional identities

$$\mathcal{A}(\theta_j(z)) = \mathcal{A}(z), \qquad j = 1, 2, \tag{9.7}$$

where the Möbius maps $\{\theta_j(z)|j = 1, \ldots, M\}$ are familiar from §3.6.

In Chapter 6 it was established that the functional form of such a radial slit mapping is

$$\mathcal{A}(z) = \frac{\omega(z, a)\omega(z, 1/a)}{\omega(z, b)\omega(z, 1/b)}, \tag{9.8}$$

where $a, b \in D_z$ and A are all real parameters and $\omega(.,.)$ is the prime function associated with the triply connected circular domain D_z and satisfying identities (6.1)–(6.4). It is instructive to check that (9.8) satisfies conditions (9.7). We find

$$\mathcal{A}(\theta_j(z)) = \frac{\omega(\theta_j(z), a)\omega(\theta_j(z), 1/a)}{\omega(\theta_j(z), b)\omega(\theta_j(z), 1/b)}, \qquad j = 1, 2, \tag{9.9}$$

and, on use of the prime function identities (6.3)–(6.4),

$$\mathcal{A}(\theta_j(z)) = e^{2\pi \mathrm{i}(v_j(a) + v_j(1/a) - v_j(b) - v_j(1/b))}\mathcal{A}(z), \qquad j = 1, 2. \tag{9.10}$$

However, in Exercise 4.7 it was established that for reflectionally symmetric domains of this type,

$$\overline{v_j}(z) = -v_j(z) + d_j, \qquad j = 1, 2, \tag{9.11}$$

for some real constants $\{d_j|j = 1, 2\}$ which can be set to vanish. Hence (9.10) becomes

$$\mathcal{A}(\theta_j(z)) = e^{2\pi \mathrm{i}(v_j(a) - \overline{v_j}(1/a) - v_j(b) + \overline{v_j}(1/b))}\mathcal{A}(z), \qquad j = 1, 2, \tag{9.12}$$

or, on use of the fact that

$$\overline{v_j}(1/z) = v_j(z), \qquad j = 1, 2, \tag{9.13}$$

we confirm that $\mathcal{A}(z)$ does indeed satisfy (9.7). It is also clear that $\mathcal{A}(z)$ as given in (9.8) is meromorphic, that is, its only singularities are poles, in the doubled domain $F = D_z \cup D_z'$.

Meromorphic functions in F that satisfy the functional relations (9.7) with respect to a set of mappings $\{\theta_j(z)|j = 1, \ldots, M\}$ associated with the given circular domain D_z are known as *automorphic functions*.

Another slit mapping encountered in Chapter 6 is the unbounded circular slit mapping to three slits on the unit circle as illustrated in Figure 6.5. Let this different mapping to a complex w plane again be denoted by the same designation,

$$w = \mathcal{A}(z). \tag{9.14}$$

Since the image of C_0 under this mapping lies on the unit w circle,

$$\mathcal{A}(z)\overline{\mathcal{A}(z)} = 1, \tag{9.15}$$

implying that

$$\mathcal{A}(z) = \frac{1}{\overline{\mathcal{A}}(1/z)}, \tag{9.16}$$

where we have used the fact that $\overline{z} = 1/z$ on C_0. Since the image of the circle C_j for $j = 1, 2$ also lies on the unit w circle,

$$\mathcal{A}(z)\overline{\mathcal{A}(z)} = 1, \tag{9.17}$$

implying that

$$\mathcal{A}(z) = \frac{1}{\overline{\mathcal{A}}(S_j(z))}. \tag{9.18}$$

Both (9.16) and (9.18) can be analytically continued off the circles where they were derived, and together they imply the same relation (9.5), leading once again to (9.7).

It is again possible to check this property directly using the functional form for such a mapping $\mathcal{A}(z)$ found in Chapter 4:

$$\mathcal{A}(z) \equiv \frac{\omega(z,a)\omega(z,1/a)}{\omega(z,\overline{a})\omega(z,1/\overline{a})}, \qquad a = \mathrm{i}r, \ \ 0 < r < 1. \tag{9.19}$$

The identities (6.1)–(6.4) can be used to confirm that this expression satisfies (9.7). It is also clear that this function has two simple poles in $D_z \cup D_z'$ at $\overline{a}$ and $1/\overline{a}$. It is therefore an automorphic function.

Yet another example of a conformal slit map that is also an automorphic function is discussed in Exercise 9.1.

With these example automorphic functions in mind to motivate our inquiries, we now explore how to construct more general functions within this class. In applications it is often required to construct automorphic functions having a preordained distribution of poles. Before introducing the constructive techniques to do this it is necessary to say a bit more about Schottky groups. A preview of these groups has already been given in Exercise 3.3.

9.3 ▪ Schottky groups

In Chapter 8 we saw that a loxodromic function is invariant under the action on its argument z of the Schottky group of Möbius maps

$$\Theta \equiv \{\rho^{2n} z | n \in \mathbb{Z}\}. \tag{9.20}$$

A different Schottky group, which we continue to denote by Θ, can be associated with a triply connected circular domain with circle centers at $\{\delta_j | j = 1, 2\}$ and radii $\{q_j | j = 1, 2\}$. In contrast to the group (9.20), which is generated by $\theta_1(z) = \rho^2 z$ and its inverse $\theta_1^{-1}(z) = \rho^{-2} z$, the generators of the new group Θ are the two Möbius maps $\theta_1(z), \theta_2(z)$ and their inverses $\theta_1^{-1}(z)$ and $\theta_2^{-1}(z)$, which are also Möbius maps. Any composition of these four maps will also be a Möbius map. The Schottky group Θ is defined as the set of Möbius maps that are compositions of any of these four generating Möbius maps $\{\theta_1(z), \theta_2(z), \theta_1^{-1}(z), \theta_2^{-1}(z)\}$. The domain of discontinuity for this group [81] is the region $F = D_z \cup D_z'$ exterior to the four circles C_1, C_2, C_1', and C_2'; this can be thought of as the region where a representative set of zeros and poles all sit. We say that any function that is meromorphic in F and invariant under the action of Θ on its argument is an automorphic function with respect to the group Θ.

9.4 ▪ Properties of automorphic functions

Some general properties of automorphic functions can be obtained by considering certain strategically chosen contour integrals.

Let $\mathcal{A}(z)$ be an automorphic function satisfying

$$\mathcal{A}(\theta_j(z)) = \mathcal{A}(z), \qquad \theta_j(z) \in \Theta, \tag{9.21}$$

where Θ is some Schottky group of the kind discussed in the previous section. Suppose too that $\mathcal{A}(z)$ has Z zeros and N poles in F counted by multiplicity. To be concrete, we assume F is unbounded and that none of the poles of $\mathcal{A}(z)$ are at infinity. All of the following arguments can be extended when these assumptions are relaxed.

By the argument principle,

$$Z - N = \lim_{R\to\infty} \left[\frac{1}{2\pi \mathrm{i}} \int_{\partial F_R} \frac{\mathcal{A}'(z)}{\mathcal{A}(z)} dz \right], \tag{9.22}$$

where the integration is around the dumbbell contour ∂F_R shown in Figure 4.2 and originally introduced in Chapter 4 to establish properties of the differential $d\Pi^{z,w}(a)$. As $R \to \infty$ (9.22) becomes

$$Z - N = -\frac{1}{2\pi \mathrm{i}} \sum_{j=1}^{M} \left\{ \int_{C_j} \frac{d\mathcal{A}(z)}{\mathcal{A}(z)} + \int_{C_j'} \frac{d\mathcal{A}(z)}{\mathcal{A}(z)} \right\}, \tag{9.23}$$

since the contribution from C_R vanishes as $R \to \infty$ under the assumption that $\mathcal{A}(z)$ has no poles at infinity; we have also made use of the continuity of $\mathcal{A}(z)$ across all the barriers (L_j, L_j^+) for $j = 1, \dots, M$. If we introduce the change of variable $z \mapsto \theta_j(z)$ in the integral around C_j on the right-hand side of (9.23), where we must introduce a minus sign owing to the change in the direction of integration associated with this change of variable, we can alternatively write

$$Z - N = \sum_{j=1}^{M} \int_{C_j'} \left[\frac{d\mathcal{A}(\theta_j(z))}{\mathcal{A}(\theta_j(z))} - \frac{d\mathcal{A}(z)}{\mathcal{A}(z)} \right] = 0, \tag{9.24}$$

where, in the second equality, we have used the automorphicity property (9.21) of $\mathcal{A}(z)$. We conclude that

$$Z = N. \tag{9.25}$$

An automorphic function therefore has the same number of zeros as poles in F.

There are other constraints on the poles and zeros of an automorphic function. These can be deduced by considering other integrals taken around this same dumbbell contour ∂F_R.

Suppose that $\mathcal{A}(z)$ has N distinct simple poles in F located at positions $\{b_n | n = 1, \dots, N\}$ so that, near b_n,

$$\mathcal{A}(z) = \frac{r_n}{z - b_n} + \text{locally analytic function}, \qquad n = 1, \dots, N. \tag{9.26}$$

Hence r_n is the residue of $\mathcal{A}(z)$ at b_n. We know there will be N zeros of $\mathcal{A}(z)$ in F, so let these be simple zeros at positions $\{a_n | n = 1, \dots, N\}$. Assume, for ease of exposition, that none of the poles of $\mathcal{A}(z)$ are at infinity; we also avoid the situation where any pole or zero of $\mathcal{A}(z)$ is on the boundary of F, although these special cases can be easily incorporated into the general theory.

Consider the contour integrals

$$\lim_{R\to\infty}\left[\int_{\partial F_R} v_j(z)\frac{\mathcal{A}'(z)}{\mathcal{A}(z)}dz\right], \qquad j=1,\ldots,M, \tag{9.27}$$

where the functions $\{v_j(z)|j=1,\ldots,M\}$ are precisely those introduced in Chapter 2. By the residue theorem, and the fact that $\mathcal{A}'(z)/\mathcal{A}(z)$ has a simple pole of residue $+1$ at any simple zero a_n of $\mathcal{A}(z)$ and a simple pole of residue -1 at any simple pole b_n of $\mathcal{A}(z)$, the integral in (9.27) equals

$$2\pi\mathrm{i}\sum_{n=1}^{N}[v_j(a_n)-v_j(b_n)]. \tag{9.28}$$

On the other hand, we can add up the contributions for all the boundary components of ∂F_R. The contribution from C_R as $R\to\infty$ can be shown to vanish, while the other boundary contributions give

$$-\sum_{k=1}^{M}\left\{\int_{C_k} v_j(z)\frac{\mathcal{A}'(z)}{\mathcal{A}(z)}dz+\int_{C_k'} v_j(z)\frac{\mathcal{A}'(z)}{\mathcal{A}(z)}dz\right. \tag{9.29}$$

$$\left.+\int_{L_k} v_j(z)\frac{\mathcal{A}'(z)}{\mathcal{A}(z)}dz+\int_{L_k^+} v_j(z)\frac{\mathcal{A}'(z)}{\mathcal{A}(z)}dz\right\}. \tag{9.30}$$

This can be written

$$-\sum_{k=1}^{M}\left\{-\int_{C_k'} v_j(\theta_k(z))\frac{\mathcal{A}'(z)}{\mathcal{A}(z)}dz+\int_{C_k'} v_j(z)\frac{\mathcal{A}'(z)}{\mathcal{A}(z)}dz\right. \tag{9.31}$$

$$\left.+\int_{L_k} v_j(z)\frac{\mathcal{A}'(z)}{\mathcal{A}(z)}dz+\int_{L_k^+} v_j(z)\frac{\mathcal{A}'(z)}{\mathcal{A}(z)}dz\right\}, \tag{9.32}$$

where in the integral around C_k we have introduced the change of variable $z\mapsto\theta_k(z)$ and used the automorphicity property (9.21) of $\mathcal{A}(z)$. Now we know that the function $v_j(z)$ jumps by $+1$ as its argument travels continuously from L_j anticlockwise around C_j to meet L_j^+, but it has no jump across L_k for $k\neq j$. Moreover, both $\mathcal{A}'(z)$ and $\mathcal{A}(z)$ are continuous across all the barriers $\{(L_j,L_j^+)|j=1,\ldots,M\}$. Also, from the general theory of Chapter 3,

$$v_j(\theta_k(z))-v_j(z)=\tau_{jk}. \tag{9.33}$$

By virtue of these observations (9.32) reduces to

$$\sum_{k=1}^{M}\tau_{jk}\int_{C_k'}\frac{\mathcal{A}'(z)}{\mathcal{A}(z)}dz-\int_{L_j^+}\frac{\mathcal{A}'(z)}{\mathcal{A}(z)}dz, \tag{9.34}$$

and we conclude that

$$2\pi\mathrm{i}\sum_{n=1}^{N}[v_j(a_k)-v_j(b_k)]=\int_{L_j}d\log\mathcal{A}(z)+\sum_{k=1}^{M}\tau_{jk}\int_{C_k'}d\log\mathcal{A}(z). \tag{9.35}$$

But the quantities

$$\frac{1}{2\pi\mathrm{i}}\int_{L_j}d\log\mathcal{A}(z), \qquad \frac{1}{2\pi\mathrm{i}}\int_{C_k'}d\log\mathcal{A}(z) \tag{9.36}$$

must be integers since $\mathcal{A}(z)$ returns to its same value at each end of L_j (by its invariance under the group action) while C'_k is a closed curve. Finally, (9.35) implies that

$$\sum_{n=1}^{N}[v_j(a_n) - v_j(b_n)] = m_j + \sum_{k=1}^{M}\tau_{jk}n_k, \qquad j = 1, \dots, M, \tag{9.37}$$

for some integers m_j, n_k.

(9.37) is one statement of the M constraints on the N zeros and poles of an automorphic function.

There is another way to state these conditions. Consider a different set of M contour integrals given by

$$\lim_{R\to\infty}\int_{\partial F_R}\mathcal{A}(z)dv_j(z), \qquad j = 1, \dots, M. \tag{9.38}$$

Each integrand is a single-valued meromorphic function in F, so, by the residue theorem, the integrals must equal

$$\sum_{n=1}^{N} r_n v'_j(b_n) = 0, \qquad j = 1, \dots, M, \tag{9.39}$$

where we have used (9.26). Now consider the contributions to the integral from the various components making up the contour ∂F_R. The contribution from C_R can be shown to vanish as $R \to \infty$ on use of the assumption that $\mathcal{A}(z)$ has no poles at infinity. The differential $dv_j(z)$ is analytic and single-valued in F and therefore also in the region bounded by ∂F_R for any sufficiently large R. It also satisfies

$$dv_j(\theta_k(z)) = dv_j(z), \qquad k = 1, \dots, M, \tag{9.40}$$

which can be seen from (9.33). Since $\mathcal{A}(z)$ is automorphic with respect to the group Θ, contributions from the integrations around each C_k will be exactly canceled by the contribution to the integral from C'_k. It follows that the integral (9.38) is zero. Therefore we conclude

$$\sum_{n=1}^{N} r_n v'_j(b_n) = 0, \qquad j = 1, \dots, M. \tag{9.41}$$

(9.41) is another statement of the M constraints on the N poles and zeros of the automorphic function $\mathcal{A}(z)$. In this case the dependence of (9.41) on the zeros $\{a_n | n = 1, \dots, N\}$ is hidden in the residues $\{r_n | n = 1, \dots, N\}$.

If $N = 1$, then, to satisfy (9.41), we must have $r_1 = 0$, which effectively removes the simple pole at b_1. Consequently it is not possible to find an automorphic function with just a single simple pole in F.

9.5 ▪ The quasi-automorphic function $\mathcal{K}(z, a)$

The function $\mathcal{K}(z, a)$ is not an automorphic function, but it comes in handy for constructing them. To illustrate this we now derive an alternative representation of the radial slit map (9.8).

Pick two distinct points $a, b \in \mathbb{R}$ in a circular domain D_z and define

$$\mathcal{H}(z, b) \equiv \mathcal{K}(z, b) - \mathcal{K}(z, 1/b) - [\mathcal{K}(a, b) - \mathcal{K}(a, 1/b)]. \tag{9.42}$$

This function has simple poles at b and $1/b$, and it is clear by the choice of the term in square brackets in (9.42) that it vanishes at $z = a$. Moreover, on use of (6.88), it is easily checked that $\mathcal{H}(z,b)$ satisfies

$$\mathcal{H}(\theta_j(z), b) = \mathcal{H}(z,b), \qquad j = 1, \ldots, M. \tag{9.43}$$

We have made use of the result established in Exercise 3.7 that for reflectionally symmetric domains D_z such as that considered here,

$$\overline{v_j}(z) = -v_j(z) + d_j \tag{9.44}$$

for some set of real constants $\{d_j | j = 1, \ldots, M\}$ (that can be set to vanish). It follows on use of (3.8) that

$$\overline{v_j}(1/z) = v_j(z) = -v_j(1/z) + d_j \tag{9.45}$$

and hence, on differentiation with respect to z, that

$$z v_j'(z) = (1/z) v_j'(1/z), \qquad j = 1, \ldots, M. \tag{9.46}$$

From (9.43) we deduce that $\mathcal{H}(z,b)$ is an automorphic function with simple poles at b and $1/b$. Since $\mathcal{H}(z,b)$ vanishes at $z = a$, the identity (6.86) can be used to verify that

$$\mathcal{H}(1/a, b) = 0. \tag{9.47}$$

So $\mathcal{H}(z,b)$ also vanishes at $1/a$.

Let $\mathcal{A}(z)$ be the radial slit map defined in (9.8). We have established that it is an automorphic function. The ratio

$$\mathcal{R}(z) \equiv \frac{\mathcal{H}(z,b)}{\mathcal{A}(z)} \tag{9.48}$$

is analytic and single-valued everywhere in F since the simple poles in the numerator and denominator at b and $1/b$ cancel each other, and any poles at a and $1/a$ due to the zeros of the denominator are removed by the zeros of the numerator at those points. Moreover, $\mathcal{R}(z)$ satisfies the identity

$$\mathcal{R}(\theta_j(z)) = \mathcal{R}(z), \qquad j = 1, \ldots, M. \tag{9.49}$$

Since $\mathcal{R}(z)$ is an automorphic function with no poles in the doubled domain F, it follows that it must be the constant function. Thus the two functions $\mathcal{H}(z,b)$ and $\mathcal{A}(z)$ are proportional:

$$\mathcal{H}(z,b) = a_0 \mathcal{A}(z). \tag{9.50}$$

The following identity has therefore been derived:

$$a_0 \left[\frac{\omega(z,a)\omega(z,1/a)}{\omega(z,b)\omega(z,1/b)} \right] = \mathcal{K}(z,b) - \mathcal{K}(z,1/b) - [\mathcal{K}(a,b) - \mathcal{K}(a,1/b)], \tag{9.51}$$

where the value of the constant a_0 can be determined by evaluating both sides at any nonsingular point, for example, at $z = 1$:

$$a_0 = [\mathcal{K}(1,b) - \mathcal{K}(1,1/b) - [\mathcal{K}(a,b) - \mathcal{K}(a,1/b)]] \frac{\omega(1,b)\omega(1,1/b)}{\omega(1,a)\omega(1,1/a)}. \tag{9.52}$$

Putting everything together we arrive at the identity

$$\frac{\mathcal{K}(z,b)-\mathcal{K}(z,1/b)-[\mathcal{K}(a,b)-\mathcal{K}(a,1/b)]}{\mathcal{K}(1,b)-\mathcal{K}(1,1/b)-[\mathcal{K}(a,b)-\mathcal{K}(a,1/b)]} = \frac{\omega(1,b)\omega(1,1/b)}{\omega(1,a)\omega(1,1/a)}\frac{\omega(z,a)\omega(z,1/a)}{\omega(z,b)\omega(z,1/b)}. \tag{9.53}$$

If we now make the choice $M=2$, we retrieve the result of Exercise 6.4, where "two views of a three-slit map" were studied. There the identity (9.53) was derived in a very different way by considering the conformal map from D_z to two slits on the real axis as an instance of a radial slit map on the one hand and a parallel slit map on the other.

9.6 ▪ Two ways to construct automorphic functions

The identity (9.53) indicates the possibility of representing a given automorphic function in two different ways. This feature is not special to that particular automorphic function but applies more generally. There are two ways to seek a representation of a given automorphic function, as we now illustrate.

Consider a triply connected circular domain D_z and its associated Schottky group Θ generated by two Möbius maps $\theta_1(z)$ and $\theta_2(z)$ and their inverses. The restriction to the case of a triply connected example is made only to simplify the presentation; the extension to constructing functions that are automorphic with respect to the Schottky groups Θ associated with any higher connected domain will be obvious.

Representation as a ratio of products of prime functions: For some $N \geq 2$ consider the function

$$\mathcal{A}(z) = A\prod_{n=1}^{N}\frac{\omega(z,a_n)}{\omega(z,b_n)}, \tag{9.54}$$

where $\{a_n, b_n | n=1,\dots,N\}$ is a set of distinct complex numbers. For ease of exposition, we assume that all the poles of $\mathcal{A}(z)$ are simple. We know that $\omega(z,a)$ has a simple zero at $z=a$ and, by virtue of the identities (6.3)–(6.4), at all equivalent points under the action of the Schottky group Θ. By virtue of the identities (6.3)–(6.4),

$$\mathcal{A}(\theta_j(z)) = A\prod_{n=1}^{N}\frac{\omega(\theta_j(z),a_n)}{\omega(\theta_j(z),b_n)} = e^{2\pi\mathrm{i}\sum_1^N[v_j(a_n)-v_j(b_n)]}\prod_{n=1}^{N}\frac{\omega(z,a_n)}{\omega(z,b_n)}, \quad j=1,2. \tag{9.55}$$

It is clear that $\mathcal{A}(z)$ will be automorphic with respect to the action of θ_1 and θ_2 if the zeros and poles $\{a_n, b_n | n=1,\dots,N\}$ satisfy the two conditions

$$\sum_1^N[v_1(a_n)-v_1(b_n)] = m_1, \qquad \sum_1^N[v_2(a_n)-v_2(b_n)] = m_2 \tag{9.56}$$

for some $m_1, m_2 \in \mathbb{Z}$. Observe that (9.56) corresponds to a constraint of the general form (9.37) derived earlier, and using general considerations.

Another function with the same zeros and poles is

$$\tilde{\mathcal{A}}(z) = e^{2\pi\mathrm{i}v_k(z)}\mathcal{A}(z) = Ae^{2\pi\mathrm{i}v_k(z)}\prod_{n=1}^{N}\frac{\omega(z,a_n)}{\omega(z,b_n)}, \tag{9.57}$$

where we can pick the integer $k = 1$ or $k = 2$. On letting $z \mapsto \theta_j(z)$ for $j = 1, 2$ we find

$$\tilde{\mathcal{A}}(\theta_j(z)) = A e^{2\pi i v_k(\theta_j(z))} \prod_{n=1}^{N} \frac{\omega(\theta_j(z), a_n)}{\omega(\theta_j(z), b_n)}. \tag{9.58}$$

In view of the identity

$$v_k(\theta_j(z)) - v_k(z) = \tau_{kj}, \tag{9.59}$$

and on use of the identities (6.3)–(6.4), we find

$$\tilde{\mathcal{A}}(\theta_j(z)) = e^{2\pi i \tau_{kj}} e^{2\pi i \sum_{n=1}^{N}[v_j(a_n) - v_j(b_n)]} \underbrace{A e^{2\pi i v_k(z)} \prod_{n=1}^{N} \frac{\omega(z, a_n)}{\omega(z, b_n)}}_{\tilde{\mathcal{A}}(z)}, \qquad j = 1, 2. \tag{9.60}$$

Thus $\tilde{\mathcal{A}}(z)$ will be invariant with respect to the action of θ_1 and θ_2 if the zeros and poles $\{a_n, b_n | n = 1, \dots, N\}$ satisfy the two conditions

$$\tau_{k1} + \sum_{n=1}^{N}[v_1(a_n) - v_1(b_n)] = m_1, \qquad \tau_{k2} + \sum_{n=1}^{N}[v_2(a_n) - v_2(b_n)] = m_2 \tag{9.61}$$

for some $m_1, m_2 \in \mathbb{Z}$. Again, we observe that (9.61) are constraints of the form (9.37) derived earlier using general considerations.

The important point to note is that the poles and zeros of an automorphic function are not freely specifiable but must satisfy a set of constraints. Generally, a function that is automorphic with respect to a Schottky group Θ generated by M Möbius maps $\{\theta_j(z) | j = 1, \dots, M\}$ and their inverses must satisfy M constraints akin to (9.56) or (9.61) relevant for this $M = 2$ example.

Representation in terms of $\mathcal{K}(z, .)$: On the other hand, consider the function defined by

$$\mathcal{A}(z) = \sum_{n=1}^{N} A_n \mathcal{K}(z, b_n) + A_0, \tag{9.62}$$

where $\{A_k | k = 0, 1, \dots, M\}$ are complex constants. By (6.88),

$$\begin{aligned} \mathcal{A}(\theta_j(z)) &= \sum_{n=1}^{N} A_n \mathcal{K}(\theta_j(z), b_n) + A_0 = \sum_{n=1}^{N} A_n (2\pi i b_n v_j'(b_n) + \mathcal{K}(z, b_n)) + A_0 \\ &= \mathcal{A}(z) + 2\pi i \sum_{n=1}^{N} A_n b_n v_j'(b_n), \qquad j = 1, 2. \end{aligned} \tag{9.63}$$

Hence two conditions for $\mathcal{A}(z)$ to be automorphic with respect to the Schottky group Θ generated by θ_1 and θ_2 are

$$\sum_{n=1}^{N} A_n b_n v_1'(b_n) = 0, \qquad \sum_{n=1}^{N} A_n b_n v_2'(b_n) = 0. \tag{9.64}$$

From (9.62) and the local form of $\mathcal{K}(., .)$, it follows that near $z = b_n$,

$$\mathcal{A}(z) = -\frac{A_n b_n}{z - b_n} + \text{locally analytic function}. \tag{9.65}$$

Observe that (9.64) are constraints of the alternative general form (9.41) derived earlier using the more general considerations of §9.4.

For a given automorphic function $\mathcal{A}(z)$ it is clear that by identifying its two different representations just described, identities akin to (9.53) can be written down.

If the automorphic function has a second order pole at b_n, say, then it should be clear that a square of the relevant prime function, i.e., $\omega(z, b_n)^2$, should be included in the denominator of the representation of the form (9.54). Similarly, the representation (9.62) can be generalized by including contributions from a term $\mathcal{L}(z, b_n)$, where the function $\mathcal{L}(z, a)$ is related to a derivative of $\mathcal{K}(z, a)$ with respect to a as defined in (9.72) and studied in Exercise 9.5. For even higher order poles, additional functions obtained by successive derivatives with respect to a can be introduced.

Exercises

9.1. **(Special radial slit map as an automorphic function)** Consider the special instance of the radial slit map from a multiply connected circular domain as given in (6.79) of §6.11:

$$f(z) = -\frac{\omega(z,a)\omega(z,1/\overline{a})}{\omega(z,\overline{a})\omega(z,1/a)}, \qquad a = \mathrm{i}r,\ 0 < r < 1. \tag{9.66}$$

(a) Verify that $f(z)$ is an automorphic function with respect to the Schottky group Θ generated by $\{\theta_j(z) | j = 1, \dots, M\}$ and their inverses.

(b) Following the analysis of §9.6 find an alternative expression for $f(z)$ in terms of the quasi-automorphic function $\mathcal{K}(.,.)$.

9.2. **(The loxodromic group)** This question concerns the Schottky group Θ with a single generator $\theta_1(z) = \rho^2 z$ and its inverse $\theta_1^{-1}(z) = \rho^{-2} z$ given by (8.22).

(a) Show that the union of the images of the annular region F defined in (8.20) under all the elements of the Schottky group (8.22) is the entire extended complex plane minus the two points 0 and ∞.

(b) Show that the two points 0 and ∞ are *fixed points* of the Schottky group Θ, where the notion of a fixed point was defined in part (d) of Exercise 3.1.

Remark: This question illustrates how images of F under the action of the Schottky group Θ tessellate *almost* all the complex plane except for so-called *singular points* of the group. In this simple example of the loxodromic group with a single generator and its inverse, the singular points coincide with the two fixed points of the generating maps. For Schottky groups with more generators this singular set associated with Θ has a much more interesting geometrical structure.[30]

9.3. **(Example construction of an automorphic function)** Consider the triply connected circular domain D_z shown in Figure 9.1 with

$$q_1 = q_2 = q, \qquad \delta_1 = -\delta_2 = \delta, \tag{9.67}$$

where the parameters q and δ are real. It is required to find a function $f(z)$ which is automorphic with respect to the Schottky group Θ generated by the maps $\{\theta_j(z) | j = 1, 2\}$ (and their inverses) associated with this domain D_z. It is known that the function $f(z)$ has simple poles at $\pm b_1$ and at $\pm \mathrm{i} b_2$ and simple zeros at 0, ∞, and $\pm a$, where the parameters a, b_1, and b_2 are real.

[30]See [101].

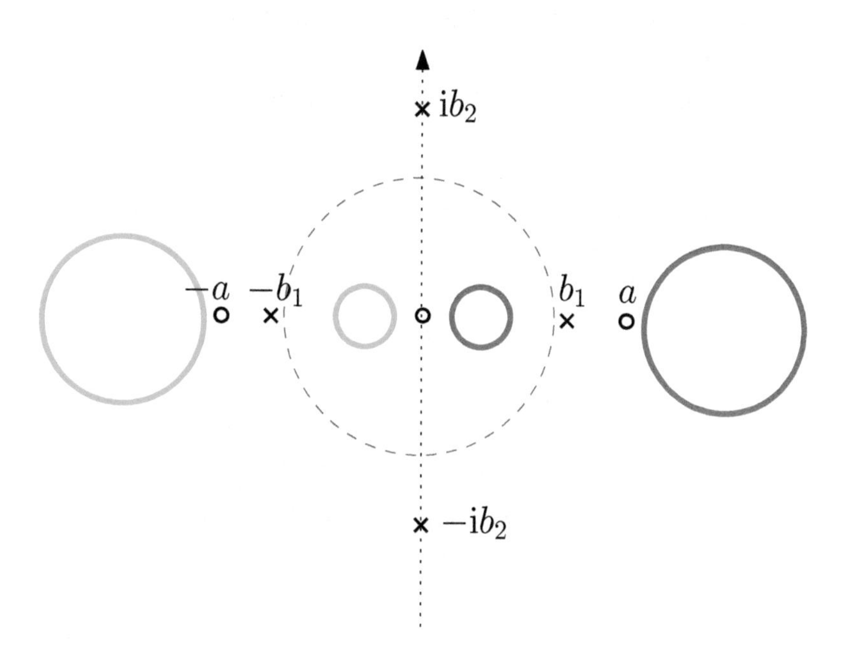

Figure 9.1. *The location of the poles and zeros in the region $F = D_z \cup D'_z$ for the automorphic function $f(z)$ of Exercise 9.3.*

(a) Show that one possible representation of $f(z)$ is given by

$$f(z) = R\left[\frac{\omega(z,0)\omega(z,\infty)\omega(z,a)\omega(z,-a)}{\omega(z,b_1)\omega(z,-b_1)\omega(z,\mathrm{i}b_2)\omega(z,-\mathrm{i}b_2)}\right], \qquad R \in \mathbb{R}, \tag{9.68}$$

where $\omega(.,.)$ is the prime function associated with D_z, and write down any constraints on the real parameters q, δ, a, b_1, and b_2.

(b) Show that an alternative representation of $f(z)$ is

$$\begin{aligned} f(z) = {} & A\left[\mathcal{K}(z,b_1) - \mathcal{K}(z,-b_1)\right] + B\left[\mathcal{K}(z,\mathrm{i}b_2) - \mathcal{K}(z,-\mathrm{i}b_2)\right] \\ & - A\left[\mathcal{K}(0,b_1) - \mathcal{K}(0,-b_1)\right] - B\left[\mathcal{K}(0,\mathrm{i}b_2) - \mathcal{K}(0,-\mathrm{i}b_2)\right], \end{aligned} \tag{9.69}$$

where A and B are real parameters, and write down any constraints on the real parameters A, B, q, δ, a, b_1, and b_2.

(c) Write down an expression for A and B in terms of R, q, δ, a, b_1, and b_2. *Hint:* You may find it useful to introduce the function defined by $\tilde{\omega}(z,a) = \omega(z,a)/(z-a)$.

9.4. **(Uniform field around circular objects)** Consider the unbounded region D_w in the complex w plane exterior to three circular discs with specified centers and radii. It is required to find an analytic function, $f(w)$ say, having a simple pole with unit residue at $w = \infty$ and satisfying the conditions that (i) the imaginary part of $f(w)$ is constant on the boundaries of the three circular discs and (ii) $f(w)$ is a single-valued, analytic function in D_w.

(a) Find the functional form of a conformal mapping,

$$w = g(z), \tag{9.70}$$

from a bounded multiply connected circular domain D_z to the unbounded circular domain D_w.

(b) Show that the composed function

$$F(z) \equiv f(g(z)) \tag{9.71}$$

is quasi-automorphic with respect to the Schottky group Θ generated by the maps $\{\theta_j(z)|j = 1, 2\}$ and their inverses associated with D_z.

(c) Hence find an expression for $F(z)$ using the function theory of this chapter.

9.5. **(The quasi-automorphic function $\mathcal{L}(z, a)$)** Define the function

$$\mathcal{L}(z, a) \equiv a \frac{\partial \mathcal{K}(z, a)}{\partial a}. \tag{9.72}$$

(a) Verify that $\mathcal{L}(z, a)$ has a second order pole at $z = a$.

(b) Verify that it satisfies the functional identity

$$\overline{\mathcal{L}}(1/z, 1/a) = \mathcal{L}(z, a). \tag{9.73}$$

(c) Verify that it satisfies the identity

$$\mathcal{L}(\theta_j(z), a) = \mathcal{L}(z, a) + 2\pi \mathrm{i} a v_j'(a) + 2\pi \mathrm{i} a^2 v_j''(a), \qquad j = 1, \ldots, M. \tag{9.74}$$

9.6. **(Identity between differentials)** Let $\alpha = ir$ where $0 < r < 1$ is a real parameter and consider the circular slit map

$$\eta(z) = \frac{\omega(z, \alpha)\omega(z, 1/\alpha)}{\omega(z, \overline{\alpha})\omega(z, 1/\overline{\alpha})}, \tag{9.75}$$

where $\omega(.,.)$ is the prime function associated with some $M + 1$ connected circular domain D_z having all circles $\{C_j | j = 1, \ldots, M\}$ centered on the real diameter $(-1, 1)$. Establish the following relation between differentials:

$$\frac{d\eta(a)}{\eta(a)} \left[\frac{\eta(a) + \eta(z)}{\eta(a) - \eta(z)} \right] = d \log \left[\frac{\omega(a, z)^2 \omega(a, 1/z)^2}{\omega(a, \alpha)\omega(a, \overline{\alpha})\omega(a, 1/\alpha)\omega(a, 1/\overline{\alpha})} \right], \tag{9.76}$$

where z is an arbitrary point in $D_z \cup D_z'$. Hence derive the functional identity

$$\frac{\eta'(a)}{\eta(a)} \left[\frac{\eta(a) + \eta(z)}{\eta(a) - \eta(z)} \right] = \frac{1}{a} \Big[2K(a, z) + 2K(a, 1/z) \\ - K(a, \alpha) - K(a, \overline{\alpha}) - K(a, 1/\alpha) - K(a, 1/\overline{\alpha}) \Big], \tag{9.77}$$

where $K(.,.)$ is the function introduced in §6.12.

Chapter 10

Polycircular arc domains

10.1 ▪ Beyond slit domains and polygons

Equipped with the new tools, themselves built within our prime function framework, from Chapters 8 and 9 we can now proceed to incorporate broader classes of planar domains, beyond slit domains and polygons, into those amenable to treatment using that same framework.

Polygons are an important class of shapes in applications not least because a common approximation of a general planar shape in practice is to mark out a discrete set of points on the boundary of the shape and to join them by straight lines. As the number of marked points increases and if they are well distributed, one gets an increasingly good polygonal approximation to the given shape. This can be done even if the shape has holes, that is, if it is multiply connected. The S–C theory of Chapter 7 is then relevant to these possibly multiply connected polygonal approximants.

A better approximation of a given shape is to use circular arcs to join a discrete set of points on the boundary. Such an approximant is no longer polygonal. Instead, it is an example of a polycircular arc domain. Such domains constitute an important class that is ubiquitous in applications. They are the focus of this chapter. Such a domain is defined to be one whose boundaries are made up of a union of circular arcs; recall that a circular arc is defined as a curve of constant curvature. This includes straight line segments which are circular arcs having zero curvature. A polygon is a special case of a polycircular arc domain where all of the boundary arcs have zero curvature.

10.2 ▪ What is a polycircular arc domain?

We encountered polycircular arc domains in Chapters 1, 5, and 6: the bounded and unbounded circular slit mappings are basic examples. Indeed, the challenge of constructing a conformal mapping to a polycircular arc domain can be exemplified by revisiting the unbounded circular slit and radial slit maps of Chapter 1. Both maps can be written in terms of the prime function of the unit disc. Using the unit disc $|\eta| < 1$ in a parametric η plane the relevant formulas for these mappings are

$$z = -\frac{\omega(\eta, a)\omega(\eta, 1/a)}{\omega(\eta, \overline{a})\omega(\eta, 1/\overline{a})}, \qquad a = \mathrm{i}r, \tag{10.1}$$

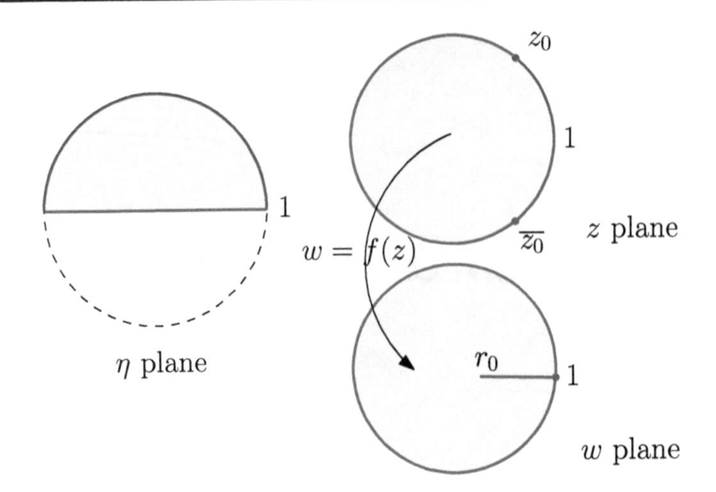

Figure 10.1. *Explicit example (10.5) of a simply connected mapping to a polycircular arc domain. Elimination of η between the unbounded circular slit and radial slit maps (10.1) and (10.2) produces a polycircular arc mapping $w = f(z)$ in (10.5) from the z plane to the w plane.*

$$w = -\frac{\omega(\eta, a)\omega(\eta, 1/\overline{a})}{\omega(\eta, \overline{a})\omega(\eta, 1/a)}, \tag{10.2}$$

where $0 < r < 1$. The image in a complex z plane is the unbounded region exterior to a circular arc on $|z| = 1$; the image in a complex w plane is the unbounded region exterior to a radial slit along the real axis and intersecting the circle $|w| = 1$ at $w = 1$, which can be shown to be invariant with respect to reflection in this circle.

Since, for the unit disc, the prime function is $\omega(\eta, a) = (\eta - a)$, it can be shown that the two ends of the circular slit on the unit circle $|z| = 1$ are at z_0 and $\overline{z_0}$, where

$$z_0 = \left(\frac{1-a}{1+a}\right)^2 = \left(\frac{1-\mathrm{i}r}{1+\mathrm{i}r}\right)^2. \tag{10.3}$$

The two ends of the radial slit in the w plane are at r_0 and $1/r_0$, where

$$r_0 = \left|\frac{a-\mathrm{i}}{\overline{a}-\mathrm{i}}\right|^2 = \left(\frac{1-r}{1+r}\right)^2. \tag{10.4}$$

As illustrated in Figure 10.1, the upper half unit disc in the η plane maps via (10.1) to the interior of the unit z circle and via (10.2) to the interior of the unit w disc minus the slit between $(r_0, 1)$.

It is possible, after some algebra, to eliminate the variable η between the two expressions (10.1) and (10.2) to find that z and w are directly related by

$$w = f(z) = z\left[\frac{(1+a^2)(1-z) + 2(1+z) - (1-a^2)[(z-z_0)(z-\overline{z_0})]^{1/2}}{(1+a^2)(1-z) - 2(1+z) - (1-a^2)[(z-z_0)(z-\overline{z_0})]^{1/2}}\right]^2. \tag{10.5}$$

This formula is an explicit example of a conformal mapping to a simply connected polycircular arc domain. It transplants the interior of the unit z disc to the interior of the unit w disc minus the slit between $(r_0, 1)$. The segment of the unit z circle between z_0 and $\overline{z_0}$ is transplanted to the radial slit on the real axis between $(r_0, 1)$; the remainder of the unit

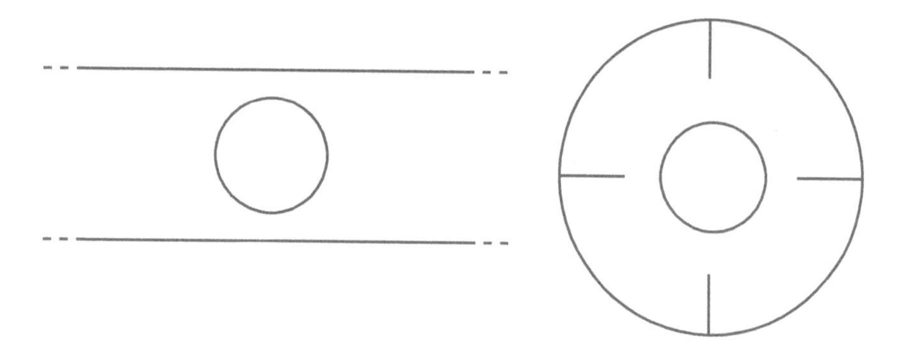

Figure 10.2. *Two doubly connected polycircular arc domains arising in applications: a "disc-in-channel" geometry (left) and a "finned annular duct" (right). Each boundary has piecewise constant curvature.*

z circle is transplanted to the unit w circle. The two square root singularities of $f(z)$ at $z = z_0, \overline{z_0}$ are the preimages of the two 90° corners at $w = 1$ on either side of the slit. All portions of the boundary of the target domain in the w plane are circular arcs.

It is easy to envisage multiply connected cases. Two doubly connected polycircular arc domains arising frequently in applications are shown in Figure 10.2: a "disc-in-channel" geometry and a "finned annular duct" configuration. Each boundary portion of these domains can be identified as a circular arc.

The S–C theory of Chapter 7 dealt with conformal mappings from a circular domain D_z to a given multiply connected polygon. That theory no longer holds as soon as one or more of the edges of a polycircular arc domain have nonzero curvature. It is natural to ask how to construct conformal mappings from the class of circular preimage domains to multiply connected polycircular arc domains for which the boundary arcs can have arbitrary curvature. This matter is addressed in this chapter, with the prime function again playing a pivotal role.

10.3 ▪ Construction of maps to polycircular arc domains

In S–C theory the derivative of the conformal mappings from circular preimage domains to multiply connected polygons can be written down explicitly in terms of the prime function of the preimage domain, usually up to a finite set of parameters. For multiply connected polycircular arc domains this is no longer possible. It turns out that the prime function can nevertheless be used to write down the ordinary differential equation determining the mapping function, again, up to a finite set of parameters.

The aim now is to derive these ordinary differential equations. In what follows it is convenient to use the term *polycircular arc map* to refer to the conformal mapping from a circular preimage region to a polycircular arc domain.

Let D_w be an $M+1$ connected polycircular arc domain with each boundary component labeled by the index $j = 0, 1, \ldots, M$. Suppose that the jth boundary component is made up of $n_j \geq 1$ circular arcs. Figure 10.3 shows an example for the triply connected case $M = 2$, where $n_0 = n_1 = n_2 = 4$. On the kth circular arc of the jth boundary of D_w, for $k = 1, \ldots, n_j$, let

$$\overline{w} = \mathcal{S}_k^{(j)}(w) = \overline{\Delta_k^{(j)}} + \frac{[Q_k^{(j)}]^2}{w - \Delta_k^{(j)}}, \qquad k = 1, \ldots, n_j, \qquad j = 0, 1, \ldots, M, \qquad (10.6)$$

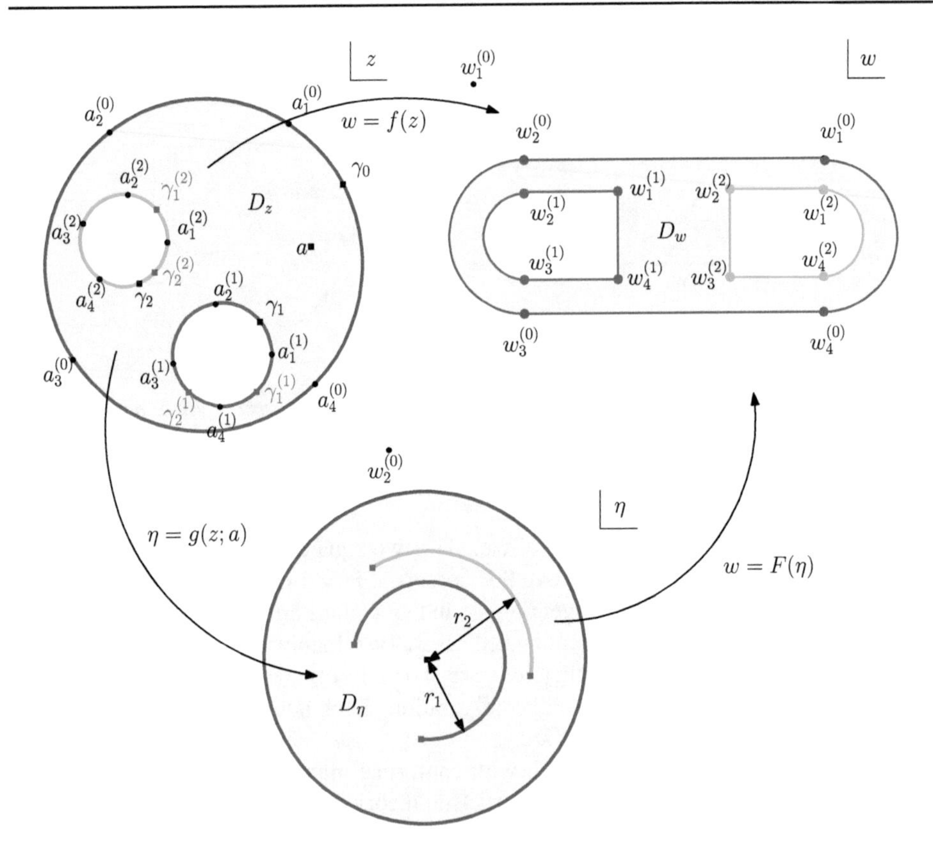

Figure 10.3. *The construction of the ordinary differential equation for the mapping from D_z to the polycircular arc domain D_w in a complex w plane proceeds via an intermediate mapping to a bounded circular slit domain in a parametric η plane.*

be the Schwarz function of this circular arc as defined in §1.9. The parameters $\{\Delta_k^{(j)}, Q_k^{(j)}\}$ are the center and radius of the circle of which this boundary portion is an arc. This Schwarz function has a meromorphic continuation off the arc, and there is no ambiguity in taking derivatives, with respect to w, of these local Schwarz functions. On differentiation of (10.6) with respect to w, and use of (10.6), we can write

$$\frac{d\mathcal{S}_k^{(j)}(w)}{dw} = -\frac{\mathcal{S}_k^{(j)}(w) - \overline{\Delta_k^{(j)}}}{w - \Delta_k^{(j)}}. \tag{10.7}$$

Differentiation of (10.7) with respect to w yields

$$\frac{d^2\mathcal{S}_k^{(j)}(w)}{dw^2} = -\frac{d\mathcal{S}_k^{(j)}(w)/dw}{w - \Delta_k^{(j)}} + \frac{\mathcal{S}_k^{(j)}(w) - \overline{\Delta_k^{(j)}}}{(w - \Delta_k^{(j)})^2} = 2\frac{\mathcal{S}_k^{(j)}(w) - \overline{\Delta_k^{(j)}}}{(w - \Delta_k^{(j)})^2}. \tag{10.8}$$

On taking a ratio of (10.7) and (10.8) we find

$$\frac{d\mathcal{S}_k^{(j)}(w)/dw}{d^2\mathcal{S}_k^{(j)}(w)/dw^2} = -\frac{1}{2}(w - \Delta_k^{(j)}), \tag{10.9}$$

with a further differentiation of (10.9) with respect to w leading to

$$-\frac{d^3\mathcal{S}_k^{(j)}(w)}{dw^3} + \frac{3}{2}\frac{(d^2\mathcal{S}_k^{(j)}(w)/dw^2)^2}{d\mathcal{S}_k^{(j)}(w)/dw} = 0. \tag{10.10}$$

It is significant that this nonlinear third order ordinary differential equation for the Schwarz function of the kth circular arc of the jth boundary does not depend explicitly on the data $\{\Delta_k^{(j)}, Q_k^{(j)}\}$; that is, the equation is independent of the indices j and k. Consequently, the local Schwarz function of *every* circular arc segment of the boundary of the polycircular arc domain satisfies the same differential equation (10.10). This important observation lies at the heart of the following construction.

As in the S–C problem, the required map $w = f(z)$ from D_z to the polycircular domain D_w can be constructed by first mapping D_z to a bounded circular slit domain D_η in a complex η domain:

$$\eta = g(z;a) = \frac{\omega(z,a)}{|a|\omega(z,1/\overline{a})}, \tag{10.11}$$

where a is some point in D_z chosen to map to $\eta = 0$. The image of each circle C_j for $j = 1, \ldots, M$ is a circular arc L_j in the η plane of radius r_j; we also let C_0 map to the unit circle L_0, of radius $r_0 = 1$, in the η plane. Recall that there are two simple zeros of the derivative $dg(z;a)/dz$ on each of the circles $\{C_j | j = 1, \ldots, M\}$, and, as done in Chapter 7, these will be denoted by

$$\{\gamma_1^{(j)}, \gamma_2^{(j)} | j = 1, \ldots, M\}. \tag{10.12}$$

Figure 10.3 shows a schematic.

Consider now the function $F(\eta)$ defined by the relation

$$w = f(z) = F(g(z;a)) = F(\eta). \tag{10.13}$$

Geometrically, $F(\eta)$ is the conformal mapping from D_η to the polycircular domain D_w. On the kth circular arc of the jth boundary we have

$$S_k^{(j)}(w) = \overline{w} = \overline{F(\eta)} = \overline{F}(r_j^2/\eta), \tag{10.14}$$

where we have used the fact that η lies on L_j so that $\overline{\eta} = r_j^2/\eta$. On substitution of the expression (10.14) into (10.10) it can be shown, after use of the chain rule and some algebra—see Exercise 10.7—that

$$\overline{\eta^2\{F(\eta),\eta\}} = \eta^2\{F(\eta),\eta\}, \qquad \eta \text{ on } L_j \text{ for } j = 0, 1, \ldots, M, \tag{10.15}$$

where we introduce the Schwarzian derivative

$$\{Z(\eta),\eta\} \equiv \frac{Z'''(\eta)}{Z'(\eta)} - \frac{3}{2}\left(\frac{Z''(\eta)}{Z'(\eta)}\right)^2, \tag{10.16}$$

and where the prime notation denotes derivative with respect to the argument of the function.

Since it will have a crucial role to play in what follows, we introduce the notation

$$T(z) \equiv \eta^2\{F(\eta),\eta\}, \tag{10.17}$$

where, as this notation suggests, we view the quantity $\eta^2\{F(\eta),\eta\}$ as a function of z. The properties of $T(z)$ will be studied in detail in §10.5.

Remark: There is an alternative, more geometrical, derivation of (10.15) that makes use of Study's formula for the curvature of the image under a conformal mapping of a circular arc and a related formula for the rate of change of this curvature with arclength in the image domain. Exercise 10.8 explores this alternative derivation. Both approaches are discussed in the original work on this problem [37].

10.4 ▪ Differential equation for the polycircular arc map

The chain rule for Schwarzian derivatives—see Exercise 10.9—can be used to show that if $f(z)$ and $\eta(z)$ are two functions of a variable z and $F(\eta)$ is defined by the composition $F(\eta(z)) = f(z)$, then

$$\left(\frac{d\eta}{dz}\right)^2 \{F(\eta), \eta\} = \{f(z), z\} - \{\eta(z), z\}. \tag{10.18}$$

This can be used to find the relationship between $\{f(z), z\}$ and $\{F(\eta), \eta\}$. In terms of $T(z)$ and $g(z, a)$ this differential equation is

$$\{f(z), z\} = \frac{1}{g(z;a)^2}\left(\frac{dg(z;a)}{dz}\right)^2 T(z) + \{g(z;a), z\}. \tag{10.19}$$

It turns out that the right-hand side of equation (10.19) can be written down in explicit form, up to a finite set of accessory parameters, using the prime function of the conformally equivalent circular domain D_z. The quantities $g(z;a)$, $dg(z;a)/dz$, and $\{g(z;a), z\}$ are, of course, readily determined in terms of the prime function from formula (10.11). It only remains to find $T(z)$.

10.5 ▪ The function $T(z)$

Let us now examine the function $T(z)$ and its properties. First note that condition (10.15) implies that

$$\overline{T(z)} = T(z), \qquad z \in C_j, \ j = 0, 1, \dots, M. \tag{10.20}$$

These imply that

$$\overline{T(\overline{z}^{-1})} = T(z), \qquad z \in C_0, \tag{10.21}$$

$$\overline{T(\overline{S_j(z)})} = T(z), \qquad z \in C_j, \ j = 1, \dots, M, \tag{10.22}$$

where, as usual, $S_j(z)$ is the Schwarz function of circle C_j. These analytic functional relations can be continued off the respective circles, implying that

$$\overline{T(\overline{S_j(z)})} = \overline{T(\overline{z}^{-1})} \tag{10.23}$$

or, on taking complex conjugates and changing variable $\overline{z} \mapsto 1/z$,

$$T(\overline{S_j}(1/z)) = T(z), \qquad j = 1, \dots, M. \tag{10.24}$$

But since from §3.6 we know that

$$\theta_j(z) \equiv \overline{S_j}(1/z), \tag{10.25}$$

then

$$T(\theta_j(z)) = T(z), \qquad j = 1, \dots, M. \tag{10.26}$$

In other words, $T(z)$ is invariant under the action of the Schottky group Θ generated by $\{\theta_j(z) | j = 1, \dots, M\}$ and their inverses.

What about the singularities of $T(z)$? Since, near any prevertex $a_k^{(j)}$, the derivative of the conformal mapping has the local form

$$\frac{df}{dz} = (z - a_k^{(j)})^{\beta_k^{(j)}} h_k^{(j)}(z) \tag{10.27}$$

for some parameter $\beta_k^{(j)}$, and some locally analytic function $h_k^{(j)}(z)$ that is nonvanishing at $a_k^{(j)}$, then

$$\frac{f''(z)}{f'(z)} = \frac{\beta_k^{(j)}}{(z - a_k^{(j)})} + \frac{dh_k^{(j)}(z)}{dz}, \tag{10.28}$$

$$\frac{f'''(z)}{f'(z)} - \left[\frac{f''(z)}{f'(z)}\right]^2 = -\frac{\beta_k^{(j)}}{(z - a_k^{(j)})^2} + \frac{d^2 h_k^{(j)}(z)}{dz^2}. \tag{10.29}$$

It follows that

$$\{f(z), z\} = -\frac{\beta_k^{(j)}}{(z - a_k^{(j)})^2} - \frac{1}{2}\left[\frac{\beta_k^{(j)}}{(z - a_k^{(j)})} + \frac{dh_k^{(j)}(z)}{dz}\right]^2 + \frac{d^2 h_k^{(j)}(z)}{dz^2} \tag{10.30}$$

$$= -\frac{\beta_k^{(j)}}{(z - a_k^{(j)})^2} - \frac{1}{2}\left[\frac{\beta_k^{(j)}}{(z - a_k^{(j)})}\right]^2 - \frac{\beta_k^{(j)}}{(z - a_k^{(j)})}\frac{dh_k^{(j)}(z)}{dz} \tag{10.31}$$

$$+ \frac{dh_k^{(j)}(z)}{dz^2} - \frac{1}{2}\left[\frac{dh_k^{(j)}(z)}{dz}\right]^2. \tag{10.32}$$

Hence, at each of the preimages $\{a_k^{(j)} | k = 1, \ldots, n_j, j = 0, 1, \ldots, M\}$, the quantity $\{f(z), z\}$ must have a second order pole of known strength

$$-\frac{1}{2}\left[\left[\beta_k^{(j)}\right]^2 + 2\beta_k^{(j)}\right] \frac{1}{(z - a_k^{(j)})^2}. \tag{10.33}$$

It follows from (10.19) that $T(z)$ has second order poles at $\{a_k^{(j)} | k = 1, \ldots, n_j, j = 0, 1, \ldots, M\}$.

The quantity $\{f(z), z\}$, which is completely independent of the intermediate map $\eta = g(z; a)$, is analytic at the points $\{\gamma_1^{(j)}, \gamma_2^{(j)} | j = 1, \ldots, M\}$, while $\{g(z; a), z\}$ has second order poles at these points since they are the simple zeros of $dg(z; a)/dz$. It follows from (10.19) that $T(z)$ must have *fourth order* poles at $\{\gamma_1^{(j)}, \gamma_2^{(j)} | j = 1, \ldots, M\}$, because these are the simple zeros of $d\eta/dz$, and the strengths of these fourth order poles of $T(z)$ must be chosen so that there are *no* singularities of the right-hand side of (10.19) at the set of points $\{\gamma_1^{(j)}, \gamma_2^{(j)} | j = 1, \ldots, M\}$.

The functional relations (10.26), together with the fact that $T(z)$ has a finite set of poles on the boundary of F, mean that $T(z)$ is an automorphic function with respect to the Schottky group Θ generated by $\{\theta_j(z) | j = 1, \ldots, M\}$ and their inverses.

Combining all the information obtained so far, the function $T(z)$ must have the following properties:

(a) It is an automorphic function with respect to the Schottky group Θ.

(b) It satisfies the functional relation $\overline{T}(z^{-1}) = T(z)$.

(c) It has second order poles at the prevertices $\{a_k^{(j)}\}$ and fourth order poles at the points $\{\gamma_1^{(j)}, \gamma_2^{(j)} | j = 1, \ldots, M\}$.

It is therefore possible, using the machinery based on the prime function introduced in Chapter 9 to construct automorphic functions, to write down a representation of the function $T(z)$ appearing in the differential equation (10.19) up to a finite set of accessory parameters. Usually, this set of parameters must be found numerically as part of the construction of the mapping function.

10.6 ▪ The Schwarz differential equation

Given that $T(z)$ can be written down, in terms of the prime function, up to a finite set of parameters, and given that $g(z;a)$ is also known explicitly in terms of the prime function, the differential equation (10.19) can be written as

$$\{f(z), z\} = \mathcal{T}(z), \tag{10.34}$$

where

$$\mathcal{T}(z) \equiv \frac{1}{g(z;a)^2}\left(\frac{dg(z;a)}{dz}\right)^2 T(z) + \{g(z;a), z\} \tag{10.35}$$

can be written down up to a finite set of parameters. We are now in a position to linearize this equation by considering two linearly independent solutions, $u_1(z)$ and $u_2(z)$, of the second order linear differential equation

$$\frac{d^2u}{dz^2} + \frac{\mathcal{T}(z)}{2}u = 0, \tag{10.36}$$

which is known up to a finite parameter set. The Wronskian $W(z) = u_1(z)u_2'(z) - u_2(z)u_1'(z)$ of two linearly independent solutions is easily shown to satisfy the first order differential equation

$$\frac{dW}{dz} = 0, \tag{10.37}$$

which can be integrated easily. The Wronskian is nonzero if $u_1(z)$ and $u_2(z)$ are linearly independent, so we choose $W = 1$. Note that if we let

$$f(z) = \frac{u_1(z)}{u_2(z)}, \tag{10.38}$$

then

$$\frac{df}{dz} = \frac{u_1'(z)}{u_2(z)} - \frac{u_2'(z)u_1(z)}{u_2(z)^2} = \frac{1}{u_2^2} \tag{10.39}$$

if $W = 1$. It is then easy to demonstrate, on taking two more derivatives, that $f(z)$ as given in (10.38) is a solution to (10.34).

10.7 ▪ The simply and doubly connected cases

The theory for simply and doubly connected polycircular arc domains is a little simpler. This is because no intermediate conformal mapping is necessary: for a simply connected domain, the preimage domain is a unit disc; for a doubly connected domain, on use of an automorphism of the disc, we can take the preimage region to be a concentric annulus in a parametric plane. This means that all the arguments of the preceding construction hold in the original preimage domains, without any need to map to an intermediate circular arc domain.

The best way to illustrate the general theory is by example. The next two sections present constructions of doubly connected domains arising frequently in applications. We focus on the doubly connected case since even the theory for this situation has only been presented quite recently [36].[31]

[31] It is interesting that, while the theory of conformal mapping to simply connected polycircular arc domains is well known and classical, the extension even to doubly connected domains has only been made in the last few years [36]. The construction of even higher connected maps, using the prime function as the primary theoretical tool and the device of an intermediate conformal slit mapping to a circular slit domain, is an even more recent development [37].

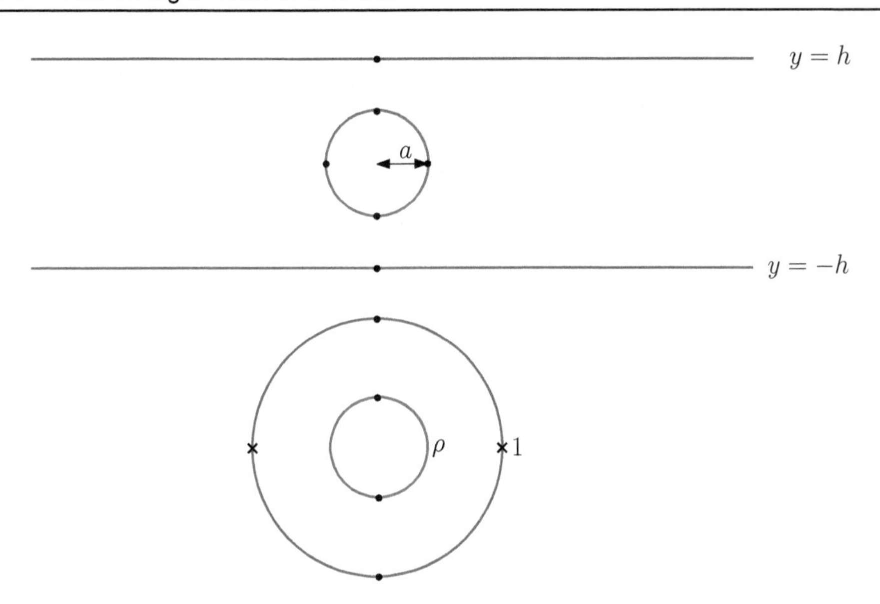

Figure 10.4. *Conformal mapping from a concentric annulus $\rho < |z| < 1$ to the exterior of a disc of radius a in a channel of width $2h$ ($h > a$) in a complex w plane. The preimages of the ends of the channel are at $z = \pm 1$.*

10.8 ▪ Channel region exterior to a disc

Consider the domain in a complex w plane comprising the interior of the channel of width $2h$ given by

$$-\infty < \mathrm{Re}[w] < \infty, \qquad -h < \mathrm{Im}[w] < h \tag{10.40}$$

and exterior to a disc of radius a centered at the origin,

$$|w| > a, \qquad a < h. \tag{10.41}$$

Such a domain is doubly connected, and, since its boundaries all have constant curvature, it is a polycircular arc domain. The challenge is to construct the conformal map

$$w = f(z) \tag{10.42}$$

from the concentric annulus

$$\rho < |z| < 1, \tag{10.43}$$

to this disc-in-channel geometry. The parameter ρ is to be determined as part of the construction. This polycircular arc domain has just two vertices as $\mathrm{Re}[w] \to \pm\infty$.

An instructive first step is to consider the simply connected mapping from the unit disc $|z| < 1$ to the channel region with no circular hole. From §1.7 this is known to be given by

$$w = \frac{2h}{\pi} \log \left[\frac{1+z}{1-z} \right] \tag{10.44}$$

and corresponds to the case $\rho = 0$. Knowledge of this map will be useful in tackling the doubly connected case when $\rho > 0$.

It is natural to seek a mapping that shares the reflectional symmetries of the target domain about the real and imaginary axes. We therefore suppose that the upper half annulus

$$\rho < |z| < 1, \qquad \mathrm{Im}[z] > 0, \tag{10.45}$$

is the preimage of the upper half of the channel exterior to the disc, that is,

$$-\infty < \mathrm{Re}[w] < \infty, \qquad 0 < \mathrm{Im}[w] < h, \qquad |w| > a. \tag{10.46}$$

The prevertices on C_0 of the two ends of the channel can be taken to be at ± 1 even for $\rho > 0$:

$$a_1^{(0)} = 1, \qquad a_2^{(0)} = -1. \tag{10.47}$$

At these prevertices it is easy to establish that

$$\beta_1^{(0)} = -1 = \beta_2^{(0)}. \tag{10.48}$$

The portion of the annulus along the positive real axis, i.e., real z satisfying

$$\rho < z < 1, \tag{10.49}$$

is the preimage of the positive real w axis outside the disc:

$$a < w < \infty. \tag{10.50}$$

Similarly,

$$-1 < z < -\rho \tag{10.51}$$

is transplanted to

$$-\infty < w < -a. \tag{10.52}$$

From the general theory presented earlier the nonlinear differential equation satisfied by the conformal map $f(z)$ has the form

$$z^2 \{f(z), z\} = T(z), \tag{10.53}$$

where $T(z)$ will now be constructed, up to a finite set of parameters, using the prime function of the annulus (10.43). The function $T(z)$

(a) must be a loxodromic function satisfying

$$T(\rho^2 z) = T(z); \tag{10.54}$$

(b) must satisfy the functional relation

$$\overline{T}(1/z) = T(z); \tag{10.55}$$

(c) must have second order poles of strength $+1/2$ at ± 1. This follows from requirement (10.33) together with use of (10.48).

It is known from the theory of Chapter 8 on loxodromic functions that one way to construct the required function $T(z)$ is to make use of the functions $K(z, a)$ and $L(z, a)$ introduced in Exercise 8.2: the function $L(z, a)$ has a second order pole at $z = a$, while $K(z, a)$ has a simple pole there. On use of the properties of $K(z, a)$ and $L(z, a)$ established in Exercise 8.2 the candidate function

$$-\frac{1}{2}\left[L(z, 1) + L(z, -1)\right] + \mathrm{i}B\left[K(z, 1) - K(z, -1)\right] + C, \tag{10.56}$$

where $B, C \in \mathbb{R}$ are constants, can be shown to satisfy all three conditions (a)–(c) listed above.

Since we seek a conformal map for which the two portions (10.49) and (10.51) of the real z axis, where $\overline{z} = z$, are transplanted to the centerline of the channel exterior to the disc (10.50) and (10.52), where $\overline{w} = w$, we expect that

$$\overline{f}(z) = f(z) \tag{10.57}$$

and hence that

$$\overline{T}(z) = T(z). \tag{10.58}$$

This implies that $B = 0$. The form for $T(z)$ is therefore given by

$$T(z) = -\frac{1}{2}\left[L(z,1) + L(z,-1)\right] + C. \tag{10.59}$$

The two real parameters ρ and C are accessory parameters. The differential equation (10.53) will be determined once these two parameters are found.

We also need to determine initial conditions for the solution of this third order differential equation. A convenient point to consider is $z = \mathrm{i}$, which, owing to the symmetries of the configuration, is the preimage of the point of the upper channel boundary closest to the cylinder:

$$f(\mathrm{i}) = \mathrm{i}h. \tag{10.60}$$

We also expect, because of the left-right symmetry shared by the preimage domain and the target domain, that the derivative of $f(z)$ is real at this point, i.e.,

$$f'(\mathrm{i}) = d \in \mathbb{R}. \tag{10.61}$$

The value of the real parameter d is not yet known; like ρ and C it must be found as part of the construction. At this point a useful device is to consider the curvature of the boundaries. The curvature of the boundary of the domain corresponding to $z = \mathrm{i}$ is zero—because the boundary is a straight line—which implies, on use of Study's formula, given in Exercise 7.1 (which expresses the boundary curvature in terms of the conformal mapping function), that

$$f''(\mathrm{i}) = \mathrm{i}d. \tag{10.62}$$

To summarize, for some choice of the three parameters ρ, C, and d, the solution of the nonlinear third order differential equation

$$z^2\left\{f(z), z\right\} = -\frac{1}{2}\left[L(z,1) + L(z,-1)\right] + C \tag{10.63}$$

for $f(z)$ will produce the conformal mapping we seek.

To solve this equation, and to find the relevant accessory parameters, we have at least two options.

One way to proceed is to linearize the ordinary differential equation (10.63) to produce the associated second order Schwarz differential equation, as discussed in §10.6.

A second option, which we will use here, is to introduce the functions

$$y_1(z) \equiv f(z), \qquad y_2(z) \equiv f'(z), \qquad y_3(z) \equiv f''(z). \tag{10.64}$$

The third order equation (10.63) can be reformulated as the first order system

$$y_1'(z) = y_2(z), \tag{10.65}$$

$$y_2'(z) = y_3(z), \tag{10.66}$$

$$y_3'(z) = \frac{T(z)}{z^2} y_2 + \frac{3}{2}\frac{y_3^2}{y_2}, \tag{10.67}$$

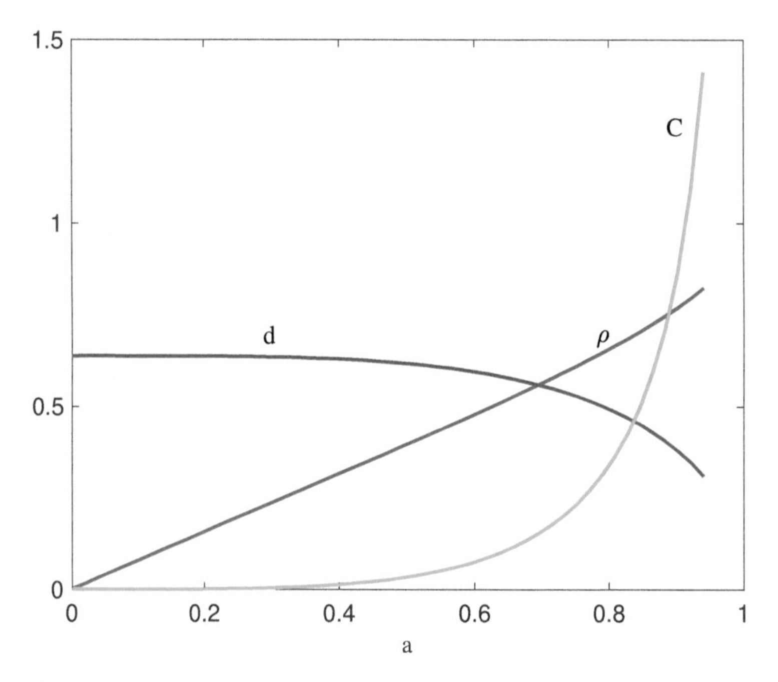

Figure 10.5. *Solution of the parameter problem for ρ, C, and d for the disc-in-channel geometry of Figure 10.4 as functions of the disc radius a for fixed channel height $h = 1$.*

with conditions at $z = \mathrm{i}$ given by

$$y_1(\mathrm{i}) = \mathrm{i}h, \qquad y_2(\mathrm{i}) = d, \qquad y_3(\mathrm{i}) = \mathrm{i}d. \tag{10.68}$$

To solve this system, and to find the accessory parameters ρ, C, and d, an iterative scheme is appropriate. Given an initial guess for ρ, C, and d, the system (10.65)–(10.68) can be solved and the values of $f(\mathrm{i}\rho)$, $f(-\rho)$, and $f(-\mathrm{i}\rho)$, say, determined. By iteration on the choice of ρ, C, and d, it must be arranged for these quantities to be given by

$$f(\mathrm{i}\rho) = \mathrm{i}, \qquad f(-\rho) = -a, \qquad f(-\mathrm{i}\rho) = -\mathrm{i}a, \tag{10.69}$$

which are expected on the grounds of symmetry.

In this nonlinear iteration, the solution (10.44) of the simply connected problem when $a = 0$ and $\rho = 0$ proves useful since it provides explicit values of the corresponding C and d for this case; a continuation process for $\rho > 0$ can then be employed. It was by means of such a scheme that the solution of the accessory parameter problem for ρ, C, and d shown in Figure 10.5 was calculated.

Once the accessory parameters are determined for a given target configuration, the required mapping function is the solution $y_1(z)$ of the (now known) system (10.65)–(10.67) with known initial conditions (10.68).

10.9 ▪ Annular duct with fins

Figure 10.2 shows a schematic of an annular duct with fins, a geometry of interest in heat transfer and turbomachinery. It is instructive to construct such a doubly connected

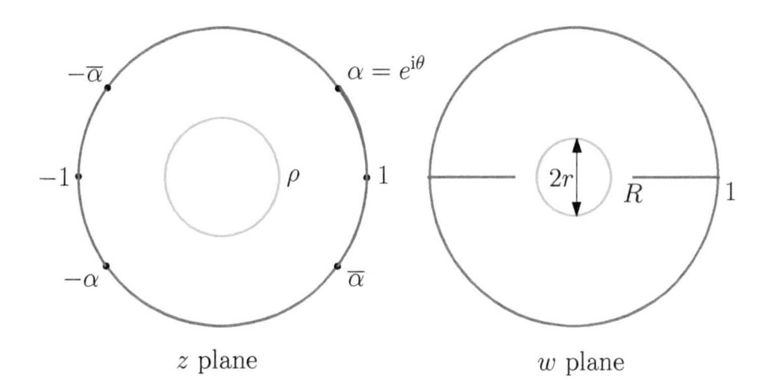

Figure 10.6. *Polycircular arc mapping of the doubly connected annulus $\rho < |z| < 1$ to a circular two-finned duct in a complex w plane.*

polycircular arc mapping. Figure 10.6 shows a schematic in which there are two fins on the outer wall. We will construct the polycircular arc mapping to this domain from a concentric annulus

$$\rho < |z| < 1. \tag{10.70}$$

Since there are two reentrant fins on the outer wall of the duct it is natural to seek an arrangement of the preimage vertices that shares the symmetries of the target. Hence the preimages of the two ends of the slits will be at ± 1 with the four $\pi/2$ corners where the fins meet the outer circular pipe boundary having preimages at $z = \pm\alpha, \pm\overline{\alpha}$, where $\alpha = e^{i\theta}$; see Figure 10.6.

Just as we did for the disc-in-channel geometry, an advisable strategy is to see if a special case involving a domain of lower connectivity might usefully be considered first. If so, one then has the opportunity to use initial conditions from that special case to "continue" the parameters to the higher connected case as the size of the holes—or the radius of an interior preimage hole—increases from zero.

We therefore first consider the case $\rho = 0$ so that the mapping in Figure 10.6 is from the simply connected unit z disc to a circular pipe in a complex w plane with two reentrant fins. This polycircular arc map from the unit disc $|z| < 1$ is denoted by

$$w = f(z). \tag{10.71}$$

Suppose the two fins lie on the positive and negative real axis, $[-1, -R], [R, 1]$, for some $0 < R < 1$. Let the preimages of the ends of the fin on the positive real axis be at $1, \alpha, \overline{\alpha}$, where $\alpha = e^{i\theta}$. By symmetry the preimages of the fin on the negative real axis will be at $-1, -\alpha, -\overline{\alpha}$. The points $z = \pm 1$ are the preimages of the ends of the fins at $\pm R$.

The differential equation to solve is

$$z^2\{f(z), z\} = T(z). \tag{10.72}$$

It can be shown that the relevant $T(z)$ is given, up to parameters, by

$$T(z) = -\frac{3}{2}\frac{z}{(z-1)^2} + \frac{3}{2}\frac{z}{(z+1)^2} + \frac{3z}{8}\left[\frac{\alpha}{(z-\alpha)^2} + \frac{\overline{\alpha}}{(z-\overline{\alpha})^2}\right]$$
$$-\frac{3z}{8}\left[\frac{\alpha}{(z+\alpha)^2} + \frac{\overline{\alpha}}{(z+\overline{\alpha})^2}\right] + iB\left[\frac{z}{z-\alpha} - \frac{z}{z-\overline{\alpha}} + \frac{z}{z+\alpha} - \frac{z}{z+\overline{\alpha}}\right] \tag{10.73}$$

where $B \in \mathbb{R}$ is a real constant. This function satisfies

$$\overline{T}(1/z) = T(z) \tag{10.74}$$

and has second order singularities of the required strength at the six prevertices $z = \pm 1$, $\pm\alpha$, $\pm\overline{\alpha}$. It is clear by inspection that this function vanishes at $z = 0$; in fact, that zero is a second order zero, as can be checked directly. The function $T(z)$ is only known up to the two real parameters B and θ. These remain to be determined. It is convenient to write

$$T(z) = -\frac{3}{2}\frac{z}{(z-1)^2} + \hat{T}(z) = -\frac{3}{2}\frac{1}{(z-1)^2} - \frac{3}{2}\frac{1}{(z-1)} + \hat{T}(z), \tag{10.75}$$

where

$$\begin{aligned}\hat{T}(z) = {} & \frac{3}{2}\frac{z}{(z+1)^2} + \frac{3z}{8}\left[\frac{\alpha}{(z-\alpha)^2} + \frac{\overline{\alpha}}{(z-\overline{\alpha})^2}\right] - \frac{3z}{8}\left[\frac{\alpha}{(z+\alpha)^2} + \frac{\overline{\alpha}}{(z+\overline{\alpha})^2}\right] \\ & + \mathrm{i}B\left[\frac{z}{z-\alpha} - \frac{z}{z-\overline{\alpha}} + \frac{z}{z+\alpha} - \frac{z}{z+\overline{\alpha}}\right].\end{aligned} \tag{10.76}$$

For the disc-in-channel geometry, integration of the ordinary differential equation (10.72) was initiated away from any vertices. In the present example the strategy is to initiate integration of the nonlinear equation (10.72) at the vertex at $z = 1$. To do so we write

$$f(z) = R + (z-1)^2 g(z) \tag{10.77}$$

for some function $g(z)$ to be determined that is analytic at $z = 1$. It is readily checked, on performing some series expansions near $z = 1$, that

$$\begin{aligned}\frac{f''(z)}{f'(z)} &= \frac{2g(z) + 4(z-1)g'(z) + (z-1)^2 g''(z)}{(z-1)[2g(z) + (z-1)g'(z)]} \\ &= \frac{1}{z-1} + \frac{3}{2}\frac{g'(z)}{g(z)} + \left[\frac{g''(z)}{2g(z)} - \frac{3}{4}\left(\frac{g'(z)}{g(z)}\right)^2\right](z-1) + \cdots.\end{aligned} \tag{10.78}$$

It then follows that, near $z = 1$,

$$z^2\{f(z), z\} = -\frac{3}{2}\frac{1}{(z-1)^2} - \frac{3}{2}\left[2 + \frac{g'(1)}{g(1)}\right]\frac{1}{(z-1)} + \cdots. \tag{10.79}$$

On the other hand, from (10.75), near $z = 1$ we have

$$T(z) = -\frac{3}{2}\frac{1}{(z-1)^2} - \frac{3}{2}\frac{1}{(z-1)} + \hat{T}(1) + \cdots. \tag{10.80}$$

From (10.72), and on equating the principal parts of both sides at $z = 1$, we find that

$$\frac{g'(1)}{g(1)} = -1. \tag{10.81}$$

To solve for $g(z)$ we introduce the following new variables:

$$y_1(z) = g(z), \qquad y_2(z) = g'(z), \qquad y_3(z) = \frac{f''(z)}{f'(z)} - \frac{1}{z-1}. \tag{10.82}$$

From (10.81) initial conditions at $z = 1$ can be written as

$$y_1(1) = a, \qquad y_2(1) = -a, \qquad y_3(1) = -\frac{3}{2}, \tag{10.83}$$

where, to obtain the last condition, we have used (10.78) and (10.81). The parameter a remains to be determined.

We need to establish the system of ordinary differential equations satisfied by y_1, y_2, and y_3. Since

$$\frac{f''(z)}{f'(z)} = y_3(z) + \frac{1}{(z-1)}, \tag{10.84}$$

it follows immediately on substitution into (10.72) that

$$y_3'(z) = \left[\frac{3}{2}\frac{1}{(z-1)^2} + \frac{T(z)}{z^2}\right] + \frac{y_3^2}{2} + \frac{y_3}{z-1}, \tag{10.85}$$

which gives y_3' in terms of y_1, y_2, and y_3. Now from (10.78) we find

$$\frac{f''(z)}{f'(z)} = y_3(z) + \frac{1}{(z-1)} = \frac{2g(z) + 4(z-1)g'(z) + (z-1)^2 g''(z)}{(z-1)[2g(z) + (z-1)g'(z)]} \tag{10.86}$$

$$= \frac{2y_1 + 4(z-1)y_2 + (z-1)^2 y_2'}{(z-1)[2y_1 + (z-1)y_2]}, \tag{10.87}$$

which can be rearranged to give an expression for y_2' in terms of y_1, y_2, and y_3:

$$y_2' = y_2 y_3 + \frac{2y_1 y_3}{(z-1)} - \frac{3y_2}{(z-1)}. \tag{10.88}$$

Collecting these results, we arrive at the nonlinear system

$$y_1' = y_2, \tag{10.89}$$

$$y_2' = y_2 y_3 + \frac{2y_1 y_3}{(z-1)} - \frac{3y_2}{(z-1)}, \tag{10.90}$$

$$y_3' = \left[\frac{3}{2}\frac{1}{(z-1)^2} + \frac{T(z)}{z^2}\right] + \frac{y_3^2}{2} + \frac{y_3}{z-1}. \tag{10.91}$$

This system appears to be singular at $z = 1$, but in fact the singularities are removable by virtue of the initial conditions (10.83). Indeed, it is important to evaluate the equations in the limit $z \to 1$ to establish the initial values $y_1'(1)$, $y_2'(1)$, and $y_3'(1)$. An expansion of each equation leads to

$$y_1'(1) = y_2(1) = -a, \tag{10.92}$$

$$y_2'(1) = \frac{1}{4}(3y_2(1)y_3(1) + 2y_1(1)y_3'(1)), \tag{10.93}$$

$$0 = \hat{T}(1) + \frac{y_3(1)^2}{2} - \frac{3}{2}. \tag{10.94}$$

This set of equations is illuminating: the final equation turns out not to give a condition on $y_3'(1)$ but to provide instead a constraint on the two unknown parameters B and θ appearing in $\hat{T}(1)$, namely,

$$\hat{T}(1) = \frac{3}{8}, \tag{10.95}$$

which, on use of (10.76), leads to

$$B = \frac{-\frac{3}{8}\left[\frac{\alpha}{(1-\alpha)^2} + \frac{\overline{\alpha}}{(1-\overline{\alpha})^2}\right] + \frac{3}{8}\left[\frac{\alpha}{(1+\alpha)^2} + \frac{\overline{\alpha}}{(1+\overline{\alpha})^2}\right]}{+\mathrm{i}\left[\frac{1}{1-\alpha} - \frac{1}{1-\overline{\alpha}} + \frac{1}{1+\alpha} - \frac{1}{1+\overline{\alpha}}\right]}. \tag{10.96}$$

Since $\alpha = e^{\mathrm{i}\theta}$, this formula provides an expression for B in terms of θ.

Since $y_3'(1)$ is undetermined, we set it to be

$$y_3'(1) = b, \tag{10.97}$$

where b is to be found. Equation (10.93) then implies

$$y_2'(1) = \frac{1}{4}\left[\frac{9a}{2} + 2ab\right], \tag{10.98}$$

where we have used (10.83).

In summary, we have to solve the nonlinear first order system (10.89)–(10.91) subject to initial conditions (10.83) and with initial values of the derivatives given by

$$y_1'(1) = -a, \qquad y_2'(1) = \frac{1}{4}\left[\frac{9a}{2} + 2ab\right], \qquad y_3'(1) = b. \tag{10.99}$$

There are three unknown parameters: a, b, and θ.

By all the symmetries of the system we expect that

$$f(0) = 0, \qquad f(\mathrm{i}) = \mathrm{i}. \tag{10.100}$$

We can therefore find the three unknowns by iterating on the solution of the system until the following three real equations are satisfied:

$$\mathrm{Re}[f(0)] = 0 = R + y_1(0), \qquad f(\mathrm{i}) = \mathrm{i} = R + (\mathrm{i}-1)^2 g(\mathrm{i}). \tag{10.101}$$

The second equation has a real and an imaginary part, so (10.101) constitutes three real equations for the three real unknowns. To satisfy these equations we integrate the system for (y_1, y_2, y_3) from $z = 1$ to the two points $z = 0, \mathrm{i}$. This is easily done using standard integration routines since these two endpoints of integration are not prevertices.

There is the opportunity for a useful check on the solution to this parameter problem, and the clue to this has already been presaged by the very first example of this chapter.

Consider *another* annulus in a parametric η plane:

$$q < |\eta| < 1. \tag{10.102}$$

This is not to be confused with the annulus $\rho < |z| < 1$ of Figure 10.6. From Chapter 5 it is known that the unbounded circular slit map from the annulus (10.102) to a complex z plane and the radial slit map to a complex w plane are

$$z = \mathcal{C}(\eta) = -\frac{\omega(\eta,\gamma)\omega(\eta,1/\gamma)}{\omega(\eta,\overline{\gamma})\omega(\eta,1/\overline{\gamma})}, \qquad \gamma = \mathrm{i}\sqrt{q}, \tag{10.103}$$

$$w = \mathcal{R}(\eta) = -\frac{\omega(\eta,\gamma)\omega(\eta,1/\overline{\gamma})}{\omega(\eta,\overline{\gamma})\omega(\eta,1/\gamma)}, \tag{10.104}$$

where $\omega(.,.)$ is now the prime function for the new annulus (10.102). These formulas are exactly the same as in (10.1) and (10.2), the only difference being that, in those formulas, the prime function $\omega(.,.)$ was that associated with the unit disc. As illustrated in Figure 10.7 the map (10.103) transplants the upper half annulus

$$q < |\eta| < 1, \qquad \mathrm{Im}[\eta] > 0, \tag{10.105}$$

to the interior of the unit z disc with the two upper half semicircles C_0 and C_1 being transplanted to two circular arcs on the unit z circle; on the other hand, the map (10.104) transplants this same upper half semiannulus (10.105) to the interior of the unit disc in a w plane with two equal-length radial slits excised, with those slits emanating from the unit circle into the interior of the disc.

The inference is that the sequence of maps

$$z \mapsto \eta = \mathcal{C}^{-1}(z) \mapsto w = \mathcal{R}(\eta) = \mathcal{R}(\mathcal{C}^{-1}(z)) \tag{10.106}$$

is equivalent to the polycircular arc mapping from the simply connected z domain, with $\rho = 0$, in Figure 10.6 to the simply connected finned domain in the w plane that we seek, i.e.,

$$f(z) = \mathcal{R}(\mathcal{C}^{-1}(z)). \tag{10.107}$$

Consequently, by picking a value of the parameter q and finding the associated endpoint $\alpha = e^{\mathrm{i}\theta}$ of the circular arc slit and the endpoint R of the radial slit, as given by the images of formulas (10.103) and (10.104), we have an alternative route to finding the required correspondence between R and θ. This provides an independent check on the solution to the parameter problem arising in a direct construction of the mapping from the z disc to the polycircular arc domain in the w plane.

Having solved the case of a simply connected circular duct with fins, that is, the case $\rho = 0$ in Figure 10.6, the analysis for $\rho \neq 0$ is a straightforward adaptation where the numerical solution of the generalized accessory parameter problem for $\rho > 0$ is aided by a continuation process from the $\rho = 0$ case.

To construct the function $T(z)$ when $\rho > 0$ we again make use of the function $K(z, a)$ introduced in §5.12 and $L(z, a)$ associated with the annulus $\rho < |z| < 1$ introduced in Exercise 8.2:

$$K(z,a) \equiv z\frac{\partial}{\partial z}\log\omega(z,a), \qquad L(z,a) \equiv z\frac{\partial K(z,a)}{\partial z}, \tag{10.108}$$

where $\omega(.,.)$ is the prime function associated with the annulus $\rho < |z| < 1$. In the limit $\rho \to 0$,

$$K(z,a) \to \frac{z}{z-a}, \qquad L(z,a) \to -\frac{za}{(z-a)^2}. \tag{10.109}$$

On use of this fact, it is easy to deduce that the required functional form of $T(z)$ when $\rho \neq 0$ is

$$\begin{aligned} T(z) =& \frac{3}{2}\left(L(z,1) + L(z,-1)\right) - \frac{3}{8}\left[L(z,\alpha) + L(z,\overline{\alpha}) + L(z,-\alpha) + L(z,-\overline{\alpha})\right] \\ &+ \mathrm{i}B\left[K(z,\alpha) - K(z,\overline{\alpha}) + K(z,-\alpha) - K(z,-\overline{\alpha})\right] + C, \end{aligned} \tag{10.110}$$

where we have added a constant C; this constant is known to be zero when $\rho = 0$, but not necessarily when $\rho > 0$. On use of (10.109) it can be checked that (10.110) with $C = 0$ is consistent with (10.73) in the simply connected limit $\rho \to 0$.

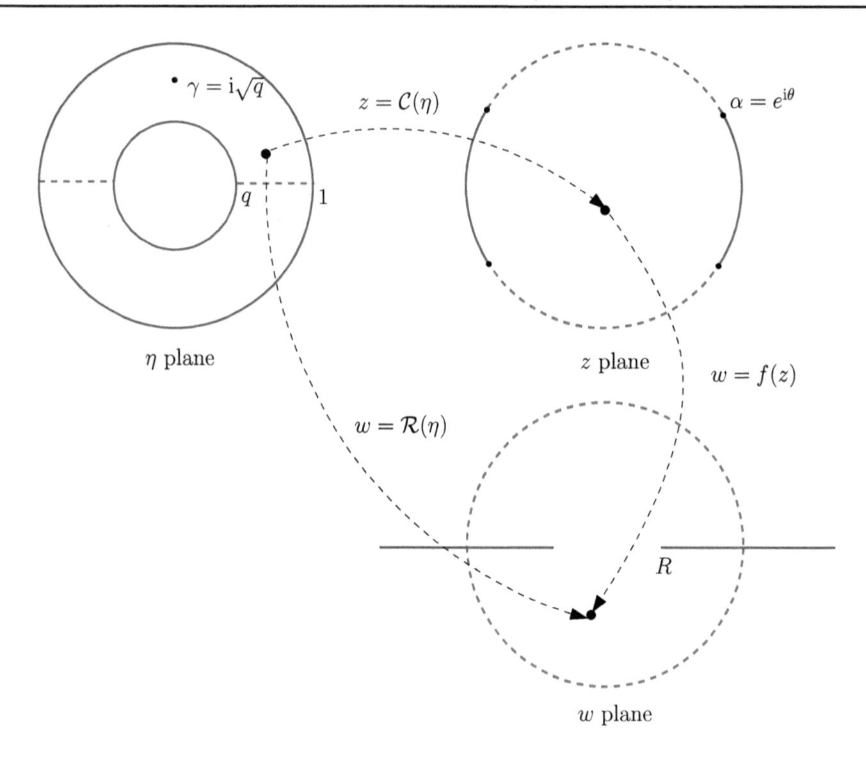

Figure 10.7. *A construction to check the solution to the parameter problem for the mapping in Figure 10.6 when $\rho = 0$. The schematic shows the "two-slit" generalization of the "one-slit" construction of Figure 10.1: elimination of the parameter η between the unbounded circular slit map (10.103) and radial map (10.104) from the upper half annulus $q < |\eta| < 1$ produces the required polycircular arc mapping from the z plane to the w plane.*

The transformation to the new system involving the set $\{y_1, y_2, y_3\}$ carries over exactly as in the simply connected case; the initial conditions (10.83) are also the same. We define $\hat{T}(z)$ via the relation

$$T(z) = -\frac{3}{2}\frac{z}{(z-1)^2} + \hat{T}(z). \tag{10.111}$$

Then the condition

$$\hat{T}(1) = \frac{3}{8} \tag{10.112}$$

provides a formula relating B to θ, ρ, and C:

$$B = \frac{\frac{3}{8} - \frac{3}{2}[\hat{L}(1,1) + L(1,-1)] + \frac{3}{8}[L(1,\alpha) + L(1,\overline{\alpha}) + L(1,-\alpha) + L(1,-\overline{\alpha})] - C}{\mathrm{i}(K(1,\alpha) - K(1,\overline{\alpha}) + K(1,-\alpha) - K(1,-\overline{\alpha}))}, \tag{10.113}$$

where $\hat{L}(z,1)$ is defined by

$$L(z,1) = -\frac{z}{(z-1)^2} + \hat{L}(z,1). \tag{10.114}$$

There are five unknown real parameters: a, b, θ, ρ, and C. These can be found by insisting that

$$\mathrm{Re}[f(\rho)] = r, \qquad \mathrm{Im}[f(\mathrm{i}\rho)] = r, \qquad f(\mathrm{i}) = \mathrm{i}, \tag{10.115}$$

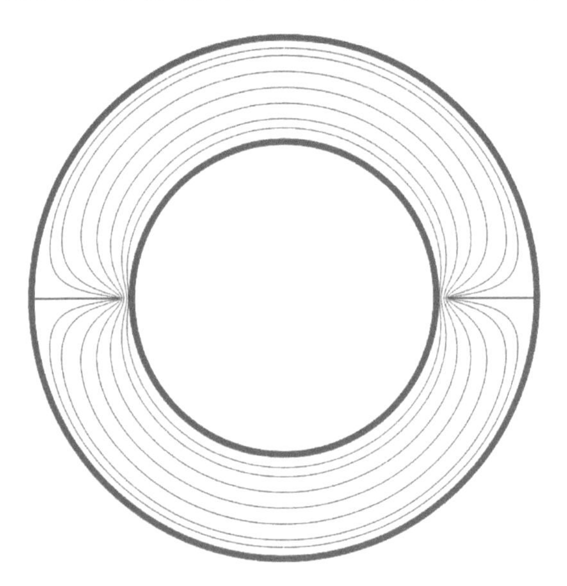

Figure 10.8. *Selection of images, under the polycircular conformal mapping from the annulus (10.70) to an annular duct with two fins, of several concentric circles* $|z| = \text{constant}$ *for* $r_d = 0.6, R = 0.65$.

which are expected by the symmetry of the preimage and target configurations and which provide four real conditions. These must be supplemented by the condition that the mapping is single-valued in the annulus.

Figure 10.8 shows the images under such a conformal mapping of concentric circles in the preimage annulus (10.70).

Exercises

10.1. This exercise concerns the polycircular arc mapping (10.5).

(a) Show that the inverse of the mapping in equation (10.1) is

$$\eta = \frac{1+a^2}{2a}\left[\frac{1-z}{1+z}\right] - \frac{1-a^2}{2a(1+z)}[(z-z_0)(z-\overline{z_0})]^{1/2}, \tag{10.116}$$

where z_0 is given in (10.3).

(b) Verify from (10.1) and (10.2) that

$$w = z\left[\frac{\eta+1/a}{\eta-1/a}\right]^2. \tag{10.117}$$

(c) Combine parts (a) and (b) to confirm the validity of formula (10.5).

10.2. **(Polycircular arc map to a triply connected domain)** This exercise concerns the polycircular arc domain D_w comprising the region inside a unit disc in a w

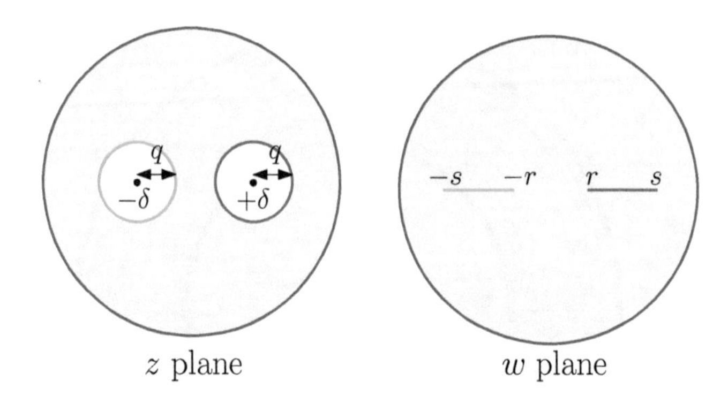

Figure 10.9. *The mapping from a circular preimage domain D_z to a triply connected polycircular arc domain D_w comprising a unit disc with two slits $[-s, -r]$ and $[r, s]$ excised along the real diameter (Exercise 10.2).*

plane exterior to two slits $[-s, -r]$ and $[r, s]$ on the real diameter, as shown in Figure 10.9.

(a) Using the general theory presented in this chapter, find the functional form of the relevant automorphic function $T(z)$ and write down the differential equation to be solved for the conformal mapping

$$w = f(z) \tag{10.118}$$

from a triply connected circular region D_z to D_w.

(b) The same mapping can be constructed using other considerations of previous chapters. Consider the sequence of conformal mappings

$$z \mapsto \eta = \mathcal{F}(z) \mapsto w = \mathcal{G}(\eta) = \mathcal{G}(\mathcal{F}(z)), \tag{10.119}$$

where $\mathcal{F}$ is the Cayley-type mapping

$$\mathcal{F}(z) = -\frac{\omega(z, +1)}{\omega(z, -1)}, \tag{10.120}$$

$\omega(.,.)$ is the prime function of the circular domain D_z, and $\mathcal{G}$ is the Cayley mapping

$$\mathcal{G}(\eta) = \frac{1-\eta}{1+\eta}. \tag{10.121}$$

Show that

$$w = f(z) = \mathcal{G}(\mathcal{F}(z)) \tag{10.122}$$

is the desired conformal mapping described in part (a).

(c) Use the result in part (b) to check the construction carried out in part (a).

10.3. **(Analytical formulas for polycircular arc maps)** By adapting the result of part (b) in Exercise 10.2 find an explicit formula for the polycircular arc mapping from a triply connected circular preimage domain to the region exterior to a unit disc and two slits on either side of it, as shown in Figure 10.10.

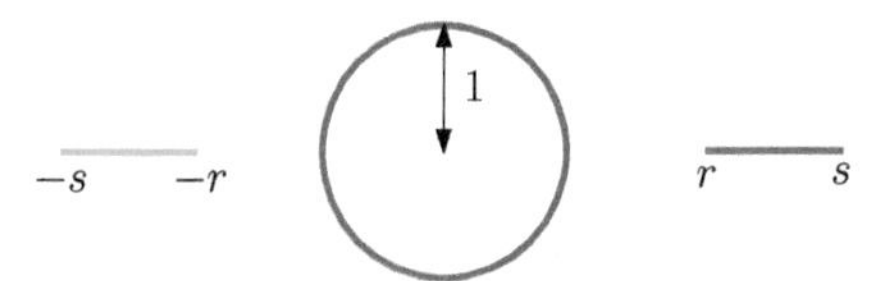

Figure 10.10. *The unbounded polycircular arc domain exterior to the unit disc and two finite-length slits on the real axis to either side (Exercise 10.3).*

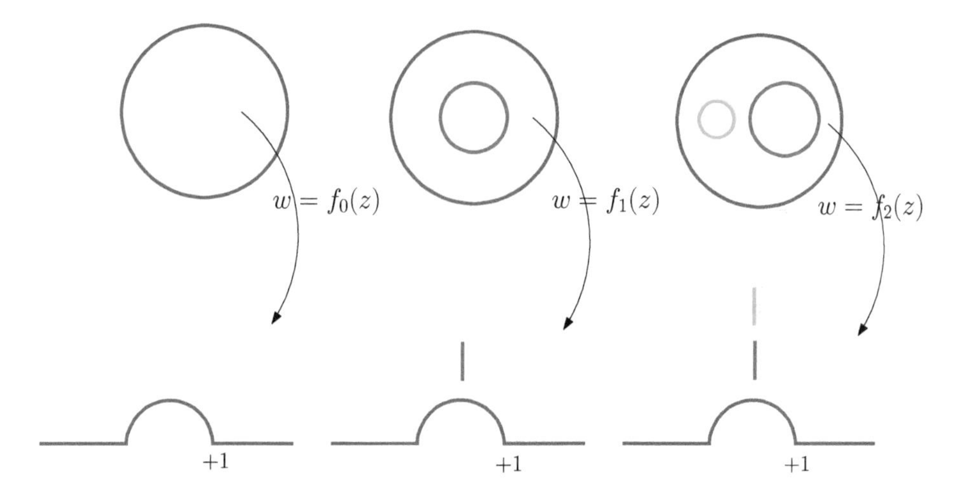

Figure 10.11. *Polycircular arc mappings $f_j(z)$ for $j = 0$, 1, and 2 to simply, doubly, and triply connected domains in a complex w plane (Exercise 10.4).*

10.4. **(Analytical formulas for polycircular arc maps)** This exercise concerns the three polycircular arc mappings shown in Figure 10.11. The domain involved in each case is $(j+1)$ connected for $j = 0$, 1, and 2. We will show that analytical expressions for these multiply connected polycircular arc mappings can be constructed using the prime function for each circular preimage domain.

(a) Find the analytical form of the simply connected mapping $w = f_0(z)$ from the interior of the unit z disc to the unbounded region in the upper half w plane exterior to a semicircular hump of unit radius, as shown in Figure 10.11. This can be done by the following sequence of mappings: (i) first transplant the unit disc to the first quadrant by a sequence of elementary transformations; (ii) transplant the first quadrant to a lower half semidisc using an inverse Cayley map; (iii) perform a simple inversion of this lower half semidisc to the simply connected region shown in Figure 10.11.

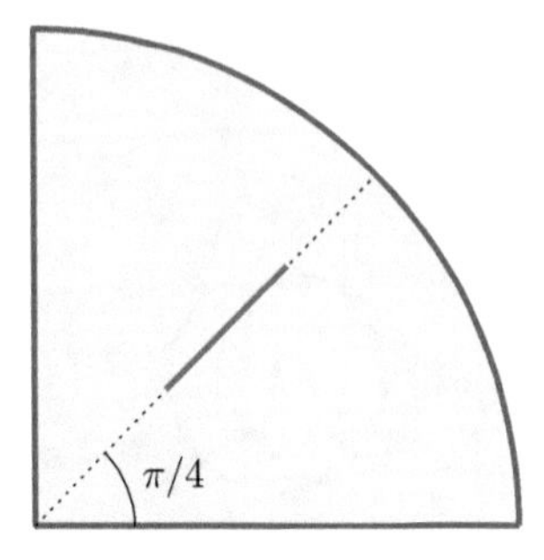

Figure 10.12. *The doubly connected polycircular arc domain comprising a sector of the disc with a slit (Exercise 10.5).*

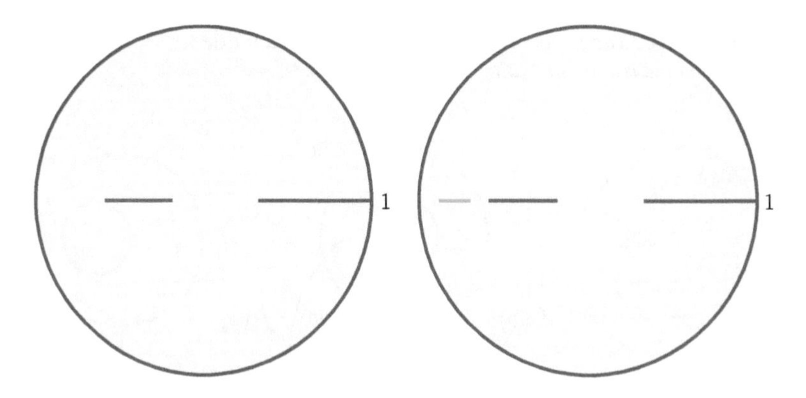

Figure 10.13. *The doubly connected (left) and triply connected (right) polycircular arc domains comprising a unit disc with an attached reentrant fin and one (left) or two (right) extra slits. (Exercise 10.6).*

(b) Find analytical expressions for the generalized mappings $f_1(z)$ and $f_2(z)$ by repeating the construction in part (a) but, in part (i), by first mapping the circular domains in the z plane to the first quadrant, with j slits along the bisecting ray having argument equal to $\pi/4$, as done, for the doubly connected case, in Exercise 7.3.

(c) Argue that the same formula for these polycircular slit mappings works for any finite number of slits along the imaginary axis above the semicircular hump.

10.5. **(Polycircular arc map to a sector of a disc with a slit)** Use the analytical formula for $f_1(z)$ derived in Exercise 10.4 to find an analytical expression, in terms of the prime function, for the concentric annulus $\rho < |z| < 1$ for the polycircular arc mapping to the quadrant with a slit along its bisector, as shown in Figure 10.12.

10.6. **(Polycircular arc map to disc with reentrant fin and slits)** Use the results of Exercise 7.5 to find analytical formulas, in terms of the prime function, for the polycircular arc mappings from circular preimage regions to the interior of a unit disc with an attached reentrant fin with additional slits along the real diameter, as shown in Figure 10.13.

10.7. **(Derivation of the key relation (10.15))** This exercise takes the reader through the principal algebraic steps needed to arrive at (10.15).

(a) Show, using the chain rule, that

$$\frac{d\mathcal{S}_k^{(j)}(w)}{dw} = -\frac{\overline{\eta}}{\eta}\frac{\overline{F'(\eta)}}{F'(\eta)}, \tag{10.123}$$

where $\mathcal{S}_k^{(j)}(w)$ is defined in (10.6) and η is the bounded circular slit map (10.11).

(b) Show that the second derivative is given by

$$\frac{d^2\mathcal{S}_k^{(j)}(w)}{dw^2} = \frac{2\overline{\eta}}{\eta^2}\frac{\overline{F'(\eta)}}{F'(\eta)^2} + \frac{\overline{\eta}^2}{\eta^2}\frac{\overline{F''(\eta)}}{F'(\eta)^2} + \frac{\overline{\eta}}{\eta}\frac{\overline{F'(\eta)}F''(\eta)}{F'(\eta)^3}. \tag{10.124}$$

(c) Show that the third derivative is given by

$$\frac{d^3\mathcal{S}_k^{(j)}(w)}{dw^3} = -\frac{6\overline{\eta}}{\eta^3}\frac{\overline{F'(\eta)}}{F'(\eta)^3} - \frac{6\overline{\eta}^2}{\eta^3}\frac{\overline{F''(\eta)}}{F'(\eta)^3} - \frac{6\overline{\eta}}{\eta^2}\frac{\overline{F'(\eta)}F''(\eta)}{F'(\eta)^4}$$
$$-\frac{3\overline{\eta}^2}{\eta^2}\frac{\overline{F''(\eta)}F''(\eta)}{F'(\eta)^4} - \frac{\overline{\eta}^3}{\eta^3}\frac{\overline{F'''(\eta)}}{F'(\eta)^3} + \frac{\overline{\eta}}{\eta}\frac{\overline{F'(\eta)}F'''(\eta)}{F'(\eta)^4} - \frac{3\overline{\eta}}{\eta}\frac{\overline{F'(\eta)}F''(\eta)^2}{F'(\eta)^5}.$$

(d) Hence verify that (10.10) leads to (10.15).

10.8. **(Alternative derivation of (10.15))** This exercise explores an alternative derivation of the key relation (10.15) that is the natural generalization of the derivation of multiply connected S–C maps via the "pre-Schwarzian," as explored in Exercise 7.2. The function $F(\eta)$ here is defined in (10.13); it is the conformal mapping from a bounded circular slit domain D_η to the polycircular domain of interest D_w in a complex w plane.

(a) Derive Study's formula

$$\kappa = \pm\frac{1}{|\eta F'(\eta)|}\text{Re}\left[1 + \frac{\eta F''(\eta)}{F'(\eta)}\right] \tag{10.125}$$

for the curvature κ of the image curve of any circular boundary in D_η under the action of $F(\eta)$ (cf. Exercise 7.1).

(b) Show that the curvature κ changes with arclength s in the image domain according to

$$\frac{d\kappa}{ds} = -\frac{1}{|\eta F'(\eta)|^2}\text{Im}\left[\eta^2\{F(\eta),\eta\}\right], \tag{10.126}$$

where $\{F(\eta),\eta\}$ is the Schwarzian derivative defined in (10.16).

(c) Hence show that, for a polycircular arc domain, (10.15) must hold, where $T(\eta)$ is defined in (10.17).

10.9. **(Chain rule for Schwarzian derivatives)** Given two analytic functions $w(z)$ and $z = f(t)$, establish that the Schwarzian derivative of the composition $W(t) \equiv w(f(t))$ is

$$\{W(t),t\} = \{w(z),z\}\left[\frac{df}{dt}\right]^2 + \{f(t),t\}. \tag{10.127}$$

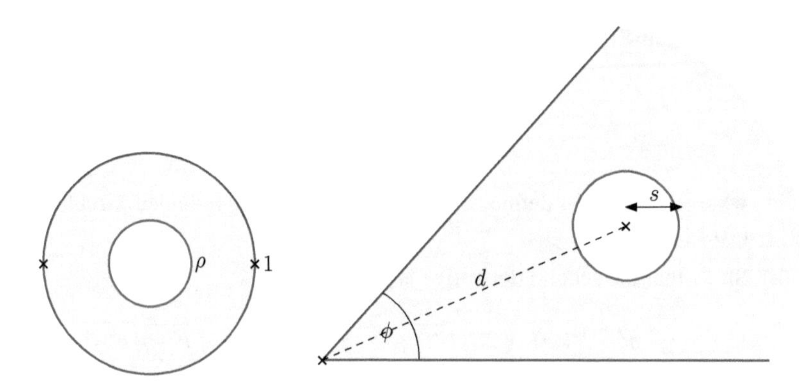

Figure 10.14. *Mapping of a concentric annulus to the polycircular arc domain comprising the interior of a wedge of opening angle ϕ and exterior to an excised circular disc on the bisector (Exercise 10.10).*

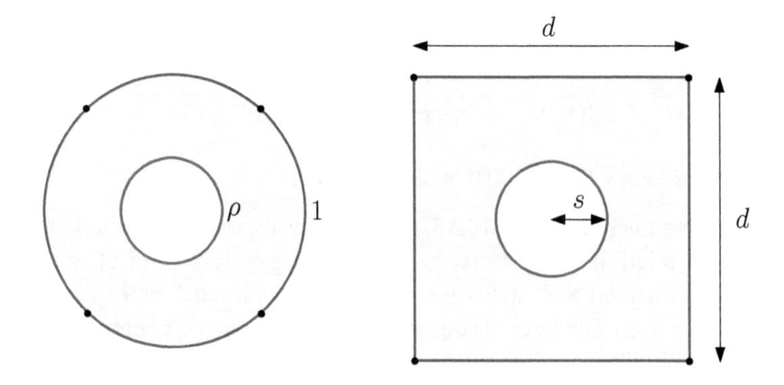

Figure 10.15. *Mapping of a concentric annulus to a doubly connected polycircular arc domain comprising the region interior to a d-by-d square exterior to a centralized circular disc of radius s (Exercise 10.11).*

10.10. **(Region exterior to a disc in an infinite wedge)** A natural extension of the problem of mapping a concentric annulus to the region exterior to a disc in a channel shown in Figure 10.4 is to construct the mapping from the same annular preimage domain to the unbounded region in a wedge of opening angle ϕ containing a circular disc of radius s situated at a distance d from the wedge apex along the angle bisector; see the schematic in Figure 10.14. Use the techniques of this chapter to construct such a conformal mapping.

10.11. **(Region exterior to a disc in a square)** Another natural extension of the disc-in-channel geometry shown in Figure 10.4 is the case of a disc of radius s in a d-by-d square, as shown in Figure 10.15. Use the techniques of this chapter to construct the conformal mapping from a concentric annulus to the shaded region in the square outside the circular disc shown in Figure 10.15.

Chapter 11
Quadrature domains

11.1 ▪ Beyond slit, polygonal, and polycircular arc domains

From the slit domains arising naturally as we built up the basic theory of the prime function in Chapters 1–6, we moved, in Chapter 7, to their natural extensions: more general polygonal domains, including multiply connected cases.

Then we expanded the class of domains amenable to our techniques when we contemplated the class of polycircular arc domains in Chapter 10, again including the multiply connected situation.

But we can go even further. In this chapter, we add the fascinating class of quadrature domains to the compendium of planar regions amenable to treatment within our prime function framework. Unlike the polygonal and polycircular domains treated so far, the quadrature domains we consider will have smooth boundaries with no corners. Their boundary shapes are not so easily characterized as being made up of a concatenation of simpler curves such as circular arcs. Interestingly, quadrature domains can, however, be constructed by a geometrical process of merging circular discs, as we will explain in this chapter.[32]

11.2 ▪ Why study quadrature domains?

Quadrature domains are a class of planar quadrature domains possessing several interesting mathematical features. This chapter will explain their characteristics and show how to construct them. The theory of quadrature domains is not well known among applied scientists outside a small community of workers who have studied the Laplacian growth, or Hele-Shaw, problem that will be discussed in Chapter 26 in Part II; the relevance of quadrature domains is well known there. As it turns out, this is but one of many applications where these domains are useful; see [27, 5] and Chapters 23, 24, and 27, for example. Such studies are likely just a small sample of the multifarious applications of these domains yet to be discovered. Here we give a nontechnical introduction to quadrature domains for the benefit of applied scientists and with emphasis on their explicit construction.[33]

---

[32]We will discover in Part II that several physical processes lead to precisely this kind of "merging" taking place in the natural world; see Chapters 26 and 27.

[33]More technical treatments on the theory of quadrature domains can be found elsewhere [86].

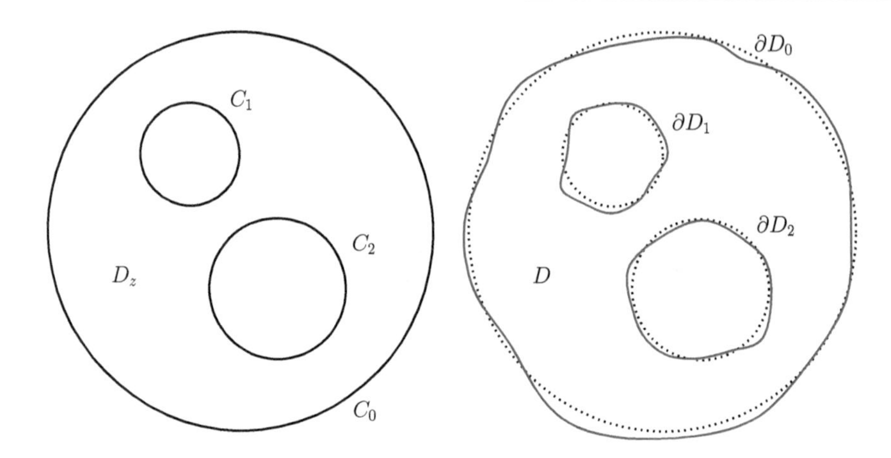

Figure 11.1. *A circular domain D_z and a deformation D for which the Schwarz functions of all boundary curves are meromorphic continuations of each other.*

11.3 ▪ Distinct boundaries with a common Schwarz function

One of the main reasons that multiply connected quadrature domains turn out to be useful in applications concerns a feature of the Schwarz functions of their boundary components.

Consider a circular domain D_z, of the kind introduced in Chapter 2, comprising the unit disc with M circular discs excised. A schematic is shown to the left of Figure 11.1. The Schwarz function of C_0 is

$$\overline{z} = S_0(z) = \frac{1}{z}, \tag{11.1}$$

and the Schwarz functions $\{S_j | j = 1, \ldots, M\}$ of the inner boundaries $\{C_j | j = 1, \ldots, M\}$ are

$$\overline{z} = S_j(z) = \overline{\delta_j} + \frac{q_j^2}{z - \delta_j}, \qquad j = 1, \ldots, M. \tag{11.2}$$

These Schwarz functions are all Möbius maps and, as such, are meromorphic functions. Indeed, each $S_j(z)$ has a simple pole at δ_j, which is the center of C_j. Clearly the Schwarz functions of the different boundary components are distinct.

It is natural to ask if there might exist some deformation of this circular domain, D say, as shown schematically to the right of Figure 11.1, for which the Schwarz function $\mathcal{S}_j(w)$ of any boundary component $\partial D_j, j = 0, 1, \ldots, M$, i.e.,

$$\overline{w} = \mathcal{S}_j(w), \qquad w \in \partial D_j, \tag{11.3}$$

can be *meromorphically* continued to any other boundary component ∂D_k ($k \neq j$) and be coincident there with the Schwarz function of that boundary. This means that we have a common Schwarz function for all boundary components, and we can write

$$\overline{w} = \mathcal{S}_j(w) = \mathcal{S}(w), \qquad z \in \partial D_j, \tag{11.4}$$

where the function $\mathcal{S}(w)$ is meromorphic in D. Expressed differently, all boundary components have a common Schwarz function, and that function has a distribution of isolated poles inside the domain. Ideally, we would hope that D might be in some sense "close" to our original domain D_z.

Such multiply connected domains D exist. They are known as *quadrature domains*. It also turns out to be possible to approximate any given domain by a quadrature domain, as we explore in §11.13.[34] We will discuss ways to characterize quadrature domains in terms of quadrature identities and give a geometrical construction by merging circular discs. The circular disc is the simplest example of a planar quadrature domain.

11.4 ▪ The circular disc as a quadrature domain

Let D denote a circular disc of radius r centered at the origin. Suppose $h(z)$ is analytic in D and integrable over it so that the two-dimensional integral

$$\int\int_D h(z)dxdy \tag{11.5}$$

makes sense. By the complex form of the Stokes theorem,

$$\int\int_D h(z)dxdy = \frac{1}{2\mathrm{i}}\int_{|z|=r} h(z)\overline{z}dz. \tag{11.6}$$

We also know that on $|z| = r$,

$$\overline{z} = \frac{r^2}{z}, \tag{11.7}$$

where the right-hand side is the Schwarz function of the circle $|z| = r$. Consequently,

$$\int\int_D h(z)dxdy = \frac{1}{2\mathrm{i}}\int_{|z|=r} h(z)\overline{z}dz = \frac{1}{2\mathrm{i}}\int_{|z|=r} \frac{h(z)r^2}{z}dz = \pi r^2 h(0), \tag{11.8}$$

where, in the last equality, we have used the residue theorem to collect the residue contribution of the (meromorphic in D) integrand at $z = 0$.

We are left with the final result that, for any function $h(z)$ analytic in the disc D,

$$\int\int_D h(z)dxdy = \pi r^2 h(0). \tag{11.9}$$

This formula tells us that the two-dimensional integral of any analytic function $h(z)$ over this disc is equal to πr^2 times its value at the center of the disc. (11.9) is an example of a *quadrature identity*. It resembles a quadrature rule for approximating a two-dimensional integral, in this case, using just a single quadrature point at the origin. It is remarkable because it is not an approximation but an identity that holds for a large class of admissible functions.

It is worth noting that (11.9) is a statement of the *mean value theorem* [61] in complex analysis since we can rewrite it as

$$h(0) = \frac{\int\int_D h(z)dxdy}{\int\int_D dxdy} \tag{11.10}$$

and see that the value of $h(z)$ at the center of the disc is its weighted average, or mean value, over the support of the disc.

[34]The boundaries of the two domains are said to be "close" with respect to some Hausdorff measure of the proximity of two sets [86].

It is natural to think of (11.9) as a geometrical property associated with the circular disc. One is led to ask if there are any other planar domains, besides the simple circular disc, for which a quadrature identity akin to (11.9) also holds. The class of quadrature domains described in §11.2—that is, those domains whose boundary components have a common Schwarz function that is meromorphic in the domain—happens to satisfy quadrature identities that are a generalization of (11.9), as we now show.

11.5 ▪ Quadrature identities

Suppose that D is a bounded $(M+1)$-connected planar domain in the complex w plane, such as that shown to the right of Figure 11.1 in a triply connected case $M = 2$, with the special feature (11.4) that the common Schwarz function $\mathcal{S}(w)$ of all the boundary components has N simple poles inside D at $\{w_n | n = 1, \ldots, N\}$ having residues $\{r_n^2 \in \mathbb{C} | n = 1, \ldots, N\}$. This restriction to simple poles is made to simplify the presentation, but it is not necessary. Also, the choice of the notation r_n^2 for the complex-valued residue contribution is made in analogy with (11.9); however, note that, in the general case, the constant r_n need not be real.

Suppose some function $h(w)$ is analytic and single-valued in D and assumed to be integrable over D so that all of the following steps hold. By the Stokes theorem,

$$\int\!\!\int_D h(w) dA_w = \frac{1}{2\mathrm{i}} \int_{\partial D_0} \mathcal{S}_0(w) h(w) dw - \frac{1}{2\mathrm{i}} \sum_{j=1}^{M} \int_{\partial D_j} \mathcal{S}_j(w) h(w) dw, \quad (11.11)$$

where

$$\overline{w} = \mathcal{S}_j(w), \qquad w \in \partial D_j, \qquad j = 0, 1, \ldots, M. \quad (11.12)$$

Here $\mathcal{S}_j(w)$ is the Schwarz function of the boundary component ∂D_j. However, recall that we have supposed that the domain D is such that (11.4) holds. Use of this property in (11.11) implies

$$\int\!\!\int_D h(w) dA_w = \frac{1}{2\mathrm{i}} \int_{\partial D} \mathcal{S}(w) h(w) dw, \quad (11.13)$$

where ∂D denotes the entire boundary of D traversed with the domain sitting to the left. The important point is that we are now integrating the *same* function over the entire boundary of D. Since, by assumption, that function is meromorphic in D, we can use the residue theorem to evaluate the integral by collecting residues at the N simple poles of the integrand at the points $\{w_n | n = 1, \ldots, N\}$:

$$\int\!\!\int_D h(w) dA_w = \frac{1}{2\mathrm{i}} \int_{\partial D} \mathcal{S}(w) h(w) dw = \sum_{n=1}^{N} \pi r_n^2 h(w_n). \quad (11.14)$$

Formula (11.14) reveals that a two-dimensional integral of an arbitrary analytic single-valued function $h(z)$ over the domain D can be reduced to a finite sum over the point set $\{w_n | n = 1, \ldots, N\}$ inside D. (11.14) is another example of a *quadrature identity*, generalizing the simpler one (11.9) relevant to a circular disc. We see that a quadrature identity can be associated to a given quadrature domain. N is called the *order* of the quadrature domain. The set of $2N$ complex-valued parameters $\{w_n, \pi r_n^2 | n = 1, \ldots, N\}$ is called the *quadrature data*.

11.6 ▪ Conformal mappings and automorphic functions

What do the conformal mappings from the canonical class of multiply connected circular domains to multiply connected quadrature domains look like? It turns out that they are given by automorphic functions with respect to the Schottky group Θ associated with the circular preimage domain.

To see this, consider the conformal map

$$w = f(z) \tag{11.15}$$

from a preimage circular domain D_z to a given multiply connected quadrature domain D. As usual, let D_z' denote the reflection of D_z in the unit circle C_0. We suppose that the Schwarz functions of the boundary curves of D satisfy (11.4). If the unit circle C_0 is transplanted by $f(z)$ to ∂D_0, then the condition

$$\overline{w} = \mathcal{S}_0(w) = \mathcal{S}(w) \tag{11.16}$$

implies

$$\overline{f(z)} = \overline{f}(1/\overline{z}) = \mathcal{S}(f(z)), \tag{11.17}$$

where we have used the fact that $z = 1/\overline{z}$ on C_0. If C_j is transplanted by $f(z)$ to ∂D_j, for $j = 1, \ldots, M$, then the condition

$$\overline{w} = \mathcal{S}_j(w) = \mathcal{S}(w) \tag{11.18}$$

implies

$$\overline{f(z)} = \overline{f(\overline{S_j(z)})} = \mathcal{S}(f(z)), \tag{11.19}$$

where we have used the fact that $z = \overline{S_j(z)}$ on C_j. This relation between analytic functions can be (meromorphically) continued off C_j into D. The relations (11.17) and (11.19) together imply that

$$\overline{f(1/\overline{z})} = \overline{f(\overline{S_j(z)})}, \qquad j = 1, \ldots, M. \tag{11.20}$$

On taking a complex conjugate and changing the complex variable $\overline{z} \mapsto 1/z$, we find

$$f(\theta_j(z)) = f(z), \qquad \theta_j(z) \equiv \overline{S_j}(1/z), \qquad j = 1, \ldots, M. \tag{11.21}$$

From (11.17) we deduce that if $\mathcal{S}(w)$ has a simple pole at w_a whose preimage is at z_a so that

$$w_a = f(z_a), \tag{11.22}$$

then $\overline{f(1/\overline{z})}$ has a simple pole at z_a implying that $f(z)$ has a simple pole at $1/\overline{z_a}$.

In this way we see that the conformal mapping $f(z)$ to a quadrature domain D from a conformally equivalent circular domain D_z is an automorphic function in the sense defined in the Chapter 9; that is, the mapping function is invariant under the action on its argument of the Schottky group Θ generated by the Möbius maps $\{\theta_j(z) | j = 1, \ldots, M\}$ and their inverses. For a bounded domain, the mapping is analytic in D_z, and all of its poles are located in D_z'. Unbounded quadrature domains can also be considered, of course, and Exercise 11.7 discusses an example case.

These observations mean that we can use the prime function machinery for loxodromic and automorphic functions developed in Chapters 8 and 9 to construct conformal maps from circular preimage domains to multiply connected quadrature domains using the given quadrature data to fix the zeros, poles, and other parameters in the conformal mapping function.[35] The next section illustrates how this can be done.

[35] Use of the prime function to construct quadrature domains corresponding to a given quadrature identity was first proposed by Crowdy and Marshall [53].

11.7 ▪ Quadrature domains by merging circular discs

We have seen how the circular disc can be viewed as one of the simplest quadrature domains associated with the quadrature identity (11.9). One way to construct more complicated quadrature domains of higher order is by a geometrical continuation process involving the "merging" of circular discs. What this means will become clear in the examples to follow. After the merger, quadrature domains with noncircular boundaries emerge. This construction helps us to build up some geometrical intuition about quadrature domains.

Merging two discs: Consider two circular discs both of radius r with centers at $\pm d$ on the real axis in a w plane. Let the discs be denoted by D_1 and D_2. If $d > r$, the domain $D = D_1 \cup D_2$ is disconnected, and, clearly, if $h(w)$ is analytic in D, then

$$\int\int_D h(w)dA_w = \int\int_{D_1} h(w)dA_w + \int\int_{D_2} h(w)dA_w = \pi r^2 h(d) + \pi r^2 h(-d). \tag{11.23}$$

Owing to the disconnected nature of the domain, this result is really just two instances of (11.9).

It is of interest to ask, however, if there exists a domain D satisfying the same identity (11.23) when $d < r$. Then we might imagine that any such domain D is formed by the merging of the two discs D_1 and D_2.

To explore whether such a domain exists, consider the conformal mapping

$$w = f(z) = \frac{Rz}{z^2 - a^2}, \qquad a > 1, \qquad a, R \in \mathbb{R}, \tag{11.24}$$

from the unit z disc $|z| < 1$. It is clear that $z = 0$ is transplanted to $w = 0$ and, since $a > 1$, this rational function $f(z)$ is analytic in the unit z disc. Now, by the Stokes theorem,

$$\int\int_D h(w)dA_w = \frac{1}{2\mathrm{i}} \int_{\partial D} \overline{w} h(w)dw. \tag{11.25}$$

Since $|z| = 1$ corresponds to ∂D, we can transfer the line integral to the boundary of the unit z disc:

$$\int\int_D h(w)dA_w = \frac{1}{2\mathrm{i}} \int_{|z|=1} \overline{f(z)} h(f(z)) \frac{df}{dz} dz. \tag{11.26}$$

On use of the facts that, on $|z| = 1$ where $\overline{z} = 1/z$,

$$\overline{f(z)} = \frac{Rz}{1 - z^2 a^2}, \qquad \frac{df}{dz} = -\frac{R(z^2 + a^2)}{(z^2 - a^2)^2}, \tag{11.27}$$

(11.26) becomes

$$\int\int_D h(w)dA_w = -\frac{R^2}{2\mathrm{i}} \int_{|z|=1} \frac{zh(f(z))}{1 - z^2 a^2} \frac{(z^2 + a^2)}{(z^2 - a^2)^2} dz. \tag{11.28}$$

It is easy to check that the integrand is meromorphic in the unit disc, with two simple poles at $z = \pm 1/a$. On collecting residues at these two poles the residue theorem leads to

$$\int\int_D h(w)dA_w = \frac{\pi R^2(1 + a^4)}{2(1 - a^4)^2} h(f(1/a)) + \frac{\pi R^2(1 + a^4)}{2(1 - a^4)^2} h(f(-1/a)). \tag{11.29}$$

Since $f(-1/a) = -f(1/a)$, this has precisely the form (11.23) if we make the identifications

$$d = f(1/a) = \frac{Ra}{1 - a^4}, \qquad r^2 = \frac{R^2(1 + a^4)}{2(1 - a^4)^2}. \tag{11.30}$$

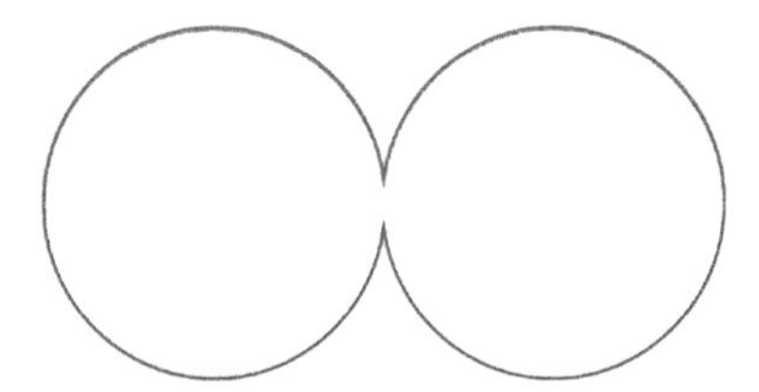

Figure 11.2. *A simply connected quadrature domain formed by the merging of two equal circular discs.*

For given d and r, (11.30) provides two equations for the conformal mapping parameters R and a.

Figure 11.2 shows an example of a simply connected domain in which the two circular discs have only just merged, forming a thin neck region between them where the boundary has two near-cusp points of high curvature.

Merging three discs: Suppose now that we consider merging three circular discs, D_1, D_2, and D_3, all of radius r. The centers of the discs will be denoted by d_1, d_2, and d_3. This case is more interesting since now two topologically distinct arrangements are possible.

One geometrical configuration is to center all three discs on the real axis with

$$d_1 = 0, \qquad d_2 = -d_3 = d. \tag{11.31}$$

If $d > 2r$, the domain will be disconnected with quadrature identity

$$\begin{aligned}\int\int_D h(w)dA_w &= \int\int_{D_1} h(w)dA_w + \int\int_{D_2} h(w)dA_w + \int\int_{D_3} h(w)dA_w \\ &= \pi r^2 h(d_1) + \pi r^2 h(d_2) + \pi r^2 h(d_3) \qquad (11.32)\\ &= \pi r^2 h(0) + \pi r^2 h(d) + \pi r^2 h(-d). \qquad (11.33)\end{aligned}$$

By analogy with the two-disc case, when $d < 2r$ we expect the discs will merge into a simply connected quadrature domain with quadrature identity

$$\int\int_D h(w)dA_w = \pi r^2 h(0) + \pi r^2 h(d) + \pi r^2 h(-d). \tag{11.34}$$

Let the conformal mapping from the unit z disc to this quadrature domain be

$$w = f(z) = \frac{Rz(z^2 - b^2)}{z^2 - a^2}, \qquad a, b, R \in \mathbb{R}, \tag{11.35}$$

with $a, b > 1$. This is a rational function with two simple poles outside the unit disc and with three simple zeros, two of which are also outside the unit disc. Again, $z = 0$ is transplanted to $w = 0$. On $|z| = 1$, where $\overline{z} = 1/z$,

$$\overline{w} = \overline{f(z)} = \overline{f}(1/z) = \frac{R(1 - z^2 b^2)}{z(1 - z^2 a^2)}, \tag{11.36}$$

which has three simple poles inside the unit z disc at $z = 0, \pm 1/a$. Intuitively, it is useful to think of each of these three simple poles as being "associated with" the center of each of the three merged discs. This is because, from (11.22), we must have

$$d_1 = f(0) = 0, \qquad d_2 = f(1/a) = d, \qquad d_3 = f(-1/a) = -d. \tag{11.37}$$

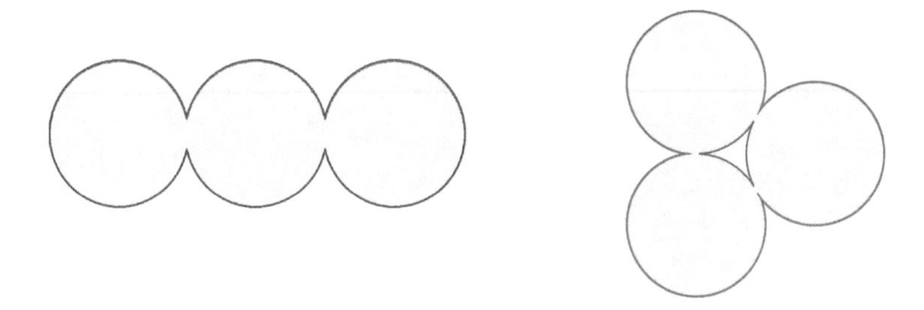

Figure 11.3. *Two ways to produce a quadrature domain by merging three equally sized circular discs: the domain on the left is simply connected; that to the right is doubly connected.*

This provides a single real constraint on the three real parameters R, a, and b. Two further real constraints emerge by insisting that the residues of $\overline{f}(1/z)f'(z)$ at $z = 0$ and $z = 1/a$ are equal to r^2, as required for the quadrature identity (11.34) to hold; the oddness of the mapping (11.35) under $z \mapsto -z$ guarantees that the residues at $\pm 1/a$ are always equal for all choices of the parameters. Exercise 11.1 explores these relations between parameters.

The schematic to the left in Figure 11.3 shows an example where three equally sized discs in a linear array have just merged, forming two thin neck regions between the now-merged discs.

Another configuration is to center the three discs at $d_1 = d$, $d_2 = de^{2\pi \mathrm{i}/3}$, and $d_3 = de^{4\pi \mathrm{i}/3}$. If $d > 2r/\sqrt{3}$, then the domain $D = D_1 \cup D_2 \cup D_3$ is disconnected, with the quadrature identity

$$\int\int_D h(w)dA_w = \int\int_{D_1} h(w)dA_w + \int\int_{D_2} h(w)dA_w + \int\int_{D_3} h(w)dA_w$$
$$= \pi r^2 h(d_1) + \pi r^2 h(d_2) + \pi r^2 h(d_3). \quad (11.38)$$

In this case, for $d = 2r/\sqrt{3} - \epsilon$ where $\epsilon > 0$ is sufficiently small, we expect the resulting merged domain to be doubly connected, with a single hole enclosed by the three now-merged discs forming an annular region around the hole.

Since this annular configuration after merging is doubly connected, we consider finding a conformal mapping to it from a concentric annulus $\rho < |z| < 1$. We know from the considerations of §11.6 that such a mapping must be a loxodromic function, which means we can deploy the theory of Chapter 8 to find it. From there we know that there are (at least) two possible choices of representation for such a function. Let us represent this loxodromic function as a ratio of products of prime functions since this has the advantage of making the poles and zeros of the function explicit. Consequently, consider

$$w = f(z) = Rz\frac{\omega(z, a/\rho^{2/3})\omega(z, \sigma_3 a/\rho^{2/3})\omega(z, \sigma_3^2 a/\rho^{2/3})}{\omega(z, a)\omega(z, \sigma_3 a)\omega(z, \sigma_3^2 a)}, \qquad R, a \in \mathbb{R}, \quad (11.39)$$

where $1 < a < 1/\rho$ and the constant σ_3 is a cubic root of unity, i.e.,

$$\sigma_3 = e^{2\pi \mathrm{i}/3}, \quad (11.40)$$

and where $\omega(.,.)$ is the prime function associated with the concentric annulus $\rho < |z| < 1$. It is straightforward to use the properties (5.4)–(5.6) of this prime function to confirm that the function in (11.39) is indeed loxodromic. The poles and zeros in the z plane clearly have a 3-fold rotational symmetry. Moreover, the function (11.39) also has the property that $w \mapsto \sigma_3 w$ as $z \mapsto \sigma_3 z$, which means that the image in the w plane will share this 3-fold rotational symmetry. This mapping is also such that the common Schwarz function of both boundaries has three simple poles inside the domain at

$$d_1 = f(1/a) = d, \qquad d_2 = f(\sigma_3/a) = d\sigma_3, \qquad d_3 = f(\sigma_3^2/a) = d\sigma_3^2. \tag{11.41}$$

By the 3-fold rotational symmetry of the mapping, this places a single real constraint on the two real parameters R and a. A second constraint comes from insisting that the residue of $\overline{f}(1/z)f'(z)$ at $z = 1/a$ equals r^2 so that the quadrature identity (11.38) holds.

A typical doubly connected quadrature domain of this type is shown to the right in Figure 11.3, displaying three thin neck regions between the three now-merged circular discs.

Merging four discs: Moving on to the merging of four circular discs of equal size, even more topologically distinct arrangements are possible. If we align all the disc centers on the real axis with

$$d_1 = -d_2 = d, \qquad d_3 = -d_4 = 3d, \tag{11.42}$$

then for $r > d$ we expect to find, after merging, a simply connected quadrature domain corresponding to a quadrature identity of the form

$$\int\int_D h(w) dA_w = \pi r^2 \left[h(d_1) + h(d_2) + h(d_3) + h(d_4)\right]. \tag{11.43}$$

The conformal mapping from the unit disc is given by a rational function, and finding it is the subject of Exercise 11.2.

Another possibility is to arrange the centers to be equally spaced around a circle:

$$d_i = de^{\pi ik/2}, \qquad k = 0, 1, 2, 3. \tag{11.44}$$

Motivated by the merging of a ring of 3 discs just discussed, we consider the conformal mapping from the concentric annulus $\rho < |z| < 1$ given by the loxodromic function

$$w = f(z) = Rz\frac{\omega(z, a/\rho^{2/4})\omega(z, \sigma_4 a/\rho^{2/4})\omega(z, \sigma_4^2 a/\rho^{2/4})\omega(z, \sigma_4^3 a/\rho^{2/4})}{\omega(z, a)\omega(z, \sigma_4 a)\omega(z, \sigma_4^2 a)\omega(z, \sigma_4^3 a)}, \qquad R, a \in \mathbb{R}, \tag{11.45}$$

where $1 < a < 1/\rho$ and

$$\sigma_4 = e^{2\pi i/4}. \tag{11.46}$$

Now $\omega(.,.)$ is the prime function associated with the concentric annulus $\rho < |z| < 1$ discussed in Chapter 5. This is a straightforward generalization of the 3-disc case, and we leave the reader to verify that (11.45) is a representation of the required mapping. A typical doubly connected quadrature domain of this type is shown to the left in Figure 11.4.

A third possibility is to place the centers at

$$d_1 = d_2 = d, \qquad d_3 = -d_4 = i\mathcal{D}, \qquad d, \mathcal{D} \in \mathbb{R}, \tag{11.47}$$

and to choose the parameter r to be sufficiently large that the discs have merged, as shown to the right in Figure 11.4. After merging, these discs form a triply connected quadrature domain with the same quadrature identity (11.43).

Consequently, let $w = f(z)$ be the conformal mapping to such a domain from a triply connected circular domain D_z of the kind considered in Chapter 6. Let $\omega(.,.)$ denote the prime function associated with such a domain, and let

$$f(z) = \frac{R\omega(z, 0)\omega(z, \infty)\omega(z, a)\omega(z, -a)}{\omega(z, b_1)\omega(z, -b_1)\omega(z, ib_2)\omega(z, -ib_2)}, \tag{11.48}$$

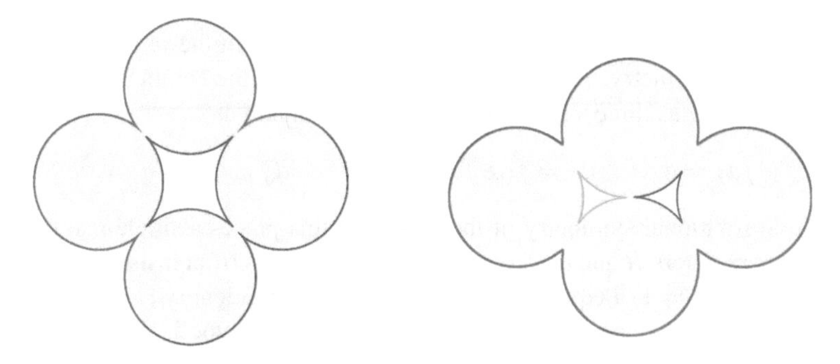

Figure 11.4. *A doubly connected quadrature domain formed by the merging of four equal circular discs. The functional form of the conformal mapping from a concentric annulus $\rho < |z| < 1$ is given in (11.45). Also shown is a triply connected quadrature domain whose conformal mapping from a triply connected circular preimage is given by (11.48).*

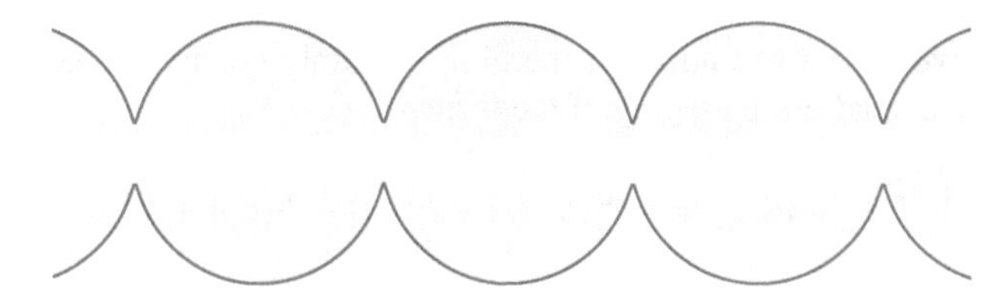

Figure 11.5. *The image domain D, in a complex w plane, of the cut annulus $\rho < |z| < 1$ under the image of (11.118) for various values of ρ together with some of its periodic translates in the x direction.*

where R, a, b_1, and b_2 are real parameters and $a, b_1, b_2 > 1$. Provided these parameters satisfy certain constraints that ensure that this function is automorphic with respect to the Schottky group Θ associated with the circular preimage D_z, this function provides the required conformal mapping from D_z to a quadrature domain of the kind shown to the right in Figure 11.4, which has been plotted using such a function.

Exercise 9.3 is concerned with the construction of an automorphic function of precisely the form (11.48), as well as the construction of an alternative representation of the same automorphic function written in terms of the function $\mathcal{K}(z, a)$. The reader is referred there to be guided through the details.[36] Looking ahead, in §11.9 we will discuss these two different constructions for general quadrature domains. Moreover in §11.10–§11.12 we will also see how the representation in terms of the function $\mathcal{K}(z, a)$ emerges from other considerations involving the so-called Bergman kernel functions of the domains.

A periodic array of discs: We can continue this construction for any number N of discs of equal size; see Exercise 11.4.

Given N equally sized circular discs, it is always possible to concatenate them on a line and consider merging them into a simply connected quadrature domain of order N. What is the limit of this procedure as $N \to \infty$? The resulting domain would resemble what is shown in Figure 11.5. Such a construction is explored in Exercise 11.9. This example is interesting because while this periodic domain has no holes and is simply connected, the function theory associated with a doubly connected preimage annulus turns out to be the most natural way to parametrize this domain. In Part II, we will see other situations in which a similar circumstance arises.

[36]This example was first considered in [53], where more details can be found.

11.8 ▪ The inverse problem

The geometrical construction of quadrature domains by merging circular discs helps provide insights into the matter of reconstructing a quadrature domain, possibly multiply connected, from its quadrature identity. For a given number of disconnected discs, the quadrature identity is determined at the outset. But notice that, as we decide on the various ways in which we can configure a given number of discs to produce a quadrature domain by merging, we do not let the quadrature identity itself change. Consequently, we might end up with either a simply connected quadrature domain or a higher connected one, depending on how we arranged the discs to be merged. If the merged domain produces multiply connected domains, then "holes" appear in the merged domain whose size varies according to the degree of merging. This gives us the clue that, for multiply connected domains, complete information on the geometry of the domain is *not* available from a given quadrature identity and that it is necessary to supplement the quadrature data with additional information about the size of the holes. We explore this matter further in the next section, where we count the parameters in a representation of the conformal mapping function.

11.9 ▪ Reconstructing a domain from its quadrature identity

In §11.7 we showed how to derive explicit forms of the conformal map from a circular domain to a quadrature domain given its associated quadrature identity. In all the examples considered, this identity had the general form

$$\int\int_{D_w} h(w)dA_w = \sum_{n=1}^{N} \pi r_n^2 h(w_n), \tag{11.49}$$

where $h(w)$ is analytic and single-valued in D. Those conformal maps which, in view of the results of §11.6, are known to be automorphic functions were written down as ratios of products of prime functions. However, we know from §9.6 that we can alternatively write these same mappings in terms of the function $\mathcal{K}(z,a)$. For comparison, it is instructive to effect this construction for a quadrature domain having a quadrature identity of the form (11.49). For a simply connected domain it is known that the quadrature identity contains sufficient information to uniquely determine the associated quadrature domain. This is not true, however, for multiply connected domains, where additional information associated with the holes is needed to uniquely determine the domain.

To illustrate this, suppose we seek to reconstruct a triply connected domain with quadrature identity (11.49), where we assume that N is sufficiently large that such a domain exists. It will be conformally equivalent to a circular preimage domain D_z with two interior holes centered at δ_1 and δ_2 with radii q_1 and q_2. This represents six real degrees of freedom. However, three real degrees of freedom are available in the Riemann mapping theorem, two of which can be fixed by fixing a choice of δ_1. This leaves a rotational degree of freedom still to be specified.

Now let

$$w = f(z) = \sum_{n=1}^{N} A_n \mathcal{K}(z, b_n) + A_0, \tag{11.50}$$

where the coefficients $\{A_n | n = 1, \ldots, N\}$ must satisfy the conditions

$$\sum_{n=1}^{N} A_n v_1'(b_n) = 0, \qquad \sum_{n=1}^{N} A_n v_2'(b_n) = 0 \tag{11.51}$$

so that $f(z)$ will be automorphic with respect to the Schottky group Θ generated by $\theta_1(z), \theta_2(z)$ and their inverses. These conditions arise on use of the properties (6.88) of $\mathcal{K}(.,.)$. The formulas (11.51) constitute two complex, or four real, conditions on the parameters in the mapping function (11.50).

Now the Schwarz function is

$$\overline{w} = \overline{f}(1/z) = \sum_{n=1}^{N} \overline{A_n} \overline{\mathcal{K}}(1/z, \overline{b_n}) + \overline{A_0}. \tag{11.52}$$

But from (6.86)

$$\overline{\mathcal{K}}(1/z, \overline{b_n}) = 1 - \mathcal{K}(z, 1/\overline{b_n}); \tag{11.53}$$

hence

$$\overline{w} = \overline{f}(1/z) = -\sum_{n=1}^{N} \overline{A_n} \mathcal{K}(z, 1/\overline{b_n}) + \sum_{n=1}^{N} \overline{A_n} + \overline{A_0}, \tag{11.54}$$

which has simple poles at $z = 1/\overline{b_n}, n = 1, \ldots, N$. We know that w_n are the locations of the poles of the Schwarz function, so we must have

$$f(1/\overline{b_n}) = w_n, \qquad n = 1, \ldots, N. \tag{11.55}$$

This provides N complex constraints, or $2N$ real conditions, on the parameters in the mapping function.

Now since

$$\mathcal{K}(z, 1/b) \sim -\frac{(1/b)}{(z - 1/b)} + \text{a locally analytic function} \tag{11.56}$$

and since

$$w - w_n = f'(1/\overline{b_n})(z - 1/\overline{b_n}) + \cdots, \tag{11.57}$$

the residues of the Schwarz function at $w = w_n$ are, for each n,

$$\overline{A_n}(1/\overline{b_n}) f'(1/\overline{b_n}) = r_n^2, \tag{11.58}$$

implying, on inspection of (11.54), that

$$A_n = \frac{\overline{r_n^2}}{1/b_n \overline{f'}(1/b_n)}, \qquad n = 1, \ldots, N. \tag{11.59}$$

This appears to give another N complex constraints on the parameters in the mapping function; however it is natural to fix the outstanding rotational degree of freedom in the mapping theorem by insisting that A_1, say, is purely real and ignoring the imaginary part of equation (11.59) for $n = 1$. Nothing is lost in making this assumption since the N equations (11.59) are not in fact independent: this can be seen by setting $h(w) = 1$ in the quadrature identity (11.49), leaving a purely real quantity on the left-hand side, from which it is seen that the quantity

$$\sum_{n=1}^{N} r_n^2 \tag{11.60}$$

is necessarily real. This means that the set (11.59) really only constitutes $2N - 1$ real conditions.

Having used up the three available degrees of freedom in the mapping theorem the available parameters in the mapping (11.50) are

$$\{A_n | n = 1, \ldots, N\}, \qquad \{b_n | n = 1, \ldots, N\}, \qquad A_0, \qquad \delta_2, \qquad \{q_j | j = 1, 2\}, \tag{11.61}$$

which, since A_1 has been taken to be real, constitutes a total of $(2N-1)+2N+2+2+2 = 4N+5$ real parameters.

The conditions (11.51), (11.55), and (11.59) to be imposed on the mapping constitute a total of $4 + 2N + (2N - 1) = 4N + 3$ constraints. This leaves two conditions still to be imposed, and it is natural to think of these freedoms to be associated with specifying the areas of the two holes.

This same feature carries over to quadrature domains of higher connectivity: if the connectivity of the domain is $M + 1$, then, in addition to the data appearing in the quadrature identity, a unique quadrature domain can only be constructed provided that this data is supplemented by information on the area of the M holes.

11.10 ▪ Bergman kernel function

Interestingly, the functional form of the conformal mappings derived in the previous section can be derived in a different way. This alternative method makes use of the *Bergman kernel function* [66, 90] and will be explained in detail in §11.12. To set the stage, this section will give some background on the Bergman kernel function, which is defined to be

$$\mathcal{B}(z, \overline{a}) \equiv -4\frac{\partial^2 G(z, a)}{\partial z \partial \overline{a}}, \tag{11.62}$$

where $G(z, a)$ is the first-type Green's function familiar from earlier chapters. Since $G(z, a)$ is a harmonic function in both z and a, $\mathcal{B}(z, \overline{a})$ is analytic in z and antianalytic in a, as the notation reflects.

The Bergman kernel function has the following symmetry: on swapping a and z,

$$\mathcal{B}(a, \overline{z}) = -4\frac{\partial^2 G(a, z)}{\partial a \partial \overline{z}}, \tag{11.63}$$

but since $G(a, z) = G(z, a)$,

$$\mathcal{B}(a, \overline{z}) = -4\frac{\partial^2 G(z, a)}{\partial a \partial \overline{z}} = \overline{\mathcal{B}(z, \overline{a})}. \tag{11.64}$$

We can also define the modified Bergman kernel function

$$\mathcal{B}_0(z, \overline{a}) \equiv -4\frac{\partial^2 G_0(z, a)}{\partial z \partial \overline{a}}, \tag{11.65}$$

where $G_0(z, a)$ is the modified Green's function considered in Chapter 2. Since $G_0(a, z) = G_0(z, a)$,

$$\mathcal{B}_0(a, \overline{z}) = -4\frac{\partial^2 G_0(z, a)}{\partial a \partial \overline{z}} = \overline{\mathcal{B}_0(z, \overline{a})}. \tag{11.66}$$

From (2.47) we can write down explicit expressions for the Bergman kernel function in multiply connected circular domains:

$$\mathcal{B}(z, \overline{a}) = \mathcal{B}_0(z, \overline{a}) + \sum_{j,n=1}^{M} \mathcal{P}_{jn} v_n'(z) \overline{v_j'(a)}, \tag{11.67}$$

where, from (4.108),

$$\mathcal{B}_0(z,\overline{a}) = -4\frac{\partial^2 G_0(z,a)}{\partial z \partial \overline{a}} = -\frac{1}{\pi}\frac{\partial^2}{\partial z \partial \overline{a}} \log \omega(z, 1/\overline{a}). \tag{11.68}$$

The simplicity of this formula relating the modified Bergman kernel function to the prime function should be noted.[37]

Both of these kernel functions have interesting transformation properties under conformal mapping of the two independent variables. It follows from (11.62) that the Bergman kernel function $\mathcal{B}^{(w)}(w,\overline{\alpha})$ in some domain D_w obtained from a multiply connected circular domain D_z by a conformal mapping

$$w = f(z), \qquad \alpha = f(a), \tag{11.69}$$

is given by

$$\mathcal{B}^{(w)}(w,\overline{\alpha}) = \frac{1}{f'(z)}\frac{1}{\overline{f'(a)}}\mathcal{B}(z,a). \tag{11.70}$$

Similarly, it follows from (11.65) that the modified Bergman kernel function satisfies

$$\mathcal{B}_0^{(w)}(w,\overline{\alpha}) = \frac{1}{f'(z)}\frac{1}{\overline{f'(a)}}\mathcal{B}_0(z,a). \tag{11.71}$$

11.11 ▪ Reproducing formulas

Both the Bergman and modified Bergman kernel functions have associated *reproducing formulas*. Suppose $f(z)$ is analytic in D_z and continuous up to the boundary. Consider the integral

$$\int\int_{D_\epsilon} \mathcal{B}(z,\overline{a}) f(a) dA_a = \int\int_{D_\epsilon} -4\frac{\partial^2 G(z,a)}{\partial z \partial \overline{a}} f(a) dA_a, \tag{11.72}$$

where D_ϵ is the same domain as D_z but with a small circular disc centered at $z = a$ with radius $\epsilon \ll 1$ removed. The integral (11.72) can be written as

$$\int\int_{D_\epsilon} \frac{\partial}{\partial \overline{a}}\left(-4\frac{\partial G(z,a)}{\partial z} f(a)\right) dA_a = \frac{1}{2\mathrm{i}}\int_{\partial D_\epsilon} -4\frac{\partial G(z,a)}{\partial z} f(a) da, \tag{11.73}$$

where we have used the complex form of the Stokes theorem, which is valid because the integrand is regular in D_ϵ. But since $G(a,z) = 0$ when a is on ∂D_z for any z, we have $\partial G(a,z)/\partial z = 0$ when a is on ∂D_z, so

$$\frac{\partial G(a,z)}{\partial z} = \frac{\partial G(z,a)}{\partial z} = 0, \qquad a \in \partial D_z, \tag{11.74}$$

and the only contribution to the integral comes from the boundary of the small circular disc of radius ϵ. Now since

$$\frac{\partial G(z,a)}{\partial z} = -\frac{1}{4\pi(z-a)} + \text{a locally analytic function}, \tag{11.75}$$

[37]This section is based on the work in [43], which appears to be the first to combine Bergman kernel functions with the theory of the prime function of circular domains to construct quadrature domains. Extensions to other classes of generalized quadrature domains have been made in [96].

the boundary integral contribution around the small circular disc of radius ϵ tends, as $\epsilon \to 0$, to

$$\frac{1}{2\pi \mathrm{i}} \int_{|a-z|=\epsilon} \frac{f(a)}{a-z} da = f(z), \tag{11.76}$$

where we have incorporated the fact that $|a-z| = \epsilon$ is traversed in the anticlockwise direction. Hence we have shown that

$$f(z) = \int\int_{D_z} \mathcal{B}(z, \overline{a}) f(a) dA_a. \tag{11.77}$$

The same arguments lead to the reproducing formula for $f(z)$ provided it is now analytic *and* single-valued in D_z:

$$f(z) = \int\int_{D_z} \mathcal{B}_0(z, \overline{a}) f(a) dA_a. \tag{11.78}$$

11.12 ▪ Quadrature domains and Bergman kernel functions

We can now show another way to arrive at a representation of the conformal mapping as a linear combination of parametric derivatives of $\log \omega(z, a)$. Using the reproducing formula

$$f(w) = \int\int_{D_w} \mathcal{B}_0^{(w)}(w, \overline{\alpha}) f(\alpha) dA_\alpha \tag{11.79}$$

for any function $f(w)$ that is analytic and single-valued in D_w and making the choice $f(w) = 1$ leads to

$$1 = \int\int_{D_w} \mathcal{B}_0^{(w)}(w, \overline{\alpha}) dA_\alpha. \tag{11.80}$$

We can take a complex conjugate of this equation to find

$$1 = \int\int_{D_w} \overline{\mathcal{B}_0^{(w)}(w, \overline{\alpha})} dA_\alpha = \int\int_{D_w} \mathcal{B}_0^{(w)}(\alpha, \overline{w}) dA_\alpha, \tag{11.81}$$

where we have used the symmetry (11.66) of the modified Bergman kernel. The formula we have produced has an integrand that is now analytic in the variable α. At this point we suppose that D_w is a quadrature domain with the quadrature identity (11.49); then (11.81) becomes

$$1 = \sum_{n=1}^{N} \pi r_n^2 \mathcal{B}_0^{(w)}(\alpha_n, \overline{w}). \tag{11.82}$$

We can now use the transformation formula for the modified Bergman kernel under conformal mapping to deduce that

$$1 = \sum_{n=1}^{N} \pi r_n^2 \mathcal{B}_0(a_n, \overline{z}) \frac{1}{f'(a_n)} \frac{1}{\overline{f'(z)}} \tag{11.83}$$

or

$$\overline{f'(z)} = \sum_{n=1}^{N} \pi r_n^2 \mathcal{B}_0(a_n, \overline{z}) \frac{1}{f'(a_n)}. \tag{11.84}$$

The complex conjugate of this equation is

$$f'(z) = \sum_{n=1}^{N} \overline{\pi r_n^2 \mathcal{B}_0(a_n, \overline{z}) \frac{1}{f'(a_n)}} = \sum_{n=1}^{N} \pi \overline{r_n^2} \mathcal{B}_0(z, \overline{a_n}) \frac{1}{\overline{f'(a_n)}}. \tag{11.85}$$

However we know from (11.68) that

$$\mathcal{B}_0(z, \overline{a_n}) = -\frac{1}{\pi} \frac{\partial^2}{\partial z \partial \overline{a_n}} \log \omega(z, 1/\overline{a_n}) \tag{11.86}$$

so that on substitution into (11.85), and after integration with respect to z, we find

$$f(z) = \sum_{n=1}^{N} \frac{\overline{r_n^2}}{\overline{a_n^2 f'(a_n)}} \frac{\partial}{\partial(1/\overline{a_n})} \log \omega(z, 1/\overline{a_n}) + A_0, \tag{11.87}$$

where A_0 is a constant of integration. This can be rewritten as

$$f(z) = \sum_{n=1}^{N} \frac{\overline{r_n^2}}{\overline{a_n f'(a_n)}} \mathcal{K}(z, 1/\overline{a_n}) + A_0, \tag{11.88}$$

which is identical to expression (11.50) with coefficients given as in (11.59) once we make the identification $a_n = 1/\overline{b_n}$.

Formula (11.88) looks like an explicit expression, but this is not so because the constants premultiplying the parametric derivatives of $\log \omega(z, a)$ themselves depend on $f(z)$. Nevertheless the functional form is correct, and the coefficients can be found by ensuring that the coefficients produce the correct quadrature identity. We also know from other considerations that $f(z)$ must be invariant under the Schottky group Θ which places further constraints on the parameters appearing in the mapping. As discussed earlier, an $(M+1)$-connected quadrature domain associated to a given quadrature identity is only unique up to specification of the area of the M holes. Further discussion of this idea of using the prime function to construct multiply connected quadrature domains from their quadrature data via the Bergman kernel function can be found in [43].

11.13 ▪ Approximating a domain by a quadrature domain

The concentric annulus $\rho < |z| < 1$ has two circular boundaries: $|z| = 1$ and $|z| = \rho$. The Schwarz functions are given, respectively, by

$$S_0(z) = \frac{1}{z}, \qquad S_1(z) = \frac{\rho^2}{z}. \tag{11.89}$$

These are distinct functions; they are certainly not meromorphic continuations of each other through the annulus.

Now consider the conformal mapping from the concentric annulus $\rho < |z| < 1$ given by[38]

$$w = f(z) = z \prod_{k=1}^{N} \left[\frac{\omega(z, \sigma_N^{k-1} a/\rho^{2/N})}{\omega(z, \sigma_N^{k-1} a)} \right], \qquad 1 < a < 1/\rho, \qquad a \in \mathbb{R}, \tag{11.90}$$

[38] This class of functions has been introduced by the author and used in various contexts [58, 31, 28].

where $N \geq 3$ and $\sigma_N = e^{2\pi i/N}$ is an Nth root of unity and where $\omega(.,.)$ is the prime function of the annulus $\rho < |z| < 1$. It is easy to verify using the identity (5.6) that $f(z)$ is a loxodromic function satisfying

$$f(\rho^2 z) = f(z). \tag{11.91}$$

The Schwarz function of the image of C_0 under the map (11.90) is

$$\overline{w} = S_0(w) = \overline{f(z)} = \overline{f(1/\overline{z})}. \tag{11.92}$$

On the other hand, the Schwarz function of the image of C_1 under the same map is

$$\overline{w} = S_1(w) = \overline{f(z)} = \overline{f(\rho^2/\overline{z})} = \overline{f(1/\overline{z})} = S_0(w), \tag{11.93}$$

where we have used the functional relation (11.91) in the penultimate equality. The domain given by the image of the annulus $\rho < |z| < 1$ under the map (11.90), provided it is a univalent function, is therefore a quadrature domain since the Schwarz functions of its two boundary curves are meromorphic continuations of each other through the domain.

Since it depends on the integer $N \geq 3$, it is of interest to study the image of the annulus under the map (11.90) as $N \to \infty$. Since

$$\rho^{-2/N} = e^{-(2/N)\log\rho} \approx 1 - \frac{2}{N}\log\rho + \mathcal{O}\left(\frac{1}{N^2}\right), \tag{11.94}$$

then

$$\omega(z, \sigma_N^{k-1}a/\rho^{2/N}) \approx \omega(z, \sigma_N^{k-1}a) - \frac{2\sigma_N^{k-1}a}{N}\log\rho\frac{\partial\omega(z,a)}{\partial a} + \mathcal{O}\left(\frac{1}{N^2}\right) \tag{11.95}$$

so that, on substitution into the numerator of (11.90),

$$w = f(z) = z\prod_{k=1}^{N}\left[\frac{\omega(z, \sigma_N^{k-1}a/\rho^{2/N})}{\omega(z, \sigma_N^{k-1}a)}\right] \approx z + \mathcal{O}\left(\frac{1}{N}\right). \tag{11.96}$$

We see that as $N \to \infty$, (11.90) tends to the identity map. This means that the image of the annulus $\rho < |z| < 1$ will become closer to the same annulus in the w plane as $N \to \infty$.

This doubly connected case study demonstrates that the annulus $\rho < |z| < 1$, which is not a quadrature domain, can be *approximated* by a quadrature domain. This explicit example[39] illustrates the general fact that any given multiply connected domain can be approximated, in a sense that can be made precise, by a quadrature domain [86]. Exercise 11.10 gives another example.

11.14 ▪ The modified Schwarz potential

In the class of quadrature domains in which a meromorphic function $S(z)$ exists that coincides with the Schwarz function of each boundary component, it is possible to define the function, up to an arbitrary constant, by

$$\psi(z,\overline{z}) = z\overline{z} - \int^{z} S(z')dz' - \int^{\overline{z}} \overline{S}(z')dz'. \tag{11.97}$$

[39] This example was first set out, in the context of a physical problem, in [31].

This function is called the modified Schwarz potential in D. If $S(z)$ has only simple poles at the set of points $\{z_n | n = 1, \dots, N\}$ in D, then $\psi(z, \overline{z})$ has logarithmic singularities at these points. An important feature of $\psi(z, \overline{z})$ is the boundary property that

$$\psi = \frac{\partial \psi}{\partial n} = 0, \qquad z \in \partial D, \tag{11.98}$$

where the arbitrary constant has been chosen to ensure that the first condition holds. Thus such potentials satisfy trivial Dirichlet-type *and* Neumann-type boundary conditions on ∂D.

One fruitful application of the modified Schwarz potential is to the construction of vortical equilibria of the Euler equations of fluid dynamics. This topic is examined in Chapter 24, where it is shown to be a fertile source of equilibrium streamfunctions for vortical flows [34, 30, 33, 29, 52, 51].

Exercises

11.1. Show that for the mapping (11.35) to satisfy a quadrature identity of the form (11.34), b must be a real root of the equation

$$b^4 - \frac{b^2}{a^2}\left(\frac{1+3a^8}{1+a^4}\right) - \left(\frac{1-3a^4}{1+a^4}\right) = 0 \tag{11.99}$$

satisfying $|a| > 1$.
Hint: Set the residues of $\overline{f}(1/z)f'(z)$ at $z = 0$ and $z = 1/a$ equal to each other.

11.2. **(Merging four discs in a linear array)** Find the functional form of the rational function conformal mapping from the unit z disc to the class of quadrature domains obtained by the merging of four touching equally sized discs aligned along the real axis in a complex w plane and having a quadrature identity of the form (11.43).

11.3. **(Schwarz function with higher order poles)** The conformal mapping from a unit disc $|z| < 1$ to a simply connected quadrature domain in the complex w plane is known to be given by

$$w = f(z) = \frac{z}{(z-a)^2}, \qquad a \in \mathbb{R}, \ a > 1. \tag{11.100}$$

What is the form of the corresponding quadrature identity?

11.4. **(Merging more equally sized discs)** In §11.7 it was shown how two, three, and then four discs of equal size can "merge" together to form quadrature domains of varying connectivity. Extend these considerations by considering the various configurations formed by the merging of *five* equally sized circular discs. Examine how to construct a representation of the conformal mapping from a canonical circular domain to these various configurations. What are the possibilities for *six* equally sized circular discs?

11.5. **(Another representation of (11.90))** Use the theory of Chapter 9 to show that the conformal map (11.90) can alternatively be written as

$$A \sum_{k=0}^{N-1} \sigma_N^k \mathcal{K}(z, a\sigma_N^k), \qquad \sigma_N = e^{2\pi i/N}, \tag{11.101}$$

where

$$A = -\frac{1}{\tilde{\omega}(a,a)} \frac{\prod_{k=1}^{N} \omega(a, \sigma_N^{k-1} a/\rho^{2/N})}{\prod_{k=2}^{N} \omega(a, \sigma_N^{k-1} a)}, \tag{11.102}$$

where

$$\tilde{\omega}(z,a) = \frac{\omega(z,a)}{z-a}. \tag{11.103}$$

11.6. **(Quadrature domain boundaries as algebraic curves)** Consider the algebraic curve given by

$$w^2\overline{w}^2 - w^2 - \overline{w}^2 - 2r^2 w\overline{w} = 0, \tag{11.104}$$

where $r \geq 1$ is a real parameter.

(a) Check that points on the boundaries of two circular discs in a complex w plane centered at ± 1 and of radius 1 such that they touch at the origin are solutions of (11.104) when $r = 1$.

(b) Now assume that $r > 1$. Verify that if w and $\overline{w}$ are given by the following functions of a third variable z,

$$w = \frac{Rz}{z^2 - a^2}, \qquad \overline{w} = \frac{Rz}{1 - z^2 a^2}, \tag{11.105}$$

then these relations are consistent provided $|z| = 1$.

(c) Show that with w and $\overline{w}$ given by (11.105) then (11.104) is satisfied identically provided that the parameters are chosen so that

$$R = \pm\frac{(1 - a^4)}{a}, \qquad r^2 = \frac{1 + a^4}{2a^2}. \tag{11.106}$$

(d) Check that relations (11.106) are identical to (11.30) with $d = 1$.

(e) Hence argue that the boundary of the quadrature domain satisfying the quadrature identity

$$\int\int_D h(w) dA_w = \pi r^2 h(1) + \pi r^2 h(-1) \tag{11.107}$$

is given by points on the algebraic curve (11.104).

(f) The point $w = 0$ is also a solution of (11.104). Confirm that this is an isolated solution.

Remark: The author has used this connection with algebraic curves as the basis for a constructive method for multipolar vortex equilibria of the Euler equation [33]. More details on the connection between quadrature domains and algebraic curves are given in [86].

11.7. **(Unbounded doubly connected quadrature domain)** Consider the unbounded region D_w described by the image under a conformal map from the annulus $\rho < |z| < 1$ having the form

$$w = f(z) = R\left[\frac{\omega(z, -\sqrt{\rho})\omega(z, -1/\sqrt{\rho})\omega(z, 1/\sqrt{\rho})}{\omega(z, \sqrt{\rho})\omega(z, e^{i\theta}/\sqrt{\rho})\omega(z, e^{-i\theta}/\sqrt{\rho})}\right], \tag{11.108}$$

where R, θ, and ρ are real constants and where $\omega(.,.)$ is the prime function of the annulus. This function has three simple zeros at $-\sqrt{\rho}, \pm\sqrt{\rho}^{-1}$ and three simple poles at $\sqrt{\rho}, \sqrt{\rho}^{-1} e^{\pm i\theta}$. The point $z = \sqrt{\rho}$ inside the annulus is the preimage of the point at infinity.

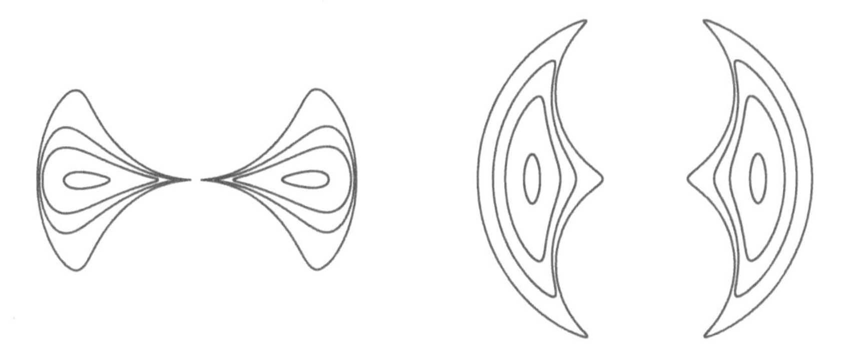

Figure 11.6. *Superposition of typical boundary shapes described by the mapping (11.108) for $\theta = \pi/3$ (left) and $\theta = 2\pi/3$ (right) and $\rho = 0.005, 0.05, 0.1,$ and 0.2 with centroids fixed. The unbounded region exterior to each boundary pair has an interpretation as an unbounded doubly connected quadrature domain.*

(a) Use the properties of the prime function summarized in §5.3 to verify that

$$f(\rho/z) = -f(z). \tag{11.109}$$

Hence argue that the images of C_0 and C_1 under this mapping are rotations of each other by angle π.

(b) Check that (11.108) satisfies

$$\overline{f}(z) = f(z). \tag{11.110}$$

Hence argue that the image domain is reflectionally symmetric about the real w axis. Figure 11.6 shows a superposition, for different choices of ρ, of the images of C_0 and C_1 under such a mapping for the two choices $\theta = \frac{\pi}{3}, \frac{2\pi}{3}$.

(c) Verify that (23.15) satisfies the functional relation

$$f(\rho^2 z) = f(z) \tag{11.111}$$

for all choices of the parameters R, θ, and ρ. Hence argue that $f(z)$ is a loxodromic function.

(d) By considering the integration of an arbitrary analytic function $h(z)$ over the region D_w, within the class of functions decaying sufficiently fast at infinity that such an integral exists, show that D_w can be viewed as an unbounded doubly connected quadrature domain.

11.8. **(Double quadrature domains)** Consider the image D_w in a complex w plane of the unit disc D_z under the rational function conformal map

$$w = f(z) = z - \frac{8a^2 z}{z^2 - a^2}, \tag{11.112}$$

where $a > 1$ is a real parameter.

(a) Confirm that D_w is a quadrature domain satisfying a quadrature identity of the form

$$\int\int_{D_w} h(w) dA_w = \pi r_0^2 h(0) + \pi r_a^2 h(w_a) + \pi r_a^2 h(-w_a), \tag{11.113}$$

where $h(z)$ is some analytic function in D_w and where

$$w_a = \frac{(9a^4 - 1)}{a(a^4 - 1)}. \tag{11.114}$$

Find the quadrature data r_0^2 and r_a^2.

(b) Show that

$$f'(z) = \left[\frac{z^2 + 3a^2}{z^2 - a^2}\right]^2 \tag{11.115}$$

and hence that, on $|z| = 1$,

$$\left|\frac{df}{dz}\right| = \frac{(3a^2 + z^2)(1 + 3a^2z^2)}{(a^2 - z^2)(1 - a^2z^2)}. \tag{11.116}$$

(c) Show that if $h(w)$ is some analytic function in D_w, then

$$\int_{\partial D_w} h(w)|dw| = 6\pi h(0) - \frac{4\pi}{a}\left(\frac{3a^4+1}{a^4-1}\right)h(w_a) + \frac{4\pi}{a}\left(\frac{3a^4+1}{a^4-1}\right)h(-w_a), \tag{11.117}$$

where ∂D_w is the boundary of D_w.

Remark: The domain D_w is special in that it admits a quadrature identity (11.113) of the kind familiar from this chapter as well as a different kind of quadrature identity (11.117) involving a boundary integral; a domain satisfying a quadrature identity of the latter form is called a *quadrature domain for arclength*. This example shows that certain special domains admit quadrature identities of both kinds. Such domains have been called *double quadrature domains* [54].

11.9. **(A periodic domain)** In Exercise 8.13 the function

$$f(z) = -\frac{\mathrm{i}}{2\pi}\log\left(\frac{z}{\sqrt{\rho}}\right) - \frac{\mathrm{i}\log\rho}{\pi}K\left(z, \frac{1}{\sqrt{\rho}}\right) \tag{11.118}$$

defined in an annulus $\rho < |z| < 1$ was introduced, with $K(z, a)$ defined in (5.77).

(a) Verify that

$$f(\rho^2 z) = f(z). \tag{11.119}$$

(b) Use the result of Exercise 5.4 to show that

$$f(\sqrt{\rho}) = 0. \tag{11.120}$$

(c) Show that for any $z \in C_0$

$$f(\rho z) = \overline{f(z)}. \tag{11.121}$$

Hint: You may find the result of Exercise 5.3 useful here.

(d) It is of interest to examine this function as a conformal mapping

$$w = f(z) \tag{11.122}$$

and to consider the image D_w of the annulus $\rho < |z| < 1$ under its action. Since $f(z)$ is not single-valued in this annulus, it is natural to introduce a logarithmic branch cut, along the negative real z axis, say. Use part (c) to argue that the region D_w enclosed by the images of C_0, C_1, and the two sides of the branch cut is as shown in Figure 11.7 where $z = \sqrt{\rho}$ maps to the origin with the images of C_0 and C_1 being symmetric with respect to reflection in the real w axis.

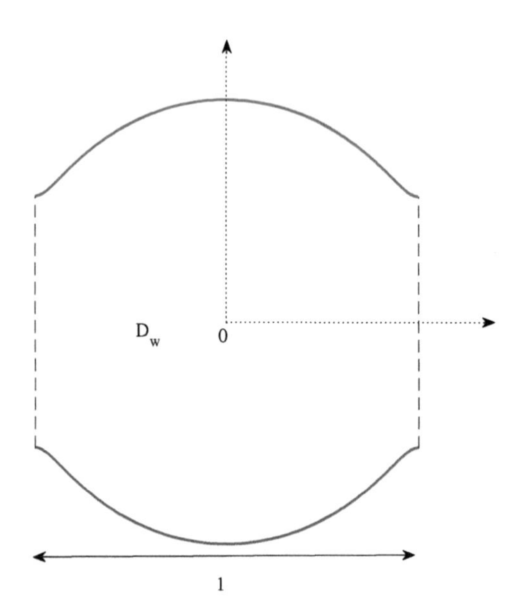

Figure 11.7. *The image domain D_w, in a complex w plane, of the cut annulus $\rho < |z| < 1$ under the image of (11.118). The point $z = \sqrt{\rho}$ maps to the origin. The images of C_0 and C_1 are reflections of each other in the real w axis.*

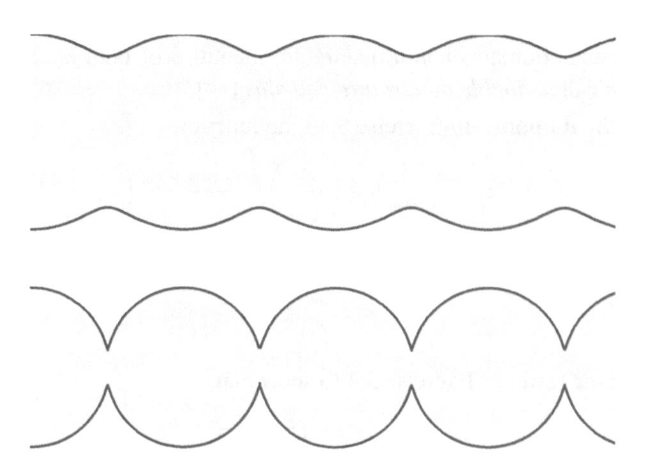

Figure 11.8. *The image domain, in a complex w plane, of the cut annulus $\rho < |z| < 1$ under the image of (11.118) for various values of ρ together with some of its periodic translates in the x direction.*

(e) Argue why this image D_w can be viewed as a single period window of a periodic domain akin to those shown in Figure 11.8 for two different values of the parameter ρ.

(f) We know that conformal mappings from a concentric annulus satisfying the functional identity in part (a) are usually associated with doubly connected quadrature domains. Explore if there is any sense in which D_w shown in Figure 11.7, or its periodic extension shown in Figure 11.8, can be interpreted as a quadrature domain.

11.10. **(Quadrature domain approximant)** In Exercise 5.7, the conformal mapping was constructed from a concentric annulus $\rho < |z| < 1$ to the unbounded region D in a complex w plane exterior to two equal-sized circular discs of radius s centered at $\pm id$.

(a) Verify that D is not a quadrature domain.

(b) Combine the result of Exercise 5.7 with formula (11.90) to find a conformal mapping from the same concentric annulus to a quadrature domain approximant of D.

Chapter 12

Cauchy transforms

12.1 ▪ Why study the Cauchy transform of a domain?

A mathematical concept that is intimately related to the theory of quadrature domains is the Cauchy transform of a domain. It has mathematical connections to the theory of the Schwarz function, it is the generating function of the geometrical moments of a given domain so it is a means of encoding the domain geometry, and it can be useful in understanding a class of unsteady free boundary problems to be considered in Chapters 26 and 27.

The theory of the Cauchy transform is rich and has many links to other areas of mathematics. We only touch on the theory here with the aim of reaching the central results—stated in §12.6 and §12.7—which we will need later in considering free boundary problems. A discussion of wider aspects of the theory of Cauchy transforms can be found in [66, 88].

12.2 ▪ Cauchy transform of a simply connected domain

Consider first a simply connected domain D in a complex z plane and let D_0 be the unbounded domain exterior to it, as shown in Figure 12.1. The *Cauchy transform* of D is defined to be the two-dimensional integral[40]

$$\mathcal{C}(z) = \frac{1}{\pi} \int\int_D \frac{dx'dy'}{z'-z}. \tag{12.1}$$

This integral, which depends only on the geometry of D, defines an analytic function of z outside D; this is clear because the integral can be thought of as a continuous superposition of analytic functions of z. If $z \in D_0$, the integrand is regular in D and we can use the complex form of the Stokes theorem, which says that

$$\int\int_D \frac{\partial \Phi}{\partial \overline{z}} dxdy = \frac{1}{2\mathrm{i}} \int_{\partial D} \Phi dz \tag{12.2}$$

[40] Some authors add a minus sign to this definition.

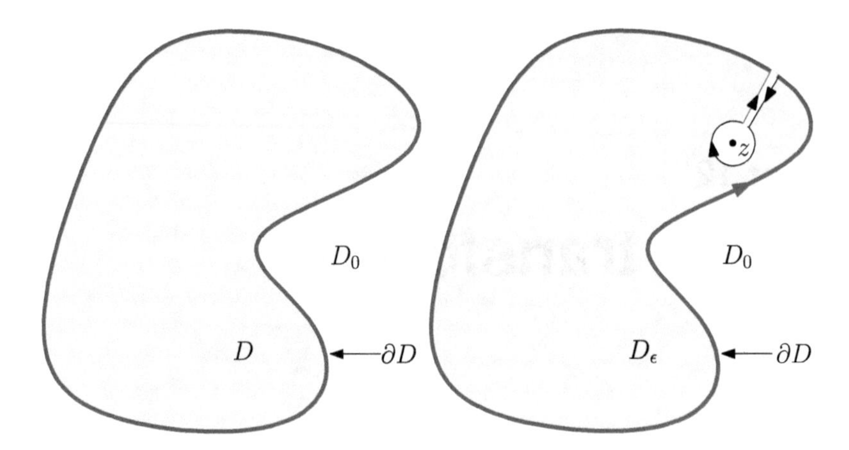

Figure 12.1. *A simply connected domain D and the domain D_0 exterior to it. To evaluate the Cauchy transform for $z \in D$, it is necessary to consider a modified domain D_ϵ where a small circle of radius ϵ centered at z is removed, with cross-cuts to the boundary ∂D introduced to form a closed contour.*

for any function Φ that is regular in D. The boundary of D is denoted by ∂D. Thus, by the Stokes theorem, for any z in D_0 the Cauchy transform (12.1) defines an analytic function:

$$\mathcal{C}(z) = C_0(z) \equiv \frac{1}{2\pi \mathrm{i}} \int_{\partial D} \frac{\overline{z'} dz'}{z' - z}, \qquad z \in D_0. \tag{12.3}$$

We can now think about the analytic continuation of this function into D. It is important to emphasize that the analytic continuation of $C_0(z)$ inside D is not equal to the function defined by the integral (12.1) for points z inside D. Indeed, for points $z \in D$ the function defined by the integral (12.1) is not even an analytic function.

To see this, for $z \in D$, consider another domain D_ϵ in which D is modified by the addition of the "keyhole contour" depicted in Figure 12.1; a small circle of radius ϵ surrounds the point $z \in D$ with z being its center. This small circle is joined to ∂D by two cross-cuts (that sit on top of each other even though, for clarity, they are shown slightly displaced in Figure 12.1). Calculation of the integral around this new closed contour, where the two cross-cuts are traversed in opposite directions, gives, in the limit $\epsilon \to 0$,

$$\mathcal{C}(z) = \frac{1}{2\pi \mathrm{i}} \int_{|z'|=1} \frac{\overline{z'} dz'}{z' - z} - \overline{z}. \tag{12.4}$$

This is because the contributions from the cross-cuts cancel each other and the nonanalytic term $-\overline{z}$ is collected from the integration around the small circle centered at z. Introducing the new analytic function

$$C_I(z) \equiv \frac{1}{2\pi \mathrm{i}} \int_{\partial D} \frac{\overline{z'} dz'}{z' - z}, \qquad z \in D, \tag{12.5}$$

defined for points $z \in D$, we deduce that the Cauchy transform is given in D and D_0 by

$$\mathcal{C}(z) = \begin{cases} C_0(z), & z \in D_0, \\ C_I(z) - \overline{z}, & z \in D, \end{cases} \tag{12.6}$$

with the condition that $\mathcal{C}(z)$ is continuous everywhere in the plane, including at the boundary ∂D, implying the relation

$$C_0(z) = C_I(z) - \overline{z}, \qquad z \in \partial D, \tag{12.7}$$

or

$$\overline{z} = C_I(z) - C_0(z), \qquad z \in \partial D. \tag{12.8}$$

We can verify that these results are consistent with the Plemelj formulas [61].

If we consider the limit of $C_0(z)$ as $z \to \partial D^+$—that is, as we approach ∂D from outside D—we have

$$C_0(z) \to \mathcal{P}\left\{\frac{1}{2\pi \mathrm{i}} \int_{\partial D} \frac{\overline{z'} dz'}{z' - z}\right\} - \frac{\overline{z}}{2}, \tag{12.9}$$

where $\mathcal{P}$ denotes a principal part integral, while, as $z \to \partial D^-$, meaning that we approach ∂D from inside D, we find

$$C_I(z) \to \mathcal{P}\left\{\frac{1}{2\pi \mathrm{i}} \int_{\partial D} \frac{\overline{z'} dz'}{z' - z}\right\} + \frac{\overline{z}}{2}. \tag{12.10}$$

On subtraction of (12.9) from (12.10), we arrive at

$$C_I(z) - C_0(z) = \overline{z}, \qquad z \in \partial D, \tag{12.11}$$

which coincides with (12.8).

By way of example, let us compute the Cauchy transform for the unit disc.

For $z \in D_0$, we find

$$C_0(z) = \frac{1}{2\pi \mathrm{i}} \int_{|z'|=1} \frac{\overline{z'} dz'}{z' - z} = \frac{1}{2\pi \mathrm{i}} \int_{|z'|=1} \frac{dz'}{z'(z' - z)}, \tag{12.12}$$

where, to obtain the second equality, we have used the fact that $\overline{z'} = 1/z'$ for points with $|z'| = 1$. The integrand is now meromorphic inside the curve so the residue theorem applies, and, on picking up the residue at $z' = 0$, we find

$$C_0(z) = -\frac{1}{z}, \qquad |z| > 1. \tag{12.13}$$

This function is analytic everywhere outside the disc and decays like $1/z$ as $z \to \infty$, as expected. The analytic continuation of this function into D is meromorphic inside D_z with a single simple pole at $z = 0$.

For $z \in D$, we find

$$C_I(z) = \frac{1}{2\pi \mathrm{i}} \int_{|z|=1} \frac{\overline{z'} dz'}{z' - z} = \frac{1}{2\pi \mathrm{i}} \int_{|z|=1} \frac{dz'}{z'(z' - z)} = 0, \tag{12.14}$$

where, now, the two residue contributions inside the disc at $z' = 0$ and $z' = z$ add up to zero. We deduce that the Cauchy transform for the unit disc is given by

$$\mathcal{C}(z) = \begin{cases} -\dfrac{1}{z}, & z \in D_0, \\ -\overline{z}, & z \in D. \end{cases} \tag{12.15}$$

This expression confirms that the Cauchy transform is continuous on the boundary of the disc $|z| = 1$.

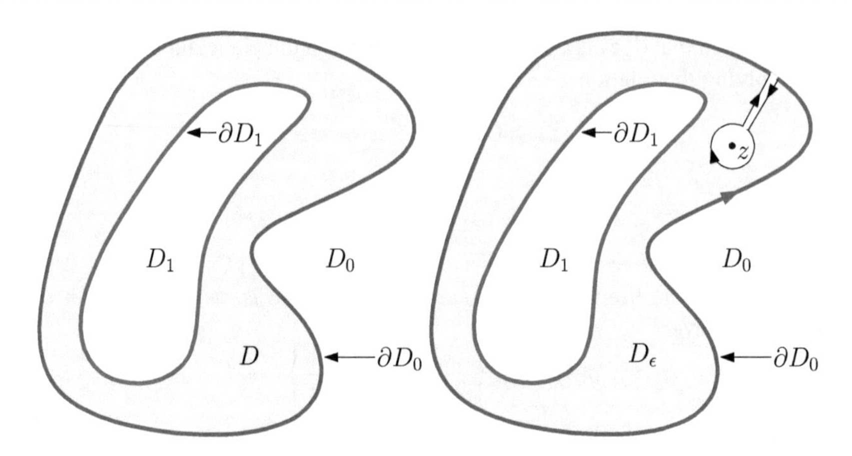

Figure 12.2. *A doubly connected domain D, the domain D_0 exterior to it, and the enclosed hole D_1 (left). To evaluate the Cauchy transform for $z \in D$, it is necessary to consider a modified domain D_ϵ where a small circle of radius ϵ centered at z is removed with a cross-cut to ∂D_0 introduced to form a closed contour.*

12.3 ▪ Cauchy transform of a doubly connected domain

The notion of a Cauchy transform can be extended to a doubly connected domain.

First, let us consider a bounded doubly connected annular domain D such as that shown on the left in Figure 12.2. Let D_0 be the exterior of D, and let D_1 be the hole in D. It is convenient to use ∂D_0 and ∂D_1 to denote the outer and inner boundaries of the annular domain, respectively. As before, the Cauchy transform of D is

$$\mathcal{C}(z) = \frac{1}{\pi} \int\int_D \frac{dx'dy'}{z' - z}. \tag{12.16}$$

For z in either D_0 or D_1 the Cauchy transform (12.16) defines an analytic function which we denote as follows:

$$\mathcal{C}(z) = \begin{cases} C_0(z), & z \in D_0, \\ C_1(z), & z \in D_1. \end{cases} \tag{12.17}$$

One can contemplate taking the analytic continuation of $C_0(z)$ into D and on into D_1; similarly, one can think of analytically continuing $C_1(z)$ through D into D_0. However this analytic continuation of $C_0(z)$ into D_1 does not generally coincide with the analytic function $C_1(z)$; nor will the analytic continuation of $C_1(z)$ into D_0 generally coincide there with $C_0(z)$. (There are, however, special domains D for which these continuations *do* coincide, as we will see in §12.5.) Both functions, however, can be expressed by the line integrals

$$C_0(z) = \frac{1}{2\pi \mathrm{i}} \int_{\partial D} \frac{\overline{z'}dz'}{z' - z}, \qquad z \in D_0, \tag{12.18}$$

$$C_1(z) = \frac{1}{2\pi \mathrm{i}} \int_{\partial D} \frac{\overline{z'}dz'}{z' - z}, \qquad z \in D_1, \tag{12.19}$$

on use of the complex form of the Stokes theorem (12.2).

When $z \in D$, just as we did in the simply connected case of Figure 12.1, we can integrate around the boundary of a modified domain D_ϵ, as sketched on the right in Figure 12.2. Again, D has been modified by a circular disc of radius ϵ centered at the singularity at z and with an associated keyhole contour. Calculation of the integral over this modified domain in the limit $\epsilon \to 0$ leads to

$$\mathcal{C}(z) = \frac{1}{2\pi\mathrm{i}} \int_{\partial D} \frac{\overline{z'} dz'}{z' - z} - \overline{z}. \tag{12.20}$$

This follows by applying the Stokes theorem (12.2) over the modified domain D_ϵ where the integrand is now regular, computing the boundary integral contribution from the radius-ϵ circle, and taking the limit $\epsilon \to 0$. It is this last boundary integral contribution that produces the $-\overline{z}$ in (12.20). Putting all this together yields

$$\mathcal{C}(z) = \begin{cases} C_0(z), & z \in D_0, \\ C_I(z) - \overline{z}, & z \in D, \\ C_1(z), & z \in D_1, \end{cases} \tag{12.21}$$

where

$$C_I(z) \equiv \frac{1}{2\pi\mathrm{i}} \int_{\partial D} \frac{\overline{z'} dz'}{z' - z} \tag{12.22}$$

is an analytic function for $z \in D$.

The Cauchy transform is a continuous function everywhere in the plane, including at the two boundaries ∂D_0 and ∂D_1. Hence,

$$C_0(z) = C_I(z) - \overline{z} \qquad \text{on } \partial D_0 \tag{12.23}$$

and

$$C_1(z) = C_I(z) - \overline{z} \qquad \text{on } \partial D_1. \tag{12.24}$$

Again, it is easy to verify that these results are consistent with the Plemelj formulas. If we consider the limit of $C_0(z)$ as $z \to \partial D_0^+$, we have

$$C_0(z) \to \mathcal{P}\left\{\frac{1}{2\pi\mathrm{i}} \int_{\partial D} \frac{\overline{z'} dz'}{z' - z}\right\} - \frac{\overline{z}}{2}, \tag{12.25}$$

where $\mathcal{P}$ denotes a principal part integral while, as $z \to \partial D_0^-$,

$$C_I(z) \to \mathcal{P}\left\{\frac{1}{2\pi\mathrm{i}} \int_{\partial D} \frac{\overline{z'} dz'}{z' - z}\right\} + \frac{\overline{z}}{2} \tag{12.26}$$

so that, on subtraction,

$$C_I(z) - C_0(z) = \overline{z}, \tag{12.27}$$

which coincides with (12.23). Similarly, as $z \to \partial D_1^-$,

$$C_1(z) \to \mathcal{P}\left\{\frac{1}{2\pi\mathrm{i}} \int_{\partial D} \frac{\overline{z'} dz'}{z' - z}\right\} - \frac{\overline{z}}{2}; \tag{12.28}$$

hence on subtraction of (12.28) from (12.26),

$$C_I(z) - C_1(z) = \overline{z}, \tag{12.29}$$

which is consistent with (12.24).

An explicit example is to let D be the concentric annulus

$$\rho < |z| < 1. \tag{12.30}$$

Direct calculation shows that if $z \in D_0$,

$$C_0(z) = \frac{1}{2\pi \mathrm{i}} \int_{\partial D} \frac{\overline{z'} dz'}{z' - z} = \frac{1}{2\pi \mathrm{i}} \int_{|z'|=1} \frac{dz'}{z'(z'-z)} - \frac{1}{2\pi \mathrm{i}} \int_{|z'|=\rho} \frac{\rho^2 dz'}{z'(z'-z)} \tag{12.31}$$

$$= -\frac{1}{z} + \frac{\rho^2}{z}, \tag{12.32}$$

where, since $z \in D_0$, we pick up the residues from both contour integrals at $z' = 0$. Similarly, for $z \in D$,

$$C_I(z) = \frac{1}{2\pi \mathrm{i}} \int_{|z'|=1} \frac{dz'}{z'(z'-z)} - \frac{1}{2\pi \mathrm{i}} \int_{|z'|=\rho} \frac{\rho^2 dz'}{z'(z'-z)} = \frac{\rho^2}{z}, \tag{12.33}$$

where, since $z \in D$, we now pick up two residues in the first contour integral at $z' = 0$ and $z' = z$ and only a single residue in the second at $z' = 0$. Finally, for $z \in D_1$,

$$C_1(z) = \frac{1}{2\pi \mathrm{i}} \int_{|z'|=1} \frac{dz'}{z'(z'-z)} - \frac{1}{2\pi \mathrm{i}} \int_{|z'|=\rho} \frac{\rho^2 dz'}{z'(z'-z)} = 0, \tag{12.34}$$

where, since $z \in D_1$, we pick up two residues in both integrals at $z' = 0$ and $z' = z$. The Cauchy transform is therefore given by

$$\mathcal{C}(z) = \begin{cases} -\dfrac{1}{z} + \dfrac{\rho^2}{z}, & z \in D_0, \\ \dfrac{\rho^2}{z} - \overline{z}, & z \in D, \\ 0, & z \in D_1. \end{cases} \tag{12.35}$$

This reduces to (12.15) when $\rho \to 0$.

Finally, it is of interest to relate these considerations to the notion of the Schwarz function of each boundary curve. Suppose

$$S_j(z) = \overline{z}, \qquad z \in \partial D_j, \ j = 0, 1, \tag{12.36}$$

denotes the Schwarz function of the jth boundary. Equations (12.23) and (12.24) imply

$$C_0(z) = C_I(z) - S_0(z) \qquad \text{on } \partial D_0 \tag{12.37}$$

and

$$C_1(z) = C_I(z) - S_1(z) \qquad \text{on } \partial D_1. \tag{12.38}$$

On use of (12.32)–(12.34) and the facts that

$$S_0(z) = \frac{1}{z}, \qquad S_1(z) = \frac{\rho^2}{z}, \tag{12.39}$$

it is easy to check that these various functions satisfy (12.37)–(12.38).

12.4 ▪ Cauchy transforms of multiply connected domains

With these simply and doubly connected examples in mind, it should be clear now how to extend the notion of a Cauchy transform to a general multiply connected domain. The

Cauchy transform $\mathcal{C}(z)$ of a bounded $(M+1)$-connected domain D is the continuous function satisfying

$$\mathcal{C}(z) = \begin{cases} C_0(z), & z \in D_0, \\ C_I(z) - \overline{z}, & z \in D, \\ C_j(z), & z \in D_j,\ j = 1, \ldots, M, \end{cases} \tag{12.40}$$

where the exterior of D is denoted by D_0 and the M holes are denoted by D_j for $j = 1, \ldots, M$.

By the continuity of $\mathcal{C}(z)$ on the boundary ∂D_0, we have

$$C_0(z) = C_I(z) - \overline{z} = C_I(z) - S_0(z), \qquad z \in \partial D_0, \tag{12.41}$$

where

$$\overline{z} = S_0(z), \qquad z \in \partial D_0, \tag{12.42}$$

is the Schwarz function $S_0(z)$ of ∂D_0. On the other hand, continuity of $\mathcal{C}(z)$ on the boundary ∂D_j for $j = 1, \ldots, M$ gives

$$C_j(z) = C_I(z) - \overline{z} = C_I(z) - S_j(z), \qquad z \in \partial D_j, \tag{12.43}$$

where

$$\overline{z} = S_j(z), \qquad z \in \partial D_j, \tag{12.44}$$

is the Schwarz function $S_j(z)$ of ∂D_j.

Since the Cauchy transform of a domain is related to the Schwarz functions of its boundary components, and since the latter functions encode information on the shape of each boundary curve, it is clear that information about the analytic functions defined by the Cauchy transform outside a bounded multiply connected domain, and in each of its holes, provides a good deal of information about the shape of the $M+1$ boundaries of D. As it turns out, for a multiply connected domain, knowledge of the Cauchy transform is not quite enough to fully pin down the shape of the domain, a feature we will see in more detail in the next section. This is intimately related to the feature observed in Chapter 11 that the shape of a multiply connected quadrature domain cannot be reconstructed from its quadrature data alone but must be supplemented with information concerning the holes.

In any case, the fact that the Cauchy transform encodes shape information turns out to be particularly useful in certain free boundary problems involving an evolving domain where it can be straightforward to ascertain the time evolution of the Cauchy transform directly from the problem statement; this is studied in §12.6. This evolution of the Cauchy transform can then be used to reconstruct the shape of the time-evolving domain. The problem becomes an inverse problem: to reconstruct the shape of a domain from its Cauchy transform.

12.5 ▪ Cauchy transforms of quadrature domains

From Chapter 11 we know that multiply connected quadrature domains are distinguished by the property that the Schwarz function of each boundary component is the meromorphic continuation through the domain D of any other boundary component. This means that there exists a function $S(z)$, having only poles in D, such that

$$\overline{z} = S(z), \qquad z \in \partial D_j,\ j = 0, 1, \ldots, M. \tag{12.45}$$

The Cauchy transform of a multiply connected domain is the continuous function satisfying

$$\mathcal{C}(z) = \begin{cases} C_0(z), & z \in D_0, \\ C_I(z) - \overline{z}, & z \in D, \\ C_j(z), & z \in D_j,\ j = 1, \dots, M. \end{cases} \tag{12.46}$$

By the continuity of $C(z)$ on the boundary ∂D_0, we have

$$C_0(z) = C_I(z) - \overline{z} = C_I(z) - S(z), \qquad z \in \partial D_0, \tag{12.47}$$

while continuity of $C(z)$ on the boundary ∂D_j for $j = 1, \dots, M$ gives

$$C_j(z) = C_I(z) - \overline{z} = C_I(z) - S(z), \qquad z \in \partial D_j. \tag{12.48}$$

By analytic continuation off these boundaries into D we deduce that

$$C_0(z) = C_j(z), \qquad j = 1, \dots, M. \tag{12.49}$$

This means that, for a bounded quadrature domain, the analytic functions defined by the Cauchy transform outside a bounded quadrature domain, and in each hole, are all analytic continuations of the same function. Moreover, since $S(z)$ has only a finite collection of poles inside D, it follows from (12.47) that the continuation of $C_0(z)$, and hence every $C_j(z)$, into D is meromorphic. It follows that $C_0(z)$ (and hence each of the functions $C_j(z)$ for $j = 1, \dots, M$) must be a rational function: it is analytic outside D as well as inside all the holes, and it has only poles inside D.

The fact that there is a rational function associated with a multiply connected quadrature domain turns out to be useful in applications, not least in problems where the boundaries of a domain evolve in time according to some physical law or by some imposed rule.

12.6 ▪ Evolution of the Cauchy transform

In certain physical problems the rational character of the Cauchy transform of a multiply connected domain can be shown to be preserved by the dynamical evolution. If we allow for the situation in which some planar domain $D(t)$ is evolving in time t, then its Cauchy transform will also evolve. It is of interest to ascertain how the Cauchy transform of a time-evolving domain $D(t)$ will change with time t.

For $j = 0, 1, \dots, M$ consider the evolution of

$$C_j(z,t) = \frac{1}{2\pi \mathrm{i}} \int_{\partial D(t)} \frac{\overline{z'} dz'}{z' - z}, \tag{12.50}$$

where z sits in the jth hole. By the product rule, the time derivative is the sum of several components,

$$\frac{\partial}{\partial t} C_j(z,t) = \frac{1}{2\pi \mathrm{i}} \int_{\partial D(t)} \left\{ \frac{\overline{z'_t} dz'}{z' - z} + \frac{\overline{z'} dz'_t}{z' - z} - \frac{\overline{z'} z'_t dz'}{(z' - z)^2} \right\}, \tag{12.51}$$

where we use the shorthand notation z_t for the derivative, with respect to time, of points z on the boundary curve. We now apply integration by parts on the second term:

$$\frac{\partial}{\partial t} C_j(z,t) = \frac{1}{2\pi \mathrm{i}} \int_{\partial D(t)} \frac{\overline{z'_t} dz'}{z' - z} \underbrace{- \frac{d\overline{z'} z'_t}{z' - z} + \frac{\overline{z'} z'_t dz'}{(z' - z)^2}}_{\text{from 2nd term}} - \frac{\overline{z'} z'_t dz'}{(z' - z)^2} \tag{12.52}$$

$$= \frac{1}{2\pi \mathrm{i}} \int_{\partial D(t)} \frac{\overline{z'_t} dz'}{z' - z} - \frac{d\overline{z'} z'_t}{z' - z} = -\frac{1}{\pi} \int_{\partial D(t)} \frac{\mathrm{Im}[z'_t d\overline{z'}]}{z' - z}. \tag{12.53}$$

This general formula will be useful later, where a rule for determining the quantity z_t is dictated by the physical circumstances of the problem.

12.7 ▪ A special class of free boundary problems

Suppose the Cauchy transform $C_j(z,t)$ defined in (12.50) for z in the jth hole in a multiply connected domain $D(t)$ evolves according to the partial differential equation

$$\frac{\partial C_j(z,t)}{\partial t} + \frac{\partial I_j(z,t)}{\partial z} = 0, \tag{12.54}$$

where $I_j(z,t)$ is the inhomogeneous Cauchy transform

$$I_j(z,t) = \frac{1}{\pi} \int\int_{D(t)} \frac{\sigma(z',t)}{z'-z} dx'dy', \tag{12.55}$$

and where $\sigma(z,t)$ is known to be an analytic, single-valued function of z in $D(t)$. Suppose too that, at the initial instant $t = 0$, the domain $D(0)$ is a quadrature domain where the Schwarz function of each boundary has only simple poles at $\{z_k(0) \in D(0)|k = 1,\dots,N\}$ inside $D(0)$, i.e.,

$$C_j(z,0) = \sum_{k=1}^{N} \frac{R_k(0)}{z - z_k(0)}, \qquad j = 0,1,\dots,M. \tag{12.56}$$

The restriction to simple poles can easily be lifted, but we focus on this case here for definiteness.

Consider the difference of two such functions for $j = p, q$, say, with $p \neq q$. Since $D(0)$ is a quadrature domain,

$$C_p(z,0) - C_q(z,0) = 0, \qquad I_p(z,0) - I_q(z,0) = 0. \tag{12.57}$$

The analytic continuation of $C_j(z,t)$ and $I_j(z,t)$ into $D(t)$, at each instant, can be expressed for each $j = 0,1,\dots,M$ by

$$C_I(z,t) - C_j(z,t) = S_j(z,t), \qquad I_I(z,t) - I_j(z,t) = \sigma(z,t)S_j(z,t), \tag{12.58}$$

where $C_I(z,t)$ and $I_I(z,t)$ are analytic in $D(t)$ and $S_j(z,t)$ denotes the Schwarz function of the jth boundary component. Since it is a relation between analytic functions for z in the jth hole, equation (12.54) also governs the evolution of the analytic continuation of $C_j(z,t)$ into $D(t)$ for each $j = 0,1,\dots,M$. Hence the difference $C_p(z,t) - C_q(z,t)$ can be shown to satisfy

$$\frac{\partial}{\partial t}(C_p(z,t) - C_q(z,t)) + \frac{\partial}{\partial z}\left[\sigma(z,t)(C_p(z,t) - C_q(z,t))\right] = 0 \tag{12.59}$$

for values of $z \in D(t)$. The unique solution of this homogeneous partial differential equation with analytic and single-valued coefficients and vanishing initial data is

$$C_p(z,t) - C_q(z,t) = 0, \tag{12.60}$$

implying, since this is valid for any choices of p and q, that the domain remains a quadrature domain under evolution.

Since quadrature domains are preserved under evolution, it remains to determine how the quadrature data evolves. We can now remove the subscript on (12.54) and write

$$\frac{\partial C(z,t)}{\partial t} + \frac{\partial I(z,t)}{\partial z} = 0. \tag{12.61}$$

It is easy to establish that this can be rewritten as

$$\frac{\partial C(z,t)}{\partial t} + \frac{\partial}{\partial z}(\sigma(z,t)C(z,t)) + \Sigma(z,t) = 0, \tag{12.62}$$

where

$$\Sigma(z,t) \equiv \frac{1}{\pi}\int\int_{D(t)} \left\{ \frac{\sigma(z',t)-\sigma(z,t)}{z'-z} \right\} dA_z \tag{12.63}$$

is analytic in $D(t)$ since the pole of the integrand at $z' = z$ is removable. Equivalently,

$$\frac{\partial C(z,t)}{\partial t} + \sigma(z,t)\frac{\partial C(z,t)}{\partial z} + \frac{\partial \sigma(z,t)}{\partial z}C(z,t) + \Sigma(z,t) = 0. \tag{12.64}$$

This partial differential equation for $C(z,t)$ has coefficient functions that are known a priori to be analytic in $D(t)$. It follows from the general theory of linear partial differential equations that there can be no creation of new singularities, and, indeed, if we substitute the local form

$$C(z,t) = \frac{R_k(t)}{z - z_k(t)} + \text{a locally analytic function} \tag{12.65}$$

for z near $z_k(t)$, we arrive at the ordinary differential equations

$$\frac{dz_k}{dt} = \sigma(z_k,t), \qquad \frac{dR_k(t)}{dt} = 0. \tag{12.66}$$

These equations govern the evolution of each simple pole $z_k(t)$ of $C(z,t)$ and its associated residue $R_k(t)$.

In the multiply connected case the equations (12.66) are not enough to fully determine the evolution of $D(t)$; some additional information is needed on the evolution of the areas of the holes in the domain. In any particular problem this is given by additional physical conditions associated with the holes.

We end this short chapter here and refer the reader to the exercises, where these general ideas are explored, and to Part II, where their usefulness is exemplified in the context of specific physical problems.

Exercises

12.1. **(Cauchy transform as generating function of geometrical moments)** Suppose D is a bounded simply connected domain. Show that as $|z| \to \infty$,

$$C(z) = \frac{1}{\pi}\int\int_D \frac{dA'_z}{z'-z} \sim -\frac{1}{\pi}\sum_{n=0}^{\infty}\frac{M_n}{z^{n+1}}, \tag{12.67}$$

where

$$M_n \equiv \int\int_D z^n dA_z, \qquad n \geq 0. \tag{12.68}$$

The Cauchy transform of a bounded domain is therefore a generating function for its complex moments (12.68).

12.2. **(Cauchy transform of an ellipse)** Let D be the interior of an ellipse with semimajor axis of length a along the real axis and semiminor axis of length b along the imaginary axis in a complex z plane.

(a) Show that a conformal mapping of the form

$$z = \frac{\alpha}{\zeta} + \beta\zeta \qquad (12.69)$$

transplants the interior of the unit disc $|\zeta| < 1$ in the complex ζ plane to the *exterior* of D, and find the appropriate values of the real parameters α and β.

(b) Use the conformal mapping in part (a) to find the Cauchy transform of D at all points in the plane.

12.3. **(Hele-Shaw flow driven by sources/sinks)** The velocity potential ϕ for the velocity $\mathbf{u}$ of a multiply connected region of fluid $D(t)$ in a two-dimensional Hele-Shaw cell, i.e.,

$$\mathbf{u} = \nabla\phi, \qquad (12.70)$$

satisfies, at each instant,

$$\nabla^2\phi = \sum_{k=1}^{N} m_k(t)\delta(z - z_k), \qquad z \in D(t), \qquad (12.71)$$

where $\{m_k(t)|k = 1, \ldots, N\}$ is the set of time-varying injection rates of a set of N sources located at positions $\{z_k|k = 1, \ldots, N\}$ which are fixed in time. The example of a time-evolving triply connected domain $D(t)$ with two holes ($M = 2$) pressurized at $\Phi_1(t)$ and $\Phi_2(t)$ and driven by three point sources ($N = 3$) is shown in Figure 12.3.

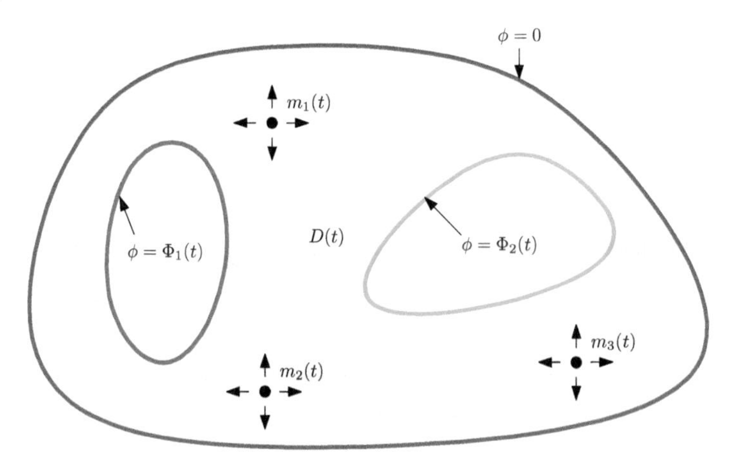

Figure 12.3.

On each of the boundaries of the fluid region $D(t)$ a Dirichlet-type condition holds:

$$\phi = \Phi_j(t), \qquad z \in \partial D_j(t), \qquad j = 0, 1, \ldots, M, \qquad (12.72)$$

where $\Phi_j(t)$ is a given function of time that is independent of position and $\partial D_j(t)$ denotes the jth component of the boundary of a multiply connected region $D(t)$. Without loss of generality, we can set

$$\Phi_0(t) = 0. \qquad (12.73)$$

The evolution of the boundary of $D(t)$ is determined by the following rule: the normal velocity V_n of any point on the boundary is

$$V_n = \mathbf{u}.\mathbf{n} = \nabla\phi.\mathbf{n}, \tag{12.74}$$

where $\mathbf{n}$ denotes the unit normal at the boundary point. Show using (12.53) that, with this evolution rule, the Cauchy transform $C_j(z,t)$ defined in (12.50) is given by

$$C_j(z,t) = -\sum_{k=1}^{N} \frac{Q_k(t)}{\pi} \frac{1}{z - z_k} + C_j(z,0), \tag{12.75}$$

where $C_j(z,0)$ is the initial Cauchy transform for z in the jth hole and

$$Q_k(t) \equiv \int_0^t m_k(t')dt'. \tag{12.76}$$

Argue from this result that if the initial fluid domain $D(0)$ is a triply connected quadrature domain, then $D(t)$ remains a triply connected quadrature domain (at least for sufficiently short times).

12.4. **(Hele-Shaw flow in a rotating cell)** Another version of the Hele-Shaw problem discussed in Exercise 12.3 is to have the fluid motion driven by steady rotation of the Hele-Shaw cell in which the fluid is placed (rather than by a distribution of sources/sinks). In this case it can be shown[41] that the free boundary problem involves a time-evolving domain $D(t)$ where the Cauchy transform of $D(t)$ evolves according to the partial differential equation (12.54)–(12.55) with the particular choice

$$\sigma(z,t) = \Omega z, \tag{12.77}$$

where Ω is some constant that depends on the fixed rotation rate of the cell.

(a) Show that

$$I_j(z,t) = \frac{\Omega}{\pi}\mathcal{A}(t) + \Omega z C_j(z,t), \tag{12.78}$$

where $C_j(z,t)$ is defined in (12.50) and $\mathcal{A}(t)$ denotes the total area of the domain $D(t)$.

(b) Hence show that

$$\frac{\partial C_j(z,t)}{\partial t} + \Omega z \frac{\partial C_j(z,t)}{\partial z} + \Omega C_j(z,t) = 0. \tag{12.79}$$

(c) It has been argued in §12.7 that this free boundary problem will admit quadrature domain solutions with

$$C_j(z,0) = \hat{C}_0(z), \qquad C_j(z,t) = C(z,t) \qquad \forall j = 0,1,\ldots,M, \tag{12.80}$$

for some initial condition $\hat{C}_0(z)$ that has only poles in $D(0)$ and where

$$\frac{\partial C(z,t)}{\partial t} + \Omega z \frac{\partial C(z,t)}{\partial z} + \Omega C(z,t) = 0. \tag{12.81}$$

By using the method of characteristics or otherwise, show that the solution of this partial differential equation satisfying the initial condition (12.80) is

$$C(z,t) = e^{-\Omega t}\hat{C}_0(ze^{-\Omega t}). \tag{12.82}$$

[41] See Chapter 26 for more details, or the treatment in [31].

(d) Show from (12.82) that if

$$\hat{C}_0(z) = \sum_{n=1}^{N} \frac{r_n}{z - z_n} \tag{12.83}$$

for some set of constants $\{r_n, z_n | n = 1, \ldots, N\}$, where each $z_n \in D(0)$, then

$$C(z,t) = \sum_{n=1}^{N} \frac{r_n}{z - z_n e^{\Omega t}}. \tag{12.84}$$

Hence confirm from this case study that the singularities of the Cauchy transform satisfy the ordinary differential equations (12.66).[42]

[42]This problem is mathematically equivalent to free boundary evolution in a rotating Hele-Shaw cell, as discussed in Chapter 26.

Chapter 13

Schwarz problems in multiply connected domains

13.1 ▪ Boundary value problems

Boundary value problems involving analytic functions are ubiquitous in the applied sciences, and there is a wide range of techniques for solving them. One of the most common boundary value problems arising in applications is the Dirichlet problem. It is the problem of finding a harmonic function in some domain given its values on the domain boundary. It is natural to ask about the solution to this problem in the class of multiply connected circular domains D_z. Can the solution be expressed in terms of the prime function associated with D_z?

The solution of a Dirichlet problem depends on the regularity of the given boundary data. We want to avoid complicating this presentation with functional analysis technicalities, so, roughly speaking, we will first present integral representations of the solution of the Dirichlet problem when the given boundary data is sufficiently smooth.[43] Then, in §13.8, we show how other aspects of our framework (e.g., the Cayley-type maps considered in Chapters 5 and 6) can be used, in conjunction with these integral expressions, to allow us to incorporate less regularity in the boundary data, including discontinuous data. Our objective is to show the key role played by the prime function in all cases. The analysis in this chapter is based on the original presentation in [20].

13.2 ▪ The Dirichlet problem in a circular domain

Consider the following boundary value problem—the Dirichlet problem—for a single-valued field ϕ in a multiply connected circular domain D_z. It is required to find the solution ϕ satisfying

$$\nabla^2\phi = 0, \qquad z \in D_z, \tag{13.1}$$

with

$$\phi = \Phi_j(z), \qquad z \in C_j,\ j = 0, 1, \dots, M, \tag{13.2}$$

where the set of functions $\{\Phi_j(z) | j = 0, 1, \dots, M\}$ constituting the boundary data for this problem is given.

[43] The reader is referred to the treatise by Bell [66] for a clear presentation of these technical issues.

13.3 ▪ An integral formula

Since we seek to find a harmonic function ϕ, it is useful to consider Green's second identity. An application of this identity using ϕ and the first-type Green's function encountered in Chapter 2 implies that, for each $j = 1, \dots, M$,

$$\int\int_{D_z} \left[\phi\nabla^2 G(z,a) - G(z,a)\nabla^2\phi\right] dA_z = \int_{\partial D_z} \left[\phi\frac{\partial G(z,a)}{\partial n_z} - G(z,a)\frac{\partial\phi}{\partial n_z}\right] ds_z. \tag{13.3}$$

Formula (13.3) leads to the integral representation

$$\phi(a) = -\int_{\partial D_z} \phi\frac{\partial G(z,a)}{\partial n_z} ds_z. \tag{13.4}$$

We know from (2.48) that

$$G(z,a) = G_0(z,a) + \sum_{k=1}^{M} \sigma_k(z)\Sigma_k(a), \tag{13.5}$$

where $G_0(z,a)$ is the modified Green's function and

$$\Sigma_k(z) = \text{Im}[v_k(z)], \qquad k = 1, \dots, M. \tag{13.6}$$

On substitution of (13.5) into (13.4) we arrive at the representation

$$\phi(a) = -\int_{\partial D_z} \phi\frac{\partial G_0(z,a)}{\partial n_z} ds_z - \sum_{k=1}^{M} \Sigma_k(a)\int_{\partial D_z} \phi\frac{\partial\sigma_k(z)}{\partial n_z} ds_z. \tag{13.7}$$

This can be written as

$$\phi(a) = -\int_{\partial D_z} \phi\frac{\partial G_0(z,a)}{\partial n_z} ds_z + \sum_{k=1}^{M} A_k\Sigma_k(a), \tag{13.8}$$

where the real constants $\{A_k | k = 1, \dots, M\}$ are

$$A_k = -\int_{\partial D_z} \phi\frac{\partial\sigma_k(z)}{\partial n_z} ds_z, \qquad k = 1, \dots, M. \tag{13.9}$$

This integral formula will be useful, so we rewrite it in terms of the prime function. We know from Exercise 1.21 that

$$\frac{\partial G_0}{\partial n_z} ds_z = \text{Re}[-d\mathcal{G}_0(z,a)]. \tag{13.10}$$

This means that (13.8) can be written as

$$\text{Re}[w(a)] = \text{Re}\left[\int_{\partial D_z} \phi d\mathcal{G}_0(z,a)\right] - \text{Re}\left[\sum_{k=1}^{M} A_k \text{i} v_k(a)\right]. \tag{13.11}$$

It was a central result of Chapter 4 that $\mathcal{G}_0(z,a)$ can be expressed in terms of the prime function of D_z as

$$\mathcal{G}_0(z,a) = \frac{1}{2\pi\text{i}} \log\left(\frac{\omega(z,a)}{|a|\omega(z,1/\overline{a})}\right) \tag{13.12}$$

so that

$$d\mathcal{G}_0(z,a) = \frac{1}{2\pi\mathrm{i}}\left[d\log\omega(z,a) - d\log\omega(z,1/\overline{a})\right]. \tag{13.13}$$

Notice that we can write

$$\begin{aligned}\mathrm{Re}\left[\int_{\partial D_z}\phi d\mathcal{G}_0(z,a)\right] &= \mathrm{Re}\left[\frac{1}{2\pi\mathrm{i}}\int_{\partial D_z}\phi\left[d\log\omega(z,a) - d\log\omega(z,1/\overline{a})\right]\right]\\ &= \mathrm{Re}\left[\frac{1}{2\pi\mathrm{i}}\int_{\partial D_z}\phi\left[d\log\omega(z,a) + d\log\overline{\omega}(\overline{z},1/a)\right]\right],\end{aligned} \tag{13.14}$$

where, in the second equality, we have used the fact that the real part of a complex quantity is the same as the real part of its complex conjugate. Then (13.11) becomes

$$\mathrm{Re}[w(a)] = \mathrm{Re}\left[\frac{1}{2\pi\mathrm{i}}\int_{\partial D_z}\phi\left[d\log\omega(z,a) + d\log\overline{\omega}(\overline{z},1/a)\right] - \sum_{k=1}^{M}A_k\mathrm{i}v_k(a)\right].$$

The functions on both sides of this expression whose real parts are being taken are analytic functions of the variable a in D_z. It can therefore be concluded that

$$w(a) = \frac{1}{2\pi\mathrm{i}}\int_{\partial D_z}\phi\left[d\log\omega(z,a) + d\log\overline{\omega}(\overline{z},1/a)\right] - \sum_{k=1}^{M}A_k\mathrm{i}v_k(a) + \mathrm{i}c, \tag{13.15}$$

where c is a real constant. Inspection reveals that this is an integral representation of a function analytic in D_z purely in terms of its real part on the boundary ∂D_z.

13.4 ▪ The Schwarz problem in D_z

Returning to the Dirichlet problem stated in §13.2, we can think of the harmonic function ϕ as the real part of some function $w(z)$ that is analytic in D_z, i.e.,

$$\phi(z) = \mathrm{Re}[w(z)]. \tag{13.16}$$

The problem stated in §13.2 will be solved if we can find the function $w(z)$ analytic in D_z satisfying the boundary condition

$$\mathrm{Re}[w(z)] = \Phi_j(z), \qquad z \in C_j,\ j = 0,1,\ldots,M, \tag{13.17}$$

where $\{\Phi_j(z)|j = 0,1,\ldots,M\}$ is a given set of real functions. We refer to this as the Schwarz problem for the analytic function $w(z)$ in D_z. Its solution is clearly only defined up to addition of a pure imaginary constant.

For now we simply assume that the given set of functions $\{\Phi_j(z)|j = 0,1,\ldots,M\}$ has sufficient regularity on all the boundaries that all integrals in the following formulas exist. In §13.8 we present complementary techniques that can help if the boundary data is less smooth and has discontinuities, for example.

The solution of the Schwarz problem can now be immediately inferred from the integral formula (13.15):

$$\begin{aligned}w(a) &= \frac{1}{2\pi\mathrm{i}}\int_{C_0}\Phi_0\left[d\log\omega(z,a) + d\log\overline{\omega}(\overline{z},1/a)\right]\\ &- \frac{1}{2\pi\mathrm{i}}\sum_{j=1}^{M}\int_{C_j}\Phi_j\left[d\log\omega(z,a) + d\log\overline{\omega}(\overline{z},1/a)\right] - \sum_{k=1}^{M}A_k\mathrm{i}v_k(a) + \mathrm{i}c,\end{aligned} \tag{13.18}$$

where

$$A_k = -\left[\int_{C_0} \Phi_0 \frac{\partial \sigma_k(z)}{\partial n_z} ds_z - \sum_{j=1}^{M} \int_{C_j} \Phi_j \frac{\partial \sigma_k(z)}{\partial n_z} ds_z \right], \qquad k = 1, \ldots, M, \quad (13.19)$$

and where c is a real constant.

The expression (13.18) is an integral formula for the solution of the Schwarz problem. On taking the real part of this integral expression we arrive at the solution of the Dirichlet problem of §13.2.

13.5 ▪ The modified Schwarz problem in D_z

It is clear from inspection of (13.15), in particular from the appearance of the functions $\{v_k(a)|k = 1, \ldots, M\}$ in that expression, that the solution of the Schwarz problem is analytic in D_z but not necessarily single-valued unless all the constants $\{A_k|k = 1, \ldots, M\}$ vanish. This is because the functions $\{v_k(a)|k = 1, \ldots, M\}$ are not single-valued in D_z, while we know that $\omega(z, a)$ and $\overline{\omega}(\overline{z}, 1/a)$ are single-valued functions of a.

This prompts us to consider the so-called modified Schwarz problem, which also arises regularly in applications. In this problem we ask for a *single-valued* analytic function $w_s(z)$, say, whose real part has given values on the boundary of D_z,

$$\phi(z) = \mathrm{Re}[w_s(z)] = \Phi_j(z), \qquad z \in C_j, \ j = 0, 1, \ldots, M. \quad (13.20)$$

If we set

$$w_s(z) = \phi + \mathrm{i}\psi, \quad (13.21)$$

so that ψ is the harmonic conjugate to ϕ in D_z, then if $w_s(z)$ is to be single-valued the following M conditions must hold:

$$\int_{C_k} d\psi = \int_{C_k} \frac{\partial \psi}{\partial s_z} ds_z = 0, \qquad k = 1, \ldots, M. \quad (13.22)$$

The conditions that the constants $\{A_k|k = 1, \ldots, M\}$ appearing in (13.15) must vanish, on the other hand, implies from (13.9) the following conditions on the set of data $\{\Phi_j(z, \overline{z})|j = 0, 1, \ldots, M\}$:

$$\begin{aligned}\int_{\partial D_z} \phi \frac{\partial \sigma_k(z)}{\partial n_z} ds_z &= \int_{\partial C_0} \Phi_0(z, \overline{z}) \frac{\partial \sigma_k(z)}{\partial n_z} ds_z - \sum_{j=1}^{M} \int_{\partial C_j} \Phi_j(z, \overline{z}) \frac{\partial \sigma_k(z)}{\partial n_z} ds_z \\ &= 0, \qquad k = 1, \ldots, M. \end{aligned} \quad (13.23)$$

A solution to the modified Schwarz problem will only exist if these conditions on the given data are satisfied. In this event, formula (13.15) then reduces to the simpler form[44]

$$w(a) = \frac{1}{2\pi \mathrm{i}} \int_{\partial D_z} \phi \left[d \log \omega(z, a) + d \log \overline{\omega}(\overline{z}, 1/a)\right] + \mathrm{i}c. \quad (13.24)$$

Consequently, on use of (13.20), we can write

$$\begin{aligned} w(a) &= \frac{1}{2\pi \mathrm{i}} \int_{C_0} \Phi_0 \left[d \log \omega(z, a) + d \log \overline{\omega}(\overline{z}, 1/a)\right] \\ &- \frac{1}{2\pi \mathrm{i}} \sum_{j=1}^{M} \int_{C_j} \Phi_j \left[d \log \omega(z, a) + d \log \overline{\omega}(\overline{z}, 1/a)\right] + \mathrm{i}c. \end{aligned} \quad (13.25)$$

This is an integral formula for the solution of the modified Schwarz problem.

[44] This expression in terms of the prime function first appeared in [20].

It is not immediately apparent that the M conditions (13.22) on the solution are equivalent to the M conditions (13.23) on the data. However, on substitution of ϕ and the function $\sigma_k(z)$ for $k = 1, \ldots, M$ from Chapter 2 into Green's second identity we find

$$\int\int_{D_z} [\phi(z)\nabla^2\sigma_k(z) - \sigma_k(z)\nabla^2\phi(z)]\, dA_z = \int_{\partial D_z} \left[\phi(z)\frac{\partial\sigma_k(z)}{\partial n_z} - \sigma_k(z)\frac{\partial\phi(z)}{\partial n_z}\right] ds_z, \tag{13.26}$$

which, since both $\phi(z)$ and $\sigma_k(z)$ are harmonic in D_z, implies

$$\int_{\partial D_z} \phi(z)\frac{\partial\sigma_k(z)}{\partial n_z} ds_z = \int_{\partial D_z} \sigma_k(z)\frac{\partial\phi(z)}{\partial n_z} ds_z = \int_{C_k} \frac{\partial\phi(z)}{\partial n_z} ds_z = 0, \quad k = 1, \ldots, M. \tag{13.27}$$

In the second equality we have used the known boundary values of $\sigma_k(z)$ on ∂D_z, and, in the last equality, we have used the Cauchy–Riemann equations and conditions (13.22). Finally, use of the boundary conditions (13.20) in the left-hand side of (13.27) shows the equivalence of (13.22) and (13.23).

There is an alternative way to arrive at (13.24) more directly using Green's identity and the modified Green's function $G_0(z, a)$. A swap of $G_0(z, a)$ for $G(z, a)$ in (13.3) leads to

$$\int\int_{D_z} [\phi\nabla^2 G_0(z, a) - G_0(z, a)\nabla^2\phi]\, dA_z = \int_{\partial D_z} \left[\phi\frac{\partial G_0(z, a)}{\partial n_z} - G_0(z, a)\frac{\partial\phi}{\partial n_z}\right] ds_z. \tag{13.28}$$

This produces the integral representation

$$\phi(a) = -\int_{\partial D_z} \left[\phi\frac{\partial G_0(z, a)}{\partial n_z} - G_0(z, a)\frac{\partial\phi}{\partial n_z}\right] ds_z, \tag{13.29}$$

where we note that the second term in the integral no longer vanishes since, while it vanishes on C_0, $G_0(z, a)$ is generally nonzero on the interior circles $\{C_j | j = 1, \ldots, M\}$. But $G_0(z, a)$ is nevertheless constant on these circles, as we recall from (2.50). Expression (13.29) then leads to

$$\phi(a) = -\int_{\partial D_z} \phi(z)\frac{\partial G_0(z, a)}{\partial n_z} ds_z + \sum_{k=1}^{M} \mathrm{Im}[v_k(a)] \int_{C_k} \frac{\partial\phi(z)}{\partial n_z} ds_z. \tag{13.30}$$

But

$$\int_{C_k} \frac{\partial\phi}{\partial n_z} ds_z = 0, \qquad k = 1, \ldots, M, \tag{13.31}$$

which follows from the Cauchy–Riemann equations for ϕ and ψ and the use of conditions (13.22). Hence,

$$\phi(a) = -\int_{\partial D_z} \left[\phi\frac{\partial G_0(z, a)}{\partial n_z} ds_z\right]. \tag{13.32}$$

On use of (13.10) this can be written as

$$\begin{aligned} \mathrm{Re}[w(a)] &= \mathrm{Re}\left[\int_{\partial D_z} \phi d\mathcal{G}_0(z, a)\right] \\ &= \mathrm{Re}\left[\frac{1}{2\pi\mathrm{i}} \int_{\partial D_z} \phi\left[d\log\omega(z, a) + d\log\overline{\omega}(\bar{z}, 1/a)\right]\right], \end{aligned} \tag{13.33}$$

where, in the second equality, we have used (13.13). Finally, since the quantities in square brackets on both sides of this equation are analytic functions of a in D_z, we retrieve the integral formula (13.24).

13.6 ▪ The Poisson integral formula

The expression (13.24), when applied to the special case where D_z is the unit disc, has a special name: it is the *Poisson integral formula.*

Suppose that D_z is the unit disc so that we wish to find the analytic function $w(z)$ such that

$$\mathrm{Re}[w(z)] = \Phi_0(z), \qquad z \in C_0. \tag{13.34}$$

Formula (13.15) reduces to

$$w(a) = \frac{1}{2\pi \mathrm{i}} \int_{C_0} \Phi_0(z)\,[d\log\omega(z,a) + d\log\overline{\omega}(\overline{z}, 1/a)] + \mathrm{i}c, \qquad c \in \mathbb{R}. \tag{13.35}$$

On use of the fact that $\overline{z} = 1/z$ on C_0 we can write this as

$$w(a) = \frac{1}{2\pi \mathrm{i}} \int_{C_0} \Phi_0(z)\,[d\log\omega(z,a) + d\log\overline{\omega}(1/z, 1/a)] + \mathrm{i}c. \tag{13.36}$$

Now on use of the identity (1.8) this becomes

$$w(a) = \frac{1}{2\pi \mathrm{i}} \int_{C_0} \Phi_0(z) \left[d\log\omega(z,a) + d\log\left(-\frac{\omega(z,a)}{za}\right)\right] + \mathrm{i}c \tag{13.37}$$

$$= \frac{1}{2\pi \mathrm{i}} \int_{C_0} \Phi_0(z)\,[2d\log\omega(z,a) - d\log z] + \mathrm{i}c. \tag{13.38}$$

Since, for the unit disc, $\omega(z,a) = (z-a)$, this becomes

$$w(a) = \frac{1}{2\pi \mathrm{i}} \int_{C_0} \Phi_0(z) \left(\frac{z+a}{z-a}\right) \frac{dz}{z} + \mathrm{i}c, \qquad c \in \mathbb{R}. \tag{13.39}$$

On swapping the roles of variables a and z this becomes

$$w(z) = \frac{1}{2\pi \mathrm{i}} \int_{C_0} \Phi_0(a) \left(\frac{a+z}{a-z}\right) \frac{da}{a} + \mathrm{i}c, \qquad c \in \mathbb{R}, \tag{13.40}$$

which is the familiar form of the Poisson integral formula [61].

13.7 ▪ The Villat integral formula

When the general formula (13.24) is applied to the special case in which D_z is the concentric annulus $\rho < |z| < 1$, it also has a special designation: it is the *Villat integral formula* [122].

To explore this case suppose that D_z is the concentric annulus $\rho < |z| < 1$ and we wish to find the *single-valued* analytic function $w(z)$ such that

$$\mathrm{Re}[w(z)] = \begin{cases} \Phi_0(z), & z \in C_0, \\ \Phi_1(z), & z \in C_1. \end{cases} \tag{13.41}$$

This is the modified Schwarz problem in a concentric annulus.

Since the domain is doubly connected, with $M = 1$, from (13.23) the solution exists provided the data satisfies the single constraint

$$\int_{\partial C_0} \Phi_0(z) \frac{\partial \sigma_1(z)}{\partial n_z} ds_z = \int_{\partial C_1} \Phi_1(z) \frac{\partial \sigma_1(z)}{\partial n_z} ds_z, \tag{13.42}$$

where $\sigma_1(z)$ is given in (5.8).

The general integral formula (13.15) for the solution in this case becomes

$$w(a) = \frac{1}{2\pi \mathrm{i}} \int_{C_0} \Phi_0(z) \left[d \log \omega(z,a) + d \log \overline{\omega}(1/z, 1/a)\right] - \frac{1}{2\pi \mathrm{i}} \int_{C_1} \Phi_1(z) \left[d \log \omega(z,a) + d \log \overline{\omega}(\rho^2/z, 1/a)\right] + \mathrm{i}c, \qquad c \in \mathbb{R}. \quad (13.43)$$

On use of the identities (5.5) and (5.6) it can be shown that

$$\overline{\omega}(\rho^2/z, 1/a) = \frac{\omega(z,a)}{a^2}. \quad (13.44)$$

Expression (13.43) then becomes

$$w(a) = \frac{1}{2\pi \mathrm{i}} \int_{C_0} \Phi_0(z) \left[2d \log \omega(z,a) - d \log z\right] - \frac{1}{2\pi \mathrm{i}} \int_{C_1} \Phi_1(z) \left[2d \log \omega(z,a)\right] + \mathrm{i}c.$$

Since

$$d \log \omega(z,a) = \frac{K(z,a)}{z} dz = \frac{\mathcal{K}(a,z)}{z} dz, \quad (13.45)$$

where we have used (5.78), on swapping the roles of variables a and z we can write either

$$w(z) = \frac{1}{2\pi \mathrm{i}} \int_{C_0} \Phi_0(a) \left[2K(a,z) - 1\right] \frac{da}{a} - \frac{1}{2\pi \mathrm{i}} \int_{C_1} \Phi_1(a) [2K(a,z)] \frac{da}{a} + \mathrm{i}c \quad (13.46)$$

or

$$w(z) = \frac{1}{2\pi \mathrm{i}} \int_{C_0} \Phi_0(a) \left[2\mathcal{K}(z,a) - 1\right] \frac{da}{a} - \frac{1}{2\pi \mathrm{i}} \int_{C_1} \Phi_1(a) [2\mathcal{K}(z,a)] \frac{da}{a} + \mathrm{i}c. \quad (13.47)$$

Formulas (13.46) and (13.47) are both equivalent to the classical Villat integral formula re-expressed in terms of the prime function of the annulus.

13.8 ▪ Discontinuous and singular boundary data

It is common in applications to encounter problems of Dirichlet type where the given boundary data is discontinuous or has singularities at certain boundary points. It is then necessary to supplement use of the integral formulas of §13.4 and 13.5 with additional techniques that take explicit account of such boundary point singularities.

Let us consider a problem with discontinuous boundary data. A useful trick is to exploit the linearity of the Dirichlet-type boundary value problem. The idea is to consider a decomposition of the form

$$\phi = \phi_s + \tilde{\phi}, \quad (13.48)$$

where the function ϕ_s is some harmonic function in the domain D_z that has the same jumps as the given boundary data, leaving us to find another harmonic function $\tilde{\phi}$ having smoother boundary data.[45] This latter step can be done using the integral formulas of §13.4 and §13.5.

As an example consider the problem of finding a harmonic function ϕ in the unbounded doubly connected region exterior to a circular object in the upper half plane, as shown in Figure 13.1. This circular object has "two faces": on one face the function ϕ must vanish,

[45] This is the strategy of singularity subtraction.

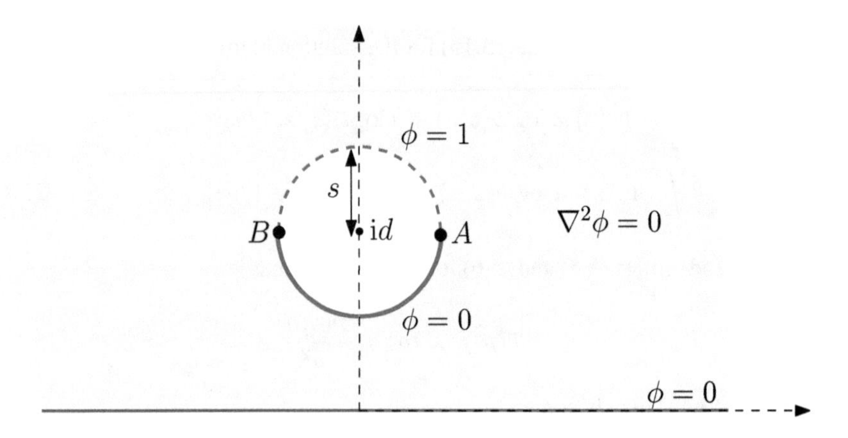

Figure 13.1. *A circular, two-faced particle of radius s at distance d above a wall.*

while on the other face ϕ must have the value unity.[46] On the infinite wall ϕ is taken to vanish. Figure 13.1 shows a schematic of the geometry and the boundary conditions. It is clear that at two boundary points, marked A and B in Figure 13.1, there is a jump in the boundary values of ϕ.

We now seek a harmonic function ϕ_s that explicitly takes account of these jumps in the boundary data. Since this boundary value problem is conformally invariant, which is easily verified, it is natural to let

$$\phi_s = \text{Im}[W(\zeta)], \tag{13.49}$$

where the objective is to find the complex potential $W(\zeta)$ as a function of a preimage ζ variable taking values in a concentric annulus

$$\rho < |\zeta| < 1. \tag{13.50}$$

The required conformal mapping function, $z = f(\zeta)$ say, was derived in Exercise 5.8, where a formula for ρ in terms of d and s is given. It only remains to find $W(\zeta)$.

To do this, consider the (multiple of a) Cayley-type mapping given by

$$S(\zeta; a, \overline{a}) = \frac{1}{a} R(\zeta; a, \overline{a}) = \frac{1}{a} \frac{\omega(\zeta, a)}{\omega(\zeta, \overline{a})}, \tag{13.51}$$

where a is a point on C_1 in the upper half ζ plane and $R(\zeta; a, \overline{a})$ is one of the Cayley-type maps considered in §5.10. It follows from (5.60) that

$$\overline{S(\zeta; a, \overline{a})} = \begin{cases} \dfrac{a}{\overline{a}} S(\zeta; a, \overline{a}), & z \in C_0, \\ \\ S(\zeta; a, \overline{a}), & z \in C_1. \end{cases} \tag{13.52}$$

This means that $S(\zeta; a, \overline{a})$ has piecewise constant argument on C_0 and C_1. If we think of the mapping

$$\chi = S(\zeta; a, \overline{a}), \tag{13.53}$$

[46] In physics, particles with "two faces" like this are known as Janus particles [8, 9].

then the concentric annulus (13.50) is mapped to the lower half χ plane, and it can be checked that the function

$$\tilde{W}(\zeta) \equiv -\frac{1}{\pi}\log\chi = -\frac{1}{\pi}\log S(\zeta;a,\overline{a}) \tag{13.54}$$

has piecewise constant imaginary part on the boundary of the annulus. In particular,

$$\mathrm{Im}[\tilde{W}(\zeta)] = \begin{cases} \dfrac{\arg[a]}{\pi}, & z \in C_0, \\ 0 \text{ or } 1, & z \in C_1, \end{cases} \tag{13.55}$$

where we have taken the principal branch of the logarithm. The point is that, as we pass through a and $\overline{a}$ on C_1, the value of $\tilde{W}(\zeta)$ jumps between 0 and 1. We now take a to be the preimage, under the conformal mapping $z = f(\zeta)$, of the point A shown in Figure 13.1, i.e.,

$$A = f(a). \tag{13.56}$$

The imaginary part of $\tilde{W}(\zeta)$ now has the required jump in its boundary values at the points A and B. But it does not yet take the correct value on C_0. This is easily fixed using $v_1(\zeta)$. Indeed, the required complex potential is

$$W(\zeta) = \tilde{W}(\zeta) + \frac{\arg[a]}{\pi}(v_1(\zeta) - 1). \tag{13.57}$$

It is easy to check that the imaginary part of this function now satisfies all the correct boundary conditions. From (13.49) this means we have found ϕ_s.

In this case, the original boundary problem is fully solved and there is no additional part $\tilde{\phi}$ of the solution to additionally compute. This is not always the case. In [9] the author solves boundary value problems similar to that considered above and, having found a function akin to (13.57) accounting for the jump in the boundary data, goes on to compute an additional contribution $\tilde{\phi}$ having smoother boundary data.

In Chapter 27 of Part II we will see another situation—in the problem of a steady stagnant-cap surfactant-laden bubble in strain—involving boundary data with certain singularities where techniques like those just illustrated can be deployed.

Exercises

13.1. **(Alternative to the Villat formula)** Suppose it is required to find a function $w(z)$ that is analytic and single-valued in the annulus $\rho < |z| < 1$ satisfying the boundary conditions

$$\mathrm{Re}[w(z)] = \begin{cases} d(z), & z \in C_0, \\ e(z), & z \in C_1. \end{cases} \tag{13.58}$$

One option is to use the Villat formula (13.46) or (13.47). In practice it can be convenient to bypass use of this formula using the following ideas. Since it is analytic and single-valued in the annulus, the function $w(z)$ has a convergent Laurent series

$$w(z) = a_0 + \sum_{n=1}^{\infty} a_n z^n + \sum_{n=1}^{\infty} \frac{a_{-n}\rho^n}{z^n}. \tag{13.59}$$

The imaginary part of a_0 can be chosen at will—it is not determined by the boundary conditions (13.58) for $w(z)$.

Suppose that, letting $z = e^{i\theta}$ on $|z| = 1$ and $z = \rho e^{i\theta}$ on $|z| = \rho$, we have the representations

$$d(z) = \sum_{n=-\infty}^{\infty} d_n e^{in\theta}, \qquad e(z) = \sum_{n=-\infty}^{\infty} e_n e^{in\theta}, \tag{13.60}$$

where $d_{-n} = \overline{d_n}$ and $e_{-n} = \overline{e_n}$.

(a) Show, on substitution of (13.59) into the boundary condition (13.58) on $|\zeta| = 1$ and consideration of the coefficients of ζ^n for $n \geq 1$, that

$$a_n + \overline{a_{-n}}\rho^n = 2d_n, \qquad n \geq 1. \tag{13.61}$$

(b) Show similarly, from the boundary condition (13.58) on $|\zeta| = \rho$, that

$$a_n\rho^n + \overline{a_{-n}} = 2e_n, \qquad n \geq 1. \tag{13.62}$$

(c) Hence show that

$$a_n = \frac{2(d_n - \rho^n e_n)}{1 - \rho^{2n}}, \qquad a_{-n} = \frac{2\overline{e_n} - 2\rho^n \overline{d_n}}{1 - \rho^{2n}}, \qquad n \geq 1. \tag{13.63}$$

(d) By consideration of the constant terms in the Laurent expansions of the boundary conditions show that

$$\text{Re}[a_0] = d_0 \tag{13.64}$$

and, for consistency, that the data must satisfy

$$e_0 = d_0. \tag{13.65}$$

Verify that this consistency condition is equivalent to (13.42).

13.2. **(Explicit solution of a Schwarz problem)** Consider the problem of finding the function $u(x, y)$ harmonic in the triply connected circular domain D_z shown in Figure 13.2, where the two interior circles C_1 and C_2 are centered on the real axis. On the boundaries of D_z, u is piecewise constant—$u = 1$ on C_0, $u = 0$ on C_2—while on C_1 the value of u is 1 on the arc of C_1 between two points a and $\overline{a}$, as shown in Figure 13.2, and vanishes on the remainder of the circle C_1. By considering the class of Cayley-type maps introduced in Chapter 6, and the functions $v_1(z)$ and $v_2(z)$, find an expression for the solution $u(x, y)$.

13.3. **(Explicit solution of a Schwarz problem)** Consider the problem of finding the function $u(x, y)$ harmonic in the unbounded region in a complex z plane exterior to three intervals of finite length along the real axis given by

$$[-s, -r], \qquad [-1, +1], \qquad [+r, +s] \tag{13.66}$$

for some $1 < r < s$. The boundary conditions on $u(x, y)$ are as follows:

$$u(x, y) = \begin{cases} 1, & z \in [-s, -r], \\ 1, & z \in [-1, 0], \\ 0, & z \in [0, 1], \\ 0, & z \in [+r, +s]. \end{cases} \tag{13.67}$$

Therefore the value of $u(x, y)$ changes halfway along the middle interval. Find an expression for the solution $u(x, y)$. *Hint:* Think about conformal mapping, conformal invariance, and Exercise 13.2.

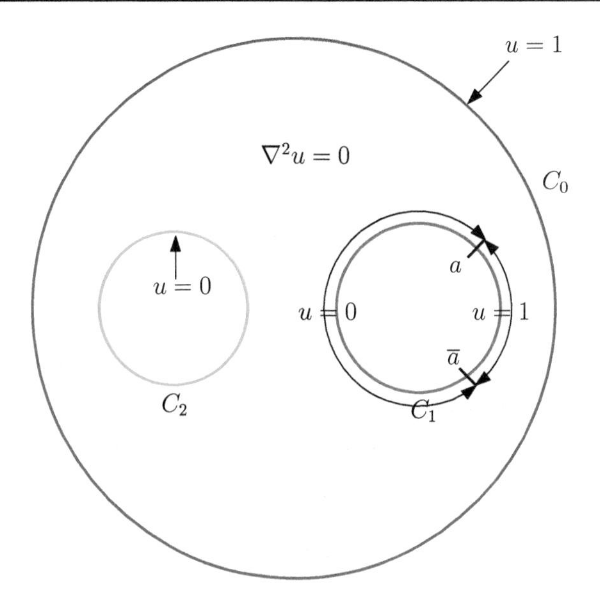

Figure 13.2. *A Dirichlet problem in a triply connected circular domain (Exercise 13.2).*

13.4. **(Moving objects in potential flow)** Another harmonic function $\phi(x, y)$ in the 3-slit domain described in Exercise 13.3 is the velocity potential of a two-dimensional velocity field $\mathbf{u}$ related by

$$\mathbf{u} = \nabla\phi. \tag{13.68}$$

The boundary conditions on this velocity field are

$$\mathbf{u}.\mathbf{n} = \mathbf{U}_j.\mathbf{n}, \qquad j = 0, 1, 2, \tag{13.69}$$

where $\mathbf{U}_j$ are a set of constant vectors and it is also stipulated that

$$\int_{\text{slit } j} \frac{\partial\phi}{\partial n} ds = 0, \qquad j = 0, 1, 2, \tag{13.70}$$

where each integral is taken around each of the three slits. Find an integral expression for ϕ with the integrand expressed in terms of an appropriate prime function.

13.5. **(Circular slit map and a connection between the Poisson and Villat integral formulas)** This exercise demonstrates a connection between two different ways to approach a certain problem of Schwarz type for an analytic function $w(z)$. It makes use of a special instance of the unbounded circular slit mappings introduced in Part I to show a connection between the Poisson and Villat integral formulas. We will use the notation C_0^+ to denote the upper half unit circle, $|z| = 1, \text{Im}[z] > 0$, and C_0^- to denote the lower half unit circle, $|z| = 1, \text{Im}[z] < 0$, with a similar notation, C_1^+ and C_1^-, for the upper and lower halves of C_1. Suppose we seek a function $w(z)$ analytic in the upper half annulus $\rho < |z| < 1, \text{Im}[z] > 0$ satisfying

$$\text{Re}[w(z)] = \begin{cases} \phi_0(z), & z \in C_0^+, \\ \phi_1(z), & z \in C_1^+, \end{cases} \tag{13.71}$$

with

$$\text{Re}[w(z)] = 0 \qquad \text{on } \overline{z} = z. \tag{13.72}$$

(a) Make use of (13.72) to argue that

$$\mathrm{Re}[w(z)] = \begin{cases} \phi_0(z), & z \in C_0^+, \\ -\phi_0(\overline{z}), & z \in C_0^-, \\ \phi_1(z), & z \in C_1^+, \\ -\phi_1(\overline{z}), & z \in C_1^-. \end{cases} \tag{13.73}$$

(b) Use the result of part (a) to show that

$$\int_{C_0} \mathrm{Re}[w(z)]\frac{dz}{z} = \int_{C_1} \mathrm{Re}[w(z)]\frac{dz}{z} = 0. \tag{13.74}$$

Hence verify that the condition on the data for $w(z)$ to be single-valued in the annulus is satisfied.

(c) Hence use the Villat integral formula (13.46) to deduce that

$$w(z)= \frac{1}{2\pi \mathrm{i}} \int_{C_0^+} \phi_0(a)\,[2K(a,z)-1]\,\frac{da}{a} \tag{13.75}$$

$$-\frac{1}{2\pi \mathrm{i}} \int_{C_0^-} \phi_0(1/a)[2K(a,z)-1]\frac{da}{a}$$

$$-\frac{1}{2\pi \mathrm{i}} \int_{C_1^+} \phi_1(a)\,[2K(a,z)]\,\frac{da}{a} \tag{13.76}$$

$$+\frac{1}{2\pi \mathrm{i}} \int_{C_1^-} \phi_1(\rho^2/a)[2K(a,z)]\frac{da}{a} + \mathrm{i}c, \tag{13.77}$$

where c is some real constant.

(d) Show that the integral in part (c) can be rewritten as

$$w(z)= \frac{1}{2\pi \mathrm{i}} \int_{C_0^+} \phi_0(a)\,[2K(a,z)+2K(a,1/z)]\,\frac{da}{a} - \frac{1}{2\pi \mathrm{i}} \int_{C_0^+} 2\phi_0(a)\frac{da}{a}$$

$$-\frac{1}{2\pi \mathrm{i}} \int_{C_1^+} \phi_1(a)\,[2K(a,z)+2K(a,1/z)]\,\frac{da}{a} + \mathrm{i}c. \tag{13.78}$$

(e) Now introduce the conformal mapping from the upper half annulus in the z plane to the interior of the unit disc $|\eta| < 1$ in a complex η plane known, from (5.38) of Chapter 5, to be given by

$$\eta = f(z) = \frac{\omega(z,\alpha)\omega(z,1/\alpha)}{\omega(z,\overline{\alpha})\omega(z,1/\overline{\alpha})}, \tag{13.79}$$

where $\alpha = \mathrm{i}r$ for $0 < r < 1$. Suppose the images of the two semicircles C_0^+ and C_1^+ on the unit circle in the η plane are denoted by L_0 and L_1, respectively, and introduce the composed function

$$W(\eta) \equiv w(f^{-1}(\eta)). \tag{13.80}$$

Show that $W(\eta)$ is analytic in the unit disc $|\eta| < 1$ with boundary values

$$\mathrm{Re}[W(\eta)] = \begin{cases} \Phi_0(\eta), & \eta \in L_0, \\ \Phi_1(\eta), & \eta \in L_1, \\ 0, & |\eta| = 1, \eta \notin L_0, L_1, \end{cases} \tag{13.81}$$

where

$$\Phi_0(\eta) \equiv \phi_0(f^{-1}(\eta)), \qquad \Phi_1(\eta) \equiv \phi_1(f^{-1}(\eta)). \tag{13.82}$$

(f) Use the Poisson integral formula (13.40) to show that

$$W(\eta) = \frac{1}{2\pi \mathrm{i}} \int_{L_0} \frac{d\tilde{\eta}}{\tilde{\eta}} \frac{\tilde{\eta}+\eta}{\tilde{\eta}-\eta} \Phi_0(\tilde{\eta}) + \frac{1}{2\pi \mathrm{i}} \int_{L_1} \frac{d\tilde{\eta}}{\tilde{\eta}} \frac{\tilde{\eta}+\eta}{\tilde{\eta}-\eta} \Phi_1(\tilde{\eta}) + \mathrm{i}d, \quad (13.83)$$

where $d \in \mathbb{R}$ is a real constant.

(g) Introduce the changes of variable

$$\eta = \frac{\omega(z,\alpha)\omega(z,1/\alpha)}{\omega(z,\overline{\alpha})\omega(z,1/\overline{\alpha})}, \qquad \tilde{\eta} = \frac{\omega(a,\alpha)\omega(a,1/\alpha)}{\omega(a,\overline{\alpha})\omega(a,1/\overline{\alpha})}, \quad (13.84)$$

and use the identity (9.77) established in Exercise 9.6 to write (13.83) as

$$\begin{aligned} w(z) =& \frac{1}{2\pi \mathrm{i}} \int_{C_0^+} [2K(a,z) + 2K(a,1/z)]\, \phi_0(a) \frac{da}{a} \\ &- \frac{1}{2\pi \mathrm{i}} \int_{C_1^+} [2K(a,z) + 2K(a,1/z)]\, \phi_1(a) \frac{da}{a} + C + \mathrm{i}d, \end{aligned} \quad (13.85)$$

where the constant C is

$$\begin{aligned} C =& -\frac{1}{2\pi \mathrm{i}} \int_{C_0^+} [K(a,\alpha) + K(a,\overline{\alpha}) + K(a,1/\alpha) + K(a,1/\overline{\alpha})]\, \phi_0(a) \frac{da}{a} \\ &+ \frac{1}{2\pi \mathrm{i}} \int_{C_1^+} [K(a,\alpha) + K(a,\overline{\alpha}) + K(a,1/\alpha) + K(a,1/\overline{\alpha})]\, \phi_1(a) \frac{da}{a}. \end{aligned}$$

(h) Show that for a on C_0,

$$K(a,\alpha) + K(a,1/\overline{\alpha}) = K(a,\alpha) + 1 - K(1/a,\overline{\alpha}) = 1 + K(a,\alpha) - K(\overline{a},\overline{\alpha}). \quad (13.86)$$

Hence deduce that the real part of the left-hand side is unity.

(i) Show that for a on C_1,

$$\begin{aligned} K(a,\alpha) + K(a,1/\overline{\alpha}) &= K(a,\alpha) + 1 - K(1/a,\overline{\alpha}) \\ &= K(a,\alpha) + 1 - (K(\rho^2/a,\overline{\alpha}) + 1) \\ &= K(a,\alpha) - K(\overline{a},\overline{\alpha}). \end{aligned} \quad (13.87)$$

Hence deduce that the real part of the left-hand side is zero.

(j) Use parts (h) and (i) to show that

$$C = -\frac{1}{2\pi \mathrm{i}} \int_{C_0^+} 2\phi_0(a) \frac{da}{a} + \mathrm{i}e, \quad (13.88)$$

where e is a real constant.

(k) Substitute (13.88) into (13.85) to retrieve the result (13.78) of part (d) and obtained directly using the Villat formula.

13.6. (**Alternate point trapezoidal rule**) The integral representations derived in this chapter usually require numerical evaluation. The integrals are of Cauchy type with integrands that are singular on the boundaries. There is a handy numerical integration technique, a variant of the usual trapezoidal rule, that can be used to

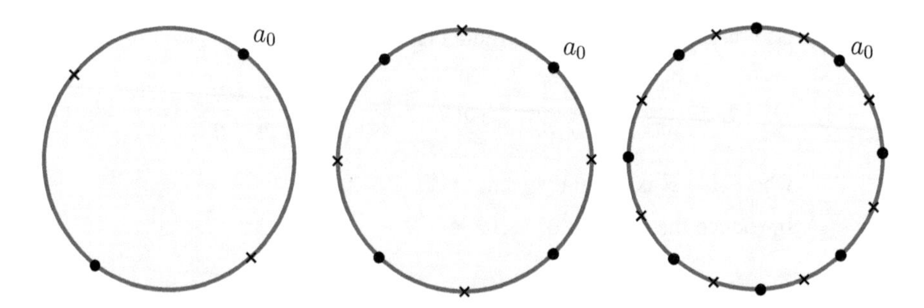

Figure 13.3. *Given a point a_0 on C_0 two sets of N equally spaced points on C_0 are identified. The cases $N = 2, 4,$ and 8 are shown. The set of N points indicated by the crosses is used in the "alternate-point trapezoidal rule" to evaluate the principal part integral (13.90) at a_0.*

evaluate such Cauchy-type integrals. This question guides the reader through this "alternate-point trapezoidal rule."

Remark: In Chapter 14, which focuses on numerical matters and where we will learn how to compute the prime function for $M \geq 1$, we will revisit this integration procedure to evaluate the integral representations found in this chapter for the higher connected cases $M \geq 1$. Here we focus on $M = 0$, where evaluation of the prime function is straightforward.

(a) Consider the problem of evaluating the Poisson integral formula (13.40) at some point a_0 on the unit circle C_0. The integrand is clearly singular for any $z = a_0$, where $a_0 \in C_0$. Use the Plemelj formulas [61] for Cauchy-type integrals to deduce that

$$w(a_0) = \Phi_0(a_0) + \mathcal{I}(a_0) + \mathrm{i}c, \tag{13.89}$$

where

$$\mathcal{I}(a_0) \equiv \mathcal{P}\left\{\frac{1}{2\pi\mathrm{i}}\int_{C_0} \Phi_0(a)\left(\frac{a+a_0}{a-a_0}\right)\frac{da}{a}\right\} \tag{13.90}$$

is a principal part integral.

(b) Write a numerical algorithm that picks an integer $N > 1$ and identifies $2N$ equally spaced points on C_0 that include the chosen point a_0. Figure 13.3 shows the cases $N = 2$ and $N = 4$. Split your set of $2N$ points into two sets of N points that "alternate," as indicated by the set of dots and crosses in Figure 13.3.

(c) Pick some test function that you know to be analytic in the unit disc D_z, and identify its real part on C_0. Evaluate the latter function in (13.90) numerically using the usual N-point trapezoidal rule based on the quadrature points corresponding to the crosses from part (b), i.e., the set of N points that does *not* include a_0, as shown schematically in Figure 13.3.

(d) The numerical procedure explained in part (c) is the alternate-point trapezoidal rule for the computation of the principal part integral (13.90). For your choice of test function in part (c), check that when you substitute the numerical result from part (c) into the right-hand side of (13.89) you retrieve your function on C_0 to within a purely imaginary constant.

Chapter 14

Computing the prime function

14.1 ▪ The value of special functions

During the course of Part I a wide variety of formulas for useful functions have been derived in terms of the prime function of a multiply connected circular domain. This was done after the theoretical work of establishing the existence of the prime function, and ascertaining its properties, had been done. The many and varied subsequent formulas then followed from these basic properties. This is the abiding value of special functions.

But how do we *evaluate* all these formulas?

It is a noteworthy feature of our special function framework that as soon as we have a fast, accurate, efficient, and robust way to compute any required prime function, we immediately have a fast, accurate, efficient, and robust way to compute all the many other functions expressible in terms of it. Our framework therefore embodies a significant economization of effort. It focuses the scientific computation and numerical analysis challenges in one place, namely, on how to compute the prime function effectively. The same is true of every special function.

14.2 ▪ How do we evaluate all these formulas?

We know the prime function $\omega(z, a)$ associated with a given multiply connected circular domain D_z to be a well-defined quantity. But what is its numerical value for given D_z, z, and a?

The simply connected case, when $M = 0$ and D_z is just the unit disc, offers no difficulty: the prime function is just $\omega(z, a) = z - a$. Any calculator or computer can evaluate this immediately.

For an $(M + 1)$-connected domain with $M \geq 1$ it turns out that there is a classical infinite-product representation of its prime function over the Schottky group Θ generated by the Möbius maps $\{\theta_j, \theta_j^{-1} | j = 1, \ldots, M\}$. This representation, which dates back to the 19th century [64], is reviewed in §14.3.

There is also an infinite-sum representation over the lattice $\mathbb{Z}^M$ deriving from a theoretical connection to the theory of theta functions. This is the topic of §14.4.

We report these two classical results for completeness. It is important to point out, however, that a freely available online resource is available, developed by the author and coworkers, where novel numerical schemes for evaluating the prime function have been devised. The idea behind these will be discussed in §14.6. Current versions of these numerical routines are available at

`github.com/ACCA-Imperial`[47]

The currently available codes rely on new representations of the prime function that avoid both the product over the Schottky group Θ and the sum over $\mathbb{Z}^M$ and rely instead on the theoretical connections of the prime function to the objects of potential theory that are evident from the theoretical development of earlier chapters [45, 40]. At the time of writing, this repository is being regularly updated, and readers are encouraged to contribute to the ongoing development of this software resource.

14.3 ▪ Infinite-product formula for the prime function

A classical infinite-product formula, over the Schottky group Θ, for evaluating the prime function dates back to the earliest work on the prime function [64]. It is

$$\omega(a,b) = (a-b) \prod_{\theta \in \Theta''} \frac{(\theta(a)-b)(\theta(b)-a)}{(\theta(a)-a)(\theta(b)-b)}, \tag{14.1}$$

where the set Θ'' denotes all elements of the Schottky group Θ, excluding the identity and all inverses. This means that if the term associated with $\theta \in \Theta$ is included in the product, then the corresponding term associated with θ^{-1} must be omitted.

We can sketch out a formal derivation of formula (14.1) that builds on the theory developed in Chapters 2–4. The differential $d\Pi^{z,w}(a)$ used in Chapter 4 to construct the prime function has the property that it is invariant under the action of the Schottky group Θ:

$$d\Pi^{z,w}(\theta(a)) = d\Pi^{z,w}(a), \qquad \theta \in \Theta. \tag{14.2}$$

It has a simple pole of residue $+1$ at z and a simple pole of residue -1 at w. Relation (14.2) provides the analytic continuation of the differential $d\Pi^{z,w}(a)$ into the images of the region F under the action of the Schottky group elements and, in particular, shows that it must have simple poles of residues ± 1 at $\theta(z)$ and $\theta(w)$ where $\theta \in \Theta$. It was also established in Chapter 4 that the differential $d\Pi^{z,w}(a)$ has the normalization

$$\int_{C_j} d\Pi^{z,w}(a) = 0, \qquad j = 1, \ldots, M. \tag{14.3}$$

With these facts in mind, and provided it converges, it is easy to check that a representation of such a differential is

$$d\Pi^{z,w}(a) = \frac{da}{a-z} - \frac{da}{a-w} + \sum_{\theta \in \Theta'} \left\{ \frac{d\theta(a)}{\theta(a)-z} - \frac{d\theta(a)}{\theta(a)-w} \right\}, \tag{14.4}$$

where Θ' denotes all elements of the Schottky group Θ, excluding the identity element; the term associated with the identity has clearly been singled out and written separately. The question of convergence is not a trivial one, and we will not discuss it here; it clearly depends on properties of the group Θ or, equivalently, the geometry of the circular domain

[47] If this link becomes inactive, readers should refer to the author's institutional website for information.

D_z. Assuming it has suitable convergence properties we can formally rewrite the sum (14.4) by explicitly separating out the inverse mappings:

$$d\Pi^{z,w}(a) = \frac{da}{a-z} - \frac{da}{a-w} + \sum_{\theta\in\Theta''} \left\{ \frac{d\theta(a)}{\theta(a)-z} - \frac{d\theta(a)}{\theta(a)-w} \right\} \tag{14.5}$$

$$+ \left\{ \frac{d\theta^{-1}(a)}{\theta^{-1}(a)-z} - \frac{d\theta^{-1}(a)}{\theta^{-1}(a)-w} \right\}, \tag{14.6}$$

where, as stated above, the set Θ'' excludes the identity and all inverses.

The cross-ratio $p(z, w, a, b)$ was introduced in §1.12. Consider the special choice of cross-ratio given by

$$p(z, w, \theta^{-1}(a), b) = \frac{(z-\theta^{-1}(a))(w-b)}{(z-b)(w-\theta^{-1}(a))}, \tag{14.7}$$

where b is some arbitrary fourth point and θ is some Möbius map. It was shown in Exercise 1.18 that a cross-ratio is invariant if any Möbius map acts on all four of its arguments. We can therefore deduce that

$$p(z, w, \theta^{-1}(a), b) = p(\theta(z), \theta(w), a, \theta(b)), \tag{14.8}$$

where

$$p(\theta(z), \theta(w), a, \theta(b)) = \frac{(\theta(z)-a)(\theta(w)-\theta(b))}{(\theta(z)-\theta(b))(\theta(w)-a)}. \tag{14.9}$$

Treating a' as the variable, a differential form of (14.8) is

$$\frac{d\theta^{-1}(a')}{\theta^{-1}(a')-z} - \frac{d\theta^{-1}(a')}{\theta^{-1}(a')-w} = \frac{da'}{a'-\theta(z)} - \frac{da'}{a'-\theta(w)}. \tag{14.10}$$

On use of this identity we can rewrite (14.6) as

$$d\Pi^{z,w}(a') = \frac{da'}{a'-z} - \frac{da'}{a'-w} + \sum_{\theta\in\Theta''} \left\{ \frac{d\theta(a')}{\theta(a')-z} - \frac{d\theta(a')}{\theta(a')-w} \right\} \tag{14.11}$$

$$+ \left\{ \frac{da'}{a'-\theta(z)} - \frac{da'}{a'-\theta(w)} \right\}. \tag{14.12}$$

On integration, up to an integer multiple of $2\pi\mathrm{i}$ associated with contour choices that we expect from the general considerations of Chapter 4, we find

$$\Pi^{z,w}_{a,b} = \int_b^a d\Pi^{z,w}(a') = \left[\log\left(\frac{a'-z}{a'-w}\right)\right]_b^a + \sum_{\theta\in\Theta''} \log\left[\frac{(\theta(a')-z)}{(\theta(a')-w)}\frac{(a'-\theta(z))}{(a'-\theta(w))}\right]_b^a. \tag{14.13}$$

Recall from §4.6 that the function $X(a,b)$ is defined by the double limit

$$\begin{aligned} X(a,b) &\equiv \lim_{\substack{z\to a\\ w\to b}} \left[-(z-a)(w-b)e^{-\Pi^{z,w}_{a,b}}\right] \\ &= \lim_{\substack{z\to a\\ w\to b}} \left[-(z-a)(w-b)\frac{(a-w)(b-z)}{(a-z)(b-w)} \prod_{\theta\in\Theta''} \frac{(\theta(a)-w)(a-\theta(w))}{(\theta(a)-z)(a-\theta(z))} \right. \\ &\qquad \left. \times \prod_{\theta\in\Theta''} \frac{(\theta(b)-z)(b-\theta(z))}{(\theta(b)-w)(b-\theta(w))}\right] \\ &= \left[(a-b)\prod_{\theta\in\Theta''} \frac{(\theta(a)-b)(a-\theta(b))}{(\theta(a)-a)(\theta(b)-b)}\right]^2 \qquad (14.14) \\ &= \omega(a,b)^2. \qquad (14.15) \end{aligned}$$

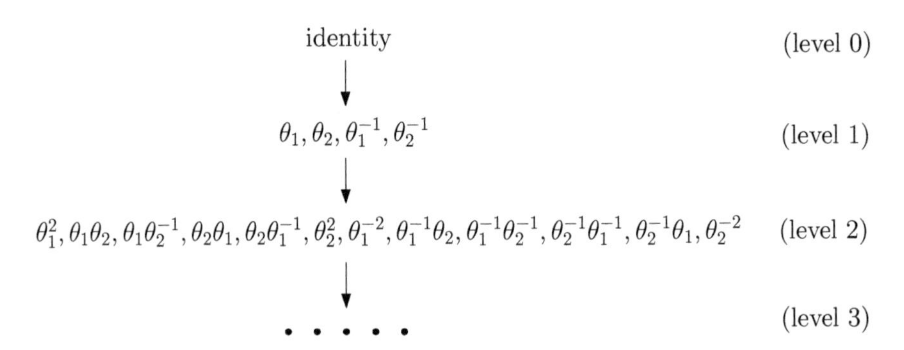

Figure 14.1. *Ordering by level of the maps in the Schottky group Θ for $M = 2$.*

It follows that

$$\omega(a,b) = X(a,b)^{1/2} = (a-b) \prod_{\theta \in \Theta''} \frac{(\theta(a)-b)(\theta(b)-a)}{(\theta(a)-a)(\theta(b)-b)} \tag{14.16}$$

which is precisely (14.1). It is an explicit infinite-product representation of $\omega(a,b)$ over "half" the Schottky group Θ, that is, over the set Θ''.

It is natural to ask how, for numerical evaluation of this infinite sum, one might truncate this infinite product. The most natural approach is to include all mappings in Θ'' up to a certain *level* of composition and then to omit all mappings of higher level.

To explain this notion of the *level of a map* we revert to the original Schottky group Θ. There is a natural ordering of its elements, all of which are Möbius maps, into levels starting with the identity map, which we call the level-zero map. The level-one maps are the $2M$ mappings in the set

$$\theta_1, \ldots, \theta_M, \theta_1^{-1}, \ldots, \theta_M^{-1}. \tag{14.17}$$

These are the basic generating maps of the Schottky group Θ. The level-two maps are all compositions of any two of the level-one maps (14.17) that do not reduce to the identity (the level-zero map). It is easy to work out that there are $2M(2M-1)$ level-two maps. The level-three maps are all compositions of any three of the level-one maps that do not reduce to a lower level map; there are $2M(2M-1)^2$ level-three maps, and so on. At level k there are $2M(2M-1)^{k-1}$ maps. Figure 14.1 shows, for $M = 2$, how this ordered "tree" of maps in Θ grows with the level.

From this rapidly growing tree we already see a disadvantage of formula (14.16). Even if this product is known to converge, which depends on the group Θ'' [62, 63], the number of mappings at each level grows exponentially with the level; recall that the number of maps at level k in the set Θ'' is $M(2M-1)^{k-1}$. The implication of this in practice is that to get even a single additional digit of accuracy in the evaluation of the prime function by adding in, say, a single additional level to a truncation of the product can necessitate a very large number of additional calculations. Some explicit numerical illustrations of this are given in [45]. This can be a punishing drawback if one requires both numerical efficiency and high accuracy.

A significant advantage of the formula (14.16), on the other hand, is that it gives the value of the prime function directly, and does not require a prior evaluation of the functions $\{v_j(z) | j = 1, \ldots, M\}$, which is a feature of alternative methods, as we will see.[48]

[48]Evaluation of the values of these first-kind integrals is equivalent, in the theory of compact Riemann surfaces, to evaluation of the so-called Abel map.

It turns out that there are analogous infinite-product representations of the functions $\{v_j(z)\}$ themselves. To see this, before delving into the general case, it is instructive to first study the case of the concentric annulus $\rho < |z| < 1$. We know from (3.35) that

$$\frac{\partial \mathcal{G}_0(\theta_1(z), a)}{\partial a} - \frac{\partial \mathcal{G}_0(z, a)}{\partial a} = v_1'(a), \tag{14.18}$$

where

$$\theta_1(z) = \rho^2 z, \qquad \theta_1^{-1}(z) = \rho^{-2} z, \tag{14.19}$$

and the group Θ is given in (8.22). We also know from (14.4) and (14.10) that we can write

$$d\Pi^{z,w}(a) = \frac{da}{a-z} - \frac{da}{a-w} + \sum_{\theta \in \Theta'} \left\{ \frac{da}{a-\theta(z)} - \frac{da}{a-\theta(w)} \right\}. \tag{14.20}$$

On use of the definition (4.2) of $d\Pi^{z,w}(a)$, we find

$$2\pi \mathrm{i} \left[\frac{\partial \mathcal{G}_0(\theta_1(z), a)}{\partial a} - \frac{\partial \mathcal{G}_0(z, a)}{\partial a} \right] da = d\Pi^{\theta_1(z), z}(a); \tag{14.21}$$

hence we can combine all these facts to deduce that

$$2\pi \mathrm{i} dv_1(a) = \frac{da}{a-\theta_1(z)} - \frac{da}{a-z} + \sum_{k \neq 0} \left\{ \frac{da}{a-\theta_1^k(\theta_1(z))} - \frac{da}{a-\theta_1^k(z)} \right\} \tag{14.22}$$

$$= \lim_{n \to \infty} \left[\frac{da}{a-\theta_1^{n+1}(z)} - \frac{da}{a-\theta_1^{-n}(z)} \right] \tag{14.23}$$

$$= \left[\frac{da}{a-A_1} - \frac{da}{a-B_1} \right], \tag{14.24}$$

where, to reach the second equality, most of the terms in a telescoping sum have canceled each other out and where A_1 and B_1 are the fixed points of the mapping $\theta_1(z)$ and its inverse, i.e.,

$$A_1 = \theta_1^{\infty}(z) = 0, \qquad B_1 = \theta_1^{-\infty}(z) = \infty. \tag{14.25}$$

It follows that

$$v_1(a) = \frac{1}{2\pi \mathrm{i}} \log a, \tag{14.26}$$

which coincides with the result (5.9) ascertained in Chapter 5 based on other considerations.

For $M \geq 2$ we can proceed similarly and, for each $j = 1, \ldots, M$, decompose the group Θ as follows:

$$\Theta = \Theta_j \times \{\theta_j^n | n \in \mathbb{Z}\}, \tag{14.27}$$

where Θ_j denotes the set of all maps in Θ without a positive or negative power of θ_j at the right-hand end (in a representation of each map as a composition of the $2M$ generating maps). Then

$$2\pi \mathrm{i} dv_j(a) = \sum_{\phi \in \Theta_j} \sum_{k=-\infty}^{\infty} \left[\frac{da}{a-\phi(\theta_j^{k+1}(z))} - \frac{da}{a-\phi(\theta_j^k(z))} \right] \tag{14.28}$$

$$= \sum_{\phi \in \Theta_j} \lim_{n \to \infty} \left[\frac{da}{a-\phi(\theta_1^{n+1}(z))} - \frac{da}{a-\phi(\theta_1^{-n}(z))} \right] \tag{14.29}$$

$$= \sum_{\phi \in \Theta_j} \frac{da}{a-\phi(A_j)} - \frac{da}{a-\phi(B_j)}, \tag{14.30}$$

where, in the second equality, we have canceled most of the terms in a telescoping sum and, in the third equality, we have introduced A_j and B_j as the fixed points of the mapping $\theta_j(z)$, i.e.,

$$A_j = \theta_j^{\infty}(z), \qquad B_j = \theta_j^{-\infty}(z). \tag{14.31}$$

It follows on integration[49] that

$$v_j(a) = \frac{1}{2\pi \mathrm{i}} \prod_{\theta \in \Theta_j} \log \left[\frac{a - \theta(A_j)}{a - \theta(B_j)} \right]. \tag{14.32}$$

The infinite-product formula (14.1) can be very convenient for solving problems in domains of low connectivity, especially given its explicit nature.

14.4 ▪ Prime function representation using theta functions

Another method of computing the prime function rests on a mathematical connection to the theory of theta functions [80, 64]. We will briefly discuss only the elements of theta function theory needed to indicate how the prime function of a typical circular domain D_z of connectivity $M+1$ can be evaluated as an infinite sum over the group $\mathbb{Z}^M$.

In Chapters 2–4 it was established that associated with any given circular domain D_z are the M functions $\{v_j(z) | j = 1, \ldots, M\}$. Given the two points z and a in D_z we can define the row vector

$$\begin{aligned} \mathbf{w} &= \left(\int_a^z dv_1(z'), \int_a^z dv_2(z'), \ldots, \int_a^z dv_M(z') \right) \\ &= (v_1(z) - v_1(a), \ldots, v_M(z) - v_M(a)). \end{aligned} \tag{14.33}$$

We also know there is an associated period matrix τ which has positive definite imaginary part. With these objects in hand it is possible to define the theta function, denoted here by θ—and not to be confused with any of the Möbius maps $\{\theta_j(z), \theta_j^{-1}(z) | j = 1, \ldots, M\}$ generating the Schottky group Θ—for the M-dimensional row vector $\mathbf{w}$, as

$$\theta(\mathbf{w}; \tau) = \sum_{\mathbf{m} \in \mathbb{Z}^M} \exp \left[\pi \mathrm{i} (\mathbf{m} + \delta) \tau (\mathbf{m} + \delta)^T + 2\pi \mathrm{i} (\mathbf{w} + \epsilon)(\mathbf{m} + \delta)^T \right], \tag{14.34}$$

where δ and ϵ can be taken to be the following special M-dimensional row vectors:[50]

$$\delta = \left(\frac{1}{2}, \underbrace{0, \ldots, 0}_{(M-1) \text{ copies}} \right), \qquad \epsilon = \underbrace{\left(\frac{1}{2}, \ldots, \frac{1}{2} \right)}_{M \text{ copies}}. \tag{14.35}$$

It is by virtue of the fact that τ has positive definite imaginary part that the sum (14.34) over the multidimensional lattice is convergent.

The prime function that has been a principal focus of this monograph can be written in terms of the theta function (14.34) as[51]

$$\omega(z, a) = \frac{\theta(\mathbf{w}; \tau)}{\sqrt{\sum_{k=1}^M b_k v_k'(z)} \sqrt{\sum_{k=1}^M b_k v_k'(a)}}, \tag{14.36}$$

[49] Similar formulas for $v_j(z)$ are featured in Baker's monograph [64].

[50] This is a special choice of odd characteristics. Other choices of odd characteristics can be made here.

[51] In this representation of the prime function there again arises the matter of which sign of the square roots to take, and which choice of characteristics.

where the set of M constants $\{b_k | k = 1, \ldots, M\}$ is given by

$$b_k = \left. \frac{\partial \theta}{\partial w_k} \right|_{\mathbf{w}=0} \tag{14.37}$$

and where w_k denotes the kth component of the row vector $\mathbf{w}$.

Remark: An important observation is that the identification (14.36) between the prime function and the theta function means that, in principle, all of the many mathematical formulas that have been written down in this monograph in terms of the prime function can now be written down in terms of the theta function defined in (14.34).

We now sketch out the key steps in establishing the identification (14.36).

Since we have made the choice (14.35) of odd characteristics, it turns out that $\theta(\mathbf{0}) = 0$ and the numerator in (14.36) is odd in $\mathbf{w}$ meaning that the right-hand side of (14.36) changes sign if z and a are swapped. Note that the denominator in (14.36) is invariant if z and a are swapped. This gives the required antisymmetry of the prime function $\omega(z, a) = -\omega(a, z)$.

By a multivariable Taylor expansion near $z = a$, we have

$$\theta(\mathbf{w}) = \theta(\mathbf{0}) + \left[\sum_{k=1}^{M} \left. \frac{\partial \theta}{\partial w_k} \right|_{\mathbf{w}=0} v_k'(a) \right] (z - a) + \cdots. \tag{14.38}$$

Therefore, near $z = a$,

$$\theta(\mathbf{w}) = \left[\sum_{k=1}^{M} \left. \frac{\partial \theta}{\partial w_k} \right|_{\mathbf{w}=0} v_k'(a) \right] (z - a) + \text{smaller terms}. \tag{14.39}$$

The denominator in (14.36), namely,

$$\sqrt{\sum_{k=1}^{M} b_k v_k'(z)} \sqrt{\sum_{k=1}^{M} b_k v_k'(a)}, \tag{14.40}$$

tends to

$$\sum_{k=1}^{M} \left. \frac{\partial \theta}{\partial w_k} \right|_{\mathbf{w}=0} v_k'(a) \tag{14.41}$$

as $z \to a$. Hence on use of (14.39) and (14.41) the function (14.36) tends to $(z - a)$ as $z \to a$. We know this to be a key property of the prime function.

The most challenging part is to confirm that the function on the right-hand side of (14.36) has the relevant transformation properties of the prime function under the action of the Schottky group, as established in Chapter 4.

If $z \mapsto \theta_j(z)$ for some $j = 1, \ldots, M$, then

$$\mathbf{w} \mapsto \mathbf{w} + \mathbf{e_j} \tau, \tag{14.42}$$

where $\mathbf{e_j}$ denotes the row vector having zeros in all elements except for 1 in the jth element. Hence each term of the form

$$\exp \left[\pi \mathrm{i} (\mathbf{m} + \delta) \tau (\mathbf{m} + \delta)^T + 2\pi \mathrm{i} (\mathbf{w} + \epsilon)(\mathbf{m} + \delta)^T \right] \tag{14.43}$$

becomes

$$\exp \left[\pi \mathrm{i} (\mathbf{m} + \delta) \tau (\mathbf{m} + \delta)^T + 2\pi \mathrm{i} (\mathbf{w} + \mathbf{e_j} \tau + \epsilon)(\mathbf{m} + \delta)^T \right]. \tag{14.44}$$

On use of properties of the exponential function, this can be written as

$$\exp\left[\pi\mathrm{i}(\mathbf{m}+2\mathbf{e_j}+\delta)\tau(\mathbf{m}+\delta)^T\right]\exp\left[2\pi\mathrm{i}(\mathbf{w}+\epsilon)(\mathbf{m}+\delta)^T\right]. \tag{14.45}$$

However, by the symmetry of the period matrix τ, we can write

$$\exp\left[\pi\mathrm{i}(\mathbf{m}+2\mathbf{e_j}+\delta)\tau(\mathbf{m}+\delta)^T\right] \tag{14.46}$$
$$= \exp\left[\pi\mathrm{i}(\mathbf{m}+\delta)\tau(\mathbf{m}+\delta)^T+\pi\mathrm{i}\mathbf{e_j}\tau(\mathbf{m}+\delta)^T+\pi\mathrm{i}(\mathbf{m}+\delta)\tau\mathbf{e_j}^T\right] \tag{14.47}$$
$$= \exp\left[\pi\mathrm{i}(\mathbf{m}+\delta+\mathbf{e_j})\tau(\mathbf{m}+\delta+\mathbf{e_j})^T-\mathrm{i}\pi\tau_{jj}\right], \tag{14.48}$$

where, in the second equality, we have in a sense "completed the square." Therefore (14.45) is

$$\begin{aligned}
&\exp\left[\pi\mathrm{i}(\mathbf{m}+\delta+\mathbf{e_j})\tau(\mathbf{m}+\delta+\mathbf{e_j})^T-\mathrm{i}\pi\tau_{jj}\right]\exp\left[2\pi\mathrm{i}(\mathbf{w}+\epsilon)(\mathbf{m}+\delta)^T\right]\\
&=\exp\left[\pi\mathrm{i}(\mathbf{m}+\delta+\mathbf{e_j})\tau(\mathbf{m}+\delta+\mathbf{e_j})^T-\mathrm{i}\pi\tau_{jj}\right]\times\\
&\exp\left[2\pi\mathrm{i}(\mathbf{w}+\epsilon)(\mathbf{m}+\delta+\mathbf{e_j})^T\right]\exp\left[-2\pi\mathrm{i}(\mathbf{w}+\epsilon)\mathbf{e_j}^T\right]\\
&=\exp\left[\pi\mathrm{i}(\mathbf{m}+\delta+\mathbf{e_j})\tau(\mathbf{m}+\delta+\mathbf{e_j})^T\right]\times\\
&\exp\left[2\pi\mathrm{i}(\mathbf{w}+\epsilon)(\mathbf{m}+\delta+\mathbf{e_j})^T\right]\exp\left[-2\pi\mathrm{i}(v_j(z)-v_j(a))-\mathrm{i}\pi\tau_{jj}\right]\exp\left[-2\pi\mathrm{i}\epsilon\mathbf{e_j}^T\right]
\end{aligned}$$

for $j=1,\ldots,M$. By our choice (14.35), we have

$$\exp\left[-2\pi\mathrm{i}\epsilon\mathbf{e_j}^T\right]=-1. \tag{14.49}$$

Since the sum over $\mathbf{m}\in\mathbb{Z}^M$ is the same as the sum over $\mathbf{m}'=\mathbf{m}+\mathbf{e_j}\in\mathbb{Z}^M$, it is clear that the sum associated with the theta function in the numerator of (14.36) acquires the multiplicative factor

$$-\exp\left[-2\pi\mathrm{i}(v_j(z)-v_j(a))-\mathrm{i}\pi\tau_{jj}\right] \tag{14.50}$$

when $z\mapsto\theta_j(z)$. Moreover, since

$$v_k(\theta_j(z))-v_k(z)=\tau_{jk}, \tag{14.51}$$

by the chain rule,

$$\theta_j'(z)v_k'(\theta_j(z))=v_k'(z), \tag{14.52}$$

implying that $\omega(z,a)$ as given in (14.36) acquires the overall multiplicative factor

$$-\frac{q_j}{1-\overline{\delta_j}z}\exp\left[-2\pi\mathrm{i}(v_j(z)-v_j(a))-\mathrm{i}\pi\tau_{jj}\right] \tag{14.53}$$

when $z\mapsto\theta_j(z)$, where here we have chosen

$$\sqrt{\theta_j'(z)}=-\frac{q_j}{1-\overline{\delta_j}z}. \tag{14.54}$$

From Chapter 4 we recognize this as precisely the multiplicative factor acquired by the prime function $\omega(z,a)$ when $z\mapsto\theta_j(z)$ for $j=1,\ldots,M$.

A final verification is to confirm that the right-hand side of (14.36) also satisfies the prime function identity (4.89). This is the topic of Exercise 14.5.

One disadvantage of using (14.36) to evaluate the prime function, at least compared to using the infinite-product representation (14.1), is that it is necessary to compute the values of the functions $\{v_j(z)|j=1,\ldots,M\}$ as a preliminary step. On the other hand, in many applications of the prime function, it is often convenient to have access to the values of the latter M functions, so this feature is not necessarily to be viewed as a disadvantage. Another drawback is that the theta function is defined over an M-dimensional lattice which must be truncated for the purposes of numerical evaluation. The multidimensionality of the indices of this sum can lead to the need to include a large number of terms for moderately large M if high accuracy is demanded.

This section has only touched the surface of exploring the connection between the theory of the prime function for multiply connected domains and the theory of theta functions. To date, this connection has really not been explored in any detail, and much more remains to be done in terms of leveraging ideas from one area for potential use in the other.

14.5 ▪ Case study: The concentric annulus

The case of the concentric annulus

$$\rho < |z| < 1 \tag{14.55}$$

is of special interest because, as we know from Chapter 5, associated with it is the simplest nontrivial example of a prime function after the elementary $\omega(z,a) = z - a$. Moreover, it turns out that the two representations of it as described in the two previous sections can be written down in a compact form, thereby providing an instructive illustration of the basic ideas.

The relevant Schottky group Θ in this case is generated by

$$\theta_1(z) = \rho^2 z, \qquad \theta_1^{-1}(z) = \rho^{-2} z. \tag{14.56}$$

On substitution of these into (14.16) we find

$$\omega(a,b) = (a-b) \prod_{n \geq 1} \frac{(\rho^{2n} a - b)(\rho^{2n} b - a)}{(\rho^{2n} a - a)(\rho^{2n} b - b)}. \tag{14.57}$$

On factoring out ab in the numerator and denominator of each term of the infinite product this can be written as

$$(a-b) \prod_{n \geq 1} \frac{(\rho^{2n}(a/b) - 1)(\rho^{2n}(b/a) - 1)}{(1-\rho^{2n})^2} \tag{14.58}$$

or as

$$\omega(a,b) = -\frac{bP(a/b,\rho)}{C^2}, \tag{14.59}$$

where it is convenient to introduce the function $P(z,\rho)$ defined by

$$P(z,\rho) \equiv (1-z) \prod_{n=1}^{\infty} (1-\rho^{2n} z)(1-\rho^{2n}/z), \qquad C = \prod_{n=1}^{\infty} (1-\rho^{2n}). \tag{14.60}$$

The explicit formulas (14.59) and (14.60) can be used to compute the prime function for the annulus. In this case it is a simple matter to truncate the infinite products in (14.60) after a suitable number of terms.

Remark: We note from (14.59) that, in this case, the prime function $\omega(z,a)$ is purely a function of the ratio z/a (and, of course, ρ). This result is consistent with Exercise 5.3, where this feature was established using independent arguments.

On the other hand, let us consider the approach via theta functions. The concentric annulus corresponds to the choices

$$\delta_1 = 0, \qquad q_1 = \rho, \qquad M = 1, \tag{14.61}$$

and we know from Chapter 4 that

$$\sigma_1(z) = \frac{\log |z|}{\log \rho}, \qquad \hat{v}_1(z) = \mathrm{i}\frac{\log z}{\log \rho}, \qquad v_1(z) = \frac{1}{2\pi \mathrm{i}} \log z, \tag{14.62}$$

and

$$\mathcal{P}_{11} = \int_{C_1} d\hat{v}_1 = \int_{|z|=\rho} \frac{\mathrm{i}}{\log \rho} \frac{dz}{z} = -\frac{2\pi}{\log \rho}, \tag{14.63}$$

so that

$$\tau_{11} = 2\mathrm{i}\mathcal{P}_{11}^{-1} = -\frac{\mathrm{i} \log \rho}{\pi}. \tag{14.64}$$

From (14.33) with $M = 1$ we identify

$$w = \frac{1}{2\pi \mathrm{i}} \log \left(\frac{z}{a} \right), \tag{14.65}$$

and from (14.34) we therefore have

$$\theta(z/a, \tau_{11}) = \sum_{m=-\infty}^{\infty} e^{\pi \mathrm{i}(m+1/2)(-\mathrm{i} \log \rho/\pi)(m+1/2)} e^{(\log(z/a)+\pi \mathrm{i})(m+1/2)} \tag{14.66}$$

$$= \sum_{m=-\infty}^{\infty} \rho^{(m+1/2)^2} (z/a)^{m+1/2} (-1)^{m+1/2}. \tag{14.67}$$

On taking a derivative with respect to w,

$$b_1 = \left. \frac{\partial \theta}{\partial w} \right|_{w=0} = 2\pi \mathrm{i} \sum_{m=-\infty}^{\infty} \left(m + \frac{1}{2} \right) \rho^{(m+1/2)^2} (-1)^{m+1/2}. \tag{14.68}$$

Now

$$v_1'(z) = \frac{1}{2\pi \mathrm{i} z}, \qquad v_1'(a) = \frac{1}{2\pi \mathrm{i} a}, \tag{14.69}$$

so the denominator in (14.36) is

$$\pm \frac{1}{\sqrt{za}} \sum_{m=-\infty}^{\infty} \left(m + \frac{1}{2} \right) \rho^{(m+1/2)^2} (-1)^{m+1/2}. \tag{14.70}$$

On division of (14.67) by (14.70) we arrive at

$$\pm\sqrt{za} \sum_{m=-\infty}^{\infty} \rho^{(m+1/2)^2} (z/a)^{m+1/2} (-1)^{m+1/2} \Bigg/ \sum_{m=-\infty}^{\infty} \left(m + \frac{1}{2} \right) \rho^{(m+1/2)^2} (-1)^{m+1/2}. \tag{14.71}$$

With the trivial observation that $\sqrt{za} = a\sqrt{z/a}$ we can write this as

$$\pm a \sum_{m=-\infty}^{\infty} \rho^{(m+1/2)^2} (z/a)^{m+1} (-1)^{m+1/2} \Bigg/ \sum_{m=-\infty}^{\infty} \left(m + \frac{1}{2} \right) \rho^{(m+1/2)^2} (-1)^{m+1/2} \tag{14.72}$$

or, on setting $n = m + 1$ in the sums,

$$\pm a \sum_{n=-\infty}^{\infty} \rho^{(n-1/2)^2}(z/a)^n(-1)^{n-1/2} \Big/ \sum_{n=-\infty}^{\infty} \left(n - \frac{1}{2}\right) \rho^{(n-1/2)^2}(-1)^{n-1/2}. \quad (14.73)$$

On canceling some common factors, we arrive at

$$\pm a \sum_{n=-\infty}^{\infty} \rho^{n(n-1)}(z/a)^n(-1)^n \Big/ \sum_{n=-\infty}^{\infty} n\rho^{n(n-1)}(-1)^n, \quad (14.74)$$

where we have used the readily established fact that

$$\sum_{n=-\infty}^{\infty} \rho^{n(n-1)}(-1)^n = 0. \quad (14.75)$$

We must choose the $+$ sign if we wish to ensure that $\omega(z,a) \sim (z-a)$ as $z \to a$; this can easily be checked by ensuring that the derivative $\partial\omega(z,a)/\partial z \to 1$ as $z \to a$. In summary, the prime function for the concentric annulus has the infinite-sum representation

$$\omega(z,a) = a \sum_{n=-\infty}^{\infty} \rho^{n(n-1)}(z/a)^n(-1)^n \Big/ \sum_{n=-\infty}^{\infty} n\rho^{n(n-1)}(-1)^n. \quad (14.76)$$

Finally, we can equate the infinite-sum representation (14.76) with the infinite-product representation (14.59) to produce the identity

$$a \sum_{n=-\infty}^{\infty} \rho^{n(n-1)}(z/a)^n(-1)^n \Big/ \sum_{n=-\infty}^{\infty} n\rho^{n(n-1)}(-1)^n = -\frac{aP(z/a,\rho)}{C^2}. \quad (14.77)$$

14.6 ▪ New methods for computing the prime function

The preface of this book made the case that the two functions

$$\omega(z,a) = z - a \qquad \text{and} \qquad \mathcal{G}_0(z,a) = \frac{1}{2\pi\mathrm{i}} \log\left(\frac{z-a}{|a|(z-1/\overline{a})}\right) \quad (14.78)$$

are intimately linked. In the *theoretical* development, the theory of Green's functions was the starting point for arriving at the prime functions. Why not, therefore, make it the starting point in any *numerical* approach to evaluating the prime functions? After all, scientists have by now invented a vast panoply of techniques for fast, accurate, effective, and robust evaluation of Green's functions and the other objects of potential theory. Why not redeploy them in the service of computing the prime function?

This matter of the numerical computation of the prime function has been reappraised in recent years by the author and his coworkers (e.g., [45, 40]). Several novel approaches to computing the prime function in multiply connected circular domains have been put forward. These endeavors have been driven by all the many uses of the prime function in solving problems in multiply connected domains, as evinced by all the formulas we have derived in terms of it in this monograph. It is the prime function, rather than the Green's function, which is the more natural object to use to express all these results.

Crowdy and Marshall [45] presented what appears to be the first numerical scheme for computing the prime function associated with a multiply connected circular domain that is not based on the classical infinite-product formula of §14.3 or the infinite-sum formula deriving from the connection with the theory of theta functions in §14.4. Instead they proposed to represent the prime function, for z and a in the doubled domain F, in the form

$$\omega(z,a) = (z-a)\tilde{\omega}(z,a), \tag{14.79}$$

where $\tilde{\omega}(z,a)$ is represented by a Fourier–Laurent expansion of the form

$$\tilde{\omega}(z,a) = A_0^{(0)} + \sum_{k=1}^{M}\sum_{n=1}^{\infty} \frac{A_n^{(k)} q_k^n}{(z-\delta_k)^n} + \sum_{k=1}^{M}\sum_{n=1}^{\infty} \frac{B_n^{(k)} q_k'^n}{(z-\delta_k')^n}, \tag{14.80}$$

where, as usual, δ_k and q_k denote the center and radius of circle C_k, respectively, and δ_k' and q_k' denote the center and radius of circle C_k', the reflection of C_k in C_0. The required simple zero of $\omega(z,a)$ at $z=a$ has been built into the representation (14.79). The form (14.80) is capable of representing a general single-valued function in a multiply connected circular domain D_z. The coefficients $A_0^{(0)}$ and $\{A_n^{(k)}, B_n^{(k)} | k=1,\ldots,M; n=1,2,\ldots\}$ of $\tilde{\omega}(z,a)$ can be determined by enforcing that the prime function satisfies the appropriate functional identities as derived in Chapter 4. In order to do this it is also necessary to determine the functions $\{v_j(z) | j=1,\ldots,M\}$. This can be done in an analogous fashion using the representation

$$v_j(z) = \frac{1}{2\pi\mathrm{i}} \log\left[\frac{z-\delta_j}{z-\delta_j'}\right] + a_0^{(j)} + \sum_{k=1}^{M}\sum_{n=1}^{\infty} \frac{a_n^{(j,k)} q_k^n}{(z-\delta_k)^n} + \sum_{k=1}^{M}\sum_{n=1}^{\infty} \frac{b_n^{(j,k)} q_k'^n}{(z-\delta_k')^n} \tag{14.81}$$

for each $j=1,\ldots,M$. The logarithmic term on the right-hand side of (14.81) enforces the conditions (2.38) and (3.13). The coefficients $a_0^{(j)}$ and $\{a_n^{(j,k)}, b_n^{(j,k)} | j,k=1,\ldots,M; n=1,2,\ldots\}$ are then found by ensuring that the imaginary parts of each $v_j(z)$ satisfy conditions (2.32). The matrix elements $\left[\mathcal{P}^{-1}\right]_{jk}$, and hence τ_{jk}, are found as part of this construction.

The scheme was found to be highly effective, especially when its numerical performance is compared to use of the classical formulas of §14.3 and §14.4. See tables 1 and 2 of [45]. It was found, for example, in computing the prime function for a triply connected domain, that adding just a handful more coefficients in the Laurent-type expansions above can be as effective (at improving accuracy) as adding several thousand more terms in a product representation over the corresponding Schottky group.

What we discern from this evidence is that the classical representations of the prime function, in the form of infinite sums or infinite products, as presented earlier in this chapter, do not necessarily constitute a competitive basis on which to develop numerical algorithms. While this observation is significant, it is no different from noticing that the infinite-product representation of $\sin z$ is not necessarily the best way to evaluate that function, even if it might have theoretical value.

Further developments in devising novel schemes for computation of the prime function have been surveyed in the review article [40]. This includes improvements of the scheme described above, as well as use of state-of-the-art boundary integral methods, such as those based on use of the so-called generalized Neumann kernel, to evaluate the prime function. Moreover, when these boundary integral techniques are coupled with fast-multipole methods, not to mention all the new quadrature rules that have been devised for integrals of

Cauchy type, they can make computation of the prime function for any given circular domain a wholly routine procedure—perhaps one that might be built into future special function packages.

The open-source software repository mentioned in §14.2 is intended to serve as a communal resource. The algorithms there are based on those described in [40]. Readers are encouraged to use this software in their work or research and to contribute to its ongoing development and improvement.[52]

In view of all the varied ways in which the prime function can be used to build so many other functions useful in applications, the challenge of further improving the efficacy of numerical schemes for evaluation of the prime function is an important one.

Exercises

14.1. **(Product of cross-ratios)** Excluding the term $(a - b)$, show that each individual term in the infinite-product representation (14.1) of the prime function $\omega(a, b)$ has the form of a cross-ratio.

14.2. **(Infinite-sum representation of $d\Pi^{z,w}(a)$)** Make use of the formula

$$\mathcal{G}_0(z, a) = \frac{1}{2\pi \mathrm{i}} \log\left(\frac{\omega(z, a)}{|a|\omega(z, 1/\overline{a})}\right) \tag{14.82}$$

together with the infinite-product representation (14.16) of the prime function to verify that

$$d\Pi^{z,w}(a) \tag{14.83}$$

as given in (14.12) corresponds to the quantity

$$2\pi \mathrm{i}\left[\frac{\partial \mathcal{G}_0(z, a)}{\partial a} - \frac{\partial \mathcal{G}_0(w, a)}{\partial a}\right] da \tag{14.84}$$

as defined in Chapter 4.

14.3. **(On the prime function for the concentric annulus)** In Chapters 2–4 we derived the prime function for *any* multiply connected circular domain from the first-type Green's function. In this way, we captured all the prime functions we needed in this book along with all the identities they satisfy. For the concentric annulus

$$\rho < |z| < 1 \tag{14.85}$$

this prime function turns out to have the representation given in (14.59) and (14.60). There is a different way to proceed, which can be helpful when first getting used to the idea of a prime function. For the concentric annulus we can contemplate *defining* its prime function $\omega(z, a)$ using the infinite product (14.59) and (14.60), that is, without using the more general approach of Chapters 2–4. This exercise explores this idea.

(a) Use standard results for proving the convergence of an infinite product to show that the defining product in (14.59) and (14.60) converges for any $\zeta \neq 0$ and for $0 \leq \rho < 1$.

[52] Users of these resources are kindly requested to cite [40], which describes the potential theoretic algorithm on which the codes are based.

(b) Suppose a lies in the annulus $\rho < |a| < 1/\rho$. Show that $z = a$ is the only simple zero of the prime function in this annulus. Verify also that the prime function is analytic and single valued as a function of both z and a everywhere in this annulus.

(c) Starting from the infinite-product definition (14.59) and (14.60) establish all the prime function identities (5.4), (5.5), (5.6), and (5.7) for the prime function of this concentric annulus.

14.4. **(Alternative derivation of sum representation when $M = 1$)** The prime function for the concentric annulus $\rho < |z| < 1$ has the infinite-sum representation (14.76); this was obtained as a special case, for $M = 1$, of the general theta function representation (14.34). There is an alternative, more direct derivation of this formula in this case that we now explore. Since $\omega(z, a)$ is analytic in the annulus $\rho < |z| < 1$, it has a convergent Laurent series there:

$$\omega(z,a) = \sum_{n=-\infty}^{\infty} b_n(a,\rho) z^n. \tag{14.86}$$

(a) Substitute (14.86) into the functional relation (5.6) and derive the recurrence relation

$$b_{n+1}(a,\rho) = -\frac{\rho^{2n}}{a} b_n(a,\rho) \qquad \forall n. \tag{14.87}$$

(b) Hence, deduce that

$$b_n(a) = c(a,\rho)\frac{(-1)^n \rho^{n^2-n}}{a^n}, \tag{14.88}$$

where $c(a, \rho)$ is independent of n.

(c) Use the condition that $\omega(z, a)/(z - a) \to 1$ as $z \to a$ to deduce that

$$c(a,\rho) = a\left[\sum_{n=-\infty}^{\infty} n\rho^{n(n-1)}(-1)^n\right]^{-1}. \tag{14.89}$$

(d) Substitute the result of part (c) into (14.86) to derive the result

$$\omega(z,a) = c(a,\rho)\sum_{n=-\infty}^{\infty} (-1)^n \rho^{n^2-n}(z/a)^n. \tag{14.90}$$

This is precisely (14.76).

14.5. **(Confirming the identity (4.89))** Confirm that the right-hand side of (14.36) relating the prime function to the theta function satisfies the identity (4.89) known to be satisfied by the prime function of a multiply connected circular domain D_z.

14.6. **(Connection with first Jacobi theta function)** The first Jacobi theta function $\Theta_1(w, q)$ is defined by the infinite sum

$$\Theta_1(w,q) = \sum_{n=-\infty}^{\infty} (-1)^{n-1/2} q^{(n+1/2)^2} e^{(2n+1)\mathrm{i}w}. \tag{14.91}$$

Show from (14.90) that

$$\omega(z,a) = -\frac{\mathrm{i}c(a,\rho)}{\rho^{1/4}}\left(\frac{z}{a}\right)^{1/2}\Theta_1\left(-\frac{\mathrm{i}}{2}\log(z/a),\rho\right). \tag{14.92}$$

From this we see that the prime function for the annulus is related to the first Jacobi theta function.

14.7. **(Connection with Weierstrass zeta function)** This exercise connects the logarithmic derivatives of the prime function of the annulus to the theory of Weierstrass.

(a) Use the results of Exercise 14.6 to show that

$$\mathcal{K}(z,a) \equiv a\frac{\partial}{\partial a}\log\omega(z,a) = \frac{1}{2} + \frac{\mathrm{i}}{2}\frac{\Theta_1'\left(-\frac{\mathrm{i}}{2}\log(z/a),\rho\right)}{\Theta_1\left(-\frac{\mathrm{i}}{2}\log(z/a),\rho\right)}, \quad (14.93)$$

$$K(z,a) \equiv z\frac{\partial}{\partial z}\log\omega(z,a) = \frac{1}{2} - \frac{\mathrm{i}}{2}\frac{\Theta_1'\left(-\frac{\mathrm{i}}{2}\log(z/a),\rho\right)}{\Theta_1\left(-\frac{\mathrm{i}}{2}\log(z/a),\rho\right)}, \quad (14.94)$$

where $\Theta_1'(w,q)$ denotes the derivative with respect to the first argument of the function and where $\mathcal{K}(z,a)$ and $K(z,a)$ were introduced in §5.12.

(b) Use the result of part (a) to confirm the identity (5.82), which was also the subject of Exercise 5.2.

(c) Now introduce the new independent variables

$$Z = -\mathrm{i}\log z, \qquad A = -\mathrm{i}\log a. \quad (14.95)$$

Check that as $z \mapsto ze^{2\pi\mathrm{i}}$ and $z \mapsto \rho^2 z$

$$Z \mapsto Z + 2\pi, \qquad Z \mapsto Z - 2\mathrm{i}\log\rho. \quad (14.96)$$

(d) Let $\mathcal{Z}(z)$ denote the Weierstrass zeta function with half periods π and $-i\log\rho$. It is known to satisfy the identity

$$\mathcal{Z}(Z) = \frac{\mathcal{Z}(\pi)Z}{\pi} + \frac{1}{2}\frac{\Theta_1'(Z/2,\rho)}{\Theta_1(Z/2,\rho)} \quad (14.97)$$

(which, given the definition (14.91), can be used to *define* the Weierstrass zeta function). Use this to verify that

$$\mathcal{K}(z,a) = \frac{1}{2} + \mathrm{i}\left(\mathcal{Z}(Z-A) - \frac{\mathcal{Z}(\pi)}{\pi}(Z-A)\right), \quad (14.98)$$

$$K(z,a) = \frac{1}{2} - \mathrm{i}\left(\mathcal{Z}(Z-A) - \frac{\mathcal{Z}(\pi)}{\pi}(Z-A)\right). \quad (14.99)$$

(e) Use the known relationship between periods of the Weierstrass zeta function

$$(-\mathrm{i}\log\rho)\mathcal{Z}(\pi) - \pi\mathcal{Z}(-\mathrm{i}\log\rho) = \frac{\mathrm{i}\pi}{2} \quad (14.100)$$

to verify the relations (5.76) and (5.81) satisfied by $\mathcal{K}(z,a)$ and $K(z,a)$ as $z \mapsto \rho^2 z$.

14.8. **(Connection with Weierstrass $\wp$ function)** The Weierstrass $\wp$ function is defined as

$$\wp(z) = -\frac{d}{dz}\mathcal{Z}(z), \quad (14.101)$$

where $\mathcal{Z}(z)$ is the Weierstrass zeta function used in Exercise 14.7. Use the results of Exercise 14.7 to establish a relation between $L(z,a)$ defined by

$$L(z,a) \equiv z\frac{\partial K(z,a)}{\partial z} \quad (14.102)$$

and the Weierstrass $\wp$ function.

14.9. **(Infinite-sum representations of $K(z,a)$ and $L(z,a)$ for the concentric annulus)** This exercise explores ways to evaluate certain functions associated with the concentric annulus

$$\rho < |z| < 1. \tag{14.103}$$

Use the infinite-product representation (14.59)–(14.60) of the prime function of this annulus to establish the following infinite-sum representations of $K(z,a)$ introduced in §5.12 and $L(z,a)$ introduced in Exercise 8.2.

(a) Show that

$$K(z,a) = -\frac{z/a}{1-z/a} - \sum_{n=1}^{\infty} \frac{\rho^{2n}(z/a)}{1-\rho^{2n}(z/a)} + \sum_{n=1}^{\infty} \frac{\rho^{2n}(a/z)}{1-\rho^{2n}(a/z)}. \tag{14.104}$$

(b) Show that

$$L(z,a) = -\frac{z/a}{(1-z/a)^2} - \sum_{n=1}^{\infty} \frac{\rho^{2n}(z/a)}{(1-\rho^{2n}(z/a))^2} - \sum_{n=1}^{\infty} \frac{\rho^{2n}(a/z)}{(1-\rho^{2n}(a/z))^2}. \tag{14.105}$$

(c) Verify that the results of (a) and (b) are consistent with the statements in Exercise 8.2(a).

14.10. **(Connection with the method of images)** The *method of images* is a construction based on the idea that if a singularity is situated in a given domain with either a straight-line or a circular arc boundary, then simple boundary conditions, such as the imaginary part of some analytic function being constant on the boundary, can be satisfied by the strategic placement of an image singularity at the reflection of that singularity in the boundary. In certain simply connected cases, this method provides an intuitive way to solve simple boundary value problems. In the multiply connected setting, where multiple boundaries are involved, it leads to a proliferation of image singularities. The locations of these image singularities turn out to be related to the image of the original singularity, and its reflection in C_0, under the action of (the nonidentity elements of) the Schottky group Θ. To see this, this exercise guides the reader through a construction, for a doubly connected annulus, of the modified Green's function $\mathcal{G}_0(z,a)$ based on these ideas. Consider a doubly connected annular domain D_z with circular boundaries C_0 and C_1 which are not necessarily concentric. Let δ_1 be the center of C_1 and let q_1 be its radius. First, recall that reflection in C_0 is given by

$$a \mapsto 1/\overline{a} \tag{14.106}$$

and reflection in C_1 is given by

$$a \mapsto \overline{S_1(a)}, \qquad S_1(z) \equiv \overline{\delta_1} + \frac{q_1^2}{z-\delta_1}, \tag{14.107}$$

where $S_1(z)$ is the Schwarz function of circle C_1. We also recall that

$$\theta_1(z) \equiv \overline{S_1}(1/z), \tag{14.108}$$

which means that the reflection of a in C_1 can be written as

$$a \mapsto \theta_1(1/\overline{a}). \tag{14.109}$$

Also, from (1.100) in Exercise 1.5, we have that

$$\theta_j^{-1}(z) = \frac{1}{\overline{\theta_j}(1/z)}. \tag{14.110}$$

This relation will also be useful in what follows. Suppose we want to find a function $w(z)$ with the following singularity near $z = a$:

$$w(z) = -\frac{\mathrm{i}}{2\pi}\log(z-a) + \text{a locally analytic function}, \tag{14.111}$$

satisfying the boundary condition

$$\mathrm{Im}[w(z)] = \text{constant on } C_0, C_1. \tag{14.112}$$

The reader will have in mind the various Green's functions associated with D_z since they satisfy conditions of precisely this kind. To construct $w(z)$ we can proceed in an iterative way and construct a series of functions

$$w_0(z),\ w_1(z),\ w_2(z),\ \cdots \tag{14.113}$$

using the method of images. To start the iteration, define

$$w_0(z) = -\frac{\mathrm{i}}{2\pi}\log(z-a). \tag{14.114}$$

The rest of this exercise will guide the reader through the construction.

(a) **Fix on C_0:** The function $w_0(z)$ has the required singularity at $z = a$, but it does not satisfy the boundary conditions. We can fix this on C_0 at least by placing a singularity of opposite sign at the reflection of a in C_0, i.e., at $1/\overline{a}$. Verify that the next iterate,

$$w_1(z) = -\frac{\mathrm{i}}{2\pi}\log(z-a) + \frac{\mathrm{i}}{2\pi}\log(z-1/\overline{a}), \tag{14.115}$$

satisfies the boundary condition (14.112) on C_0.

(b) **Fix on C_1:** The function $w_1(z)$ has the required singularity at $z = a$ and satisfies the boundary condition on C_0, but not on C_1. We can fix this by adding in *two* additional singularities, again with opposite signs, at the reflection in C_1 of each of the singularities of $w_1(z)$. Verify that the next iterate,

$$\begin{aligned} w_2(z) = &-\frac{\mathrm{i}}{2\pi}\log(z-a) + \frac{\mathrm{i}}{2\pi}\log(z-1/\overline{a}) \\ &+\frac{\mathrm{i}}{2\pi}\log(z-\theta_1(1/\overline{a})) - \frac{\mathrm{i}}{2\pi}\log(z-\theta_1(a)), \end{aligned} \tag{14.116}$$

satisfies the boundary condition (14.112) on C_1.

(c) **Fix on C_0:** Since we last fixed the boundary condition on C_0 we have added two new singularities at

$$\theta_1(1/\overline{a}),\ \theta_1(a). \tag{14.117}$$

This has destroyed the property that the condition on C_0 is satisfied. We must therefore add in opposite-signed vortices at the reflections of the points

(14.117) in C_0, i.e., at

$$\frac{1}{\overline{\theta_1(1/\overline{a})}}, \frac{1}{\overline{\theta_1(a)}}. \tag{14.118}$$

Use (14.110) to verify that these positions are given equivalently by

$$\theta_1^{-1}(a), \ \theta_1^{-1}(1/\overline{a}). \tag{14.119}$$

Hence show that the next iterate satisfying the boundary condition on C_0 is

$$\begin{aligned} w_4(z) = &-\frac{\mathrm{i}}{2\pi}\log(z-a) + \frac{\mathrm{i}}{2\pi}\log(z-1/\overline{a}) \\ &+\frac{\mathrm{i}}{2\pi}\log(z-\theta_1(1/\overline{a})) - \frac{\mathrm{i}}{2\pi}\log(z-\theta_1(a)) \\ &-\frac{\mathrm{i}}{2\pi}\log(z-\theta_1^{-1}(a)) + \frac{\mathrm{i}}{2\pi}\log(z-\theta_1^{-1}(1/\overline{a})). \end{aligned} \tag{14.120}$$

(d) **Fix on C_1:** A comparison of (14.116) and (14.120) reveals that two additional image singularities at positions

$$\theta_1^{-1}(a), \ \theta_1^{-1}(1/\overline{a}) \tag{14.121}$$

have been added since we last fixed the boundary condition on C_1. Hence, two additional image point vortices (of opposite sign) must be added to the reflections of these points in C_1, i.e., at

$$\theta_1(1/\overline{\theta_1^{-1}(a)}), \ \theta_1(1/\overline{\theta_1^{-1}(1/\overline{a})}). \tag{14.122}$$

Verify, on use of (14.110), that these points can be equivalently expressed as

$$\theta_1^2(1/\overline{a}), \ \theta_1^2(a). \tag{14.123}$$

Hence show that the next iterate satisfying the boundary condition on C_1 is

$$\begin{aligned} w_5(z) = &-\frac{\mathrm{i}}{2\pi}\log(z-a) + \frac{\mathrm{i}}{2\pi}\log(z-1/\overline{a}) \\ &+\frac{\mathrm{i}}{2\pi}\log(z-\theta_1(1/\overline{a})) - \frac{\mathrm{i}}{2\pi}\log(z-\theta_1(a)) \\ &-\frac{\mathrm{i}}{2\pi}\log(z-\theta_1^{-1}(a)) + \frac{\mathrm{i}}{2\pi}\log(z-\theta_1^{-1}(1/\overline{a})) \\ &+\frac{\mathrm{i}}{2\pi}\log(z-\theta_1^2(1/\overline{a})) - \frac{\mathrm{i}}{2\pi}\log(z-\theta_1^2(a)). \end{aligned} \tag{14.124}$$

(e) **Fix on C_0:** There are now two new image point vortices since we last fixed the boundary condition on C_0: we must therefore add in two additional image point vortices, of opposite sign, at the reflections of these points in C_0. These reflections are given by

$$1/\overline{\theta_1^2(1/\overline{a})}, \ 1/\overline{\theta_1^2(a)}. \tag{14.125}$$

Verify that, on use of (14.110), these positions can be written as

$$\theta_1^{-2}(a), \qquad \theta_1^{-2}(1/\overline{a}). \tag{14.126}$$

Hence, check that the next iterate is

$$\begin{aligned}
w_6(z) = &-\frac{\mathrm{i}}{2\pi}\log(z-a)+\frac{\mathrm{i}}{2\pi}\log(z-1/\overline{a}) \\
&+\frac{\mathrm{i}}{2\pi}\log(z-\theta_1(1/\overline{a}))-\frac{\mathrm{i}}{2\pi}\log(z-\theta_1(a)) \\
&-\frac{\mathrm{i}}{2\pi}\log(z-\theta_1^{-1}(a))+\frac{\mathrm{i}}{2\pi}\log(z-\theta_1^{-1}(1/\overline{a})) \\
&+\frac{\mathrm{i}}{2\pi}\log(z-\theta_1^{2}(1/\overline{a}))-\frac{\mathrm{i}}{2\pi}\log(z-\theta_1^{2}(a)) \\
&-\frac{\mathrm{i}}{2\pi}\log(z-\theta_1^{-2}(a))+\frac{\mathrm{i}}{2\pi}\log(z-\theta_1^{-2}(1/\overline{a})). \qquad (14.127)
\end{aligned}$$

(f) We have carried out enough steps in the iteration to see the pattern. Argue that, by repeating this process indefinitely, the final result is

$$w_\infty(z) = -\frac{\mathrm{i}}{2\pi}\log\prod_{\theta\in\Theta}\left[\frac{z-\theta_j(a)}{z-\theta_j(1/\overline{a})}\right], \qquad (14.128)$$

where Θ is the Schottky group generated by $\theta_1(z)$ and its inverse.

(g) Use the result of Exercise 3.2 to establish the two identities

$$z-\theta_j^{-1}(a) = \left(\frac{c_jz+d_j}{a_j-c_ja}\right)(\theta_j(z)-a), \qquad (14.129)$$

$$z-\theta_j^{-1}(1/\overline{a}) = \left(\frac{c_jz+d_j}{a_j-c_j/\overline{a}}\right)(\theta_j(z)-1/\overline{a}). \qquad (14.130)$$

Hence show that the function (14.128) can be rewritten as

$$\begin{aligned}
w_\infty(z) = &-\frac{\mathrm{i}}{2\pi}\log\left(\frac{z-a}{z-1/\overline{a}}\right) \\
&-\frac{\mathrm{i}}{2\pi}\left[\log\prod_{\theta\in\Theta''}\left[\frac{(z-\theta_j(a))(\theta_j(z)-a)}{(z-\theta_j(1/\overline{a}))(\theta_j(z)-1/\overline{a})}\right]\right] + \text{constant},
\end{aligned}$$

where the set Θ'' denotes all elements of the Schottky group excluding the identity and all inverses.

(h) Hence use the result (14.1) to verify that, to within a constant, the construction above has retrieved the modified Green's function

$$w_\infty(z) = \mathcal{G}_0(z,a) + \text{constant}, \qquad (14.131)$$

where

$$\mathcal{G}_0(z,a) = -\frac{\mathrm{i}}{2\pi}\log\left[\frac{\omega(z,a)}{|a|\omega(z,1/\overline{a})}\right]. \qquad (14.132)$$

(i) What would happen if, instead of starting with $w_0(z)$ and first fixing the boundary condition on C_0, we had first fixed the boundary condition on C_1? (*Hint:* Think about the other modified Green's function, $G_1(z,a)$, considered in Chapter 2.)

14.11. **(Method of images for a triply connected case)** Work out the details of how the construction given in Exercise 14.10 extends to the triply connected case where D_z has two interior boundary circles C_1 and C_2.

14.12. **(Numerical confirmation of the identity (14.77))** This exercise works towards a numerical check on the identity (14.77).

(a) Pick a value of ρ with $0 < \rho < 1$ and write a numerical procedure to evaluate the prime function for the concentric annulus (14.55) by truncating the infinite product (14.59)–(14.60) at a sufficiently large number of terms.

(b) Write another numerical routine that computes the same prime function based on the infinite-sum representation (14.76) truncated at a sufficiently large number of terms.

(c) Combine the results of your algorithms in parts (a) and (b) to confirm the identity (14.77).

14.13. **(Alternate-point trapezoidal rule)** This question is concerned with the numerical evaluation of the integral representations for the solution of the modified Schwarz problem in a concentric annulus

$$\rho < |z| < 1. \tag{14.133}$$

It is an extension, to $M = 1$, of Exercise 13.6. We are concerned mainly with evaluation of the integral expressions for points on C_0 and C_1.

(a) Develop a numerical routine to compute the function $K(z, a)$ relevant to a concentric annulus

$$\rho < |z| < 1 \tag{14.134}$$

by truncating the infinite-sum representation given in (14.104).

(b) Pick some test function that you know to be analytic and single-valued in the concentric annulus and extract the values of its real part on the boundary circles C_0 and C_1. Use this data in formula (13.46) or (13.47), together with your algorithm from part (a) and the alternate-point trapezoidal rule explained in Exercise 13.6, to compute values of your original test function on C_0 and C_1 up to an imaginary constant.

(c) Take the same boundary data and reconstruct your test function using the alternative scheme described in Exercise 13.1.

14.14. **(Numerical routine for the infinite-product representation)** Develop a numerical algorithm to compute the value of the infinite product (14.1) for the prime function for some $M \geq 2$. Do this by truncating the product so that it includes only mappings in the set Θ'' up to some predetermined level N. (*Hint:* You will need to work out a way to compute all compositions of the level-one mappings and then exclude any inverses.)

14.15. **(Numerical routine for evaluating $\{v_j(z)\}$)** Given an $(M+1)$-connected circular domain D_z develop a numerical algorithm that finds each function $\{v_j | j = 1, \ldots, M\}$ in the doubled domain F by determining the coefficients in the series representation (14.81). Find the period matrix τ associated with D_z as a subsidiary result of your algorithm.

14.16. **(Evaluating the prime function via the theta function formula (14.36))** Use the algorithm developed in Exercise 14.15 to evaluate the prime function based on the theta function formula (14.36). For a given circular domain D_z compare your numerical results against the values of the prime function obtained using your algorithm from Exercise 14.14.

14.17. **(Evaluating the prime function using downloadable codes)** Download the latest version of the numerical routines available at

`github.com/ACCA-Imperial`

and use them to check against the values of the prime function you computed using the algorithms you developed in Exercises 14.14 and 14.16.

Part II

Applications

. . . the movement away from theory and generality is the movement toward truth. All theorising is flight. We must be ruled by the situation itself and this is unutterably particular.

Under the Net
Iris Murdoch

Chapter 15

A calculus for potential theory

Problems of potential theory arise across the applied sciences, from electromagnetism and solid and fluid mechanics through to heat transfer and other steady state transport and diffusion processes. The challenge is to solve for a harmonic potential in two-dimensional domains of various shapes with boundary conditions most commonly of Dirichlet, Neumann, or mixed type.

We will show how solutions to a wide range of potential theory problems involving multiply connected planar domains can be written down explicitly using the prime function. A "calculus" geared at solving such problems can therefore be built up with the prime function serving as the basic functional building block. Consequently the solutions of a broad set of problems of potential theory can be written down in analytical form in terms of the prime function. Then, given a numerical routine to compute the prime function, such as those discussed in Chapter 14, the solutions can be reproduced and evaluated directly from the relevant calculus formula. This chapter features contour plots of the various potentials, all of which are calculated from the formulas in terms of the prime function we derive for them.

The calculus described in this chapter was first put forward by the author, in the physical context of ideal fluid dynamics, specifically vortex dynamics, in [14].

In many undergraduate texts the following complex potential for a uniform field outside a single circular object of radius r can be found:

$$h(z) = \phi + \mathrm{i}\psi = U\left[z + \frac{r^2}{z}\right], \tag{15.1}$$

where ϕ and ψ are harmonic conjugate functions having some physical significance. The imaginary part of $h(z)$ vanishes on the boundary of the circular object. The harmonic field ϕ might be an electric field potential from which an electric field can be derived, or a velocity potential describing flow in a porous medium. Here we will generalize this classical solution by finding the uniform field around *any* finite number of circular objects in an unbounded domain. It will become clear that, having adopted the prime function as our principal mathematical tool, the solution (15.1) is just the simplest case, corresponding to the choice $\omega(z, a) = (z - a)$, of a very general formula, in terms of the prime function, providing the solution to this uniform field problem for any number of objects.

The remainder of this chapter then gradually introduces more complexity into the potential theory problems, all the while showing how the generalized potentials can be

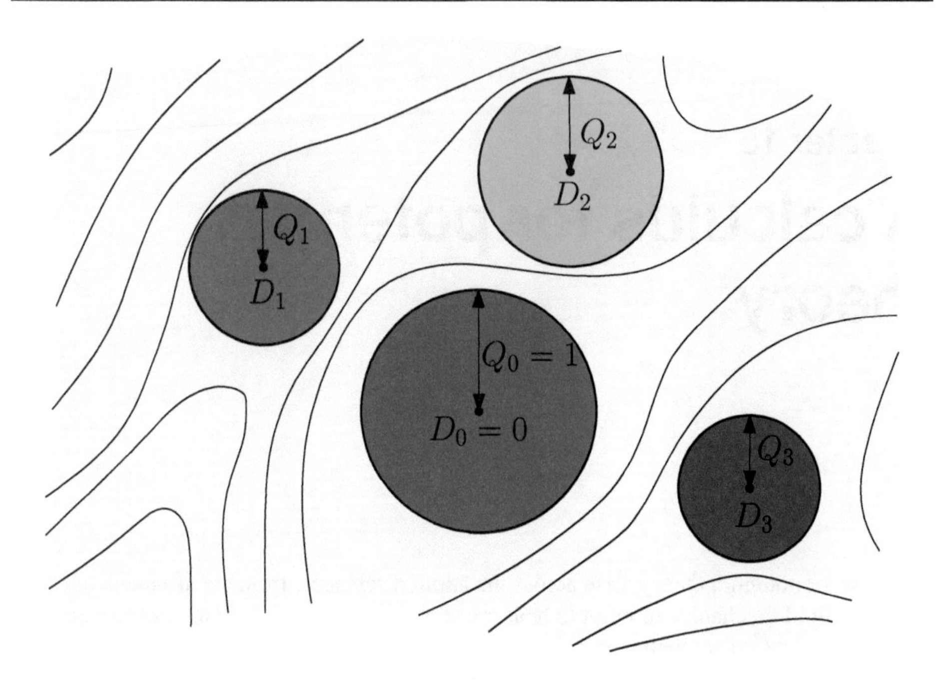

Figure 15.1. *Schematic of field lines associated with a harmonic field defined in the multiply connected domain exterior to $M+1$ circular objects. The case $M=3$ is shown.*

written down very naturally in terms of the prime function using the machinery developed in Part I.

Consider the unbounded region D_z in a two-dimensional $z = x + \mathrm{i}y$ plane exterior to $M+1$ circular objects, or discs, as shown in Figure 15.1. Let $\{\Delta_j | j = 0, \ldots, M\}$ denote the centers and $\{Q_j | j = 0, \ldots, M\}$ the radii of the circular discs $\{D_j | j = 0, \ldots, M\}$. We can take the origin to be at the center of the disc D_0, and by means of a simple rescaling, we can set the radius of this disc to unity so that

$$\Delta_0 = 0, \qquad Q_0 = 1. \tag{15.2}$$

Remark: In this chapter, unlike in Part I, D_z will denote the physical domain in which we seek to find the potential function. We will later use D_ζ to denote the preimage circular domain.

We will show first how to build the calculus in the setting of these simple geometries involving multiple circular objects. Later, this calculus will be combined with conformal mapping considerations to solve wide classes of potential theory problems in complicated multiply connected geometries.

Uniform field exterior to multiple objects: A common challenge in potential theory applications is to compute a harmonic potential Φ satisfying Laplace's equation in D_z,

$$\nabla^2 \Phi = 0, \qquad z \in D_z, \tag{15.3}$$

with Neumann-type conditions on all the domain boundaries

$$\frac{\partial \Phi}{\partial n} = 0, \qquad z \in \partial D_z, \tag{15.4}$$

and with a suitable boundary condition at infinity. We might insist that

$$\Phi \to x \qquad \text{as } x^2 + y^2 \to \infty. \tag{15.5}$$

In fluid mechanics, for example, Φ is the velocity potential associated with a two-dimensional fluid velocity field $\mathbf{u}$ given by

$$\mathbf{u} = \nabla\Phi. \tag{15.6}$$

(15.4) is the condition that the fluid not penetrate the boundaries of the fluid region. The far-field condition (15.5) means that the fluid far away from the objects is in a state of uniform flow parallel to the x axis.

In electrostatics Φ is the electric field potential where the electric field $\mathbf{E}$ is given by

$$\mathbf{E} = -\nabla\Phi \tag{15.7}$$

and (15.4) is the condition that the normal electric field components at the boundaries all vanish. The far-field condition (15.5) says that the electric field as $|z| \to \infty$ is uniform and parallel to the real axis.

These are just two physical manifestations of typical potential theory problems, all of which are amenable to solution using our mathematical framework. For definiteness, we focus on finding the ideal irrotational flow components (u, v) of an inviscid fluid. In this context to find the velocity potential Φ it is natural to seek a *complex potential*

$$h(z) = \Phi(x,y) + \mathrm{i}\Psi(x,y), \qquad z = x + \mathrm{i}y, \tag{15.8}$$

whose real part is the required function; the function Ψ, called the streamfunction, is the harmonic conjugate of Φ. The two-dimensional velocity field $\mathbf{u} = (u, v) = \nabla\Phi$ is derived from $h(z)$ by differentiation:

$$\frac{dh}{dz} = u - \mathrm{i}v. \tag{15.9}$$

By the Cauchy–Riemann equations the boundary conditions (15.4) correspond to

$$\mathrm{Im}[h(z)] = \Psi(x,y) = \text{constant}, \qquad z \in \partial D_j,\ j = 0, 1, \ldots, M. \tag{15.10}$$

The condition (15.5) on Φ implies that

$$h(z) \to z \qquad \text{as } z \to \infty. \tag{15.11}$$

In summary, as a problem in complex analysis we must find $h(z)$ satisfying the following conditions:

- $h(z)$ is analytic in D_z except at infinity, where, as $z \to \infty$,

$$h(z) \to z. \tag{15.12}$$

- On ∂D_ζ,

$$\mathrm{Im}[h(z)] = \text{constant}. \tag{15.13}$$

We know from Part I that analytic functions satisfying these conditions are not unique. This is because, while we have defined the singularities of $h(z)$ in the multiply connected domain—in this case there is only the simple pole at infinity—we have not yet specified its periods around the holes. To find a unique $h(z)$ we must additionally specify the values of the quantities

$$\int_{\partial D_j} dh(z), \qquad j = 0, 1, \ldots, M. \tag{15.14}$$

From the boundary condition (15.13) we have immediately that

$$\mathrm{Im}\left[\int_{\partial D_j} dh(z)\right] = 0, \qquad j = 0, 1, \ldots, M. \tag{15.15}$$

However we still need to specify the values of

$$\mathrm{Re}\left[\int_{\partial D_j} dh(z)\right] = 0, \qquad j = 0, 1, \ldots, M. \tag{15.16}$$

Since

$$\int_{\partial D_j} \mathbf{u}.\mathbf{dx} = \mathrm{Re}\left[\int_{\partial D_j} (u - \mathrm{i}v)(dx + \mathrm{i}dy)\right] = \mathrm{Re}\left[\int_{\partial D_j} \frac{dh}{dz}dz\right] = \mathrm{Re}\left[\int_{\partial D_j} dh(z)\right] \tag{15.17}$$

the quantities (15.16) correspond to

$$\int_{\partial D_j} \mathbf{u}.\mathbf{dx}, \qquad j = 0, 1, \ldots, M, \tag{15.18}$$

which, in the fluid dynamics problem, is called the *circulation* around the object D_j.

A problem that *does* have a unique solution, therefore, is to specify that there is zero circulation around each object in the flow. Equivalently, we insist that $h(z)$ is a single-valued analytic function in the flow domain.

A well-defined problem therefore is to find the complex potential $h(z)$ satisfying the following conditions:

- $h(z)$ is analytic in D_z except at infinity, where the far-field uniform flow condition holds:

$$h(z) \to z \qquad \text{as } z \to \infty \tag{15.19}$$

  and is also single-valued in D_z so that the zero-circulation conditions hold:

$$\mathrm{Re}\left[\int_{\partial D_j} dh(z)\right] = 0, \qquad j = 0, 1, \ldots, M. \tag{15.20}$$

- the impenetrable boundary conditions that on ∂D_z,

$$\mathrm{Im}[h(z)] = \text{constant}. \tag{15.21}$$

The simple inversion

$$z = \frac{1}{\eta}, \qquad \eta = \frac{1}{z} \tag{15.22}$$

is a conformal map that transplants the exterior of D_0 to the interior of the unit disc in a parametric η plane. Being a Möbius map, (15.22) transplants the boundaries of each of the M other circular discs $\{D_j | j = 1, \ldots, M\}$ to a circle inside the unit disc $|\eta| < 1$; we will call these $\{\tilde{C}_j | j = 1, \ldots, M\}$. By the considerations of Exercise 1.1 the circle

$$|z - \Delta_j|^2 = Q_j^2 \tag{15.23}$$

is transplanted to the circle

$$|\eta - \tilde{\delta}_j|^2 = \tilde{q}_j^2, \tag{15.24}$$

where

$$\tilde{\delta}_j = \frac{\overline{\Delta_j}}{|\Delta_j|^2 - Q_j^2}, \qquad \tilde{q}_j = \frac{Q_j}{||\Delta_j|^2 - Q_j^2|}. \tag{15.25}$$

The preimage of the point $z = \infty$ is $\eta = 0$, but, if we like, we can transfer this to any other point $\zeta = a$ in the unit disc by means of a disc automorphism. Given a nonzero point a with $|a| < 1$ the automorphism and its inverse, given by

$$\zeta = \frac{\eta + a}{1 + \overline{a}\eta}, \qquad \eta = \frac{\zeta - a}{1 - \overline{a}\zeta}, \tag{15.26}$$

move $\eta = 0$ to $\zeta = a \neq 0$. It follows from Exercise 1.7 that the image of the circle (15.24) under this mapping is

$$|\zeta - \delta_j|^2 = q_j^2, \tag{15.27}$$

where

$$\delta_j = \frac{(\tilde{\delta}_j + a)(1 + a\overline{\tilde{\delta}_j}) - a\tilde{q}_j^2}{|1 + \tilde{\delta}_j \overline{a}|^2 - \tilde{q}_j^2 |a|^2}, \qquad q_j = \tilde{q}_j \frac{1 - |a|^2}{||1 + \tilde{\delta}_j \overline{a}|^2 - \tilde{q}_j^2 |a|^2|}, \tag{15.28}$$

and where we have set $a \mapsto -a$ in formulas (1.108). On composition with (15.22), under the map

$$z = f(\zeta) \equiv \frac{1 - \overline{a}\zeta}{\zeta - a}, \tag{15.29}$$

the domain D_ζ comprising the unit ζ disc exterior to M excised circular discs of center δ_j and radius q_j is transplanted to D_z. It is useful to note that

$$z = f(\zeta) = \frac{A}{\zeta - a} - \overline{a}, \tag{15.30}$$

where

$$A = 1 - |a|^2 > 0. \tag{15.31}$$

By virtue of our choice of the rotational degree of freedom in the disc automorphism (15.26) we have ensured that the residue A of $f(\zeta)$ at $\zeta = a$ is real and positive.

Armed with the conformal map (15.30) from the circular domain D_ζ to the domain of interest a convenient way to find the required complex potential $h(z)$ is to introduce the composed analytic function

$$H(\zeta) \equiv h(f(\zeta)). \tag{15.32}$$

It can be readily checked that $H(\zeta)$ must have the following properties:

- As $\zeta \to a$,

$$H(\zeta) \sim \frac{A}{\zeta - a} + \text{a locally analytic function}. \tag{15.33}$$

- On ∂D_ζ,

$$\text{Im}[H(\zeta)] = \text{constant}. \tag{15.34}$$

- It satisfies the conditions

$$\text{Re} \int_{C_j} dH(\zeta) = 0, \qquad j = 0, 1, \ldots, M. \tag{15.35}$$

Given all the machinery based on the prime function $\omega(\zeta, a)$ in a circular domain D_ζ developed here it is now a relatively easy matter to write down explicit formulas for $H(\zeta)$ in terms of the prime function. This means that we have a parametric representation of the required solution for the flow written in terms of the prime function.

Since $H(\zeta)$ has a single simple pole in D_ζ and satisfies the condition (15.34) on all boundaries of ∂D_ζ, we are reminded of the parallel slit maps introduced in Part I. Recall that a function having a simple pole, with unit residue, at $\zeta = a$, and transplanting all boundary circles of ∂D_ζ to slits of finite length making angle θ to the positive real axis, is given by

$$\phi_\theta(\zeta, a) = e^{\mathrm{i}\theta}\left[\cos\theta\phi_0(\zeta, a) - \mathrm{i}\sin\theta\phi_{\pi/2}(\zeta, a)\right], \tag{15.36}$$

where

$$\phi_0(\zeta, a) = -\frac{1}{a}\mathcal{K}(\zeta, a) + \frac{1}{\overline{a}}\mathcal{K}(\zeta, 1/\overline{a}), \qquad \phi_{\pi/2}(\zeta, a) = -\frac{1}{a}\mathcal{K}(\zeta, a) - \frac{1}{\overline{a}}\mathcal{K}(\zeta, 1/\overline{a}),$$

with

$$\mathcal{K}(\zeta, a) \equiv a\frac{\partial\omega(\zeta, a)}{\partial a}. \tag{15.37}$$

In view of this, the parametric form of the solution for the complex potential $H(\zeta)$ satisfying (15.41)–(15.43) is deduced to be

$$z = f(\zeta) = \frac{1 - \overline{a}\zeta}{\zeta - a}, \tag{15.38}$$

$$h = A\phi_0(\zeta, a). \tag{15.39}$$

Since the flow is parallel to the x axis, we have chosen $\theta = 0$.

It is important to note that formula (15.39) is the solution for uniform flow (parallel to the real axis) past *any* number of discs. It is understood, of course, that the relevant prime function appearing in the formula for h must be the one relevant to the associated preimage domain D_ζ.

Remark: In this case it is a simple matter to invert equation (15.38) to find ζ as an explicit function of z and to substitute this expression into (15.39). In this way we can eliminate the parameter ζ if desired.

It is straightforward to generalize this to the case where the uniform flow makes an angle $\theta \neq 0$ to the positive real axis. The far-field condition on $h(z)$ becomes

$$h(z) \sim e^{-\mathrm{i}\theta} z. \tag{15.40}$$

The properties of $H(\zeta)$ in this case are as follows:

- As $\zeta \to a$,

$$H(\zeta) \sim \frac{Ae^{-\mathrm{i}\theta}}{\zeta - a} + \text{a locally analytic function.} \tag{15.41}$$

- On ∂D_ζ,

$$\mathrm{Im}[H(\zeta)] = \text{constant}. \tag{15.42}$$

- $H(\zeta)$ satisfies the circulation conditions

$$\mathrm{Re}\int_{C_j} dH(\zeta) = 0, \qquad j = 0, 1, \ldots, M. \tag{15.43}$$

A parametric form of the solution is

$$z = f(\zeta) = \frac{1 - \overline{a}\zeta}{\zeta - a}, \tag{15.44}$$

$$h = H(\zeta) = Ae^{-\mathrm{i}\theta}\phi_\theta(\zeta, a). \tag{15.45}$$

The factor $e^{-\mathrm{i}\theta}$ in front of $\phi_\theta(\zeta, a)$ ensures that $H(\zeta)$ has the behavior (15.41) since $\phi_\theta(\zeta, a)$ has unit residue at $\zeta = a$. We also recall that

$$\mathrm{Im}[e^{-\mathrm{i}\theta}\phi_\theta(\zeta, a)] = \text{constant}, \qquad \zeta \in \partial D_\zeta, \tag{15.46}$$

which ensures that (15.42) is satisfied.

Figures 15.2–15.4 show typical streamline distributions for a range of obstacles sitting in an oncoming flow from different directions. These examples range from a doubly connected fluid region (two obstacles) through to a quadruply connected situation (four obstacles), but similar plots for any number of obstacles follow in exactly the same way.

The streamline distributions in Figures 15.2–15.4 are calculated based on the formulas just derived and use of the numerical procedures to compute the prime function described in Chapter 14.

Formula (15.45) was first presented and studied in the context of uniform irrotational flow past an array of circular discs in [22].

Straining flow past multiple objects: If the complex potential $h(z)$ is not linear in z as $|z| \to \infty$ but instead has the quadratic far-field form

$$h(z) \sim \gamma z^2 \tag{15.47}$$

as $|z| \to \infty$, where $\gamma \in \mathbb{C}$, then it is said that the objects are situated in an irrotational straining flow. From (15.30) it is clear that

$$z^2 = \frac{A^2}{(\zeta - a)^2} - \frac{2A\overline{a}}{\zeta - a} + \overline{a}^2. \tag{15.48}$$

The properties of $H(\zeta)$ in this case become:

- As $\zeta \to a$,

$$H(\zeta) \sim \frac{A^2\gamma}{(\zeta - a)^2} - \frac{2A|a|e^{-\mathrm{i}\theta_a}\gamma}{\zeta - a} + \text{a locally analytic function}, \tag{15.49}$$

  where θ_a is the argument of a.

- On ∂D_ζ,

$$\mathrm{Im}[H(\zeta)] = \text{constant}. \tag{15.50}$$

- $H(\zeta)$ satisfies the circulation conditions

$$\mathrm{Re} \int_{C_j} dH(\zeta) = 0, \qquad j = 0, 1, \ldots, M. \tag{15.51}$$

The required complex potential can be found by adapting the general ideas presented in Part I.

Figure 15.2. *Streamlines around two obstacles as given by (15.45) with* $\theta = 0$.

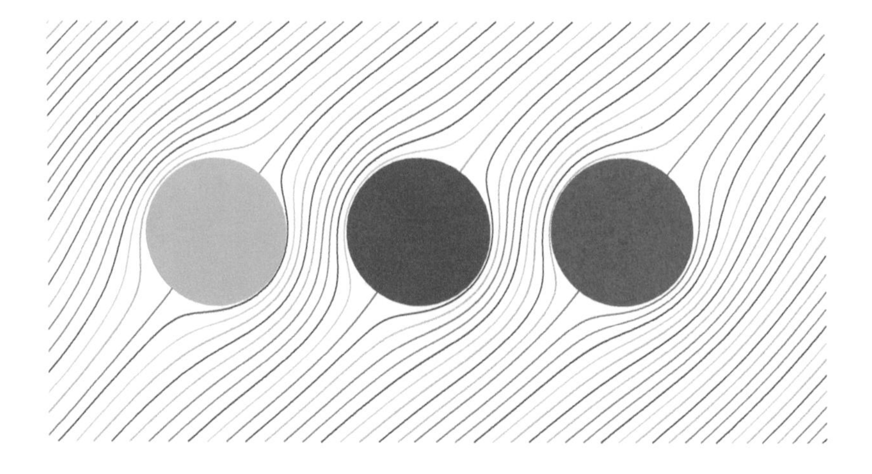

Figure 15.3. *Streamlines around three obstacles as given by (15.45) with* $\theta = \pi/4$.

Figure 15.4. *Streamlines around four obstacles as given by (15.45) with* $\theta = -\pi/4$.

Consider the two potentials

$$\Phi(\zeta, a) = \left[\frac{\partial^2}{\partial a_x \partial(\mathrm{i}a_y)}\right](-2\pi\mathrm{i}\mathcal{G}_0(\zeta, a)), \tag{15.52}$$

$$\Psi(\zeta, a) = \frac{1}{2}\left[\frac{\partial^2}{\partial a_x^2} - \frac{\partial^2}{\partial a_y^2}\right](-2\pi\mathrm{i}\mathcal{G}_0(\zeta, a)). \tag{15.53}$$

Since the real part of $-2\pi\mathrm{i}\mathcal{G}_0(\zeta, a)$ is constant on the circles $\{C_j | j = 0, 1, \ldots, M\}$, the imaginary part of $\Phi(\zeta, a)$ will be constant on these circles since $\Phi(\zeta, a)$ is a purely imaginary parametric derivative of $-2\pi\mathrm{i}\mathcal{G}_0(\zeta, a)$. On the other hand, since $\Psi(\zeta, a)$ is a purely real parametric derivative of $-2\pi\mathrm{i}\mathcal{G}_0(\zeta, a)$, the former will have constant real part on the circles $\{C_j | j = 0, 1, \ldots, M\}$.

By virtue of the identities

$$\frac{1}{2}\left[\frac{\partial^2}{\partial a_x^2} - \frac{\partial^2}{\partial a_y^2}\right] = \frac{\partial^2}{\partial a^2} + \frac{\partial^2}{\partial \overline{a}^2}, \qquad \left[\frac{\partial^2}{\partial a_x \partial(\mathrm{i}a_y)}\right] = \frac{\partial^2}{\partial a^2} - \frac{\partial^2}{\partial \overline{a}^2}, \tag{15.54}$$

both $\Phi(\zeta, a)$ and $\Psi(\zeta, a)$ will have second order poles at $\zeta = a$ of the form

$$\frac{1}{(\zeta - a)^2} + \text{a locally analytic function.} \tag{15.55}$$

Figure 15.5. *Streamline distribution associated with straining flow past two cylinders calculated using formulas (15.56)–(15.57).*

If we choose $\gamma = 1$ so that $h(z) \sim z^2$ as $|z| \to \infty$, then a parametric form of the solution for straining flow past the objects is

$$z = f(\zeta) = \frac{1 - \overline{a}\zeta}{\zeta - a}, \tag{15.56}$$

$$h = H(\zeta) = A^2\gamma\Phi(\zeta, a) - 2A|a|\gamma e^{-\mathrm{i}\theta_a}\phi_{\theta_a}(\zeta, a). \tag{15.57}$$

Here we have used the facts that

$$\mathrm{Im}[\Phi(\zeta, a)] = \text{constant}, \qquad \mathrm{Im}[e^{-\mathrm{i}\theta_a}\phi_{\theta_a}(\zeta, a)] = \text{constant}, \qquad \zeta \in \partial D_\zeta. \tag{15.58}$$

Figure 15.5 shows the streamline distribution around two equal-sized circular obstacles centered in a linear straining flow.

Adding isolated point vortices: A useful fluid dynamical model of an isolated vortex—a source of swirling motion in the flow—is the *point vortex* [14, 113]. We say that there is a point vortex of circulation Γ at position z_α in the flow if, near z_α, the complex potential has the local form

$$h(z) = \underbrace{-\frac{\mathrm{i}\Gamma}{2\pi}\log(z - z_\alpha)}_{\text{point vortex}} + \text{a locally analytic function}, \qquad \Gamma \in \mathbb{R}. \tag{15.59}$$

Adding a point vortex of circulation Γ to the flow therefore corresponds to adding a logarithmic singularity of pure imaginary strength $-\mathrm{i}\Gamma/(2\pi)$ at an isolated point in the flow.

Suppose we have a situation of uniform flow past the $M+1$ circular objects and that we wish to add a point vortex of circulation Γ_1 at some position z_1 in the uniform flow just constructed. Let the preimage of the location z_1 under the conformal mapping $f(\zeta)$ be at ζ_1 so that

$$z_1 = f(\zeta_1). \tag{15.60}$$

Let the required complex potential continue to be denoted by $h(z)$, and let

$$H(\zeta) \equiv h(f(\zeta)). \tag{15.61}$$

The requirements on $H(\zeta)$ are now as follows:

- As $\zeta \to a$,

$$H(\zeta) \to \frac{Ae^{-\mathrm{i}\theta}}{\zeta - a} + \text{a locally analytic function.} \tag{15.62}$$

- As $\zeta \to \zeta_1$,

$$H(\zeta) \to -\frac{\mathrm{i}\Gamma_1}{2\pi}\log(\zeta - \zeta_1) + \text{a locally analytic function.} \tag{15.63}$$

- On ∂D_ζ,

$$\mathrm{Im}[H(\zeta)] = \text{constant.} \tag{15.64}$$

- $H(\zeta)$ satisfies the circulation conditions

$$\mathrm{Re}\int_{C_j} dH(\zeta) = 0, \qquad j = 0, 1, \ldots, M. \tag{15.65}$$

All of the $M+1$ analytic extensions of the modified Green's functions

$$\{\mathcal{G}_j(\zeta,\zeta_1) | j = 0, 1, \ldots, M\} \tag{15.66}$$

associated with D_ζ have a logarithmic singularity, of strength $1/(2\pi\mathrm{i})$, at the point $\zeta = \zeta_1$ in D_ζ. We know from Part I that the difference between the modified Green's functions concerns their periods. It has been established that

$$\int_{C_k} d\mathcal{G}_0(z,\zeta_1) = \delta_{k0}, \qquad \int_{C_k} d\mathcal{G}_j(z,\zeta_1) = -\delta_{jk}, \;\; j = 1, \ldots, M. \tag{15.67}$$

Since any one of the modified Green's functions $\{\mathcal{G}_j(\zeta,\zeta_1)|j = 0, 1, \ldots, M\}$ has the required logarithmic singularity in D_ζ, let us focus on

$$\mathcal{G}_0(\zeta,\zeta_1) = \frac{1}{2\pi\mathrm{i}}\log\left[\frac{\omega(\zeta,\zeta_1)}{|\zeta_1|\omega(\zeta,1/\overline{\zeta_1})}\right]. \tag{15.68}$$

If we were to add to $H(\zeta)$ a contribution

$$\Gamma_1\mathcal{G}_0(\zeta,\zeta_1), \tag{15.69}$$

then, since by a Taylor expansion

$$z = f(\zeta) = z_1 + f'(\zeta_1)(\zeta - \zeta_1) + \cdots \tag{15.70}$$

so that

$$\log(z - z_1) = \log(\zeta - \zeta_1) + \log f'(\zeta_1) + \cdots, \tag{15.71}$$

and since near $\zeta = \zeta_1$

$$\Gamma_1 \mathcal{G}_0(\zeta, \zeta_1) = -\frac{\mathrm{i}\Gamma_1}{2\pi} \log(\zeta - \zeta_1) + \text{a locally analytic function} \tag{15.72}$$

so that near $z = z_1$

$$\Gamma_1 \mathcal{G}_0(\zeta, \zeta_1) = -\frac{\mathrm{i}\Gamma_1}{2\pi} \log(z - z_1) + \text{a locally analytic function,} \tag{15.73}$$

then this term contributes precisely the required point vortex singularity at z_1. It also has zero circulation around the circles $\{C_j | j = 1, \ldots, M\}$, which is as desired. But since

$$\int_{\partial C_0} d\mathcal{G}_0 = +1, \tag{15.74}$$

under the inversion (15.22), if C_0 is traversed in an anticlockwise direction, its image ∂D_0 will be traversed in a clockwise direction. This means that

$$\int_{\partial D_0} d\mathcal{G}_0 = -1. \tag{15.75}$$

Thus the contribution $\Gamma_1 \mathcal{G}_0(\zeta, \zeta_1)$ will add circulation $-\Gamma_1$ around ∂D_0, and this is not wanted. To eliminate it we incorporate the additional term

$$-\Gamma_1 \mathcal{G}_0(\zeta, a), \tag{15.76}$$

which adds a point vortex at $\zeta = a$, corresponding to the point at infinity in D_z, and which induces a circulation $+\Gamma_1$ around ∂D_0. Consequently, the two contributions to the circulation around ∂D_0 from $\Gamma_1 \mathcal{G}_0(\zeta, \zeta_1)$ and $-\Gamma_1 \mathcal{G}_0(\zeta, a)$ cancel out.

The required solution for a uniform flow with angle θ past the objects with a point vortex of circulation Γ_1 at z_1 and having zero circulation around all objects in the flow is

$$z = f(\zeta) = \frac{1 - \overline{a}\zeta}{\zeta - a}, \tag{15.77}$$

$$h = H(\zeta) = Ae^{-\mathrm{i}\theta} \phi_\theta(\zeta, a) + \Gamma_1 \left[\mathcal{G}_0(\zeta, \zeta_1) - \mathcal{G}_0(\zeta, a) \right]. \tag{15.78}$$

Figure 15.6 shows the streamline distribution associated with uniform flow past three circular cylinders in the case where there are, in addition, three point vortices located in the flow.

It is clear that to add N vortices at points $\{z_n | n = 1, \ldots, N\}$ having circulations $\{\Gamma_n | n = 1, \ldots, N\}$, the required complex potential is

$$z = f(\zeta) = \frac{1 - \overline{a}\zeta}{\zeta - a}, \tag{15.79}$$

$$h = H(\zeta) = Ae^{-\mathrm{i}\theta} \phi_\theta(\zeta, a) + \sum_{n=1}^{N} \Gamma_n \left[\mathcal{G}_0(\zeta, \zeta_n) - \mathcal{G}_0(\zeta, a) \right], \tag{15.80}$$

where $\{\zeta_n | n = 1, \ldots, N\}$ are the preimages of the vortex locations, i.e.,

$$z_n = f(\zeta_n). \tag{15.81}$$

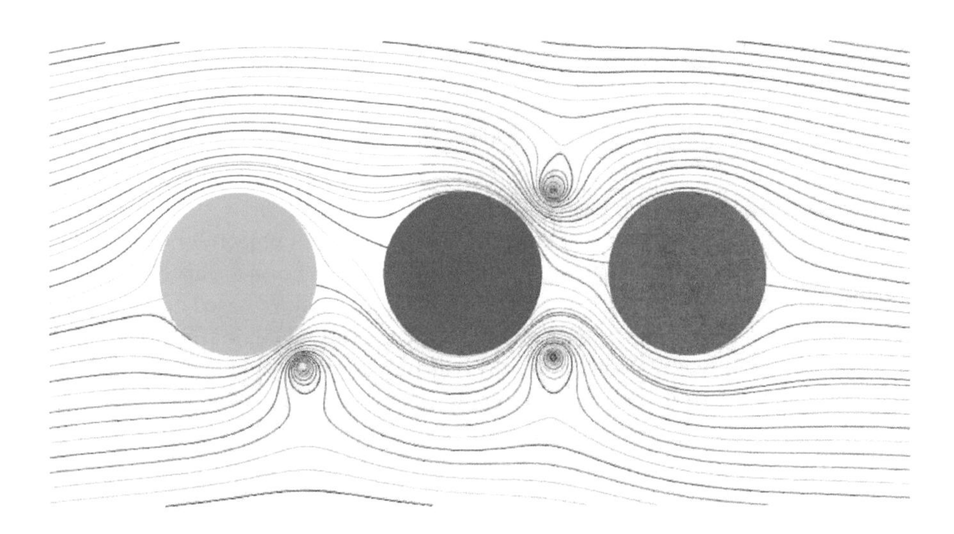

Figure 15.6. *Streamline distribution associated with uniform flow around three circular cylinders in the presence of three point vortices. There is no circulation around the cylinders. This figure was plotted using formulas (15.79)–(15.80).*

Adding circulation around the objects: If an application requires the addition of nonzero circulations around the obstacles, the functions $\{\mathcal{G}_j(\zeta,a)|j=0,1,\ldots,M\}$ can be deployed.

Suppose it is required that there should be a circulation γ_j around ∂D_j for $j = 0,1,\ldots,M$. The contribution

$$-\sum_{j=0}^{M}\gamma_j\mathcal{G}_j(\zeta,a) \tag{15.82}$$

will effect precisely this change in the round-object circulations. This can be verified using (15.67).

The complex potential for uniform flow in domain D_z, with angle θ, exterior to M circular discs with N point vortices at points $\{z_n|n=1,\ldots,N\}$ having circulations $\{\Gamma_n|n=1,\ldots,N\}$ and with circulation $\{\gamma_j|j=0,1,\ldots,M\}$ around the jth boundary of D_z is

$$z = f(\zeta) = \frac{1-\overline{a}\zeta}{\zeta - a}, \tag{15.83}$$

$$h = H(\zeta) = Ae^{-\mathrm{i}\theta}\phi_\theta(\zeta,a) + \sum_{n=1}^{N}\Gamma_n\left[\mathcal{G}_0(\zeta,\zeta_n) - \mathcal{G}_0(\zeta,a)\right] - \sum_{j=0}^{M}\gamma_j\mathcal{G}_j(\zeta,a), \tag{15.84}$$

where

$$z_n = f(\zeta_n). \tag{15.85}$$

Remark: From (2.53), the function $\mathcal{G}_j(z,a)$ is known to be given by

$$\mathcal{G}_j(z,a) \equiv \mathcal{G}_0(z,a) - v_j(z) - v_j(a) + \mathrm{i}[\mathcal{P}^{-1}]_{jj}. \tag{15.86}$$

In a remark at the end of Chapter 4 it was noted that formulas (4.108) and (4.114) can be substituted into the expression (15.86) to obtain formulas for the functions $\{\mathcal{G}_j(z,a)|j =$

$1, \ldots, M\}$ solely in terms of the prime function. If this is done, we can instead use the modified functions defined by

$$\tilde{\mathcal{G}}_j(\zeta, a) = \frac{1}{2\pi \mathrm{i}} \log \left[\frac{\omega(\zeta, a)}{\omega(\zeta, \theta_j(1/\overline{a}))} \right] \tag{15.87}$$

in place of the functions $\mathcal{G}_j(\zeta, a)$ used in, say, the solution (15.84).[53]

It is known that a combination of uniform flow past a set of objects, or "airfoils," each of which has nonzero circulation around it, can lead to a lift force on the airfoils. A detailed study of the lift on a finite collection of airfoils in a uniform flow, based on the analytical formulas above in terms of the prime function, has been carried out in [23].

Adding point sources and sinks: We say that there is a point source of strength $m > 0$ at position z_α in the flow if, near z_α, the complex potential has the local form

$$h(z) = \underbrace{\frac{m}{2\pi} \log(z - z_\alpha)}_{\text{point source}} + \text{a locally analytic function}, \qquad m \in \mathbb{R}. \tag{15.88}$$

Adding a point source of strength m to the flow therefore corresponds to adding a logarithmic singularity of purely real strength $m/(2\pi)$ at isolated points in the flow. If $m < 0$, we say that there is a point sink of strength m at z_α.

It is clear that, to respect the condition of conservation of mass in an incompressible fluid, it is only possible to find a solution comprising a source of strength m at some point in the flow if there is a compensating sink of strength m at some other point in the flow. This can be at infinity, of course, if the domain is unbounded.

Consider the function

$$\frac{m}{2\pi} \log \left[\frac{\omega(\zeta, b)\omega(\zeta, 1/\overline{b})}{\omega(\zeta, a)\omega(\zeta, 1/\overline{a})} \right]. \tag{15.89}$$

The function inside the square brackets, which is the argument of the logarithm, is recognized as a radial slit map. Such maps have constant argument, implying, on taking a logarithm as in (15.89), that the imaginary part of the function will be constant on all boundaries of ∂D_ζ. The function (15.89) also has a logarithmic singularity of strength $m/(2\pi)$ at $\zeta = b$ and a logarithmic singularity of strength $-m/(2\pi)$ at $\zeta = a$, where a is the preimage of infinity. This corresponds to a source of strength m at $z_b = f(b)$ and a sink of strength m at infinity.

A parametric form of the solution for a single source of strength m at point $z_b = f(b)$ outside the circular objects (and with a compensating sink at infinity) is

$$z = f(\zeta) = \frac{1 - \overline{a}\zeta}{\zeta - a}, \tag{15.90}$$

$$h = H(\zeta) = \frac{m}{2\pi} \log \left[\frac{\omega(\zeta, b)\omega(\zeta, 1/\overline{b})}{\omega(\zeta, a)\omega(\zeta, 1/\overline{a})} \right]. \tag{15.91}$$

Figure 15.7 shows the streamlines for a source and a sink placed in the fluid region exterior to two impenetrable circular obstacles.

A more complete study of point sources and sinks in irrotational flow around multiple objects has been given in [7], where the solutions are used to find the so-called effective size of an array of holes in a grating.

[53] It is these modified functions that were used in [23] when studying problems involving calculation of the lift on stacks of airfoils.

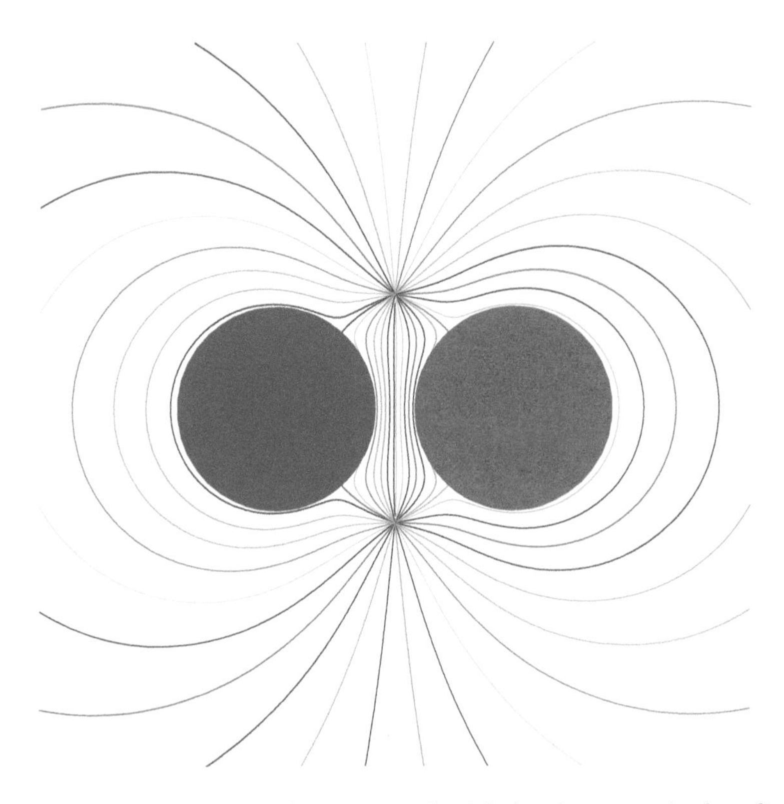

Figure 15.7. *Streamline distribution for a source and a sink placed near two circular cylinders.*

Moving objects in an irrotational flow: In many situations the objects in an ideal irrotational flow are not stationary but are translating with some velocity or rotating with some angular velocity; see [18] for a discussion of problems of "fluid stirrers" in an ideal flow, where the framework in Part I is used to find the flows. The following treatment is based on the presentation given there.

Suppose the object with boundary ∂D_j that is the image of C_j under a conformal map

$$z = f(\zeta) \tag{15.92}$$

is moving with speed

$$\mathbf{U}^{(j)} = (U_x^{(j)}, U_y^{(j)}), \qquad j = 0, 1, \ldots, M, \tag{15.93}$$

but is not rotating. The boundary condition on the fluid velocity $\mathbf{u} = (u, v)$ on ∂D_j is

$$\mathbf{u}.\mathbf{n} = \mathbf{U}^{(j)}.\mathbf{n}, \qquad j = 0, 1, \ldots, M, \tag{15.94}$$

where $\mathbf{n}$ denotes the unit normal at the surface of the moving object. In complex form, where we introduce the complex constant

$$U^{(j)} = U_x^{(j)} + \mathrm{i}U_y^{(j)}, \tag{15.95}$$

this boundary condition is

$$\mathrm{Re}\left[\frac{dh}{dz}\left(-\mathrm{i}\frac{dz}{ds}\right)\right] = \mathrm{Re}\left[\overline{U^{(j)}}\left(-\mathrm{i}\frac{dz}{ds}\right)\right], \qquad j = 0, 1, \ldots, M. \tag{15.96}$$

This boundary condition can be integrated with respect to the arclength s to give

$$\mathrm{Re}\left[-\mathrm{i}h(z)\right] = \mathrm{Re}\left[-\mathrm{i}\overline{U^{(j)}}z\right] + B_j, \qquad j = 0, 1, \ldots, M, \tag{15.97}$$

where B_j is a constant of integration.

If we now introduce the composed function

$$H(\zeta) \equiv h(f(\zeta)), \tag{15.98}$$

then the boundary condition (15.97) is

$$\mathrm{Re}\left[-\mathrm{i}H(\zeta)\right] = \mathrm{Re}\left[-\mathrm{i}\overline{U^{(j)}}f(\zeta)\right] + B_j, \qquad \zeta \in C_j. \tag{15.99}$$

Without loss of generality we can set

$$B_0 = 0. \tag{15.100}$$

Hence the boundary conditions on the unknown complex potential $H(\zeta)$ are

$$\mathrm{Re}\left[-\mathrm{i}H(\zeta)\right] = \begin{cases} \mathrm{Re}\left[-\mathrm{i}\overline{U^{(0)}}f(\zeta)\right], & \zeta \in C_0, \\ \mathrm{Re}\left[-\mathrm{i}\overline{U^{(j)}}f(\zeta)\right] + B_j, & \zeta \in C_j,\ j = 1, \ldots, M. \end{cases} \tag{15.101}$$

If we insist that the circulation around the obstacles is zero, then $H(\zeta)$ must be single-valued in the fluid region D_z with boundary conditions (15.101). Hence the problem for the function $-\mathrm{i}H(\zeta)$ is a standard modified Schwarz problem of the kind considered in Chapter 13. The solution is given by

$$\begin{aligned} H(\zeta) = {} & \frac{1}{2\pi\mathrm{i}} \int_{C_0} \left[\mathrm{Re}\left[-\mathrm{i}\overline{U^{(0)}}f(\zeta')\right]\right] \left[d\log\omega(\zeta',\zeta) + d\log\overline{\omega}(\overline{\zeta'}, 1/\zeta)\right] \\ & - \frac{1}{2\pi\mathrm{i}} \sum_{j=1}^{M} \int_{C_j} \left[\mathrm{Re}\left[-\mathrm{i}\overline{U^{(j)}}f(\zeta')\right] + B_j\right] \left[d\log\omega(\zeta',\zeta) + d\log\overline{\omega}(\overline{\zeta'}, 1/\zeta)\right] \\ & + \mathrm{i}c, \qquad c \in \mathbb{R}, \end{aligned} \tag{15.102}$$

where the constants $\{B_j | j = 1, \ldots, M\}$ are given by solutions of the linear system

$$\begin{aligned} \int_{C_0} \left[\mathrm{Re}\left[-\mathrm{i}\overline{U^{(0)}}f(\zeta)\right]\right] \frac{\partial\sigma_j(\zeta)}{\partial n_\zeta} ds_\zeta - \sum_{k=1}^{M} \int_{C_k} \left[\mathrm{Re}\left[-\mathrm{i}\overline{U^{(k)}}f(\zeta)\right] + B_k\right] \frac{\partial\sigma_j(\zeta)}{\partial n_\zeta} ds_\zeta & \\ = 0, \qquad j = 1, \ldots, M. & \end{aligned} \tag{15.103}$$

Figure 15.8 shows the instantaneous streamlines associated with two cylinders located on the real axis but moving with identical speeds in opposite directions, as indicated by the arrows.

Similar formulas to (15.102) can be derived if, in addition to translating at some specified speed, they are rotating with some specified angular velocity.

Combining the calculus with conformal mapping: The class of circular domains D_ζ is a canonical class: by this we mean that *any* given $M+1$ connected domain is conformally equivalent to such a domain. If some means of constructing the conformal map from a circular domain D_ζ to some noncircular domain D_z of physical interest can be found, either analytically or numerically, we can use the calculus just described to solve

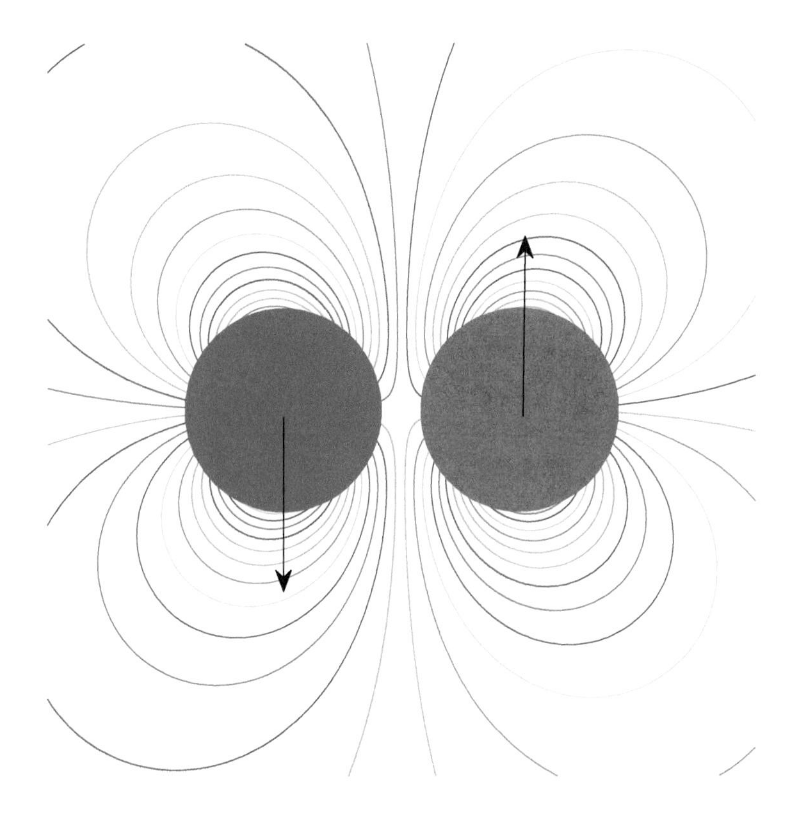

Figure 15.8. *Streamline distribution for two cylinders centered on the real axis and moving with equal speed but in opposite directions parallel to the y axis.*

problems of potential theory in these domains, and often in analytical form in terms of the prime function. This means that all the mathematical machinery presented in, say, Chapters 7 and 10 on S–C and polycircular arc mappings can now be deployed to greatly enlarge the range of problems for which analytical solutions can be written down.

Consider the problem of uniform flow with speed U and with angle θ to the positive real axis, past three parallel flat plates aligned with the imaginary y axis. Suppose that the two plates to the left and right of the central plate have the same length; this configuration, rotated by 90°, was studied in Exercise 6.15; the conformal map from a circular preimage domain D_ζ to the unbounded region exterior to these plates is a parallel slit map of the form

$$z = f(\zeta) = -\frac{1}{a}\mathcal{K}(\zeta, -a) - \frac{1}{a}\mathcal{K}(\zeta, -1/a), \tag{15.104}$$

where $\zeta = -a$ maps to infinity and $a > 0$ is a real parameter. Since we require

$$h(z) \sim Ue^{-\mathrm{i}\theta} z + \text{ a locally analytic function} \tag{15.105}$$

as $z \to \infty$, the complex potential $h(z)$ for uniform flow past these plates making angle θ with the positive real axis is, in parametric form,

$$h = H(\zeta) = Ue^{-\mathrm{i}\theta}\phi_\theta(\zeta, -a), \tag{15.106}$$

where it is easily checked that this solution satisfies (15.41)–(15.43).

Figure 15.9. *Streamlines for uniform flow (15.106) past three flat plates with $\theta = \pi/4$.*

Figure 15.9 shows the streamline distribution associated with such a flow, as computed using (15.104) and (15.106).

Channel flows are of considerable interest in applications. Consider the situation of a flat plate obstructing a uniform flow in a two-dimensional channel, as shown in Figure 15.10. The top and bottom walls of the channel are impenetrable and are therefore streamlines. Another interpretation is to dispense with the channel "walls" and view this arrangement as a single period window for uniform flow past a screen or grating aligned with the imaginary axis perforated by a periodic array of gaps; in such a flow the edges of the period window will be streamlines of the flow, assuming that the flat plate is centrally placed inside the period window.

The fluid domain is doubly connected, and the relevant conformal mapping from an annulus $\rho < |\zeta| < 1$ has been constructed in Chapter 7:

$$z = \mathrm{i}g(\zeta), \tag{15.107}$$

where $g(\zeta)$ is given in formula (7.106) and where we have rotated the configuration through $\pi/2$ so that the channel is parallel to the real z axis. If $h(z)$ is the complex potential for the flow, then on introducing

$$H(\zeta) \equiv h(\mathrm{i}g(\zeta)) \tag{15.108}$$

it is easily argued that

$$H(\zeta) = U \log\left[\frac{\omega(\zeta,+1)}{\omega(\zeta,-1)}\right]. \tag{15.109}$$

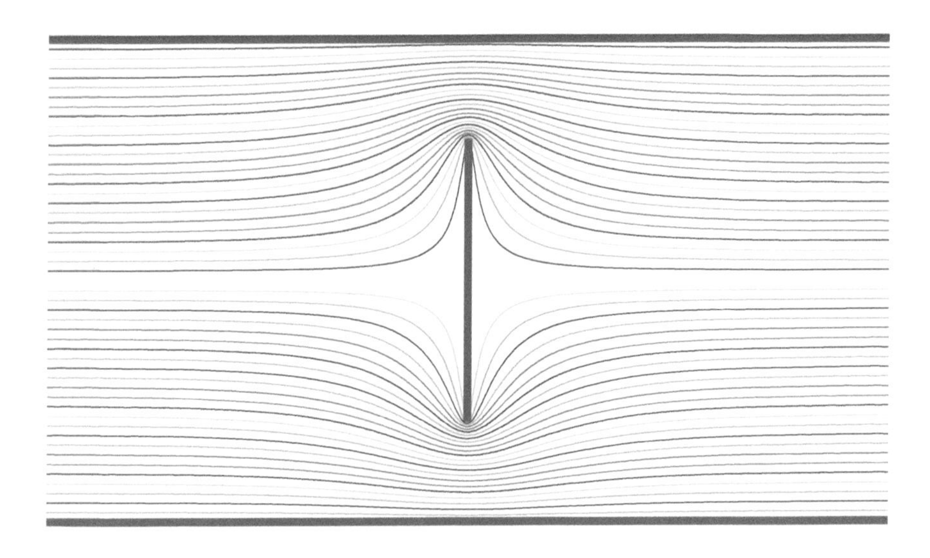

Figure 15.10. *Streamlines for uniform flow past a vertical slit in a channel computed using the S–C-type map found in Chapter 7.*

A parametric representation, in terms of parameter ζ, of the required $h(z)$ is therefore

$$z = \mathrm{i}g(\zeta), \qquad h = U \log \left[\frac{\omega(\zeta, +1)}{\omega(\zeta, -1)} \right]. \tag{15.110}$$

Figure 15.10 shows the streamline distribution in this case, plotted on the basis of formulas (15.110).

Uniform flow past a circular disc in a channel is another problem that commonly arises; equivalently, this arrangement can be viewed as a single period window for uniform flow past a periodic array of cylinders placed in a line parallel to the imaginary axis. This is another doubly connected region of fluid, and the conformal mapping from the concentric annulus $\rho < |\zeta| < 1$ to this domain is a polycircular arc mapping constructed in §10.8 of Chapter 10. Let us denote it by

$$z = \mathcal{Z}(\zeta). \tag{15.111}$$

Simply swapping the conformal mapping in the solution (15.110) we deduce that a parametric representation, in terms of parameter ζ, of the required $h(z)$ is therefore

$$z = \mathcal{Z}(\zeta), \qquad h = U \log \left[\frac{\omega(\zeta, +1)}{\omega(\zeta, -1)} \right]. \tag{15.112}$$

Figure 15.11 shows a typical streamline distribution for this case calculated using (15.112).

Finally, consider flow along a channel in which there is a periodic staggered array of reentrant fins along the channel obstructing the flow. This is expected to cause a snaking motion of the fluid along the channel. The conformal mapping from a concentric annulus to a single period of such an arrangement has already been found using the periodic version of the S–C mapping in §7.10 of Chapter 7. On use of the mapping, $\mathcal{Z}(\zeta)$ say, found there, the parametric representation

$$z = \mathcal{Z}(\zeta), \qquad h = \mathrm{i}U \log \zeta, \qquad U \in \mathbb{R}, \tag{15.113}$$

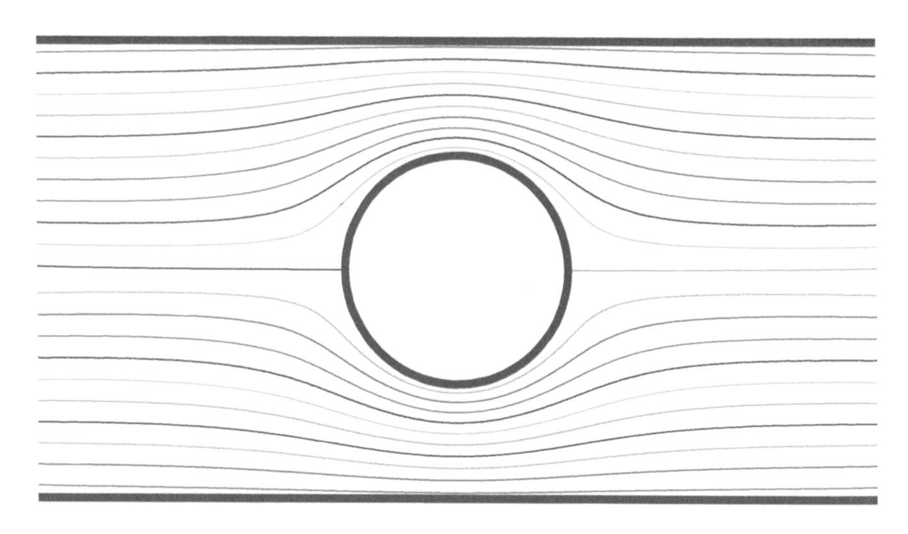

Figure 15.11. *Streamlines for uniform flow past a cylinder in a channel computed using (15.110) and the polycircular arc map found in §10.8.*

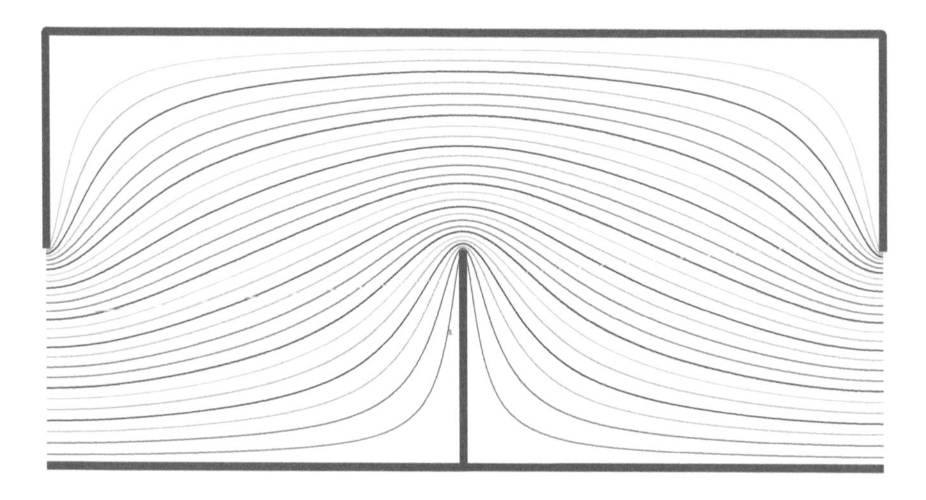

Figure 15.12. *Streamlines for uniform flow along a channel with a periodic array of staggered fins on top and bottom walls (one period window shown). The map is computed using the periodic S–C map found in §7.10.*

of the complex potential encodes the streamlines for uniform flow along this channel with periodic barriers. Notice that the complex potential $H(\zeta) \equiv h(\mathcal{Z}(\zeta))$ in this case is just a real multiple of $v_1(\zeta)$ for the concentric annulus.

Figure 15.12 shows the streamlines in one period window of such a channel flow with a staggered array of reentrant fins.

For the flows depicted in Figures 15.10, 15.11, and 15.12 it is of physical interest to calculate the associated *blockage coefficients* [11], and this can easily be done using the complex potentials just derived.

Chapter 16

Hamiltonian dynamics of point vortices

Kirchhoff–Routh theory [113] is the name given to the study of the Hamiltonian structure of the dynamics of point vortices in the two-dimensional flow of an ideal fluid. In the original formulation the domains in which the vortices move around were assumed to be either the whole two-dimensional plane without impenetrable boundaries or some simply connected subregion of the plane having a single boundary.

In 1941 Lin [94] pointed out that, in a multiply connected domain with multiple impenetrable boundaries, the Hamiltonian structure of the dynamical system survives. He used the existence of what he dubbed a "hydrodynamic Green's function" to construct the relevant Hamiltonian. Lin's approach was not constructive, and he gives no explicit calculations of any vortex trajectories. Much later, Crowdy and Marshall [48] pointed out that this "hydrodynamic Green's function" is none other than the modified Green's function $G_0(z,a)$ introduced in Part I; they went on to show that the Hamiltonians governing the point vortex dynamics can be constructed explicitly using the prime function of a conformally equivalent preimage domain D_z. They used the constructive function theory of Part I to find the trajectories of point vortices in a number of geometrically complicated domains relevant to applications [48, 49, 50]. This chapter surveys the key ideas in that work.

In ideal fluid mechanics the velocity $\mathbf{u}$ of a two-dimensional, irrotational, incompressible, and inviscid fluid in the (x, y) plane is determined by a velocity potential Φ where

$$\mathbf{u} = \nabla\Phi. \tag{16.1}$$

The irrotationality of the flow means that $\nabla \wedge \mathbf{u} = 0$ so that Φ is a harmonic function in the fluid domain. A common strategy is to consider a *complex potential* $h(z)$ where $z = x + \mathrm{i}y$ and

$$h(z) = \Phi + \mathrm{i}\Psi. \tag{16.2}$$

Ψ is the harmonic conjugate of the harmonic field Φ and is known as the *streamfunction*. Its contours give the instantaneous streamlines of the flow.

In applications, for example, in modeling a tornado or an oceanic eddy, it is often required to model isolated regions of vorticity where the assumption of irrotationality breaks down. A *point vortex* of circulation Γ at position z_a in a complex z plane has a complex potential with a singularity having the local form

$$h(z) = \underbrace{-\frac{\mathrm{i}\Gamma}{2\pi}\log(z - z_a)}_{\text{point vortex}} + \text{a locally analytic function}, \qquad \Gamma \in \mathbb{R}. \tag{16.3}$$

This logarithmic singularity of the otherwise analytic complex potential reflects the presence of the point vortex.

Suppose there are N point vortices in some multiply connected circular fluid region D_ζ where $N \geq 1$ is an integer. Let each vortex be at complex position a_j with circulation Γ_j for $j = 1, \ldots, N$. If we define $q_j = \sqrt{\Gamma_j} a_j$ and $p_j = \sqrt{\Gamma_j}\overline{a_j}$, then it can be shown that the dynamics of the vortices is given by

$$\dot{p}_j = 2\mathrm{i}\frac{\partial H^{(\zeta)}}{\partial q_j}. \tag{16.4}$$

$H^{(\zeta)}(\{a_j\})$ is a Hamiltonian for the dynamics of the vortices. This Hamiltonian is also known as the *Kirchhoff–Routh path function* since it is the function that determines the paths of the vortices.

The Kirchhoff–Routh path function is given explicitly as

$$H^{(\zeta)}(\{a_j\}) = \mathrm{Im}\left[\sum_{j=1}^{N} \Gamma_j W_B(a_j, t) + \sum_{j=1}^{N}\sum_{\substack{k=1\\k\neq j}}^{N} \frac{\Gamma_j\Gamma_k}{2}\mathcal{G}_0(a_j, a_k) + \sum_{j=1}^{N}\frac{\Gamma_j^2}{2} g(a_j, a_j)\right], \tag{16.5}$$

where $\mathcal{G}_0(\zeta, a)$ is the analytic extension of the modified Green's function associated with D_ζ introduced in Chapter 2 and $g(\zeta, a)$ is defined by

$$\mathcal{G}_0(\zeta, a) = -\frac{\mathrm{i}}{2\pi}\log(\zeta - a) + g(\zeta, a). \tag{16.6}$$

The function $g(\zeta, a)$ is analytic and single-valued at $\zeta = a$.[54] The complex potential $W_B(\zeta, t)$, which does not depend on the point vortex locations, allows us to add in the effect of some externally imposed background flow. This might be a uniform flow, a linear straining flow, or the flow due to a nonzero circulatory flow around the obstacles.

Remark: It is of no small significance that the functional forms of $W_B(\zeta, t)$ for all such background flows have been given, in terms of the prime function, in Chapter 15. These formulas can be used in expression (16.5) to add in the effects of these background flows on the dynamics of the vortices.

If the circular domain D_ζ is conformally mapped, by a function $z = f(\zeta)$, to some other conformally equivalent domain D_z such that the vortex at $a_j(t)$ is mapped to $z_j(t)$ where

$$z_j(t) = f(a_j(t)), \tag{16.7}$$

then the new Hamiltonian governing the dynamics in the domain D_z is

$$H^{(z)}(\{z_j(t)\}) = H^{(\zeta)}(\{a_j(t)\}) + \sum_{j=1}^{N}\frac{\Gamma_j^2}{4\pi}\log\left|\frac{df}{d\zeta}\right|_{\zeta=a_j(t)}. \tag{16.8}$$

The case of a single vortex, $N = 1$, is of particular interest. Since the Hamiltonian is conserved by the dynamics, and for $N = 1$ is a function only of the single vortex location, the vortex trajectories in this case are given by the level lines of the Hamiltonian. We can use the prime function to construct this Hamiltonian and, hence, the trajectories of the vortex in the given fluid domain.

Suppose a single vortex of circulation Γ is located at a point a in a multiply connected circular fluid region D_ζ in a complex ζ plane. For brevity we suppress the dependence of

[54] The imaginary part of $g(\zeta, a)$ is called the Robin function.

a on time. The objective is to find the locus of this vortex as it moves around in some given domain.

By way of example we take a triply connected domain comprising the unit ζ disc with two excised circular discs having radii and centers given by

$$q_1 = q_2 = q, \qquad \delta_1 = -\delta_2 = \delta \in \mathbb{R} \tag{16.9}$$

for some real q and δ. The Hamiltonian (16.5) is a function solely of a and is given by

$$H^{(\zeta)}(a) = \frac{\Gamma^2}{2}\mathrm{Im}[g(a,a)]. \tag{16.10}$$

We know that, in terms of the prime function of this circular domain,

$$\mathcal{G}_0(\zeta, a) = \frac{1}{2\pi \mathrm{i}} \log\left[\frac{\omega(\zeta, a)}{|a|\omega(z, 1/\overline{a})}\right]. \tag{16.11}$$

If we define the function $\hat{\omega}(\zeta, a)$ by the relation

$$\omega(\zeta, a) = (\zeta - a)\hat{\omega}(\zeta, a), \tag{16.12}$$

where we have factored out the known simple zero of $\omega(\zeta, a)$, then $g(\zeta, a)$ can be identified as

$$g(\zeta, a) = \frac{1}{2\pi \mathrm{i}} \log\left[\frac{\hat{\omega}(\zeta, a)}{|a|\omega(\zeta, 1/\overline{a})}\right]. \tag{16.13}$$

It follows immediately from (16.10) that

$$H^{(\zeta)}(a) = -\frac{\Gamma^2}{4\pi} \log\left|\frac{\hat{\omega}(a,a)}{|a|\omega(a, 1/\overline{a})}\right| = \frac{\Gamma^2}{4\pi} \log|a\omega(a, 1/\overline{a})|, \tag{16.14}$$

where we have used the fact that $\hat{\omega}(a, a) = 1$.

Figure 16.1 shows a set of trajectories for a single vortex placed in such a fluid domain D_ζ. Each line in this figure is the path on which a single vortex would move if placed in isolation in D_ζ. These trajectories should not be confused with the instantaneous streamline distributions plotted in Chapter 15.

Suppose we consider instead the situation of two circular "islands" in a semi-infinite fluid region off a coastline modeled by an infinite impenetrable wall along the real axis. If the islands are not present, a point vortex, perhaps modeling an "oceanic eddy," simply travels along the wall at constant velocity and at a fixed distance from it. The presence of the islands affects these straight line trajectories, leading to more complicated motion of the vortex.

To determine this motion, we first find a conformal mapping from a triply connected circular domain D_ζ with two circular holes to the fluid region of interest. In this case, the required conformal mapping is simply a Cayley map of the kind studied in §1.7, namely,

$$z = f(\zeta) = \mathrm{i}\left[\frac{1-\zeta}{1+\zeta}\right]. \tag{16.15}$$

This mapping transplants D_ζ to the upper half z plane exterior to two circular islands. We chose the centers δ_1, δ_2 and radii q_1, q_2 of circles C_1 and C_2 in D_ζ to be such that the circular islands in the z plane have the desired centers and radii. From (16.8) it follows that the trajectories of a vortex placed in this domain are then the level lines of

$$H^{(z)}(z_a) = H^{(\zeta)}(a) + \frac{\Gamma^2}{4\pi} \log\left|\frac{df}{d\zeta}\right|_{\zeta=a}, \tag{16.16}$$

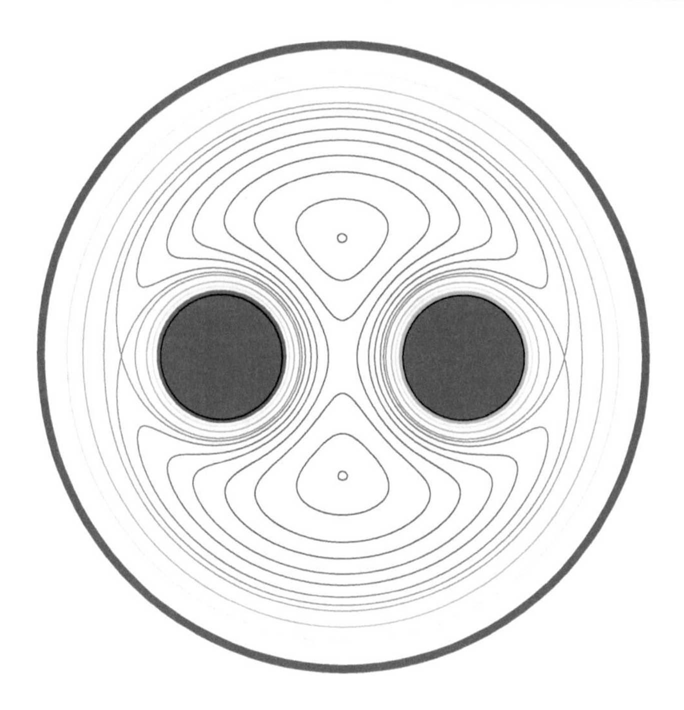

Figure 16.1. *Trajectories of a single point vortex placed in a triply connected circular domain.*

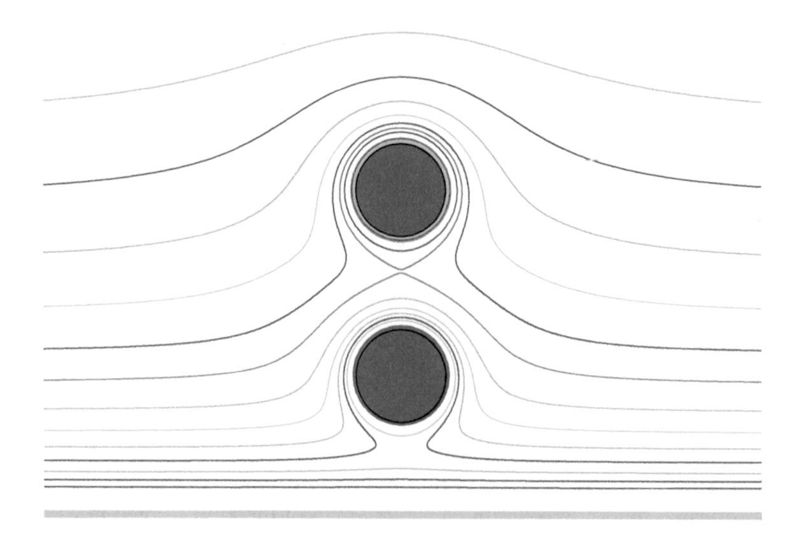

Figure 16.2. *Trajectories of a point vortex near two circular islands off an infinite coastline.*

where $z_a = f(a)$, the Hamiltonian $H^{(\zeta)}(a)$ is given in (16.14), and

$$\left.\frac{df}{d\zeta}\right|_{\zeta=a} = -\frac{2\mathrm{i}}{(a+1)^2}, \tag{16.17}$$

which follows from (16.15).

Figure 16.2 shows the example of two circular islands of unit diameter separated from each other by unit distance and with the island closest to the wall located at unit distance from it.

When the vortex is located far from the islands it will not feel their presence and will travel along the coastline at constant speed and at some fixed distance from it. Once it draws close to the offshore islands this straight line trajectory will be deflected. Physically, it is of interest to ask how far from the wall the incoming vortex must be in order to travel between the coastline and the first offshore island, or to penetrate the gap between the two offshore islands, or to skip over both offshore islands completely. The critical trajectories between these different eventualities are clear from Figure 16.2. Such matters are discussed in more detail in [49, 50].

We see how the framework of Part I can be used as a constructive tool, in combination with Lin's general theory [94], to calculate quantities of interest in the study of vortex motion. The function theory of Part I opens up the possibility of calculating the vortex motion of a wide range of geometries with relative ease. Other considerations, including the motion of multiple vortices, and vortex motion around complicated multiairfoil structures such as the Kasper wing, have been investigated in [115, 105] using the prime function machinery of Part I.

Chapter 17

Electric transport theory and the Hall effect

This chapter deals with a problem in electric transport theory and illustrates how the ideas of Part I can be used in perhaps unexpected ways. The focus here is on a problem in a simply connected domain where, indeed, well-known classical methods apply, for example, the usual form of the classical S–C formula for simply connected polygons. It turns out, however, that use of these classical methods presents mathematical and computational difficulties that, as we will show, can be overcome by thinking more broadly and embracing the mathematical ideas of Part I.

In two-dimensional electric transport theory, it is necessary to solve the equations

$$\nabla . \mathbf{j} = 0, \qquad \nabla \times \mathbf{E} = 0 \tag{17.1}$$

for an electric field $\mathbf{E} = (E_x, E_y)$ and a current $\mathbf{j} = (j_x, j_y)$ related by the transport equation

$$\begin{pmatrix} j_x \\ j_y \end{pmatrix} = \begin{bmatrix} \sigma_{xx} & -\sigma_{yx} \\ \sigma_{yx} & \sigma_{xx} \end{bmatrix} \begin{pmatrix} E_x \\ E_y \end{pmatrix}. \tag{17.2}$$

The quantities σ_{xx} and σ_{yx}, assumed here to be constant, are the longitudinal and Hall conductivities. The Hall angle is defined by

$$\delta = \tan^{-1}\left(\frac{\sigma_{xy}}{\sigma_{xx}}\right), \tag{17.3}$$

and this will enter the analysis later. From the second equation in (17.1) we deduce that the electric field is derivable from a potential ψ, i.e.,

$$\mathbf{E} = -\nabla\psi, \tag{17.4}$$

where, from the first equation in (17.1), ψ is harmonic wherever the field exists.

Let $z = x + \mathrm{i}y$, and let this harmonic function be the imaginary part of some complex potential,

$$w(z) = \phi + \mathrm{i}\psi, \tag{17.5}$$

where ϕ is the harmonic conjugate of ψ. It follows that

$$E_y + \mathrm{i}E_x = -\frac{dw}{dz} \qquad \text{or} \qquad E_x - \mathrm{i}E_y = \mathrm{i}\frac{dw}{dz}. \tag{17.6}$$

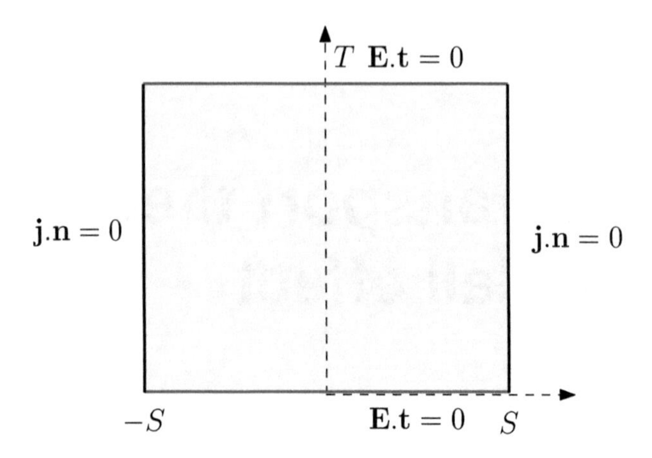

Figure 17.1. *Two-terminal conductance problem in a rectangular sample (z plane). Electrodes at the top and bottom edges of the sample force a potential difference across the sample. No current flows out of the side edges.*

If we can find dw/dz, we have determined the required electric field $\mathbf{E}$. If we introduce the complex number

$$\Sigma = \sigma_{xx} + \mathrm{i}\sigma_{yx}, \tag{17.7}$$

then it follows from (17.2) that

$$j_x + \mathrm{i}j_y = \Sigma(E_x + \mathrm{i}E_y). \tag{17.8}$$

In the two-terminal conductance problem a rectangular sample

$$-S < x < S, \qquad 0 < y < T \tag{17.9}$$

has two electrodes attached to its horizontal edges representing a source and a sink. A schematic is shown in Figure 17.1. The boundary conditions for this problem are that there is no electric field parallel to the electrodes:

$$\mathbf{E}.\mathbf{t} = 0 \qquad \text{on } y = 0, T, \tag{17.10}$$

where $\mathbf{t}$ denotes the unit tangent vector to the electrodes and there is no current flow out of the vertical sides of the sample, i.e.,

$$\mathbf{j}.\mathbf{n} = 0 \qquad \text{on } x = \pm S, \tag{17.11}$$

where $\mathbf{n}$ denotes the unit outward normal vector to the side walls.

It is useful to reformulate this problem in standard complex variable notation. Using the symbol $\mapsto$ to denote the complex form $a_x + \mathrm{i}a_y$ of a two-dimensional vector quantity (a_x, a_y), we have

$$\mathbf{t} \mapsto \frac{dz}{ds}, \qquad \mathbf{n} = -\mathrm{i}\frac{dz}{ds}, \tag{17.12}$$

where ds denotes the arclength element along a boundary. Boundary condition (17.10) can be written as

$$\mathbf{E}.\mathbf{t} \mapsto \mathrm{Re}\left[\mathrm{i}\frac{dw}{dz}\frac{dz}{ds}\right] = \mathrm{Re}\left[\mathrm{i}\frac{dw}{ds}\right] = 0, \qquad \text{on } y = 0, T, \tag{17.13}$$

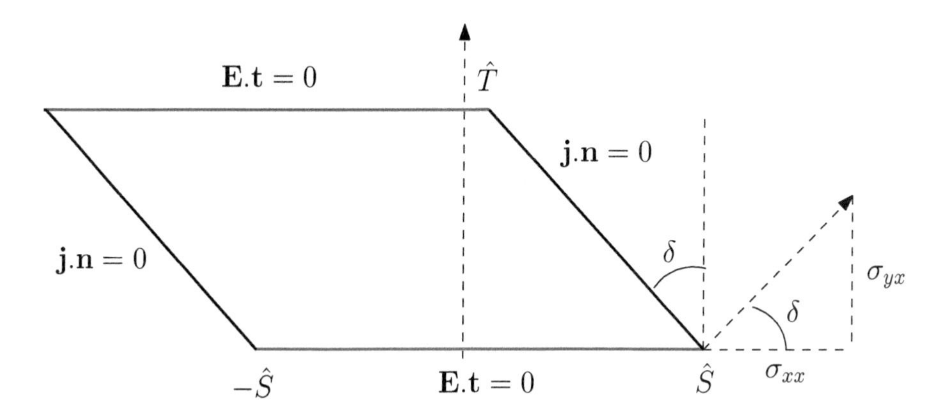

Figure 17.2. *The same two-terminal conductance problem as in Figure 17.1 but in a parallelogram sample ($\mathcal{Z}$ plane).*

while boundary condition (17.11) becomes

$$\mathbf{j}.\mathbf{n} \mapsto \text{Re}\left[\text{i}\overline{\Sigma}\frac{dw}{dz}\left(-\text{i}\frac{dz}{ds}\right)\right] = \text{Re}\left[\overline{\Sigma}\frac{dw}{ds}\right] = 0, \qquad \text{on } x = \pm S. \tag{17.14}$$

These two boundary conditions—of no tangential electric field and no normal current at a boundary ∂D of some domain D—can be written in differential form as, respectively,

$$\text{Re}\left[idw\right] = 0, \qquad \text{on } y = 0, T, \tag{17.15}$$

$$\text{Re}\left[\overline{\Sigma}dw\right] = 0, \qquad \text{on } x = \pm S. \tag{17.16}$$

These conditions can be shown to be *conformally invariant*: this means that if $w(z)$ is some complex potential satisfying either of these boundary conditions in some domain D in a complex z plane, then if we transform the given domain to another domain $\mathcal{D}$ in the complex $\mathcal{Z}$ plane via some conformal mapping

$$\mathcal{Z} = g(z), \qquad z = g^{-1}(\mathcal{Z}) \tag{17.17}$$

and define

$$\mathcal{W}(\mathcal{Z}) \equiv w(z) = w(g^{-1}(\mathcal{Z})), \tag{17.18}$$

then $\mathcal{W}(\mathcal{Z})$ is the complex potential in the $\mathcal{Z}$ plane and it will satisfy conditions analogous to (17.15) and (17.16) on the image of that portion of the original boundary where the same condition pertains. That is,

$$\text{Re}\left[\text{i}d\mathcal{W}\right] = 0 \tag{17.19}$$

on the image under $g(z)$ of the two boundary components $y = 0, T$ and

$$\text{Re}\left[\overline{\Sigma}d\mathcal{W}\right] = 0 \tag{17.20}$$

on the image under $g(z)$ of the two boundary components $x = \pm S$.

Suppose now that, in some complex $\mathcal{Z} = \mathcal{X} + \text{i}\mathcal{Y}$ plane, we take the domain $\mathcal{D}$ to be the parallelogram shown in Figure 17.2 with side walls that are perpendicular to the vector $(\sigma_{xx}, \sigma_{yx})$, and let

$$\mathcal{W}(\mathcal{Z}) = \lambda\mathcal{Z}, \tag{17.21}$$

where $\lambda \in \mathbb{R}$ is any real constant determined by the size of the voltage drop across the electrodes. This is clearly an analytic function of $\mathcal{Z}$ and therefore constitutes an admissible complex potential. The bottom and top edges of the parallelogram can be parametrized by

$$\mathcal{Z} = \mathcal{X}, \mathcal{X} + \mathrm{i}\hat{\mathcal{T}}, \qquad \mathcal{X} \in \mathbb{R}, \tag{17.22}$$

for some constant $\hat{\mathcal{T}}$ so that

$$d\mathcal{W} = d(\lambda\mathcal{Z}) = \lambda d\mathcal{X} \tag{17.23}$$

and, on these walls,

$$\mathrm{Re}\,[\mathrm{i}d\mathcal{W}] = \mathrm{Re}\,[\mathrm{i}\lambda d\mathcal{X}] = 0. \tag{17.24}$$

Hence (17.21) satisfies the boundary condition of no tangential electric field on these top and bottom walls. On the other hand, the slanting side walls of the parallelogram can be parametrized by

$$\mathcal{Z} = \pm\hat{\mathcal{S}} + \mathrm{i}\frac{\Sigma}{|\Sigma|}\mathcal{S}, \qquad \mathcal{S} \in \mathbb{R}, \tag{17.25}$$

where $\mathcal{S}$ is the arclength along the boundary. Hence, on these walls,

$$\mathrm{Re}\left[\overline{\Sigma}d\mathcal{W}\right] = \mathrm{Re}\left[\overline{\Sigma}d(\lambda\mathcal{Z})\right] = \mathrm{Re}\left[\overline{\Sigma}\mathrm{i}\frac{\lambda\Sigma}{|\Sigma|}d\mathcal{S}\right] = 0, \tag{17.26}$$

which means that (17.21) also satisfies the boundary condition of no normal current on these side walls.

The conclusion is that (17.21) is a solution of the two-terminal conductance problem in the parallelogram where electrodes are placed on the top and bottom walls and with no current through the slanting side walls.

Unfortunately, this is not the domain in which we are interested in solving the problem: we want to solve the same boundary value problem in the rectangular domain (17.9) in Figure 17.1.

However, by the conformal invariance of the boundary value problem just observed, we just need to find a conformal map (17.17), $\mathcal{Z} = g(z)$, from the rectangle (17.9) to the parallelogram. Then the required complex potential $w(z)$ in the rectangle will be

$$w(z) = \mathcal{W}(\mathcal{Z}) = \lambda\mathcal{Z} = \lambda g(z). \tag{17.27}$$

The first equality follows from the conformal invariance property just explained, the second equality follows on use of the known parallelogram solution (17.21) just established, and the third equality is just a statement of the conformal mapping (17.17) relating $\mathcal{Z}$ to z. By normalizing the potential difference across the sample, we can set $\lambda = 1$.

As a result of all these observations we see from (17.27) that solving the original problem for the complex potential $w(z)$ in the rectangular sample is equivalent to constructing the conformal mapping from the rectangle in Figure 17.1 to the parallelogram in Figure 17.2.

Since both the rectangle in Figure 17.1 and the parallelogram in Figure 17.2 are simply connected polygons, one way to find the required mapping from the rectangle to the parallelogram is to construct two S–C mappings, from a unit disc or an upper half plane in a parametric ζ plane, say, one to each of these simply connected polygons. This is exactly the approach used by Wick [125]. While this approach provides an expression for the required conformal mapping from the rectangle to the parallelogram, and hence an expression for the required complex potential determining the electric field in the rectangle, it is inconvenient in two respects. First, it involves the calculation of an inverse S–C mapping; second, the integrands of this S–C mapping become highly singular in the physically important case of $\delta \to \pi/2$, which means that the integral becomes difficult to evaluate.

Another approach based on the idea of directly mapping the rectangle to the parallelogram was put forward by Rendell and Girvin [108]; we will emulate the spirit of their method but offer a different approach. Their original analysis does not make use of the prime function and is different from what follows. The following analysis showcases the usefulness of the prime function in this problem and, in particular, the exploitation of the properties of Cayley-type maps in the same spirit of what was done in §7.10 of Chapter 7. Recent interest in the quantum Hall effect has reinvigorated interest in the analysis of Rendell and Girvin—see [60]—so the analysis here, which appears to be new, should be of interest in those applications.

Consider the map from the upper half annulus

$$\rho < |\zeta| < 1, \qquad \mathrm{Im}[\zeta] > 0, \tag{17.28}$$

to the rectangle in a z plane given by

$$-S < x < S, \qquad 0 < y < T. \tag{17.29}$$

This can be effected using an elementary map given by the logarithm:

$$z = \mathcal{Z}(\zeta) = S\left[\frac{2\log(\zeta/\rho)}{\log(1/\rho)} - 1\right], \qquad \rho = e^{-2\pi S/T}. \tag{17.30}$$

Under this mapping, the two upper half circles C_0^+ and C_1^+ are transplanted to the two *vertical* edges of the rectangle at $x = \pm S$.

Remark: In Chapter 5, the elementary logarithmic mapping (17.30) was seen to be a special case of a logarithm of the so-called annular slit mapping (5.46). It is also the simplest example of a periodic S–C map falling within the case 3 category explained in §7.7 of Chapter 7.

Now consider the function dw/dz. It is useful to think of the differential dz as one traces around the boundary of the rectangle and of dw as one traces around the boundary of the parallelogram; see Figure 17.3 for a schematic. We used exactly the same idea in §7.10 when studying the problem of a slit in a channel. A little thought reveals that this quantity has piecewise constant argument; this observation should prompt the reader to think of the Cayley-type maps studied in Part I. For the concentric annulus, recall that Cayley-type maps are functions of the form

$$\frac{\omega(z,a)}{\omega(z,b)}, \qquad a, b \in C_0 \text{ or } a, b \in C_1. \tag{17.31}$$

Given these, it is natural to attempt to construct the function dw/dz as a function of ζ. To do so, we also notice that the argument of $d\zeta$ changes by $\pi/2$ as it passes through the prevertices at $1, -1, -\rho, \rho$, with the same feature being true of dz as it passes through the vertices of the rectangle. On the other hand, the corresponding changes in argument of dw as it passes through the vertices of the parallelogram are $\pi/2+\delta, \pi/2-\delta, \pi/2+\delta, \pi/2-\delta$. From these considerations, we deduce that

$$\frac{dw}{dz} = A\left[\frac{\omega(\zeta,-1)\omega(\zeta,+\rho)}{\omega(\zeta,+1)\omega(\zeta,-\rho)}\right]^{\frac{2\delta}{\pi}}, \qquad A = e^{2i\delta}, \tag{17.32}$$

where the choice of the constant A ensures that dw/dz is real when ζ is real.

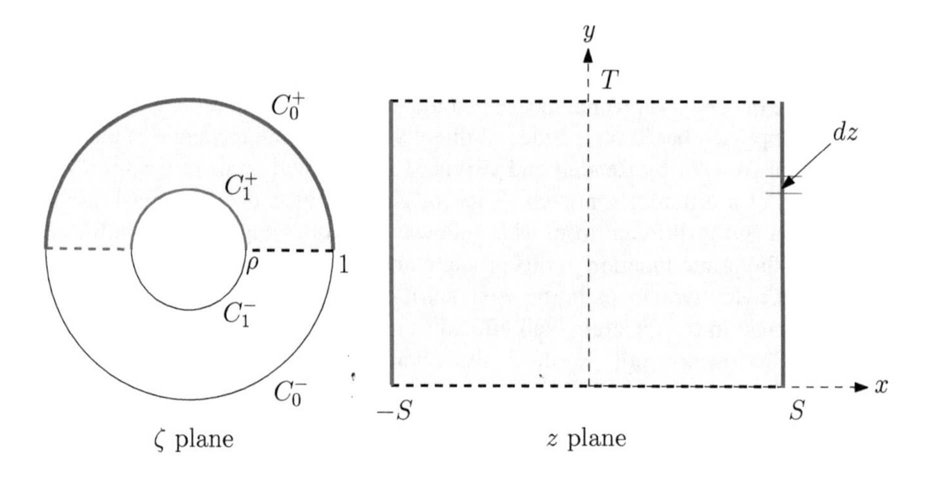

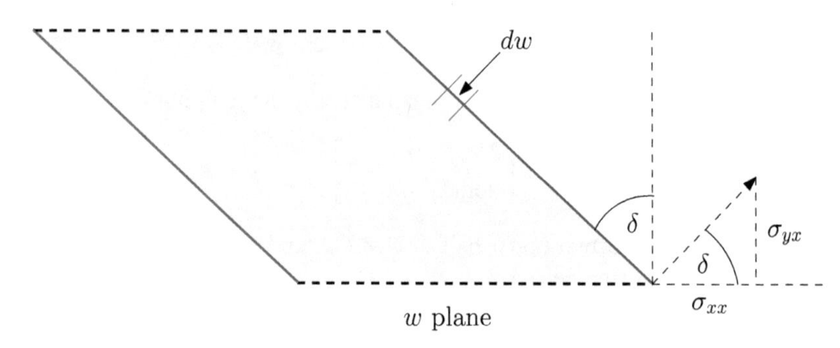

Figure 17.3. *Construction of a conformal mapping from a rectangle to a parallelogram using Cayley-type maps in an annulus.*

The solution for $w(z)$, or, equivalently, the conformal mapping taking the rectangle to the parallelogram, can now be written in parametric form as

$$z = \mathcal{Z}(\zeta) = S\left[\frac{2\log(\zeta/\rho)}{\log(1/\rho)} - 1\right], \qquad \rho = e^{-2\pi S/T}, \tag{17.33}$$

$$w = \mathcal{W}(\zeta) = \int_{\sqrt{\rho}}^{\zeta}\left[-\frac{\omega(\zeta,-1)\omega(\zeta,+\rho)}{\omega(\zeta,+1)\omega(\zeta,-\rho)}\right]^{\frac{2\delta}{\pi}} \frac{2S}{\log(1/\rho)}\frac{d\zeta}{\zeta}. \tag{17.34}$$

In the Hall effect problem, the electric field itself is related to dw/dz as in (17.6), and therefore formula (17.32) can be used to calculate the inductance of the sample.

There is another way to proceed, which is of advantage if the aspect ratio of the rectangle is too small, i.e., if $S/T \ll 0$.

Suppose we construct a *different* conformal mapping, from an upper half annulus in a parametric η plane, say,

$$\tilde{\rho} < |\eta| < 1, \qquad \mathrm{Im}[\eta] > 0, \tag{17.35}$$

to the rectangular sample, but now where C_0^+ and C_1^+ are the preimages of the *horizontal* sides of the sample at $y = 0, T$. Figure 17.4 shows a schematic for this case. Such a

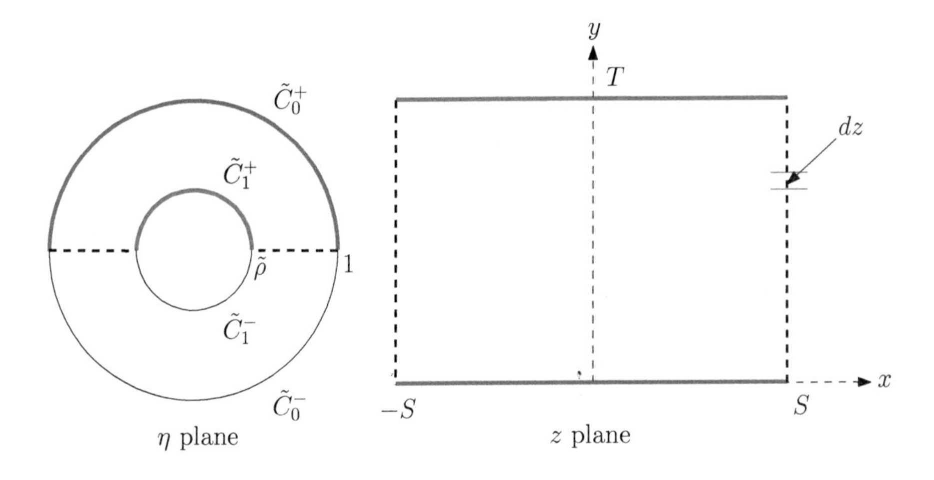

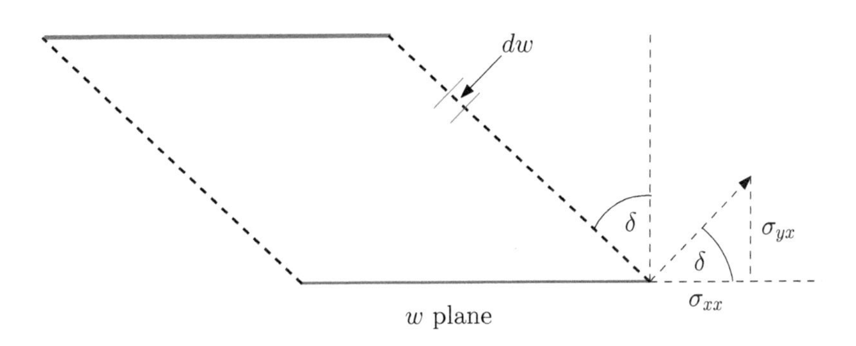

Figure 17.4. *Alternative parametrization of the same conformal mapping from the rectangle to the parallelogram as in Figure 17.3.*

mapping is found to be

$$z = \tilde{\mathcal{Z}}(\eta) = -\frac{2\mathrm{i}S}{\pi}\left[\log\eta - \frac{\mathrm{i}\pi}{2}\right], \qquad \tilde{\rho} = e^{-\pi T/(2S)}. \tag{17.36}$$

All the same arguments concerning dw/dz continue to hold, the only difference now being that the prevertices corresponding to turning angles $\pi/2+\delta, \pi/2-\delta, \pi/2+\delta, \pi/2-\delta$ have now been cyclically permuted from $1, -1, -\rho, \rho$ to $\rho, 1, -1, -\rho$. Consequently, the solution for $w(z)$, or, equivalently, the conformal mapping taking the rectangle to the parallelogram, has the alternative parametric form

$$z = \tilde{\mathcal{Z}}(\eta) = -\frac{2\mathrm{i}S}{\pi}\left[\log\eta - \frac{\mathrm{i}\pi}{2}\right], \tag{17.37}$$

$$w = \tilde{\mathcal{W}}(\eta) = \int_{\mathrm{i}}^{\eta}\left[-\frac{\tilde{\omega}(\eta,+1)\tilde{\omega}(\eta,-\tilde{\rho})}{\omega(\eta,-1)\omega(\eta,+\tilde{\rho})}\right]^{\frac{2\delta}{\pi}}\left(-\frac{2\mathrm{i}S}{\pi}\frac{d\eta}{\eta}\right), \quad \tilde{\rho} = e^{-\pi T/(2S)}. \tag{17.38}$$

An important observation is that when $S/T \gg 1$ then $\rho \to 0$; when $S/T \ll 1$ then $\tilde{\rho} \to 0$. As a general rule of thumb, it requires less effort, numerically speaking, to accurately evaluate the prime function when the radii $\{q_j | j = 1, \ldots, M\}$ of the circles $\{C_j | j = 1, \ldots, M\}$ are much smaller than unity (the radius of C_0).

This same mathematical idea—of swapping which part of the boundary the images of the circles $\{C_j | j = 1, \ldots, M\}$ correspond to—was the topic of Exercise 6.13 on "swapping the correspondences." It has also been exploited by Crowdy and Marshall [44], who use it to find two alternative representations of the conformal map from a circular preimage region to a given multiply connected quadrature domain of the kind considered in Chapter 11.

A key lesson from this chapter is therefore that the function theoretical framework of Part I can also be used, and offer advantages, even in situations not ostensibly involving multiply connected domain geometries.

Chapter 18

Laminar flow in ducts

The problem of pressure-driven flow in ducts, or pipes, of differing cross-sectional shape is an important one in fluid mechanics and heat transfer. The illustrative problems chosen here involve simply connected duct cross-sections, yet the ideas of Part I still afford various advantages in tackling them. At the very least, they offer a novel approach that differs from those usually employed. As in Chapter 17, the message is again that the mathematical framework of Part I can be of use even when the problem at hand does not ostensibly involve a multiply connected domain.

Suppose the axis of a duct carrying a viscous fluid is parallel to the Z axis in Cartesian (x, y, Z) coordinates. The hydrodynamic problem for steady unidirectional flow, $\mathbf{u} = (0, 0, w(x, y))$, where $w(x, y)$ is the axial velocity, driven by a constant pressure gradient $\partial P/\partial Z$, is

$$\nabla^2 w(x, y) = \frac{\partial P}{\partial Z} = -1, \tag{18.1}$$

where, within a suitable nondimensionalization, we can take the magnitude of the pressure gradient to be -1. The no-slip boundary condition requires that

$$w = 0 \tag{18.2}$$

on the side walls of the duct. The boundary value problem expressed in (18.1) and (18.2) is therefore Poisson's equation with Dirichlet-type boundary conditions.

Suppose the duct is rectangular: $-L/2 \leq x \leq L/2$ and $0 \leq y \leq H$. We wish to solve the boundary value problem (18.1)–(18.2) in this rectangular domain. On isolating a convenient particular solution, the general solution can be written as

$$w = \frac{y(H - y)}{2} + \text{Im}[g(z)], \tag{18.3}$$

where, now, the analytic function $g(z)$ of $z = x + iy$ is to be found. The chosen particular solution satisfies the no-slip boundary conditions on the top and bottom walls $y = 0, H$. Other choices of this particular solution are possible.

A common approach to this problem is to use Fourier series in both the x and y directions and to represent the solution as an infinite sum of separable solutions. Batchelor [65] discusses the solution of this problem using such methods. These sums can be poorly convergent, however.

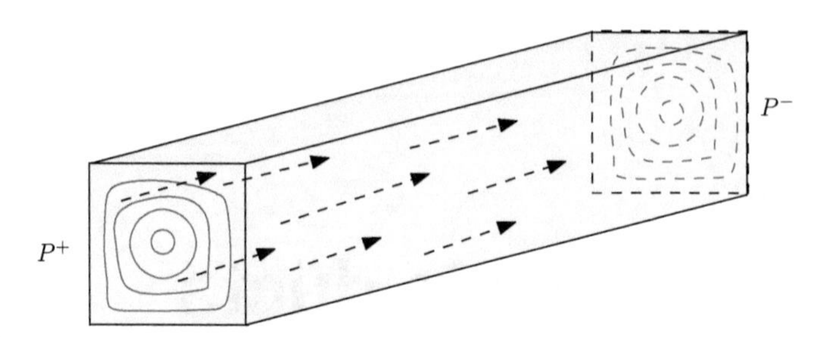

Figure 18.1. *Unidirectional flow along a rectangular duct due to a uniform pressure gradient along the channel due to the pressure difference between P^+ and P^-.*

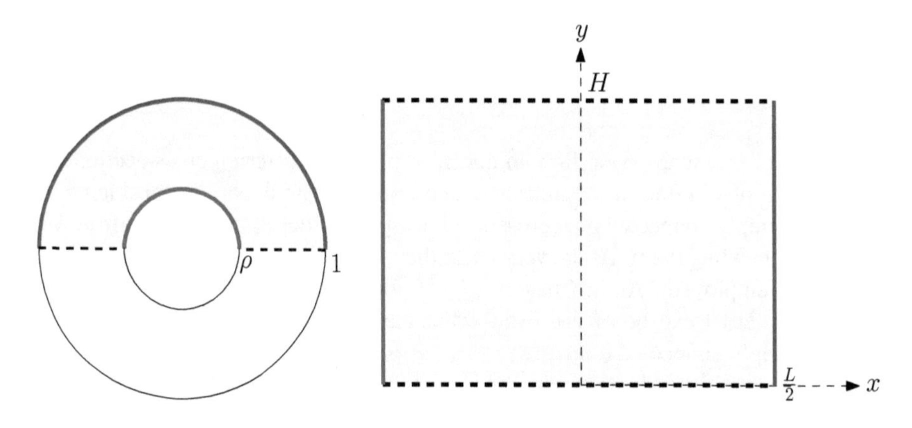

Figure 18.2. *Conformal mapping (18.5) of the upper half annulus $\rho < |\zeta| < 1, \mathrm{Im}[\zeta] > 0$, to the rectangular duct cross-section.*

We will approach the matter differently and introduce a conformal map from the concentric annulus

$$\rho < |\zeta| < 1 \tag{18.4}$$

to the rectangular cross-section of the duct. It is readily found to be

$$z = \mathcal{Z}(\zeta) = L\left[\frac{\log(\zeta/\rho)}{\log(1/\rho)}\right] - \frac{L}{2}, \qquad \rho = e^{-\pi L/H}. \tag{18.5}$$

Remark: This elementary logarithmic map, which was also deployed in Chapter 17, is a special case of a logarithm of the annular slit map (5.46).

Under this map the upper half annulus

$$\rho < |\zeta| < 1, \qquad \mathrm{Im}[\zeta] > 0, \tag{18.6}$$

is transplanted to the rectangular flow domain with the correspondences

$$\mathcal{Z}(\rho) = -\frac{L}{2}, \qquad \mathcal{Z}(1) = +\frac{L}{2}, \qquad \mathcal{Z}(-1) = +\frac{L}{2} + \mathrm{i}H, \qquad \mathcal{Z}(-\rho) = -\frac{L}{2} + \mathrm{i}H. \tag{18.7}$$

The two segments of the real ζ axis $[-1, -\rho]$ and $[\rho, 1]$ are transplanted to the no-slip walls at $y = 0, H$.

Now let

$$G(\zeta) \equiv g(\mathcal{Z}(\zeta)), \tag{18.8}$$

and write (18.3) as

$$w = \frac{(\mathcal{Z}(\zeta) - \overline{\mathcal{Z}(\zeta)})}{4\mathrm{i}} \left[H - \frac{\mathcal{Z}(\zeta) - \overline{\mathcal{Z}(\zeta)})}{2\mathrm{i}} \right] + \mathrm{Im}[G(\zeta)], \tag{18.9}$$

where it is now $G(\zeta)$ that must be determined. The latter function must be analytic in the upper half ζ annulus (18.6).

The boundary conditions on the walls at $y = 0, H$, together with (18.3), imply that

$$\mathrm{Im}[G(\zeta)] = \frac{G(\zeta) - \overline{G(\zeta)}}{2\mathrm{i}} = 0 \qquad \text{on } \overline{\zeta} = \zeta, \tag{18.10}$$

or

$$\overline{G}(\zeta) = G(\zeta). \tag{18.11}$$

The Schwarz conjugate function of $G(\zeta)$ is therefore the same as $G(\zeta)$. By the Schwarz reflection principle, we conclude that $G(\zeta)$ is analytic in the *entire* annulus $\rho < |\zeta| < 1$ and not just in the upper half annulus.

The upper half semicircles of C_0 and C_1, which we denote by C_0^+ and C_1^+, are the preimages of the two side walls at $x = \pm L/2$. Therefore

$$\mathrm{Im}[G(\zeta)] = -\frac{(\mathcal{Z}(\zeta) - \overline{\mathcal{Z}(\zeta)})}{4\mathrm{i}} \left[H - \frac{\mathcal{Z}(\zeta) - \overline{\mathcal{Z}(\zeta)})}{2\mathrm{i}} \right], \qquad \zeta \in C_0^+, C_1^+. \tag{18.12}$$

Now suppose that $\zeta \in C_0^-, C_1^-$ where this notation means the lower half semicircles of C_0 and C_1. Then $\overline{\zeta} \in C_0^+, C_1^+$. Furthermore,

$$\mathrm{Im}[G(\zeta)] = -\mathrm{Im}[\overline{G(\zeta)}] = -\mathrm{Im}[G(\overline{\zeta})] = \frac{(\mathcal{Z}(\overline{\zeta}) - \overline{\mathcal{Z}(\overline{\zeta})})}{4\mathrm{i}} \left[H - \frac{\mathcal{Z}(\overline{\zeta}) - \overline{\mathcal{Z}(\overline{\zeta})})}{2\mathrm{i}} \right], \tag{18.13}$$

where, in the first equality, we have used the fact that the imaginary parts of complex conjugate quantities are the negative of each other, the second equality follows on use of (18.11), and the third equality follows from (18.12) and is valid because $\overline{\zeta} \in C_0^+, C_1^+$. We conclude that

$$\mathrm{Im}[G(\zeta)] = \begin{cases} -\dfrac{(\mathcal{Z}(\zeta) - \overline{\mathcal{Z}(\zeta)})}{4\mathrm{i}} \left[H - \dfrac{\mathcal{Z}(\zeta) - \overline{\mathcal{Z}(\zeta)})}{2\mathrm{i}} \right], & \zeta \in C_0^+, C_1^+, \\[2ex] \dfrac{(\mathcal{Z}(\overline{\zeta}) - \overline{\mathcal{Z}(\overline{\zeta})})}{4\mathrm{i}} \left[H - \dfrac{\mathcal{Z}(\overline{\zeta}) - \overline{\mathcal{Z}(\overline{\zeta})})}{2\mathrm{i}} \right], & \zeta \in C_0^-, C_1^-. \end{cases} \tag{18.14}$$

On introducing

$$\mathcal{G}(\zeta) = -\mathrm{i}G(\zeta), \tag{18.15}$$

(18.14) can be written in the form

$$\mathrm{Re}[\mathcal{G}(\zeta)] = \begin{cases} -\dfrac{(\mathcal{Z}(\zeta) - \overline{\mathcal{Z}(\zeta)})}{4\mathrm{i}} \left[H - \dfrac{\mathcal{Z}(\zeta) - \overline{\mathcal{Z}(\zeta)})}{2\mathrm{i}} \right], & \zeta \in C_0^+, C_1^+, \\[2ex] \dfrac{(\mathcal{Z}(\overline{\zeta}) - \overline{\mathcal{Z}(\overline{\zeta})})}{4\mathrm{i}} \left[H - \dfrac{\mathcal{Z}(\overline{\zeta}) - \overline{\mathcal{Z}(\overline{\zeta})})}{2\mathrm{i}} \right], & \zeta \in C_0^-, C_1^-. \end{cases} \tag{18.16}$$

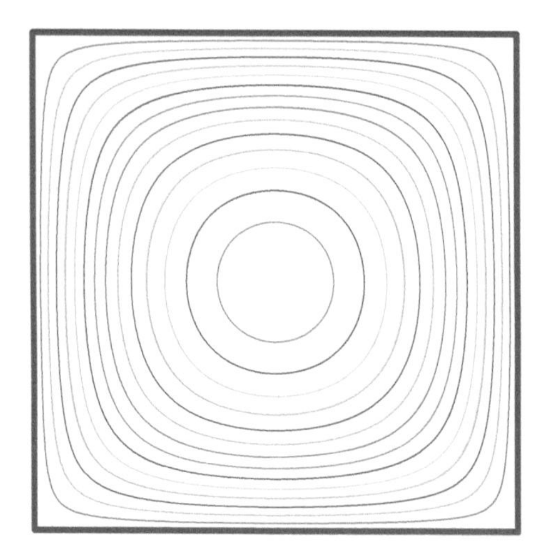

Figure 18.3. *Contours of constant longitudinal velocity for flow along a square duct.*

Since we know that $G(\zeta)$, and hence $\mathcal{G}(\zeta)$, are analytic in the entire annulus $\rho < |\zeta| < 1$, (18.16) is a standard Schwarz problem for $\mathcal{G}(\zeta)$, as considered in Chapter 13. We observe the important fact that the data on the right-hand side is continuous at $\zeta = \pm 1, \pm\rho$, where C_0^+ meets C_0^- and C_1^+ meets C_1^-. This means that it is not necessary to invoke any of the additional measures expounded in §13.8 for nonsmooth data in a problem of Schwarz type. The solution for $\mathcal{G}(\zeta)$ can therefore be written as

$$\mathcal{G}(\zeta) = A_1 v_1(\zeta) + \hat{\mathcal{G}}(\zeta), \tag{18.17}$$

where A_1 is a constant and $\hat{\mathcal{G}}(\zeta)$ is analytic and single-valued in the annulus and can be represented by the Villat integral formula. In fact, by the symmetry of the flow it can be shown that $A_1 = 0$. Hence, up to an unimportant purely imaginary constant,

$$\begin{aligned}\mathcal{G}(\zeta) = & \frac{1}{2\pi \mathrm{i}} \int_{C_0^+} \phi_0^+ \left[2d\log\omega(z,\zeta) - d\log z\right] + \frac{1}{2\pi \mathrm{i}} \int_{C_0^-} \phi_0^- \left[2d\log\omega(z,\zeta) - d\log z\right] \\ & - \frac{1}{2\pi \mathrm{i}} \int_{C_1^+} \phi_1^+ \left[2d\log\omega(z,\zeta)\right] - \frac{1}{2\pi \mathrm{i}} \int_{C_1^-} \phi_1^- \left[2d\log\omega(z,\zeta)\right],\end{aligned} \tag{18.18}$$

where $\omega(.,.)$ is the prime function for the annulus, and where ϕ_0^+, ϕ_0^- denote the values of $\mathrm{Re}[\mathcal{G}(\zeta)]$ on C_0^+, C_0^-, as given on the right-hand side of (18.16), and similarly for ϕ_1^+, ϕ_1^-. Since it follows from (18.3) that

$$w(x,y) = \frac{y(H-y)}{2} + \mathrm{Re}[\mathcal{G}(\zeta)], \tag{18.19}$$

we have derived an explicit integral representation of the required velocity field.

Figure 18.3 shows the contours of constant longitudinal velocity for flow in a square duct with $L = H$ so that $\rho = e^{-\pi}$. These have been computed on the basis of formula (18.19).

Suppose now that the duct takes the form of a circular pipe with two reentrant fins at diametrically opposite locations on the interior of the pipe wall; see the middle schematic

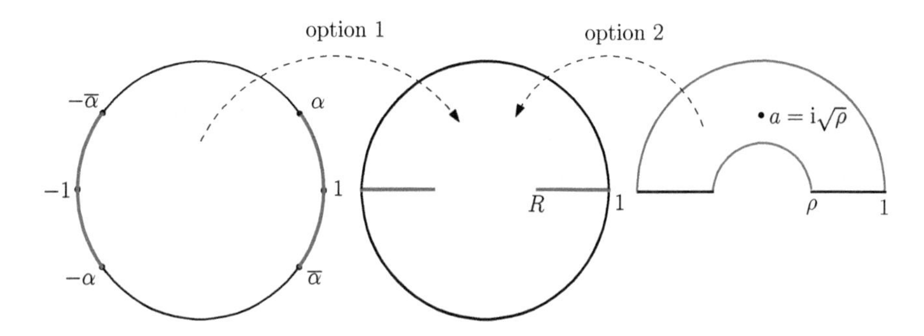

Figure 18.4. *Two options for computing flow in a two-finned circular duct: a mapping from a unit disc (option* 1*) or from an upper half annulus (option* 2*).*

in Figure 18.4. The challenge is again to solve (18.1)–(18.2). Now a convenient form of the general solution in this case is

$$w = \frac{1}{4}(1 - z\overline{z}) + \mathrm{Im}[g(z)], \tag{18.20}$$

where the analytic function $g(z)$ is to be found. We have taken a different form of the particular solution that is more convenient for this duct geometry.

There are two options for tackling this problem, options 1 and 2.

Since the cross-section of the duct is simply connected, option 1 is to use a polycircular arc mapping from a disc to the flow domain constructed using the methods of Chapter 10. Figure 10.6 with $\rho = 0$ will capture the geometry of interest here. The left two schematics in Figure 18.4 illustrate this conformal mapping approach.

Option 2 is to make use of the special form of radial slit map from the upper half annulus

$$\rho < |\zeta| < 1, \qquad \mathrm{Im}[\zeta] > 0, \tag{18.21}$$

considered in (6.79), i.e.,

$$z = \mathcal{Z}(\zeta) = -\frac{\omega(\zeta, a)\omega(\zeta, 1/\overline{a})}{\omega(\zeta, \overline{a})\omega(\zeta, 1/a)}, \qquad a = \mathrm{i}\sqrt{\rho}, \tag{18.22}$$

where $\omega(.,.)$ is the prime function of the annulus and the adjustable parameter a has been chosen to ensure that both fins are of the same length. The right two schematics in Figure 18.4 illustrate this alternative approach.

In either case, as will be seen, the boundary value problem for $w(x, y)$ can be reduced to a standard Schwarz problem for an analytic function in some domain. While option 1 appears simpler in that it involves mapping from the simply connected unit disc, where the well-known Poisson integral formula can be used to express the solution of the relevant Schwarz problem, the conformal mapping required is given by the solution of a third order nonlinear differential equation that must be solved numerically. While this is certainly possible—indeed the relevant differential equation was constructed and solved explicitly in §10.9 of Chapter 10—it seems preferable to make use of the mapping from the doubly connected annulus, not least because the conformal mapping (18.22) is available as an explicit function in terms of the prime function of the annulus. Moreover, solving the resulting Schwarz problem in the annulus is no more challenging since the relevant solution can be expressed explicitly using the Villat integral formula. We saw in Chapter 13 that the latter is nothing other than the natural generalization of the Poisson integral formula to the annulus.

In what follows we present the details for solution via option 2; the reader may find it instructive to carry out the analysis for option 1 and to compare the two methods. It is appropriate to remark that Exercise 13.5 was concerned with elucidating the connection between these two distinct approaches.

To proceed we introduce

$$G(\zeta) \equiv g(\mathcal{Z}(\zeta)) \tag{18.23}$$

and write (18.20) as

$$w = \frac{1}{4}(1 - z\overline{z}) + \mathrm{Im}[G(\zeta)]. \tag{18.24}$$

We now emulate all of the preceding analysis of the rectangular duct to find an integral representation for $G(\zeta)$.

The function $G(\zeta)$ is known to be analytic in the upper half annulus

$$\rho < |\zeta| < 1, \qquad \mathrm{Im}[\zeta] > 0. \tag{18.25}$$

However the boundary condition on the two portions of the real axis $[-1, -\rho]$ and $[\rho, 1]$, together with the chosen functional form (18.24), imply that

$$\overline{G(\zeta)} = \overline{G}(\zeta) = G(\zeta) \quad \text{on } \overline{\zeta} = \zeta. \tag{18.26}$$

By the Schwarz reflection principle, we deduce that $G(\zeta)$ is analytic in the *entire* annulus $\rho < |\zeta| < 1$.

The semicircles C_0^+ and C_1^+ in the upper half plane are the preimages of the two fins. Therefore the no-slip boundary condition (18.2) implies

$$\mathrm{Im}[G(\zeta)] = \frac{1}{4}(\mathcal{Z}(\zeta)\overline{\mathcal{Z}(\zeta)} - 1), \qquad \zeta \in C_0^+, C_1^+. \tag{18.27}$$

Now suppose that $\zeta \in C_0^-, C_1^-$; then $\overline{\zeta} \in C_0^+, C_1^+$ and

$$\mathrm{Im}[G(\zeta)] = -\mathrm{Im}[\overline{G(\zeta)}] = -\mathrm{Im}[G(\overline{\zeta})] = -\frac{1}{4}(\mathcal{Z}(\overline{\zeta})\overline{\mathcal{Z}(\overline{\zeta})} - 1), \tag{18.28}$$

where, in the first equality, we have used the fact that the imaginary parts of complex conjugate quantities have opposite signs, the second equality follows on use of (18.26), and the third equality follows from (18.27) and is valid because $\overline{\zeta} \in C_0^+, C_1^+$. We conclude that

$$\mathrm{Im}[G(\zeta)] = \begin{cases} \frac{1}{4}(\mathcal{Z}(\zeta)\overline{\mathcal{Z}(\zeta)} - 1), & \zeta \in C_0^+, C_1^+, \\ -\frac{1}{4}(\mathcal{Z}(\overline{\zeta})\overline{\mathcal{Z}(\overline{\zeta})} - 1), & \zeta \in C_0^-, C_1^-, \end{cases} \tag{18.29}$$

or, on introduction of

$$\mathcal{G}(\zeta) = -\mathrm{i}G(\zeta), \tag{18.30}$$

we have

$$\mathrm{Re}[\mathcal{G}(\zeta)] = \begin{cases} \frac{1}{4}(\mathcal{Z}(\zeta)\overline{\mathcal{Z}(\zeta)} - 1), & \zeta \in C_0^+, C_1^+, \\ -\frac{1}{4}(\mathcal{Z}(\overline{\zeta})\overline{\mathcal{Z}(\overline{\zeta})} - 1), & \zeta \in C_0^-, C_1^-. \end{cases} \tag{18.31}$$

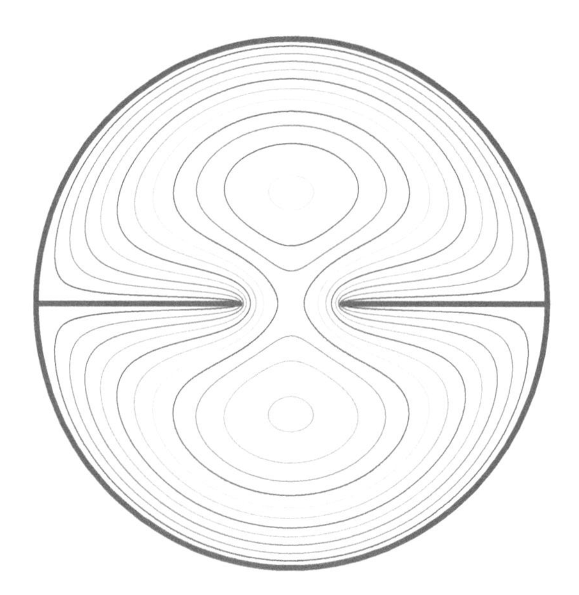

Figure 18.5. *Constant-velocity contours in a two-finned duct with $R = 0.2$ reproducing Figure 4*(a) *of Hu and Chang [92].*

This is a standard Schwarz problem for $\mathcal{G}(\zeta)$ as considered in Chapter 13. Its solution can be written as

$$\mathcal{G}(\zeta) = A_1 v_1(\zeta) + \hat{\mathcal{G}}(\zeta), \tag{18.32}$$

where $\hat{\mathcal{G}}(\zeta)$ is analytic and single-valued in the annulus and can be represented by the Villat integral formula. By the symmetry of the configuration, $A_1 = 0$ and the solution for $\mathcal{G}(\zeta)$ is given in the form of a Villat integral representation akin to (18.18), the only difference being that the data $\phi_0^{\pm}, \phi_1^{\pm}$ is now given by the right-hand side of (18.31).

Figure 18.5 shows a reproduction, using the method just described, of Figure 4(a) from Hu and Chang [92] which was found using quite different techniques.

It should be clear how to extend this analysis via option 2 to any finite number of reentrant fins. The only adjustment is to increase the connectivity of the preimage circular domain D_ζ and to employ the same expression (18.22) for the radial slit map, now using the relevant prime function. To solve the modified Schwarz problem defined in D_ζ that then arises from generalizing the above analysis, it is simply necessary to deploy the relevant integral representation for its solution as given in Chapter 13.

Again we see that, given the prime function machinery developed in Part I, problems involving simply connected geometries having no "holes" can be solved, in many cases more easily, using the prime function associated with higher connected geometries. Readers are therefore encouraged to think creatively even when solving problems in ostensibly simply connected domains.

Chapter 19

Torsion of hollow prismatic rods

For a rod with a uniform simply connected cross-section along its length, it is known that the boundary value problem governing the torsion of prismatic rods is identical to the laminar flow of viscous fluids in ducts, as considered in Chapter 18 [124]. For a hollow rod having a multiply connected cross-section, differences arise that are associated with requirements of single-valuedness of a so-called warping function.

In the laminar duct flow problems of Chapter 18, we showed how to use the theory of the generalized Poisson kernel of Chapter 13 to find integral expressions for the flows. In this chapter we solve the problem of torsion in a special class of hollow rods whose cross-sections are quadrature domains of the kind considered in Chapter 11. In such a case, the theory of Part I can be used to solve the problem in closed form without the use of the generalized Poisson kernel.

In the Saint-Venant theory of torsion, a hollow rod, with axis aligned with the Z axis in Cartesian coordinates (x, y, Z), and having a uniform doubly connected cross-section along its axis, experiences a torque at one of its ends, the precise details of the application of which are not important, with the boundary of the rod away from its ends being free of stress. Figure 19.1 shows a schematic of the physical set-up. There are two mathematical approaches to solving for the stresses in the rod: one uses the displacements and results in a problem of Neumann type for a harmonic warping function; the other uses a stress formulation and is the one we will adopt here.

The Prandtl stress function Ψ satisfies, in each (x, y) cross-section D, the partial differential equation

$$\nabla^2\Psi = -1, \qquad z \in D, \qquad \nabla^2 = \frac{\partial^2}{\partial x^2} + \frac{\partial^2}{\partial y^2} \tag{19.1}$$

with boundary condition

$$\Psi = \begin{cases} 0, & z \in \partial D_0, \\ \gamma, & z \in \partial D_1, \end{cases} \tag{19.2}$$

where γ is some constant to be determined; it cannot be arbitrarily specified but is found from the requirement that an associated warping function is single-valued. Physically this means that the rod is free of any screw dislocations. If we write the general solution of (19.1) in the complex-variable form

$$\Psi = -\frac{z\overline{z}}{4} + \mathrm{Re}[f(z)], \qquad z = x + \mathrm{i}y, \tag{19.3}$$

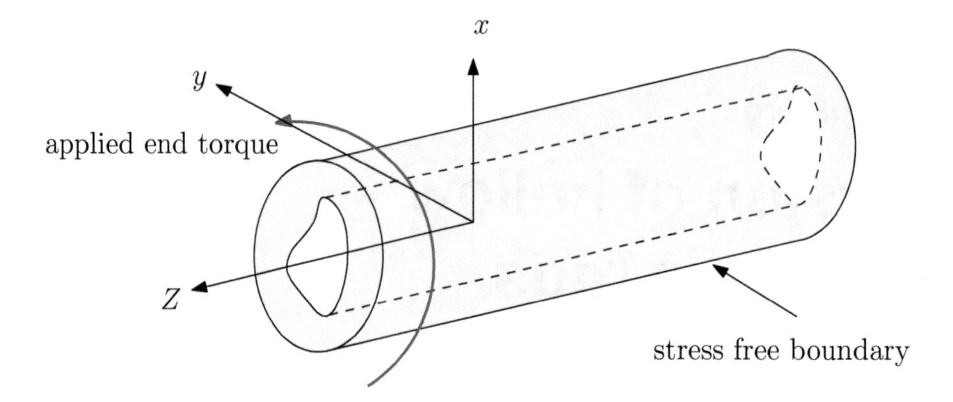

Figure 19.1. *Schematic of the torsion problem in a hollow prismatic duct with uniform doubly connected cross-section.*

where $f(z)$ is an analytic function of $z = x + iy$ in D, then the condition that the warping function is single-valued is equivalent to $f(z)$ being a single-valued analytic function in each doubly connected cross-section D. The boundary condition (19.2) then implies

$$\mathrm{Re}[f(z)] = \begin{cases} \dfrac{z\overline{z}}{4}, & z \in \partial D_0, \\ \dfrac{z\overline{z}}{4} + \gamma, & z \in \partial D_1. \end{cases} \tag{19.4}$$

Since the cross-section is doubly connected, we know from Chapter 4 that we can introduce

$$z = \mathcal{Z}(\zeta) \tag{19.5}$$

to be the conformal mapping from a concentric annulus

$$\rho < |\zeta| < 1 \tag{19.6}$$

to the duct cross-section D. The value of ρ will be determined by the chosen shape of the cross-section. We can now introduce

$$F(\zeta) \equiv f(\mathcal{Z}(\zeta)), \tag{19.7}$$

and the boundary condition (19.4) then implies that

$$\mathrm{Re}[F(\zeta)] = \begin{cases} \phi_0(\zeta) \equiv \dfrac{\mathcal{Z}(\zeta)\overline{\mathcal{Z}}(1/\zeta)}{4}, & \zeta \in C_0, \\ \phi_1(\zeta) \equiv \dfrac{\mathcal{Z}(\zeta)\overline{\mathcal{Z}}(\rho^2/\zeta)}{4} + \gamma, & \zeta \in C_1. \end{cases} \tag{19.8}$$

This is a modified Schwarz problem in the concentric annulus for the function $F(\zeta)$, which is analytic and single-valued in the annulus (19.6). The value of γ will follow from the compatibility condition for the existence of such a single-valued function, as discussed in Chapter 13.

An integral form of the solution for $F(\zeta)$ can be written down using the Villat formula discussed in Chapter 13. Indeed,

$$F(\zeta) = \mathcal{I}(\zeta) + \mathrm{i}d, \tag{19.9}$$

where

$$\mathcal{I}(\zeta)=\frac{1}{2\pi\mathrm{i}}\int_{C_0}\phi_0(a)\,[2K(a,\zeta)-1]\,\frac{da}{a}-\frac{1}{2\pi\mathrm{i}}\int_{C_1}\phi_1(a)[2K(a,\zeta)]\frac{da}{a},\qquad(19.10)$$

where $d\in\mathbb{R}$ is a constant. The function $K(.,.)$ is the logarithmic derivative of the prime function of the concentric annulus introduced in §5.12.

Given the conformal mapping from a concentric annulus to the cross-section, formulas (19.3), (19.9), and (19.10) constitute an integral representation of the stress function for a general hollow rod with doubly connected cross-section.

However, for a special class of rod cross-sections, this integral expression for $F(\zeta)$ can be circumvented and a more explicit expression found. This is the case when the rod cross-section is a doubly connected quadrature domain of the kind discussed in Chapter 11. In this way, this application ties together two separate elements of the general theory in Part I—from Chapters 11 and 13—and it is instructive to give details of this construction.

Specifically, consider the prismatic rod D with a doubly connected cross-section shown to the left in Figure 11.4, comprising four near-circular pipes in an annular arrangement enclosing a hollow zone. The conformal mapping to this doubly connected domain from a concentric annulus $\rho<|\zeta|<1$ is given by (11.90) with $N=4$. It was shown in Exercise 11.5 that (11.90) can alternatively be written in terms of the $\mathcal{K}(\zeta,.)$ function; equivalently, on use of (5.82), it can be written in terms of $K(\zeta,.)$ as

$$z=\mathcal{Z}(\zeta)=\sum_{j=1}^{4}A_jK(\zeta,a_j),\qquad(19.11)$$

where

$$A_j=A\Omega_4^{j-1},\qquad\Omega_4=e^{\pi\mathrm{i}/2},\qquad a_j=a\Omega_4^{j-1},\qquad(19.12)$$

with $1<a<1/\rho$ and

$$A=\frac{1}{\tilde{\omega}(a,a)}\frac{\prod_{k=1}^{4}\omega(a,\Omega_4^{k-1}a/\rho^{1/2})}{\prod_{k=2}^{4}\omega(a,\Omega_4^{k-1}a)},\qquad\tilde{\omega}(z,a)=\frac{\omega(z,a)}{z-a}.\qquad(19.13)$$

Since this domain is a quadrature domain, the conformal mapping satisfies

$$\mathcal{Z}(\zeta)=\mathcal{Z}(\rho^2\zeta).\qquad(19.14)$$

That is, the mapping is a loxodromic function of the kind discussed in Chapter 8. Conditions (19.4) can be written as

$$F(\zeta)+\overline{F}(1/\zeta)=\frac{\mathcal{Z}(\zeta)\overline{\mathcal{Z}}(1/\zeta)}{2},\qquad\zeta\in C_0,\qquad(19.15)$$

$$F(\zeta)+\overline{F}(\rho^2/\zeta)=\frac{\mathcal{Z}(\zeta)\overline{\mathcal{Z}}(\rho^2/\zeta)}{2}+2\gamma,\qquad\zeta\in C_1.\qquad(19.16)$$

These relations between analytic functions can be continued off the two circles C_0 and C_1, and, on use of (19.14) and subtraction of (19.15) from (19.16), together with use of (19.14), we conclude that

$$F(\rho^2\zeta)=F(\zeta)+2\overline{\gamma}.\qquad(19.17)$$

If we search for admissible $F(\zeta)$ within the class of functions satisfying (19.17), then it is enough to consider the single functional relation

$$F(\zeta)+\overline{F}(1/\zeta)=\frac{\mathcal{Z}(\zeta)\overline{\mathcal{Z}}(1/\zeta)}{2}.\qquad(19.18)$$

The function $F(\zeta)$ must be analytic in $\rho < |\zeta| < 1$, and, moreover, the right-hand side can be seen to be a loxodromic function with simple poles in the fundamental region $\rho < |\zeta| < 1/\rho$ at $\{a_j, 1/\overline{a_j}|j = 1, \ldots, 4\}$. Putting all these deductions together, we infer that $F(\zeta)$ is a quasi-loxodromic function.

We therefore propose that

$$F(\zeta) = \sum_{j=1}^{4} B_j K(\zeta, a_j) + \mathcal{D}, \tag{19.19}$$

where the simple poles $\{a_j|j = 1, \ldots, 4\}$ are all outside the annulus $\rho < |\zeta| < 1$ and where the constants $\{B_j|j = 1, \ldots, 4\}$ and $\mathcal{D}$ remain to be found. In order to satisfy (19.17) we must have

$$-\sum_{j=1}^{4} B_j = 2\overline{\gamma}, \tag{19.20}$$

where we have used the identity (5.81). The constants $\{B_j|j = 1, \ldots, 4\}$ can be determined by equating residues at the simple poles $\{a_j|j = 1, \ldots, 4\}$:

$$B_j = \frac{A_j \overline{\mathcal{Z}}(1/a_j)}{2}, \qquad j = 1, \ldots, 4. \tag{19.21}$$

With $\{B_j|j = 1, \ldots, 4\}$ determined in this way, (19.20) provides an equation for γ. It only remains to find $\mathcal{D}$. This can be done by evaluating the identity (19.18) at some chosen point, say $\zeta = 1$. This leads to

$$\mathcal{D} + \overline{\mathcal{D}} = \frac{\mathcal{Z}(1)\overline{\mathcal{Z}}(1)}{2} - \sum_{j=1}^{4} B_j K(1, a_j) - \sum_{j=1}^{4} \overline{B_j} K(1, \overline{a_j}), \tag{19.22}$$

which determines the real part of $\mathcal{D}$. The imaginary part of $\mathcal{D}$ is not determined by the boundary value problem for $F(\zeta)$, and its value is not important.

With the constants $\{B_j|j = 1, \ldots, 4\}$ determined by formula (19.21) and $\mathcal{D}$ determined, up to an unimportant imaginary constant, by (19.22), the function $F(\zeta)$ as given by (19.19) is known explicitly in terms of the prime function and its derivatives, and the solution is complete.

The explicit formula (19.19) of the solution for $F(\zeta)$ gives an alternative representation of the Villat integral expression (19.9). They are the same up to the choice of a purely imaginary constant.

Chapter 20

Laminar convective heat transfer

Steady heat transfer in uniform ducts with multiply connected cross-sections provides another challenge in which the methods of Part I can be applied. It is an area where methods of complex analysis have historically played an important role. The treatise by Shah and London [118] documents many of these contributions up to the late 1970s. To illustrate this application we build on the solutions developed in Chapters 18 and 19 and calculate the temperature field associated with steady advection and diffusion of heat by the laminar flow along the special class of annular ducts given by quadrature domains considered in the problems of torsion of Chapter 19.

Annular ducts, or "pipes," arise most commonly in heat transfer applications, so we will focus on ducts having a doubly connected cross-section. Generalization of our approach to cross-sections of higher connectivity will be clear.

For pedagogical purposes we will choose the duct to have a cross-sectional shape that is a doubly connected quadrature domain. This is because, for such ducts, it is possible to find an explicit form for the axial velocity of the fluid along the duct, just as we did for the torsion problem in Chapter 19, which then facilitates solution of the advection-diffusion problem for the temperature field, which is the principal focus of this chapter. This allows us to exemplify several aspects of the theory of Part I all at once.

Consider pressure-driven axial flow in the Z direction of a Cartesian (x, y, Z) plane,

$$\mathbf{u} = (0, 0, w(x, y)), \tag{20.1}$$

in a pipe of uniform cross-section D having boundary ∂D. Exactly such problems were the focus of Chapter 18. The axial velocity, with the strength of the pressure gradient normalized to unity, satisfies

$$\nabla^2 w = -1, \qquad (x, y) \in D, \tag{20.2}$$

with the no-slip boundary condition implying that

$$w = 0, \qquad (x, y) \in \partial D. \tag{20.3}$$

One of the canonical problems of laminar convective heat transfer is to find the steady state temperature field $T(x, y)$ satisfying

$$\nabla^2 T = cw, \tag{20.4}$$

where the constant c is related to the axial thermal gradient in the direction perpendicular to the (x, y) plane in which $T(x, y)$ is being determined. (20.4) is often called the energy equation. A typical choice of boundary condition is to take

$$\frac{\partial T}{\partial n} = q_0 \quad \text{on outer boundary}, \qquad \frac{\partial T}{\partial n} = q_1 \quad \text{on inner boundary}, \tag{20.5}$$

where q_0 and q_1 are, respectively, uniform heat fluxes through the outer and inner duct boundaries.

These boundary value problems for $T(x, y)$ only have a solution provided a compatibility condition is satisfied. This is usual for harmonic boundary value problems with boundary conditions of Neumann type. Integration of the governing equation (20.4) over the duct cross-section D leads to

$$\int\int_D \nabla^2 T dA = c \int\int_D w dA = cQ, \tag{20.6}$$

where we define

$$Q = \int\int_D w dA \tag{20.7}$$

as the axial fluid flux along the duct. On use of the divergence theorem (20.6) becomes

$$\int_{\partial D} \frac{\partial T}{\partial n} ds = q_0 \mathcal{P}_1 + q_0 \mathcal{P}_1 = cQ, \tag{20.8}$$

where $\mathcal{P}_0$ is the perimeter of the outer duct boundary, $\mathcal{P}_1$ is the perimeter of the inner duct boundary, and we have used the boundary condition (20.5). Relation (20.8) relates the parameters c, q_0, q_1, $\mathcal{P}_0$, $\mathcal{P}_1$, and Q relevant to any solution for a steady state temperature distribution $T(x, y)$ in D.

In Chapter 18 we saw that solving the hydrodynamic problem for pressure-driven flow through a doubly connected duct can be reduced to a standard Schwarz problem in the concentric annulus. We can write the general solution of (20.2) in the form

$$w = -\frac{z\overline{z}}{4} + \text{Re}[f(z)], \tag{20.9}$$

where $f(z)$ is an analytic function of the variable $z = x + iy$ in D. The boundary condition (20.3) implies

$$\text{Re}[f(z)] = \frac{z\overline{z}}{4} \tag{20.10}$$

on ∂D. Let

$$z = \mathcal{Z}(\zeta) \tag{20.11}$$

be the conformal mapping from a concentric annulus

$$\rho < |\zeta| < 1 \tag{20.12}$$

to the duct cross-section D. If we introduce

$$F(\zeta) \equiv f(\mathcal{Z}(\zeta)), \tag{20.13}$$

then (20.10) implies that

$$\text{Re}[F(\zeta)] = \begin{cases} \dfrac{\mathcal{Z}(\zeta)\overline{\mathcal{Z}}(1/\zeta)}{4}, & \zeta \in C_0, \\ \dfrac{\mathcal{Z}(\zeta)\overline{\mathcal{Z}}(\rho^2/\zeta)}{4}, & \zeta \in C_1. \end{cases} \tag{20.14}$$

This is a standard Schwarz problem for $F(\zeta)$ in the concentric annulus. An integral representation of $F(\zeta)$ can be written down using the methods of Chapter 13.

For a special class of duct cross-sections, this integral expression for $F(\zeta)$ can be circumvented and a more explicit expression found. This is the case when the duct cross-section is a doubly connected quadrature domain of the kind discussed in Chapter 11. The following analysis is similar to that presented in the torsion problem in Chapter 19, but it differs from it in several ways; not least is that this physical problem no longer demands that $F(\zeta)$ be single-valued in the annulus.

Specifically, consider the duct region D comprising four near-circular pipes given by (11.90). It was shown in Chapter 19 that the conformal map to such a domain from a concentric annulus $\rho < |\zeta| < 1$ is

$$z = \mathcal{Z}(\zeta) = \sum_{j=1}^{4} A_j K(\zeta, a_j), \tag{20.15}$$

where a_j and A_j are given in (19.12)–(19.13). Since this domain is a quadrature domain, the conformal mapping is a loxodromic function satisfying

$$\mathcal{Z}(\zeta) = \mathcal{Z}(\rho^2\zeta). \tag{20.16}$$

Later we will need the result

$$\zeta \mathcal{Z}'(\zeta) = \sum_{j=1}^{4} A_j L(\zeta, a_j), \tag{20.17}$$

where we use the prime notation to denote the derivative $\mathcal{Z}'(\zeta) = d\mathcal{Z}/d\zeta$ and where $L(\zeta, a)$ was introduced in Exercise 8.2 and has a second order pole at $\zeta = a$. Also of use in the analysis to follow will be the function

$$M(\zeta, a) \equiv \zeta \frac{\partial L(\zeta, a)}{\partial \zeta} \tag{20.18}$$

introduced in Exercise 8.3 and which has a third order pole at $\zeta = a$.

It is possible to solve the Schwarz problem (20.14) in an explicit form not involving any integral expressions. Conditions (20.10) can be written

$$F(\zeta) + \overline{F}(1/\zeta) = \frac{\mathcal{Z}(\zeta)\overline{\mathcal{Z}}(1/\zeta)}{2}, \qquad \zeta \in C_0, \tag{20.19}$$

$$F(\zeta) + \overline{F}(\rho^2/\zeta) = \frac{\mathcal{Z}(\zeta)\overline{\mathcal{Z}}(\rho^2/\zeta)}{2}, \qquad \zeta \in C_1, \tag{20.20}$$

from which we infer, on analytic continuation of these expressions off the two boundaries, that

$$F(\rho^2\zeta) = F(\zeta). \tag{20.21}$$

For this we have used (20.16). $F(\zeta)$ is invariant under the action of the Schottky group generated by $\theta_1(\zeta) = \rho^2\zeta$ and its inverse. We note that $F(\zeta)$ is not necessarily a loxodromic function since it may not be a single-valued meromorphic function in the doubled domain $\rho < |\zeta| < 1/\rho$. If we search for a candidate $F(\zeta)$ within the class of functions satisfying (20.21), then it is enough to consider the functional relation

$$F(\zeta) + \overline{F}(1/\zeta) = \frac{\mathcal{Z}(\zeta)\overline{\mathcal{Z}}(1/\zeta)}{2}. \tag{20.22}$$

The function $F(\zeta)$ must be analytic in $\rho < |\zeta| < 1$. Moreover, the right-hand side of (20.22) is known to have simple poles at $\{a_j, 1/\overline{a_j}|j = 1, \dots, 4\}$. We therefore propose that

$$F(\zeta) = \sum_{j=1}^{4} B_j K(\zeta, a_j) + C v_1(\zeta) + D, \tag{20.23}$$

where the simple poles $\{a_j|j = 1, \dots, 4\}$ are all outside the annulus $\rho < |\zeta| < 1$, while

$$v_1(\zeta) = \frac{1}{2\pi \mathrm{i}} \log \zeta, \tag{20.24}$$

and where the constants $\{B_j|j = 1, \dots, 4\}$, C, and D remain to be found. In order to satisfy (20.21) we must have

$$-\sum_{j=1}^{4} B_j + \frac{C}{2\pi \mathrm{i}} \log \rho^2 = 0. \tag{20.25}$$

The constants $\{B_j|j = 1, \dots, 4\}$ can be determined by equating residues at the simple poles $\{a_j|j = 1, \dots, 4\}$:

$$B_j = \frac{A_j \overline{\mathcal{Z}}(1/a_j)}{2}, \qquad j = 1, \dots, 4. \tag{20.26}$$

With the constants $\{B_j|j = 1, \dots, 4\}$ found, and C given by (20.25), it only remains to find D. This can be done by evaluating the identity (20.22) at some chosen point, say $\zeta = 1$. This leads to

$$D + \overline{D} = \frac{\mathcal{Z}(1)\overline{\mathcal{Z}}(1)}{2} - \sum_{j=1}^{4} B_j K(1, a_j) - \sum_{j=1}^{4} \overline{B_j} K(1, \overline{a_j}), \tag{20.27}$$

which determines the real part of D. The imaginary part of D is not determined by the boundary value problem for w.

This concludes the solution of the hydrodynamic problem for the axial flow in the duct. It is useful now to consider the integral of $f(z)$, namely

$$\mathcal{F}(z) \equiv \int^{z} f(z') dz', \tag{20.28}$$

which will arise later in the analysis of the thermal problem. The following steps constitute an instructive exercise in the theory of loxodromic functions considered in Chapter 8.

On integrating by parts, another expression for $\mathcal{F}(z)$ is

$$\mathcal{F}(z) = z f(z) - I(\zeta), \tag{20.29}$$

where

$$I(\zeta) \equiv \int^{\zeta} J(\zeta') \frac{d\zeta'}{\zeta'}, \qquad J(\zeta) \equiv \mathcal{Z}(\zeta) \left[\zeta \frac{dF}{d\zeta} \right]. \tag{20.30}$$

From (20.15) and (20.23), we find that

$$J(\zeta) = \left[\sum_{j=1}^{4} A_j K(\zeta/a_j) \right] \left[\sum_{j=1}^{4} B_j L(\zeta, a_j) + \frac{C}{2\pi \mathrm{i}} \right], \tag{20.31}$$

where we have used the fact that

$$\zeta \frac{dF}{d\zeta} = \sum_{j=1}^{4} B_j L(\zeta, a_j) + \frac{C}{2\pi \mathrm{i}}. \tag{20.32}$$

Inspection of (20.31) reveals that $J(\zeta)$ is a loxodromic function with a third order pole at each of the points $\{a_j | j = 1, \ldots, 4\}$. Consequently, from the theory of Chapter 8 we know that we can rewrite it as

$$J(\zeta) = \sum_{j=1}^{4} [R_j M(\zeta, a_j) + S_j L(\zeta, a_j) + T_j K(\zeta, a_j)] + U \tag{20.33}$$

for a set of constants $\{R_j, S_j, T_j\}$ and U and where we must insist that

$$\sum_{j=1}^{4} T_j = 0 \tag{20.34}$$

in order that $J(\zeta)$ is loxodromic. Recall from Exercises 8.2 and 8.3 that both $L(z, a)$ and $M(z, a)$ are loxodromic functions with second order and third order poles at $z = a$, respectively. It is readily shown, directly from their functional definitions, that

$$K(\zeta) = \frac{1}{\zeta - 1}, \qquad L(\zeta) = -\frac{1}{(\zeta-1)^2} - \frac{1}{\zeta - 1} + \text{a locally analytic function}, \tag{20.35}$$

$$M(\zeta) = \frac{2}{(\zeta-1)^3} + \frac{3}{(\zeta-1)^2} + \frac{1}{\zeta-1} + \text{a locally analytic function}. \tag{20.36}$$

It follows that, for ζ near a_k, we can write

$$\begin{aligned} J(\zeta) = {} & R_k \left[\frac{2a_k^3}{(\zeta - a_k)^3} + \frac{3a_k^2}{(\zeta - a_k)^2} + \frac{a_j}{\zeta - a_k} \right] + S_k \left[-\frac{a_k^2}{(\zeta - a_k)^2} - \frac{a_k}{\zeta - a_k} \right] \\ & + \frac{T_k a_k}{\zeta - a_k} + \text{a locally analytic function} \\ = {} & \frac{\mathcal{A}_3^{(k)}}{(\zeta - a_k)^3} + \frac{\mathcal{A}_2^{(k)}}{(\zeta - a_k)^2} + \frac{\mathcal{A}_1^{(k)}}{\zeta - a_k} + \text{a locally analytic function,} \end{aligned} \tag{20.37}$$

where the constants $\mathcal{A}_1^{(k)}$, $\mathcal{A}_2^{(k)}$, and $\mathcal{A}_3^{(k)}$ can be computed analytically or numerically using the formulas

$$\mathcal{A}_j^{(k)} = \frac{1}{2\pi \mathrm{i}} \int_{|\zeta - a_k| = \epsilon} J(\zeta)(\zeta - a_j)^{j-1} d\zeta, \qquad j = 1, 2, 3, \tag{20.38}$$

where $|\zeta - a_k| = \epsilon$ is some circle of sufficiently small radius ϵ centered at a_k. We can then determine that, for $k = 1, \ldots, 4$, the constants appearing in (20.33) are given by

$$R_k = \frac{\mathcal{A}_3^{(k)}}{2a_k^3}, \qquad S_k = \frac{3\mathcal{A}_3^{(k)}}{2a_k^3} - \frac{\mathcal{A}_2^{(k)}}{a_k^2}, \qquad T_k = \frac{1}{a_k} \left[\mathcal{A}_1^{(k)} - \frac{\mathcal{A}_2^{(k)}}{a_k} + \frac{\mathcal{A}_3^{(k)}}{a_k^2} \right]. \tag{20.39}$$

The constant U can be found by evaluating both sides of (20.33) at $\zeta = 1$:

$$U = J(1) - \sum_{k=1}^{4} [R_k M(1, a_k) + S_k L(1, a_k) + T_k K(1, a_k)]. \tag{20.40}$$

With the representation of $J(\zeta)$ given in (20.33) now fully determined, it follows that

$$I(\zeta) = \int^{\zeta} J(\zeta')\frac{d\zeta'}{\zeta'} = \sum_{j=1}^{4} R_j L(\zeta, a_j) + S_j K(\zeta, a_j) + T_j \log \omega(\zeta, a_j) + U \log \zeta, \tag{20.41}$$

where we have set an arbitrary constant of integration to zero. This is done without loss of generality since $\mathcal{F}(z)$ is only determined up to an arbitrary constant. An important observation is that this function is not single-valued in the annulus $\rho < |\zeta| < 1$ unless $U = 0$.

With these mathematical preliminaries complete, we turn to solving the energy equation (20.4) for the temperature $T(x, y)$ with boundary conditions (20.5).

In terms of the variables $(z, \overline{z})$, and on substitution of the form (20.9), (20.4) becomes

$$\frac{\partial^2 T}{\partial z \partial \overline{z}} = -\frac{z\overline{z}}{16} + \frac{f(z) + \overline{f(z)}}{8}. \tag{20.42}$$

This can be integrated with respect to $\overline{z}$ to give

$$\frac{\partial T}{\partial z} = -\frac{z\overline{z}^2}{32} + \frac{1}{8}\left(\overline{z}f(z) + \overline{\mathcal{F}(z)}\right) + \frac{g'(z)}{2}, \tag{20.43}$$

where $g'(z)$ is some (as yet unknown) analytic function in D. A second integration with respect to z yields

$$T = -\frac{z^2\overline{z}^2}{64} + \frac{1}{8}\left(\overline{z}\mathcal{F}(z) + z\overline{\mathcal{F}(z)}\right) + \mathrm{Re}[g(z)] \tag{20.44}$$

$$= -\frac{z^2\overline{z}^2}{64} + \frac{1}{8}\left(|z|^2(f(z) + \overline{f(z)}) - \overline{z}I(\zeta) - z\overline{I(\zeta)}\right) + \mathrm{Re}[g(z)], \tag{20.45}$$

where we have used the fact that $T(x, y)$ must be real-valued. A useful step now is to let

$$g(z) = -\frac{z\overline{I}(\zeta)}{4} + \tilde{g}(z), \tag{20.46}$$

where $\overline{I}(\zeta)$ is the Schwarz conjugate of $I(\zeta)$. Then

$$T = -\frac{z^2\overline{z}^2}{64} + \frac{1}{8}\left(|z|^2(f(z) + \overline{f(z)}) - \overline{z}I(\zeta) - z\overline{I(\zeta)} - z\overline{I}(\zeta) - \overline{z}I(\overline{\zeta})\right) + \mathrm{Re}[\tilde{g}(z)].$$

This can be decomposed as

$$T = T_p + \tilde{T}, \tag{20.47}$$

where

$$T_p = -\frac{z^2\overline{z}^2}{64} + \frac{1}{8}\left(|z|^2(f(z) + \overline{f(z)}) - z(\overline{I}(\overline{\zeta}) + \overline{I}(\zeta)) - \overline{z}(I(\zeta) + I(\overline{\zeta}))\right) \tag{20.48}$$

and

$$\tilde{T} = \mathrm{Re}[\tilde{g}(z)]. \tag{20.49}$$

It can easily be checked, on use of (20.41), that T_p is single-valued in D and, since we require T to be single-valued there too, $\tilde{T} = \mathrm{Re}[\tilde{g}(z)]$ must also be single-valued in D. It is also easy to verify that

$$\nabla^2 T_p = cw. \tag{20.50}$$

It only remains to find $\tilde{g}(z)$, and, to do this, we introduce the composed function

$$G(\zeta) \equiv \tilde{g}(\mathcal{Z}(\zeta)). \tag{20.51}$$

The boundary condition (20.5) on the outer boundary implies that

$$\frac{\partial \hat{T}}{\partial n} = q_0 - \frac{\partial T_p}{\partial n} \tag{20.52}$$

or, equivalently, that

$$\mathrm{Re}\left[2\frac{\partial \hat{T}}{\partial z}\left(-\mathrm{i}\frac{dz}{ds}\right)\right] = q_0 - \mathrm{Re}\left[2\frac{\partial T_p}{\partial z}\left(-\mathrm{i}\frac{dz}{ds}\right)\right], \tag{20.53}$$

where

$$\frac{\partial T_p}{\partial z} = -\frac{z\bar{z}^2}{32} + \frac{1}{8}\left(\bar{z}(f(z)+\overline{f(z)}) + z\bar{z}f'(z) - (\bar{I}(\zeta)+\bar{I}(\bar{\zeta})) - \bar{z}\frac{I'(\zeta)}{\mathcal{Z}'(\zeta)} - z\frac{\bar{I}'(\zeta)}{\mathcal{Z}'(\zeta)}\right). \tag{20.54}$$

Now $\partial T_p/\partial z$ is single-valued in D, and since $\partial T/\partial z$ must also be single-valued in D, then $\tilde{g}'(z)$ must also be single-valued there. The boundary value problem for $\tilde{g}'(z)$ can be written as

$$\mathrm{Re}\left[\tilde{g}'(z)\left(-\mathrm{i}\frac{dz}{ds}\right)\right] = \begin{cases} q_0 - \mathrm{Re}\left[2\dfrac{\partial T_p}{\partial z}\left(-\mathrm{i}\dfrac{dz}{ds}\right)\right] & \text{on outer boundary,} \\ q_1 - \mathrm{Re}\left[2\dfrac{\partial T_p}{\partial z}\left(-\mathrm{i}\dfrac{dz}{ds}\right)\right] & \text{on inner boundary.} \end{cases} \tag{20.55}$$

Now on C_0 it is readily checked that

$$\frac{dz}{ds} = \frac{\mathrm{i}\zeta\mathcal{Z}'(\zeta)}{|\mathcal{Z}'(\zeta)|}, \qquad \zeta \in C_0, \tag{20.56}$$

and on C_1,

$$\frac{dz}{ds} = -\frac{\mathrm{i}\zeta\mathcal{Z}'(\zeta)}{\rho|\mathcal{Z}'(\zeta)|}, \qquad \zeta \in C_1. \tag{20.57}$$

From the chain rule it follows that

$$\tilde{g}'(z) = \frac{G'(\zeta)}{\mathcal{Z}'(\zeta)}. \tag{20.58}$$

Hence (20.55) is

$$\mathrm{Re}\left[\zeta G'(\zeta)\right] = \begin{cases} \phi_0(\zeta) \equiv q_0|\mathcal{Z}'(\zeta)| - \mathrm{Re}\left[2\zeta\mathcal{Z}'(\zeta)\dfrac{\partial T_p}{\partial z}\right], \\ \phi_1(\zeta) \equiv -q_1\rho|\mathcal{Z}'(\zeta)| - \mathrm{Re}\left[2\zeta\mathcal{Z}'(\zeta)\dfrac{\partial T_p}{\partial z}\right]. \end{cases} \tag{20.59}$$

With $\tilde{g}'(z)$ required to be single-valued in D, $\zeta G'(\zeta)$ must be analytic and single-valued in the annulus $\rho < |\zeta| < 1$. We therefore recognize (20.59) as a modified Schwarz problem for $\zeta G'(\zeta)$ in D of the kind considered in Chapter 13. From the presentation there it is known that there is a condition on this modified Schwarz problem for $\zeta G'(\zeta)$ to be single-valued in the annulus $\rho < |\zeta| < 1$. It is

$$\int_{C_0} \left[|\mathcal{Z}'(\zeta)| \left[q_0 - \mathrm{Re}\left[2\frac{\partial T_p}{\partial z}\left(-\mathrm{i}\frac{dz}{ds}\right)\right]\right]\right] \frac{d\zeta}{\mathrm{i}\zeta}$$

$$= \int_{C_1} \left[-\rho|\mathcal{Z}'(\zeta)| \left[q_1 - \mathrm{Re}\left[2\frac{\partial T_p}{\partial z}\left(-\mathrm{i}\frac{dz}{ds}\right)\right]\right]\right] \frac{d\zeta}{\mathrm{i}\zeta}. \tag{20.60}$$

It is not obvious that this will be satisfied. However, this condition can be shown to be equivalent to

$$\int_{\partial D_0} q_0 ds + \int_{\partial D_1} q_1 ds = \int_{\partial D} \left[\mathrm{Re}\left[2\frac{\partial T_p}{\partial z}\left(-\mathrm{i}\frac{dz}{ds}\right)\right]\right] ds, \tag{20.61}$$

which can be written as

$$q_0\mathcal{P}_0 + q_1\mathcal{P}_1 = \mathrm{Re}\left[\int_{\partial D}\left[2\frac{\partial T_p}{\partial z}\left(-\mathrm{i}\frac{dz}{ds}\right)\right] ds\right] \tag{20.62}$$

$$= 4\mathrm{Re}\left[\int_{\partial D} \frac{1}{2\mathrm{i}}\frac{\partial T_p}{\partial z} dz\right] = \mathrm{Re}\left[\int_D \nabla^2 T_p\right] = cQ. \tag{20.63}$$

This is precisely the compatibility condition (20.8). We have therefore confirmed that the modified Schwarz problem (20.59) is solvable for the function $\zeta G'(\zeta)$ that is analytic and single-valued in D provided the parameters c, q_0, q_1, $\mathcal{P}_0$, $\mathcal{P}_1$, and Q satisfy (20.8).

Finally, the Villat formula of Chapter 13 can now be used:

$$\zeta G'(\zeta) = \mathcal{I}(\zeta) + \mathrm{i}d, \tag{20.64}$$

where

$$\mathcal{I}(\zeta) = \frac{1}{2\pi\mathrm{i}} \int_{C_0} \phi_0(a)\left[2K(a,\zeta) - 1\right]\frac{da}{a} - \frac{1}{2\pi\mathrm{i}}\int_{C_1} \phi_1(a)[2K(a,\zeta)]\frac{da}{a}, \tag{20.65}$$

and where $d \in \mathbb{R}$ is a constant. In fact, we must set $d = 0$ in order that the temperature T is single-valued in D. As a function of ζ, we can write

$$\frac{\partial T_p}{\partial z} = -\frac{\mathcal{Z}(\zeta)\overline{\mathcal{Z}(\zeta)}^2}{32} + \frac{1}{8}\left[\overline{\mathcal{Z}(\zeta)}\left(F(\zeta) + \overline{F(\zeta)} + \mathcal{Z}(\zeta)\frac{\zeta F'(\zeta)}{\zeta \mathcal{Z}'(\zeta)}\right)\right. \tag{20.66}$$

$$\left. -(\overline{I}(\zeta) + \overline{I}(\overline{\zeta})) - \overline{\mathcal{Z}(\zeta)}\frac{I'(\zeta)}{\mathcal{Z}'(\zeta)} - \mathcal{Z}(\zeta)\frac{\overline{I}'(\zeta)}{\mathcal{Z}'(\zeta)}\right]. \tag{20.67}$$

It follows that

$$G(\zeta) = \int^{\zeta} \mathcal{I}(\zeta')\frac{d\zeta'}{\zeta'}. \tag{20.68}$$

This completes the determination of T, up to an arbitrary constant:

$$T = -\frac{z^2\overline{z}^2}{64} + \frac{1}{8}\left(|z|^2(f(z) + \overline{f(z)}) - (z+\overline{z})(I(\zeta) + \overline{I(\zeta)})\right) + \mathrm{Re}[G(\zeta)]. \tag{20.69}$$

Figure 20.1 makes use of this analytical expression in terms of the prime function and its derivatives to calculate the temperature contours in a typical quadrature domain duct cross-section in the cases where a nonzero heat flux exists separately on the outer and inner boundary. The prime function is calculated using the methods of Chapter 14.

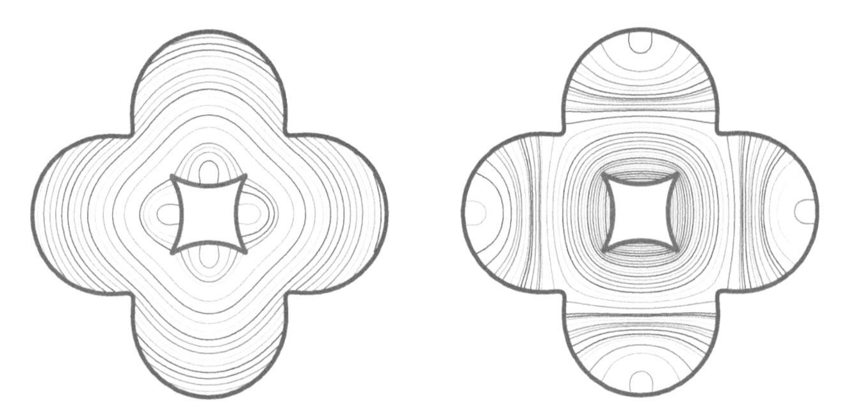

Figure 20.1. *Temperature contours in a cross-section of a doubly connected duct with a nonzero heat flux on the outer boundary only (left) and the inner boundary only (right). The cross-section corresponds to* $\rho = 0.3$ *and* $a = 1.4$.

Figure 20.2. *Temperature contours in a cross-section of a doubly connected duct with equal heat fluxes on both inner and outer boundaries.*

Figure 20.2 shows the temperature profile when both boundaries inject equal heat fluxes into the duct.

To the best of the author's knowledge, no previous studies have provided analytical solutions to the energy equation for heat transfer in a doubly connected duct. From this example calculation it should be clear how to use aspects of the mathematical framework of Part I to generalize the analysis to ducts with cross-sections of higher connectivity.

Chapter 21
Mixed-type boundary value problems

Boundary value problems for, say, harmonic or biharmonic fields, are said to be of *mixed type* if the boundary conditions on distinct parts of the boundary of the domain are different. Such problems arise in a variety of physical settings. To illustrate how the methods of Part I come into play in this context, we study some canonical mixed boundary value problems arising in the field of microfluidics. In studying so-called superhydrophobic surfaces, the challenge is to determine the flow, and an important diagnostic quantity known as the *hydrodynamic slip length*, over hybrid surfaces comprising a mixture of no-slip and no-shear regions.

A paradigmatic problem is that of longitudinal flow along a surface with unidirectional grooves. In this problem we consider the motion of a fluid in the semi-infinite region $y > 0$ above a wall located at $y = 0$ in Cartesian (x, y, Z) coordinates. This wall is punctuated by a finite collection of no-shear slots running parallel to the Z axis; physically, we envisage that these slots are made up of free surface menisci spanning some grooves etched into the surface. The key point is that such menisci are commonly modeled as being free of shear. The flow direction is longitudinal, meaning that the fluid flows parallel to the slots, and has the functional form

$$\mathbf{u} = (0, 0, w(x, y)), \tag{21.1}$$

where, as $y \to \infty$, we insist that

$$w(x, y) \to y. \tag{21.2}$$

This means that the far-field flow is one of simple shear over the surface.

If there is no pressure gradient driving the flow, then the Navier–Stokes equations governing the motion of Newtonian viscous fluids tell us that the velocity field $w(x, y)$ is harmonic in the fluid region:

$$\nabla^2 w = 0, \qquad y > 0, \qquad \nabla^2 = \frac{\partial^2}{\partial x^2} + \frac{\partial^2}{\partial y^2}. \tag{21.3}$$

On the solid wall portions of the axis $y = 0$, we must impose the boundary condition

$$w = 0. \tag{21.4}$$

This is the no-slip condition. On the other hand, on the portions of the plane $y = 0$ occupied by the no-shear slots, we impose

$$\frac{\partial w}{\partial y} = 0. \tag{21.5}$$

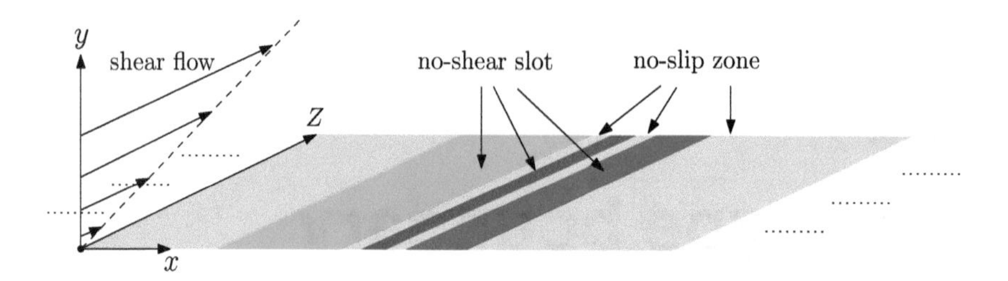

Figure 21.1. *Schematic of longitudinal shear flow over a no-slip surface endowed with three no-shear slots of differing widths (shown in green, blue, and red).*

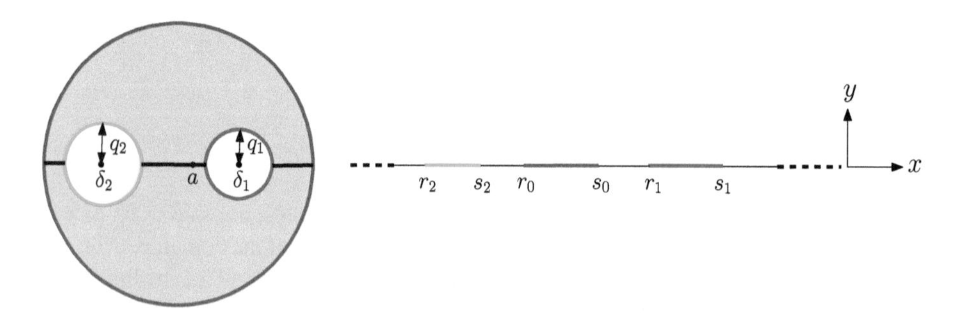

Figure 21.2. *Conformal mapping (21.7) of a triply connected circular domain D_ζ to the unbounded region exterior to three slits $[r_j, s_j]$ for $j = 0, 1, 2$ along the real axis.*

This is the no-shear condition. This mixture of Dirichlet-type and Neumann-type boundary conditions on different portions of the boundary $y = 0$ is typical of a mixed-type boundary value problem.

Suppose there are $M + 1$ no-shear slots on $y = 0$ where $M \geq 0$. Figure 21.1 shows a typical case, $M = 2$, where there are three no-shear slots, each having a different width, and with no-slip zones in between them. We will solve this mixed boundary value problem in analytical form, expressing the solution in terms of the prime function associated with a conformally equivalent circular domain D_ζ. Conformal slit maps will play a crucial role.

Consider a circular domain D_ζ with M circles $\{C_j | j = 1, \ldots, M\}$ centered at points $\{\delta_j \in \mathbb{R} | j = 1, \ldots, M\}$ on the real axis. The radii of these circles are $\{q_j \in \mathbb{R} | j = 1, \ldots, M\}$. The conformal mapping

$$z = f(\zeta) \tag{21.6}$$

from D_ζ to the unbounded region exterior to the no-shear slots is known to us from the considerations of Part I. In fact, we know from Exercises 6.4 and 6.11 that there are at least three different ways to represent such a mapping. One way is to write it as a parallel slit mapping in which all circles $\{C_j | j = 0, 1, \ldots, M\}$ are transplanted to slits parallel to the real axis in the z plane:

$$z = f(\zeta) = A\phi_0(\zeta, a) + B, \tag{21.7}$$

where A and B are real constants and the parallel slit maps $\phi_\theta(\zeta, a)$ are defined in Chapter 6 with $\theta = 0$ and the point $a \in \mathbb{R}$ chosen to sit on the real axis in D_ζ. Since a is real, and all the circles are centered on the real ζ axis, the image slits are on the real z axis.

Let us denote the intervals occupied by the no-shear slots as

$$[r_j, s_j], \qquad j = 0, 1, \ldots, M. \tag{21.8}$$

The values of the $2(M+1)$ real parameters $A, B, \{q_j, \delta_j \in \mathbb{R} | j = 1, \dots, M\}$ will be determined by requiring that the two points at which each of the $M+1$ circles $\{C_j | j = 0, 1, \dots, M\}$ intersect the real axis correspond to the set of $2(M+1)$ points (21.8). Near the point a the mapping (21.7) has the form

$$z = \frac{A}{\zeta - a} + \text{a locally analytic function.} \tag{21.9}$$

To find the longitudinal velocity field $w(x,y)$ we introduce a complex potential

$$h(z) = \chi(x,y) + \mathrm{i}w(x,y), \tag{21.10}$$

where $\chi(x,y)$ is the harmonic conjugate to $w(x,y)$. It is clear from (21.2) that

$$h(z) \to z \qquad \text{as } |z| \to \infty. \tag{21.11}$$

By the Cauchy–Riemann equations the boundary condition (21.5) corresponds to

$$\chi = \text{constant} \tag{21.12}$$

on the no-shear slots. If we succeed in determining $h(z)$, then the solution we seek is given by

$$w(x,y) = \mathrm{Im}[h(z)]. \tag{21.13}$$

It is useful to introduce the composition of the complex potential with the conformal mapping function:

$$H(\zeta) \equiv h(f(\zeta)). \tag{21.14}$$

Since the circles $\{C_j | j = 0, 1, \dots, M\}$ are transplanted to the no-shear slots under the mapping (21.7), thinking of $h = H(\zeta)$ itself as a conformal map of D_ζ to a complex h plane, we require

$$H(\zeta) = \frac{A}{\zeta - a} + \text{a locally analytic function} \tag{21.15}$$

with the circles $\{C_j | j = 0, 1, \dots, M\}$ transplanted by this mapping to a set of slits parallel to the imaginary h axis. However, on recalling the properties of the parallel slit maps considered in Part I, we immediately infer that

$$H(\zeta) = A\phi_{\pi/2}(\zeta, a) \tag{21.16}$$

will satisfy all requirements. Notice that we have again used the parallel slit maps $\phi_\theta(\zeta, a)$ defined in Part I but now with $\theta = \pi/2$.

In summary, a parametric form of the required complex potential is

$$z = A\phi_0(\zeta, a) + B, \tag{21.17}$$

$$h = A\phi_{\pi/2}(\zeta, a). \tag{21.18}$$

The velocity field follows from (21.13). The problem of simple shear over $M+1$ no-shear slots in an otherwise no-slip surface has therefore been solved in terms of the prime function of a conformally equivalent circular domain. The functional form of the solution is the same for any value of M, that is, for any number of slots.

A more physically interesting situation arises on supposing there to be a $2L$-*periodic array* of no-shear slots, with $M+1$ slots per period window. Figure 21.3 shows a schematic with $M = 2$ corresponding to three no-shear slots per period.

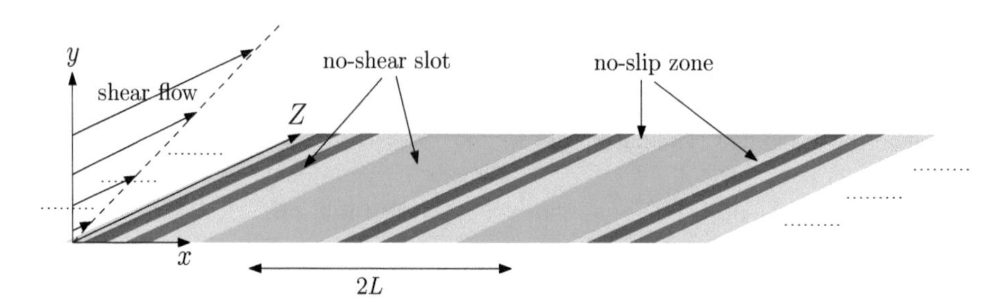

Figure 21.3. *Longitudinal shear flow over a $2L$-periodic array of no-shear slots in a no-slip surface. The case shown has three no-shear slots per period with each of the slots having a different width.*

By the periodicity of the arrangement, it is enough to consider the flow in a single period window. Now we take a circular preimage D_ζ with all circles $\{C_j | j = 1, \ldots, M\}$ aligned along the real axis, as before. We will take the upper half unit disc in this ζ plane to be transplanted to the fluid region in one period window. We assume that each circle C_j for $j = 0, 1, \ldots, M$ maps to one of the $M+1$ no-shear slots with each portion of the real ζ axis between the circles in D_ζ being transplanted to the no-slip zones between the no-shear slots.

The conformal map to a single period window of such an arrangement is

$$z = f(\zeta) = -\frac{\mathrm{i}L}{\pi} \log \eta(\zeta, a), \qquad a = \mathrm{i}r, \tag{21.19}$$

where $0 < r < 1$ is a real parameter and

$$\eta(\zeta, a) = -\frac{\omega(\zeta, a)\omega(\zeta, 1/a)}{\omega(\zeta, \overline{a})\omega(\zeta, 1/\overline{a})}. \tag{21.20}$$

This is the special case of an *unbounded circular slit map* introduced in Chapters 1, 5, and 6. The map $\eta(\zeta, a)$ transplants D_ζ to the whole z plane exterior to $M+1$ circular slits all located on the unit circle $|z| = 1$; the circles $\{C_j | j = 0, 1, \ldots, M\}$ are transplanted to the no-shear slots. Moreover, as shown in Part I, the portions of the real ζ axis between the circles $\{C_j | j = 1, \ldots, M\}$ are transplanted to the regions of the unit circle $|z| = 1$ between the images of the circles $\{C_j | j = 0, 1, \ldots, M\}$. On taking a logarithm of $\eta(\zeta, a)$, as done in (21.19), the resulting function has all the properties required of a conformal mapping to a single period window containing $M+1$ no-shear slots. (21.19) has logarithmic branch points at $\zeta = a, \overline{a}$; the branch point at $\zeta = a$ maps to $y \to \infty$ in the period window. If we encircle the branch point at $\zeta = a$ once in an anticlockwise sense, then

$$z \mapsto z + 2L, \tag{21.21}$$

which takes us into a neighboring period window. The two sides of a branch cut, in the upper unit disc, joining the points a and $\overline{a}$ will correspond to the preimages of the two sides of the period window.

To solve for the flow we again seek the complex potential defined by

$$h(z) = \chi + \mathrm{i}w. \tag{21.22}$$

We introduce the composed function

$$H(\zeta) \equiv h(f(\zeta)). \tag{21.23}$$

By the far-field condition that $h(z) \sim z$ as $y \to \infty$ we require that $H(\zeta)$ have the same logarithmic singularity at $\zeta = a$ as the conformal mapping (21.19). In contrast to (21.19), however, we require that the *imaginary* part of $H(\zeta)$ have some constant value on each of the preimage circles $\{C_j | j = 0, 1, \dots, M\}$.

A little thought reveals that the function we seek is

$$H(\zeta) = -\frac{\mathrm{i}L}{\pi} \log \tilde{\eta}(\zeta, a), \tag{21.24}$$

where

$$\tilde{\eta}(\zeta, a) = -\frac{\omega(\zeta, a)\omega(\zeta, 1/\overline{a})}{\omega(\zeta, \overline{a})\omega(\zeta, 1/a)}. \tag{21.25}$$

The latter function is the special case of an *unbounded radial slit map* introduced in Chapters 1, 5, and 6. As shown in Part I, under such a mapping the region D_ζ is transplanted to the unbounded region exterior to $M+1$ radial slits, all of which are symmetric with respect to the unit circle $|z|$ (meaning that if the edge of a slit is at some point γ inside the disc, then its other edge is at $1/\overline{\gamma}$). It is the circles $\{C_j | j = 0, 1, \dots, M\}$ that are transplanted to the radial slits; the portions of the real ζ axis between the circles $\{C_j | j = 0, 1, \dots, M\}$ are transplanted to the regions of the unit circle $|z| = 1$ between the radial slit images of the circles $\{C_j | j = 0, 1, \dots, M\}$. On taking a logarithm of the radial slit map $\tilde{\eta}(\zeta, a)$ the resulting function (21.24) will have the required logarithmic branch points at $\zeta = \pm a$, meaning that the far-field condition (21.11) will be satisfied. The imaginary part of (21.24) will clearly vanish on the portions of the real ζ axis between the circles $\{C_j | j = 0, 1, \dots, M\}$; this is because $|\tilde{\eta}(\zeta, a)| = 1$ there. Moreover, the real part of (21.24) will be constant on the circles $\{C_j | j = 0, 1, \dots, M\}$ since $\arg[\tilde{\eta}(\zeta, a)]$ is constant when ζ lies on these circles.

Figure 21.4 shows a schematic illustrating the effect of taking a logarithm of the unbounded circular slit map (21.20) and the radial slit map (21.25).

A parametric form of the sought-after longitudinal flow is therefore given by

$$z = -\frac{\mathrm{i}L}{\pi} \log \eta(\zeta, a), \qquad a = \mathrm{i}r, \tag{21.26}$$

$$h = -\frac{\mathrm{i}L}{\pi} \log \tilde{\eta}(\zeta, a), \tag{21.27}$$

where the imaginary part of h gives $w(x, y)$. Figure 21.5 shows the velocity modification from uniform shear flow over a periodic array of no-shear slots with three slots per period, as calculated from the solution (21.26)–(21.27).

A quantity of interest in applications is the *hydrodynamic slip length* [11, 12] λ defined by the condition

$$h(z) \sim z + \mathrm{i}\lambda \qquad \text{as } y \to \infty. \tag{21.28}$$

Given that we have derived the solution for the flow, it is now a simple matter to determine an expression for λ in terms of the prime function given the results (21.19) and (21.24). Observe that, by a straightforward manipulation, we can write (21.24) as

$$\begin{aligned} H(\zeta) &= -\frac{\mathrm{i}L}{\pi} \log \left[-\frac{\omega(\zeta, a)\omega(\zeta, 1/\overline{a})}{\omega(\zeta, \overline{a})\omega(\zeta, 1/a)} \right] &(21.29)\\ &= -\frac{\mathrm{i}L}{\pi} \log \left[-\frac{\omega(\zeta, a)\omega(\zeta, 1/a)}{\omega(\zeta, \overline{a})\omega(\zeta, 1/\overline{a})} \frac{\omega(\zeta, 1/\overline{a})^2}{\omega(\zeta, 1/a)^2} \right] &(21.30)\\ &= z - \frac{\mathrm{i}L}{\pi} \log \left[\frac{\omega(\zeta, 1/\overline{a})^2}{\omega(\zeta, 1/a)^2} \right]. &(21.31) \end{aligned}$$

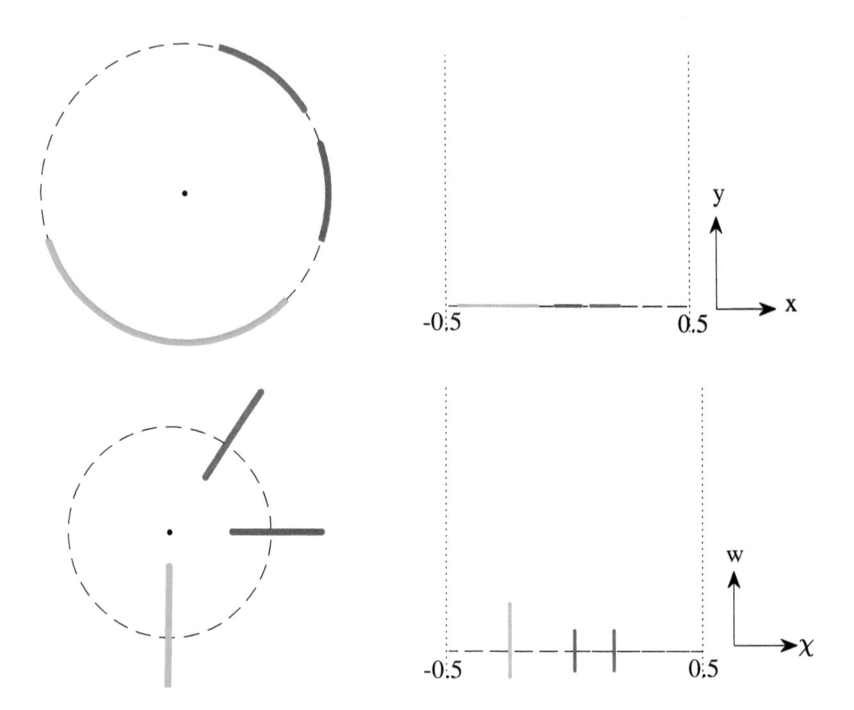

Figure 21.4. *Effect of taking a logarithm of the unbounded circular slit map (21.20) and the radial slit map (21.25) as in (21.19) and (21.24).*

From this we see immediately that as $y \to \infty$, so that $\zeta \to a$,

$$H(\zeta) \to z - \frac{\mathrm{i}L}{\pi} \log \left[\frac{\omega(a, 1/\overline{a})^2}{\omega(a, 1/a)^2} \right]. \tag{21.32}$$

It follows that

$$\lambda = -\frac{2L}{\pi} \log \left[\frac{\omega(a, 1/\overline{a})}{\omega(a, 1/a)} \right]. \tag{21.33}$$

This is the hydrodynamic slip length for any number $M + 1$ of no-shear slots written concisely in terms of the prime function of a circular domain conformally equivalent to the surface geometry. This formula was first derived in [12].

In the simply connected case $M = 0$ where $\omega(z, a) = (z - a)$, formula (21.33) becomes

$$\lambda = -\frac{2L}{\pi} \log \left[\frac{1 - r^2}{1 + r^2} \right]. \tag{21.34}$$

In Exercise 1.13 it was established that the map (21.20) transplants the unit circle $|\zeta| = 1$ to a circular slit on the unit circle in the image η plane with

$$\pi - \phi < \arg[\eta] < \pi + \phi, \tag{21.35}$$

where

$$r = \tan\left(\frac{\phi}{4}\right). \tag{21.36}$$

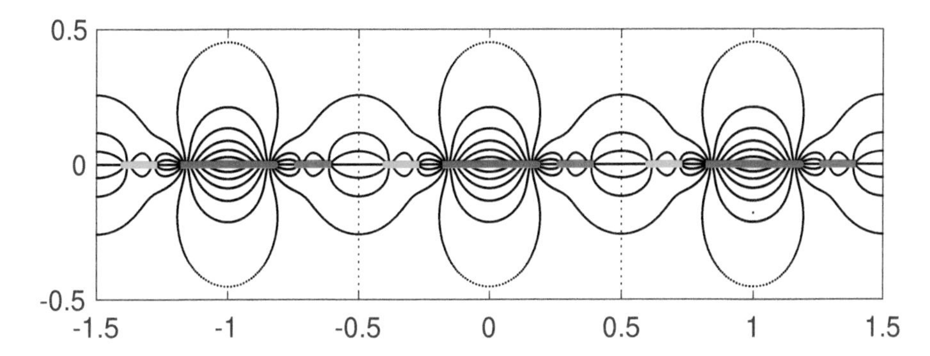

Figure 21.5. *Contours of the velocity modification from uniform shear flow over a periodic array of no-shear slots with three slots per period. In each period window a central no-shear slot is surrounded by two shorter equal-length satellite slots.*

On taking a logarithm of η we see that the quantity ϕ/π represents the so-called no-shear fraction: the fraction of the period window occupied by the no-shear slot. On substitution of (21.36) into (21.34) we find

$$\lambda = \frac{2L}{\pi} \log \sec(\phi/2). \tag{21.37}$$

It is interesting to note that this result for a single no-shear slot per period window was derived by Philip [106] in a quite different way using the theory of S–C mappings.

The prominent role played in these problems by the various conformal slit mappings introduced in Part I is remarkable. For flow over a finite set of no-shear slots in an infinite no-slip wall, the *parallel slit maps* allow us to represent the solution concisely in (21.17)–(21.18). On the other hand, for flow over a periodic array of no-shear slots, with multiple slots per period, it is the *unbounded circular slit maps* and *unbounded radial slit maps* that afford a convenient representation of the solution, as in (21.19) and (21.24).

The conformal slit maps introduced in this chapter have also been used, within a perturbation analysis, to study the effects on the hydrodynamic slip length of weak curvature of the no-shear surfaces and the effect of a subphase gas [3].

The mathematical boundary value problems considered in this chapter are important ones, and they are so fundamental that they appear in a wide range of physical contexts. To name just a few, the hydrodynamic slip lengths considered here have different interpretations in terms of so-called contact resistance and spreading resistance in heat transfer applications [59]. The same mixed boundary value problems also appear in studies of the screening effect of metallic grids on the electrostatic field in vacuum triodes; a formula akin to (21.37) was found in this context by Moizhes [100], and its generalization (21.33) derived here (and originally in [12]) will be relevant there too. The boundary value problems also arise in so-called escape problems for Brownian walkers, where the boundary condition of a domain changes depending on whether the boundary portion is an exit zone or an absorbing zone. Marshall [97] has already shown how to use the prime function machinery of Part I for escape problems in a circular disc with multiple exit holes/absorbing sections along its boundary.

Much more is possible in all these physical contexts using the framework set out in Part I. Many other applications where the conformal slit maps play an important theoretical role have been reviewed in [10].

Chapter 22

Slow viscous flow

In fluid dynamics the Reynolds number is a nondimensional quantity determining the ratio of inertial forces to viscous forces in the motion of a fluid. In the limit of zero Reynolds number the inertia of a flow is negligible and the Navier–Stokes equations governing the velocity $\mathbf{u}$ of a fluid of viscosity μ reduce to the Stokes equations.

In two dimensions, where the velocity of the fluid is $\mathbf{u} = (u, v)$, the Stokes equations take the form

$$\nabla^2 \mathbf{u} = \frac{1}{\mu} \nabla p, \qquad \nabla . \mathbf{u} = 0, \tag{22.1}$$

where $p(x, y)$ is the pressure in the fluid. Since the flow is incompressible and two-dimensional, a streamfunction $\psi(x, y)$ can be introduced that is related to the velocity via

$$(u, v) = \left(\frac{\partial \psi}{\partial y}, -\frac{\partial \psi}{\partial x} \right). \tag{22.2}$$

A complex-variable formulation of the problem can be introduced. On taking a curl of (22.1) and on use of (22.2), we find that the streamfunction satisfies the biharmonic equation

$$\nabla^4 \psi = 0. \tag{22.3}$$

Once ψ is determined, the velocity field (u, v) follows from (22.2). A representation of the general solution to (22.3) is given by

$$\psi = \mathrm{Im}[\overline{z} f(z) + g(z)], \tag{22.4}$$

where $f(z)$ and $g(z)$ are two analytic functions of the complex variable $z = x + \mathrm{i}y$. It is possible to express all physical quantities in terms of $f(z)$ and $g(z)$ and their derivatives $f'(z)$, $f''(z)$, $g'(z)$, and $g''(z)$, where primes are used to denote differentiation with respect to z:

$$\frac{p}{\mu} - \mathrm{i}\omega = 4 f'(z), \tag{22.5}$$

$$u + \mathrm{i}v = -f(z) + z\overline{f'(z)} + \overline{g'(z)}, \tag{22.6}$$

$$e_{11} + \mathrm{i}e_{12} = z\overline{f''(z)} + \overline{g''(z)}, \tag{22.7}$$

where p is the pressure, ω is the vorticity defined by

$$\omega = \frac{\partial v}{\partial x} - \frac{\partial u}{\partial y}, \tag{22.8}$$

and e_{ij} denotes the fluid rate-of-strain tensor given, in Cartesian coordinates, by

$$e_{ij} = \frac{1}{2}\left[\frac{\partial u_i}{\partial x_j} + \frac{\partial u_j}{\partial x_i}\right]. \tag{22.9}$$

For a Newtonian fluid of viscosity μ the stress tensor σ_{ij} is

$$\sigma_{ij} = -p\delta_{ij} + 2\mu e_{ij}. \tag{22.10}$$

If the complex form of a two-dimensional vector with Cartesian components (a_x, a_y) is $a_x + \mathrm{i}a_y$, then the complex form of the components of the stress tensor evaluated on some boundary with normal vector having components denoted by n_i can be shown to be

$$\sigma_{ij}n_j \mapsto 2\mu\mathrm{i}\frac{dH}{ds}, \tag{22.11}$$

where

$$H(z,\overline{z}) = f(z) + z\overline{f'(z)} + \overline{g'(z)}. \tag{22.12}$$

The drag force on an object with boundary ∂D can be found by computing the quantity

$$\int_{\partial D} \sigma_{ij}n_j ds = 2\mu\mathrm{i}\int_{\partial D} dH, \tag{22.13}$$

where the square brackets denote the change in the value of the quantity they contain on a single anticlockwise traversal of ∂D and n_j are the components of the unit normal at the boundary.

There is an additive degree of freedom in the definition of $f(z)$. To see this, note, from (22.5)–(22.7), that if

$$f(z) \mapsto f(z) + c, \qquad g'(z) \mapsto g'(z) + \overline{c}, \tag{22.14}$$

where $c \in \mathbb{C}$ is an arbitrary constant, then all physical quantities associated with the flow remain the same.

We now consider a particular Stokes flow problem which provides an instructive example of how the mathematical framework of Part I can be used.

Suppose a no-slip wall along the real axis has three gaps occupying the intervals

$$[-1,-b], \qquad [-a,+a], \qquad [+b,1] \tag{22.15}$$

for $0 < a < b < 1$, as shown in red in Figure 22.1. Lengths have been nondimensionalized with respect to L, say, where the edges of the two semi-infinite walls are at $\pm L$. This means that the no-slip wall itself occupies the intervals

$$(\infty,-1], \qquad [-b,-a], \qquad [+a,+b], \qquad [1,\infty), \tag{22.16}$$

as shown in blue in Figure 22.1.

We suppose that a flow is driven through the gaps in the wall by an imposed pressure gradient across the wall. Hasimoto [89] studied this problem for the case of two gaps in the wall, using methods different from those to be presented here. What follows is a generalization of his analysis to three and, by obvious extension, any finite number of gaps.

If we impose that the fluid pressure tends to $P > 0$ as $y \to +\infty$ and to $-P$ as $y \to -\infty$, then this means that

$$4f'(z) \to \begin{cases} +P/\mu, & y \to +\infty, \\ -P/\mu, & y \to -\infty, \end{cases} \tag{22.17}$$

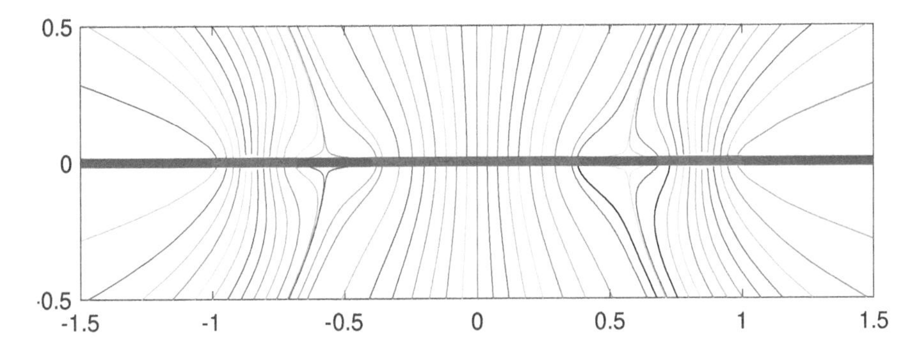

Figure 22.1. *Streamlines for pressure-driven Stokes flow through three gaps (shown in red) in an infinite wall (shown in blue).*

or, equivalently,

$$f(z) \to \begin{cases} \dfrac{Pz}{4\mu}, & y \to +\infty, \\ \\ -\dfrac{Pz}{4\mu}, & y \to -\infty. \end{cases} \tag{22.18}$$

On the no-slip wall in the intervals (22.16) the velocity must vanish, meaning that $(u, v) = (0, 0)$ and therefore that

$$u - \mathrm{i}v = -\overline{f(z)} + \overline{z}f'(z) + g'(z) = 0 \qquad \text{for } z \text{ on the wall.} \tag{22.19}$$

Once $f(z)$ has been found we can determine $g(z)$ using the boundary condition (22.19) since it implies that, on the wall where $\overline{z} = z$,

$$g'(z) = \overline{f}(z) - zf'(z), \tag{22.20}$$

and this relation can be analytically continued off the real axis.

Since it provides an instructive case study, we will present two different methods to solve this problem. Each method uses different elements of the mathematical theory of Part I. Both methods use the upper half of a circular preimage domain to correspond to the upper half region of the fluid, but the methods differ in *which* boundary portions of the real axis correspond to the preimage circles: the first method supposes the circles are the preimages of the no-slip wall portions; in the second method the circles are the preimages of the gaps in the wall. The conformal mappings to both choices of domain were studied earlier in Exercises 6.12 and 6.13.

The first solution method proceeds as follows.[55] Consider the conformal mapping

$$z = \mathcal{Z}_1(\zeta) \tag{22.21}$$

from a triply connected preimage circular domain D_ζ in a parametric ζ plane; the subscript 1 has been introduced to reflect the fact that this is the relevant conformal mapping in the first method of solution (a different mapping is introduced later for the second method). Under this mapping the unit circle C_0 is transplanted to the infinite portion of the walls

$$(\infty, -1], \qquad [1, \infty), \tag{22.22}$$

[55]This method, for the case of two gaps in the wall and for a broader range of flow types, has been used by Crowdy and Samson [55]; the example of this chapter is therefore a generalization of that analysis.

which can be thought of as joined at infinity, while two interior circles C_1 and C_2 are respectively the preimages of the two finite portions of the wall given by

$$[-b,-a], \qquad [+a,+b]. \tag{22.23}$$

By the left-right symmetry of the configuration it is natural to suppose that the two points $\zeta = \pm \mathrm{i}$ on C_0 are the preimages of points infinitely far from the wall as $y \to \pm\infty$; we can also suppose that the preimage domain D_ζ shares this left-right symmetry so that

$$\delta_1 = -\delta_2 = -\delta, \qquad q_1 = q_2 = q, \tag{22.24}$$

where the two real parameters δ and q are determined by ensuring that

$$\mathcal{Z}_1(\delta - q) = +a, \qquad \mathcal{Z}_1(\delta + q) = +b. \tag{22.25}$$

Since the image of C_0 is the real axis in the z plane, we have

$$\overline{\mathcal{Z}_1(\zeta)} = \mathcal{Z}_1(\zeta), \tag{22.26}$$

implying that

$$\overline{\mathcal{Z}_1}(1/\zeta) = \mathcal{Z}_1(\zeta). \tag{22.27}$$

Similarly, since the image of C_1 and C_2 also lies on the real axis,

$$\overline{\mathcal{Z}_1}(\phi_1(\zeta)) = \mathcal{Z}_1(\zeta), \qquad \overline{\mathcal{Z}_1}(\phi_2(\zeta)) = \mathcal{Z}_1(\zeta). \tag{22.28}$$

By continuation of these formulas off the circles on which they are derived we conclude that

$$\mathcal{Z}_1(\theta_j(\zeta)) = \mathcal{Z}(\zeta), \qquad j = 1, 2. \tag{22.29}$$

The map $\mathcal{Z}_1(\zeta)$ is invariant under the action of a Schottky group Θ generated by $\theta_1(\zeta), \theta_2(\zeta)$ and their inverses. These are the Möbius maps introduced in §3.6 of Chapter 3.

We now introduce the composition of the two complex potentials $f(z)$ and $g'(z)$ with the conformal mapping

$$F(\zeta) \equiv f(\mathcal{Z}_1(\zeta)), \qquad G(\zeta) \equiv g'(\mathcal{Z}_1(\zeta)). \tag{22.30}$$

In terms of these functions the boundary conditions (22.19) on C_j for $j = 0, 1, 2$ imply the following relations between analytic functions:

$$-\overline{F}(1/\zeta) + \overline{\mathcal{Z}_1}(1/\zeta)\frac{dF/d\zeta}{d\mathcal{Z}_1/d\zeta} + G(\zeta) = 0, \qquad \zeta \in C_0, \tag{22.31}$$

$$-\overline{F}(S_1(\zeta)) + \overline{\mathcal{Z}_1}(\theta_1(1/\zeta))\frac{dF/d\zeta}{d\mathcal{Z}_1/d\zeta} + G(\zeta) = 0, \qquad \zeta \in C_1, \tag{22.32}$$

$$-\overline{F}(S_2(\zeta)) + \overline{\mathcal{Z}_1}(\theta_2(1/\zeta))\frac{dF/d\zeta}{d\mathcal{Z}_1/d\zeta} + G(\zeta) = 0, \qquad \zeta \in C_2, \tag{22.33}$$

where we have used the fact that, on C_j,

$$\overline{\mathcal{Z}(\zeta)} = \overline{\mathcal{Z}_1}(S_j(\zeta)) = \overline{\mathcal{Z}_1}(\theta_j(1/\zeta)) \tag{22.34}$$

and where we have substituted for the Schwarz functions $\{S_j(\zeta) | j = 1, 2\}$ of the circles $\{C_j | j = 1, 2\}$ in favor of the maps $\{\theta_j(\zeta) | j = 1, 2\}$.

The relations (22.31)–(22.33) between analytic functions of ζ can be continued off the circles $\{C_j|j = 0,1,2\}$ on which they are originally derived and, together with the properties (22.29) of the mapping $\mathcal{Z}_1(\zeta)$, can be used to show that

$$F(\theta_j(\zeta)) = F(\zeta), \qquad j = 1,2. \tag{22.35}$$

Consequently, $F(\zeta)$ is also invariant with respect to the Schottky group Θ.

Since $\mathcal{Z}_1(\zeta)$ has simple poles at $\zeta = \pm\mathrm{i}$, is an automorphic function with respect to the Schottky group Θ, and must be analytic everywhere in the interior of D_ζ, it is natural to propose that

$$\mathcal{Z}_1(\zeta) = A\left[\mathcal{K}(\zeta,+\mathrm{i}) - \mathcal{K}(\zeta,-\mathrm{i})\right], \tag{22.36}$$

where A is a real constant which can be chosen to ensure that $\zeta = 1$ maps to $z = 1$. Namely,

$$\mathcal{Z}_1(\zeta) = \frac{\mathcal{K}(\zeta,+\mathrm{i}) - \mathcal{K}(\zeta,-\mathrm{i})}{\mathcal{K}(1,+\mathrm{i}) - \mathcal{K}(1,-\mathrm{i})}, \qquad A = \frac{1}{\mathcal{K}(1,+\mathrm{i}) - \mathcal{K}(1,-\mathrm{i})}. \tag{22.37}$$

(22.37) is the required mapping from D_ζ to the fluid domain. Here we are using the automorphic function theory presented in Chapter 9, where it was shown how the quasi-automorphic function $\mathcal{K}(\zeta,a)$ can be used to construct a representation of an automorphic function given the location of its poles.

Now by (6.88) we know that

$$\mathcal{K}(\theta_j(\zeta),a) = \mathcal{K}(\zeta,a) + 2\pi\mathrm{i}av_j'(a), \qquad j = 1,2. \tag{22.38}$$

We also know that

$$\overline{v_j}(1/\zeta) = v_j(\zeta) \tag{22.39}$$

and, for reflectionally symmetric domains such as this with all the centers of circles $\{C_j|j = 0,1,2\}$ on the real axis, the result of Exercise 3.7 implies that

$$\overline{v_j}(\zeta) = -v_j(\zeta) + d_j, \qquad j = 1,2, \tag{22.40}$$

for some set of constants $\{d_j|j = 1,\dots,M\}$ (which can be set to vanish without loss of generality). Hence

$$v_j(1/\zeta) = -v_j(\zeta) + d_j, \qquad j = 1,2. \tag{22.41}$$

On differentiation,

$$\frac{1}{\zeta}v_j'(1/\zeta) = \zeta v_j'(\zeta), \qquad j = 1,2. \tag{22.42}$$

This means that, for points ζ on C_0, we have

$$\zeta v_j'(\zeta) = \overline{\zeta}v_j'(\overline{\zeta}), \qquad j = 1,2. \tag{22.43}$$

We can use (22.38) and (22.43) to show that, for any A, the mapping (22.36) is indeed invariant as $\zeta \mapsto \theta_j(\zeta)$ for $j = 1,2$, as required.

We can now use the far-field boundary conditions (22.18) and the known analyticity of $F(\zeta)$ in D_ζ to propose that

$$F(\zeta) = \frac{PA}{4\mu}\mathcal{K}(\zeta,+\mathrm{i}) + \frac{PA}{4\mu}\mathcal{K}(\zeta,-\mathrm{i}) + Bv_1(\zeta) + Bv_2(\zeta), \tag{22.44}$$

where the real coefficient B is to be determined. The coefficient of $v_2(\zeta)$ has been chosen equal to that of $v_1(\zeta)$ on account of the left-right symmetry of the geometry and the imposed flow. The coefficient B must be chosen to ensure that $F(\zeta)$ satisfies the invariance property (22.35). On noting that

$$\begin{aligned} F(\theta_1(\zeta)) &= \frac{PA}{4\mu}\mathcal{K}(\theta_1(\zeta),+\mathrm{i}) + \frac{PA}{4\mu}\mathcal{K}(\theta_1(\zeta),-\mathrm{i}) + Bv_1(\theta_1(\zeta)) + Bv_2(\theta_1(\zeta)) \\ &= F(\zeta) + \frac{PA}{4\mu}\left[2\pi\mathrm{i}(\mathrm{i}v_1(\mathrm{i})) + 2\pi\mathrm{i}(-\mathrm{i}v_1(-\mathrm{i}))\right] + B(\tau_{11}+\tau_{21}) \end{aligned} \tag{22.45}$$

it is clear that B must be such that

$$\frac{PA}{4\mu}\left[2\pi\mathrm{i}(\mathrm{i}v_1(\mathrm{i})) + 2\pi\mathrm{i}(-\mathrm{i}v_1(-\mathrm{i}))\right] + B(\tau_{11}+\tau_{21}) = 0. \tag{22.46}$$

Now $F(\zeta)$ is fully determined, with $G(\zeta)$ following on use of (22.31), namely,

$$G(\zeta) = \overline{F}(1/\zeta) - \overline{\mathcal{Z}_1}(1/\zeta)\frac{dF/d\zeta}{d\mathcal{Z}_1/d\zeta}. \tag{22.47}$$

It is worth noting that, on use of (4.114) from Chapter 4, it is possible to proceed with the analysis using *only* the prime function to express the results and without use of the functions $\{v_j(\zeta)|j=1,2\}$. To do so, we pick an arbitrary point $\alpha \in D_\zeta$ and propose $F(\zeta)$ in the form

$$F(\zeta) = \frac{PA}{4\mu}\mathcal{K}(\zeta,+\mathrm{i}) + \frac{PA}{4\mu}\mathcal{K}(\zeta,-\mathrm{i}) \tag{22.48}$$

$$+\frac{B}{2\pi\mathrm{i}}\log\left[\frac{\omega(z,\theta_1(1/\overline{\alpha}))}{\omega(z,1/\overline{\alpha})}\right] + \frac{B}{2\pi\mathrm{i}}\log\left[\frac{\omega(z,\theta_2(1/\overline{\alpha}))}{\omega(z,1/\overline{\alpha})}\right], \tag{22.49}$$

where we have used (4.114) to express $v_j(\zeta)$ in terms of the prime function. This choice of $F(\zeta)$ differs from (22.44) only in an additive constant, but, as pointed out earlier, $F(\zeta)$ is only determined up to such additive constants, so this is inconsequential.

To perform some concrete calculations we restrict our attention to the special one-parameter family of domains, parametrized by r, characterized by

$$a = \frac{1}{2} - r, \qquad b = \frac{1}{2} + r, \tag{22.50}$$

where $0 < r < 1/2$. The drag force on each finite section of the wall can be computed using (22.13). By the symmetry of the flow configuration the force (F_x, F_y) on each of the two finite wall portions can be found by computing the quantity

$$2\mu\mathrm{i}\int_{C_1} 2B\,dv_1 = 4\mu\mathrm{i}B. \tag{22.51}$$

In the limit $r \to 0.5$ the three gaps in the wall become vanishingly small, and it might be expected on physical grounds that the force (per unit length in the perpendicular direction to the plane of flow) on each wall portion will tend to $(2P) \times (2r)$, which is the pressure difference across the wall multiplied by the length $2r$ of each wall portion. Figure 22.2 shows $|F_y|/(2P)$ as a function of the length $2r$ of each wall portion and reveals that $|F_y|/(2P) \to 1$ as $2r \to 1$. It also shows that placing even two very small wall portions at $x = \pm 1/2$ in the gap $[-1, 1]$ leads to quite significant forces on these walls.

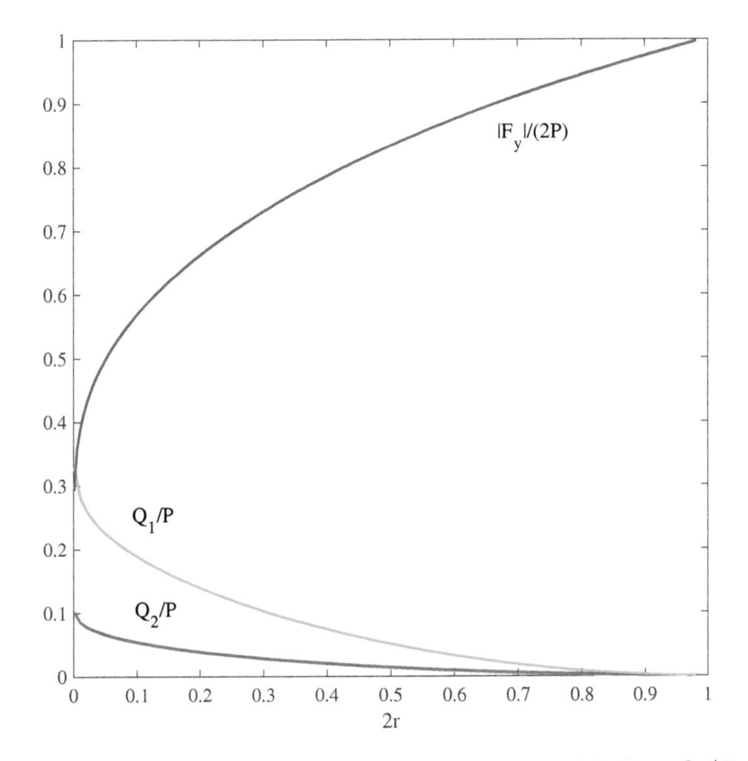

Figure 22.2. *Force on each of the occluding wall portions and the fluxes Q_1/P and Q_2/P through the gaps.*

It is also of interest to compute the magnitude of the fluxes Q_1 and Q_2 through the gaps defined by

$$Q_1 = \int_{-1/2+r}^{1/2-r} (-v)dx, \qquad Q_2 = \int_{1/2+r}^{1} (-v)dx = \int_{-1}^{-1/2-r} (-v)dx. \tag{22.52}$$

These are also plotted against the length $2r$ of the two occluding walls in Figure 22.2.

We now turn to a second approach to this same problem. By the reflectional symmetry of the flow about the real axis, it is enough to determine the flow in the upper half plane. Owing to this same symmetry it can be argued that the pressure p must be constant on the real axis; indeed, without loss of generality we can take

$$p = \text{Re}[f'(z)] = 0 \qquad \text{on the gaps.} \tag{22.53}$$

On integrating along the real axis in the gaps, we have that

$$\text{Re}[f(z)] = \text{constant} \qquad \text{on the gaps.} \tag{22.54}$$

Suppose we define the new analytic function

$$h(z) = \mathrm{i}f(z). \tag{22.55}$$

From (22.18) we have

$$h(z) \to \frac{\mathrm{i}Pz}{4\mu} \qquad \text{as } y \to \infty, \tag{22.56}$$

and, from (22.54), we have

$$\mathrm{Im}[h(z)] = \text{constant} \qquad \text{on the gaps.} \tag{22.57}$$

It can also be argued that $f(z)$, and hence $h(z)$, are single-valued on making a circuit of the gaps. Remarkably, this problem for $h(z)$ is identical to one considered in Chapter 15: uniform irrotational flow in the negative y direction past a collection of impenetrable objects given here by the three finite-length gaps (22.15) and with zero circulation around them.

We can introduce a triply connected circular domain D_η in a parametric η plane and consider the conformal map

$$z = \mathcal{Z}_2(\eta) \tag{22.58}$$

from D_η to the unbounded region exterior to the three gaps (22.15); the subscript 2 reflects the fact that this map is the one associated with the second solution method. Let the boundary circles in D_η be denoted by $\tilde{C}_0$, $\tilde{C}_1$, and $\tilde{C}_2$, and let the prime function associated with this domain be denoted by $\tilde{\omega}(.,.)$; we add the tilde to distinguish it from the prime function $\omega(.,.)$ associated with the circular region D_ζ just considered. Owing to the symmetry of the intervals about the imaginary axis, we can take the centers $\tilde{\delta}_1, \tilde{\delta}_2$ and radii $\tilde{q}_1, \tilde{q}_2$ of $\tilde{C}_1, \tilde{C}_2$ to be

$$\tilde{\delta}_1 = \tilde{\delta} = -\tilde{\delta}_2, \qquad \tilde{q}_1 = \tilde{q}_2 = \tilde{q}. \tag{22.59}$$

With this choice we anticipate that the preimage $\zeta = \alpha$ of the point at infinity is $\alpha = 0$.

The mapping of $\mathcal{Z}_2(\eta)$ can be taken to be

$$\mathcal{Z}_2(\zeta) = D\phi_0(\eta, 0) + E, \tag{22.60}$$

where D and E are two real constants and $\phi_\theta(\eta, \alpha)$ is the parallel slit mapping introduced in Chapter 6.

The values of the four real parameters D, E, $\tilde{\delta}$, and $\tilde{q}$ are chosen to ensure the four conditions

$$\mathcal{Z}_2(1) = a, \qquad \mathcal{Z}_2(-1) = -a, \qquad \mathcal{Z}_2(\tilde{\delta}_1 - \tilde{q}_1) = +1, \qquad \mathcal{Z}_2(\tilde{\delta}_1 + \tilde{q}_1) = +b. \tag{22.61}$$

Introducing the composed function

$$H(\eta) \equiv h(\mathcal{Z}_2(\eta)), \tag{22.62}$$

the solution for $H(\eta)$ is given by the parallel slit mapping

$$H(\eta) = \frac{\mathrm{i}PD}{4\mu}\phi_{\pi/2}(\eta, 0). \tag{22.63}$$

If we now introduce

$$\tilde{F}(\eta) \equiv f(\mathcal{Z}_2(\eta)), \qquad \tilde{G}(\eta) \equiv g'(\mathcal{Z}_2(\eta)), \tag{22.64}$$

it follows from (22.55) that

$$\tilde{F}(\eta) = \frac{PD}{4\mu}\phi_{\pi/2}(\eta, 0). \tag{22.65}$$

To complete the solution, $\tilde{G}(\eta)$ is found by noting that, under this conformal mapping, it is portions of the real η axis inside D_η that are transplanted to the no-slip wall where

$$-\overline{\tilde{F}(\eta)} + \overline{\mathcal{Z}_2(\eta)} \frac{d\tilde{F}/d\eta}{d\mathcal{Z}_2/d\eta} + \tilde{G}(\eta) = 0, \tag{22.66}$$

and we have used the fact that $\overline{\eta} = \eta$ on the real η axis. This relation between analytic functions can be continued off the real axis and, in particular, we deduce

$$\tilde{G}(\eta) = \overline{\tilde{F}}(\eta) - \overline{\mathcal{Z}_2}(\eta) \frac{d\tilde{F}/d\eta}{d\mathcal{Z}_2/d\eta}. \tag{22.67}$$

This completes the solution.

There is a minor technical complication in taking $\alpha \to 0$ in the definitions of $\phi_\theta(\eta, \alpha)$ for $\theta = 0, \pi/2$ since these become unbounded and must be regularized as was done, for example, in Exercises 1.15 and 1.17 in the simply connected case. Actually, there are several other ways to circumvent this difficulty as $\alpha \to 0$, and it is instructive to explore them.

One simple fix is to use an automorphism of the disc (1.36) to move the preimage of infinity at $\alpha = 0$ to a different (nonzero) point in the disc; the transformation that is useful for this was explored in Exercise 1.14. Now, however, the preimage domain will not then, in any obvious geometrical sense, share the symmetries of the target configuration. While this is inconvenient, and perhaps undesirable, it is of no real consequence.

Another option proceeds as follows. From (22.60) and (22.65) we can write

$$\mathcal{Z}_2(\eta) = -D \lim_{\alpha \to 0} \left[\frac{1}{\alpha} \tilde{\mathcal{K}}(\eta, \alpha) - \frac{1}{\overline{\alpha}} \tilde{\mathcal{K}}(\eta, 1/\overline{\alpha}) \right] + E, \tag{22.68}$$

$$\tilde{F}(\eta) = -\frac{PD}{4\mu} \lim_{\alpha \to 0} \left[\frac{1}{\alpha} \tilde{\mathcal{K}}(\eta, \alpha) + \frac{1}{\overline{\alpha}} \tilde{\mathcal{K}}(\eta, 1/\overline{\alpha}) \right], \tag{22.69}$$

where $\tilde{F}(\eta)$ is determined up to an unimportant constant. However, from Exercise 6.11, an alternative representation of the mapping $\mathcal{Z}_2(\eta)$ is known to be given by

$$\mathcal{Z}_2(\eta) = \mathcal{D} \left[\frac{\tilde{\omega}(\eta, -1)^2 + \tilde{\omega}(\eta, +1)^2}{\tilde{\omega}(\eta, -1)^2 - \tilde{\omega}(\eta, +1)^2} \right], \qquad \mathcal{D} = \frac{\tilde{\omega}(\tilde{\delta} - \tilde{q}, -1)^2 - \tilde{\omega}(\tilde{\delta} - \tilde{q}, +1)^2}{\tilde{\omega}(\tilde{\delta} - \tilde{q}, -1)^2 + \tilde{\omega}(\tilde{\delta} - \tilde{q}, +1)^2}, \tag{22.70}$$

where the constant $\mathcal{D}$ is chosen in order that $\mathcal{Z}_2(\tilde{\delta} - \tilde{q}) = +1$. Formula (22.70) can be used to evaluate the mapping function without difficulty. Moreover, a combination of (22.68), (22.69), and (22.70) means that, up to an unimportant constant, we can write

$$\begin{aligned} \tilde{F}(\eta) &= \frac{P}{4\mu} \left[-2D \lim_{\alpha \to 0} \left(\frac{1}{\alpha} \tilde{\mathcal{K}}(\eta, \alpha) \right) - \mathcal{D} \left[\frac{\tilde{\omega}(\eta, -1)^2 + \tilde{\omega}(\eta, +1)^2}{\tilde{\omega}(\eta, -1)^2 - \tilde{\omega}(\eta, +1)^2} \right] \right], \qquad (22.71) \\ &= \frac{P}{4\mu} \left[-2D \lim_{\alpha \to 0} \left(\frac{\partial \log \tilde{\omega}(\eta, \alpha)}{\partial \alpha} \right) - \mathcal{D} \left[\frac{\tilde{\omega}(\eta, -1)^2 + \tilde{\omega}(\eta, +1)^2}{\tilde{\omega}(\eta, -1)^2 - \tilde{\omega}(\eta, +1)^2} \right] \right]. \end{aligned}$$

This expression for $\tilde{F}(\eta)$ can also be readily evaluated.

The streamlines shown in Figure 22.1 have been plotted using this form of the solution. To do so, we note that

$$g(\mathcal{Z}_2(\eta)) = \int^z \tilde{G}(\eta) dz = \int^\eta \tilde{G}(\eta) \mathcal{Z}_2'(\eta) d\eta. \tag{22.72}$$

Owing to the reflectional symmetry of the domain about the real axis, we have $\overline{\tilde{F}}(\eta) = \tilde{F}(\eta)$ and $\overline{Z_2}(\eta) = Z_2(\eta)$; hence, from (22.67),

$$g(Z_2(\eta)) = \int^{\eta} \left[\tilde{F}(\eta) \frac{dZ_2(\eta)}{d\eta} - Z_2(\zeta) \frac{d\tilde{F}(\eta)}{d\eta} \right] d\eta \tag{22.73}$$

$$= -F(\eta) Z_2(\eta) + 2 \int^{\eta} \tilde{F}(\eta) \frac{dZ_2(\eta)}{d\eta} d\eta. \tag{22.74}$$

The streamfunction now follows on use of (22.4).

It can be shown from a local analysis near the edges of the walls in this problem that the functions $f(z)$ and $g'(z)$ exhibit singularities of square root type at these points. A significant advantage of both solution methods outlined above is that these square root branch points are never seen explicitly when written in terms of the uniformizing variable ζ (for method 1) or η (for method 2).

Chapter 23

Plane elasticity

There are close mathematical similarities between the problems of slow viscous flows considered in Chapter 22 and problems of plane linear elasticity [83]. The framework of Part I can therefore be exploited in the latter context, too. In this chapter solutions to a fundamental boundary value problem in plane elasticity will be determined, in analytical form in terms of the prime function, in a multiply connected setting. An important feature here is to show how the theory of quadrature domains studied in Chapter 11 can play a role in plane elasticity problems.

The stress tensor components $\{\sigma_{xx}, \sigma_{xy}, \sigma_{yy}\}$ of a two-dimensional linear elastic solid can be represented in terms of two analytic functions, or complex potentials, $\phi(z)$ and $\psi(z)$, with

$$\sigma_{xx} + \sigma_{yy} = 4\mathrm{Re}[\phi'(z)], \tag{23.1}$$

$$\sigma_{yy} - \sigma_{xx} + 2\mathrm{i}\sigma_{xy} = 2\left[\overline{z}\phi''(z) + \psi''(z)\right]. \tag{23.2}$$

It can also be shown that the quantity

$$\sigma_{ij}n_j, \tag{23.3}$$

the stresses on a curve with normal vector with components n_j, is given in complex form by

$$-\mathrm{i}\frac{dH}{ds}, \qquad H(z,\overline{z}) = \phi(z) + z\overline{\phi'(z)} + \overline{\psi'(z)}. \tag{23.4}$$

The displacements (u, v) are given, in complex form, by

$$2G(u + \mathrm{i}v) = \kappa\phi(z) - z\overline{\phi'(z)} - \overline{\psi(z)}, \tag{23.5}$$

where G is the shear modulus of the host material, and the Kolosov constant κ is given, depending on the type of problem, by

$$\kappa = \begin{cases} 3 - 4\nu, & \text{for plane strain,} \\ (3-\nu)/(1+\nu), & \text{for plane stress,} \end{cases} \tag{23.6}$$

where the constant ν is the Poisson ratio. The potentials $\phi(z)$ and $\psi(z)$ can be found by solving boundary value problems of different types.

A direct problem: two uniformly pressurized pores. One fundamental problem of elasticity involves specifying the surface traction along a given boundary contour and finding the interior stress distribution. We now show how the function theory of Part I can be used to find the stress around two uniformly pressurized pores in an unbounded elastic medium. This section is a summary of the analysis given in [5].

Consider an infinite elastic body containing two pores, with no stresses acting at infinity, and with a uniform hydrostatic pressure of magnitude p acting along the pore boundaries. We want to find the stress distribution around the pores subject to this uniform hydrostatic loading. It will be assumed that the two pores have some geometrical symmetries: they are taken to be rotations of each other through 180°, and the centroids of both pores are taken to be located on the positive and negative real axis at equal distances from the origin.

Since the imposed traction on the boundary is

$$-pn_i, \tag{23.7}$$

which, in complex form, is

$$-p\left[-\mathrm{i}\frac{dz}{ds}\right] = \mathrm{i}p\frac{dz}{ds}, \tag{23.8}$$

where ds denotes the arclength element around the boundary, on use of (23.4),

$$-\mathrm{i}\frac{dH}{ds} = \mathrm{i}p\frac{dz}{ds}, \tag{23.9}$$

or, on integration with respect to s,

$$\phi(z) + z\overline{\phi'(z)} + \overline{\psi(z)} = -z + A_j, \qquad j = 0, 1, \tag{23.10}$$

where A_j is a constant and we have taken $p = 1$. The constants A_0 and A_1 must be found as part of the solution.

For the problem of two symmetric pores, called pore 1 and pore 2, considered here, the relevant conditions on the pore boundaries can be taken to be

$$\phi(z) + z\overline{\phi'(z)} + \overline{\psi(z)} = -z + \gamma, \quad \text{on pore 1}, \tag{23.11}$$

$$\phi(z) + z\overline{\phi'(z)} + \overline{\psi(z)} = -z - \gamma, \quad \text{on pore 2}, \tag{23.12}$$

where γ is some real constant to be found. This is because the symmetry of the geometrical configuration with respect to rotation by π around the origin can be used to argue that $\phi(z)$ and $\psi(z)$ can be chosen to be odd functions:

$$\phi(-z) = -\phi(z), \qquad \psi(-z) = -\psi(z) \tag{23.13}$$

with simple zeros as $z \to \infty$. (23.13) requires that

$$\phi(0) = 0, \qquad \psi(0) = 0. \tag{23.14}$$

It turns out that this fixes all degrees of freedom in the choice of $\phi(z)$ and $\psi(z)$ and necessitates the appearance of $\pm\gamma$ in (23.12). The fact that the two constants are negatives of each other follows from (23.13) and the assumed geometrical symmetries of the pores.

To find the stress fields we know from Chapter 5 that it is enough to perform the analysis in a circular domain D_ζ given by the concentric annulus $\rho < |\zeta| < 1$. We will show that analytical solutions to this boundary value problem for the stress distribution can be found if the elastic body exterior to the two finite-area pores is described by a conformal map from the annulus $\rho < |\zeta| < 1$ having the form

$$z = f(\zeta) = R\left[\frac{\omega(\zeta, -\sqrt{\rho})\omega(\zeta, -1/\sqrt{\rho})\omega(\zeta, 1/\sqrt{\rho})}{\omega(\zeta, \sqrt{\rho})\omega(\zeta, e^{i\theta}/\sqrt{\rho})\omega(\zeta, e^{-i\theta}/\sqrt{\rho})}\right], \tag{23.15}$$

where R, θ, and ρ are real constants and where $\omega(.,.)$ is the prime function for the concentric annulus $\rho < |\zeta| < 1$.

From its form as a ratio of products of prime functions the poles and zeros of the function (23.15) are clear: there are three simple zeros at $-\sqrt{\rho}, \pm\sqrt{\rho}^{-1}$ and three simple poles at $\sqrt{\rho}, \sqrt{\rho}^{-1}e^{\pm i\theta}$. Under the mapping (23.15) C_0 and C_1 are each transplanted to one of the pore boundaries, the point $\zeta = \sqrt{\rho}$ maps to $z = \infty$, and $\zeta = -\sqrt{\rho}$ maps to the origin $z = 0$.

It can be checked that (23.15) satisfies the functional relation

$$f(\rho^2\zeta) = f(\zeta) \tag{23.16}$$

for all choices of the parameters R, θ, and ρ. Since it is also meromorphic in the doubled domain $\rho < |\zeta| < 1/\rho$, we recognize from Chapter 8 that $f(\zeta)$ qualifies as a loxodromic function and, from Chapter 11, that the image domain D_z, which is the elastic region exterior to the two pores, is an unbounded quadrature domain. Indeed, domains given by this class of mappings were the subject of Exercise 11.7, which the reader should review to ascertain some of the properties of the mapping. Figure 11.6 shows a superposition of some typical pore shapes for different values of the parameters.

We now indicate why there is a significant advantage in restricting attention to the class of quadrature domains.

Let us introduce the composition of the two potentials $\phi(z)$ and $\psi(z)$ with the conformal mapping function $f(\zeta)$:

$$\Phi(\zeta) \equiv \phi(f(\zeta)), \qquad \Psi(\zeta) \equiv \psi(f(\zeta)). \tag{23.17}$$

Since it is a one-to-one conformal mapping, $f(\zeta)$ is analytic everywhere in the annulus $\rho < |\zeta| < 1$ except for the required simple pole at $\zeta = \sqrt{\rho}$, which maps to $z = \infty$. But since $\phi(z)$ and $\psi(z)$ are analytic in the elastic medium, and decay as $z \to \infty$, the composed functions $\Phi(\zeta)$ and $\Psi(\zeta)$ will be analytic everywhere in the annulus $\rho < |\zeta| < 1$.

In terms of the functions (23.17) the complex conjugate of the boundary condition (23.11) takes the form

$$\overline{\Phi}(1/\zeta) + \overline{f}(1/\zeta)\frac{d\Phi/d\zeta}{df/d\zeta} + \Psi(\zeta) = -\overline{f}(1/\zeta) + \overline{\gamma}, \qquad \zeta \in C_0, \tag{23.18}$$

where we have used the fact that $\overline{\zeta} = 1/\zeta$ on C_0. Condition (23.18), which is now a relation between analytic functions, can be analytically continued off the boundary C_0 where it was derived.

The complex conjugate of the boundary condition (23.12) becomes

$$\overline{\Phi}(\rho^2/\zeta) + \overline{f}(\rho^2/\zeta)\frac{d\Phi/d\zeta}{df/d\zeta} + \Psi(\zeta) = -\overline{f}(\rho^2/\zeta) - \overline{\gamma}, \qquad \zeta \in C_1, \tag{23.19}$$

where we have used the fact that $\overline{\zeta} = \rho^2/\zeta$ on C_1. On use of property (23.16) of the conformal mapping function, (23.19) becomes

$$\overline{\Phi}(\rho^2/\zeta) + \overline{f}(1/\zeta)\frac{d\Phi/d\zeta}{df/d\zeta} + \Psi(\zeta) = -\overline{f}(1/\zeta) - \overline{\gamma}, \qquad \zeta \in C_1. \tag{23.20}$$

Condition (23.20), which is now a relation between analytic functions, can also be analytically continued off the boundary C_1 where it was derived.

Subtraction of (23.18) and (23.20) now leads to

$$\overline{\Phi}(\rho^2/\zeta) - \overline{\Phi}(1/\zeta) = -2\overline{\gamma}, \quad \text{or} \quad \Phi(\rho^2\zeta) = \Phi(\zeta) - 2\gamma. \tag{23.21}$$

We deduce that the function $\Phi(\zeta)$ is quasi-loxodromic with respect to the Schottky group Θ generated by the map $\theta_1(z) = \rho^2 z$ and its inverse. This is a key observation. It should be clear from the sequence of steps above that (23.21) is a consequence of the condition (23.16) pertaining to the class of quadrature domains to which we have restricted our attention.

From (23.18) we see that the analyticity properties of $\Phi(\zeta)$ throughout the complex plane will be determined by its analyticity properties in the annulus $\rho < |\zeta| < 1/\rho$. Relation (23.18) can be used, together with the known analyticity of $\Phi(\zeta)$ and $\Psi(\zeta)$ in $\rho < |\zeta| < 1$, to deduce that the only possible singularities of $\Phi(\zeta)$ in $1 < |\zeta| < 1/\rho$ are those inherited from $f(\zeta)$. But (23.15) shows that $f(\zeta)$ has just two simple pole singularities there at the points $e^{\pm i\theta}/\sqrt{\rho}$.

In view of this, as well as condition (23.21), it is natural to conclude that

$$\Phi(\zeta) = AK(\zeta, e^{i\theta}/\sqrt{\rho}) + BK(\zeta, e^{-i\theta}/\sqrt{\rho}) + C \tag{23.22}$$

for some constants A, B, and C and where K was introduced in (5.77). The function (23.22) has the required simple poles at $e^{\pm i\theta}/\sqrt{\rho}$ by construction. On use of the property (5.81) of $K(\zeta)$, $\Phi(\zeta)$ satisfies (23.21) provided that

$$-A - B = -2\gamma. \tag{23.23}$$

This gives the value of γ once A and B have been found; these latter two constants can be found by equating the residues of the simple poles at $e^{\pm i\theta}/\sqrt{\rho}$ on each side of equation (23.18). The constant C follows from ensuring that $\phi(0) = 0$, or $\Phi(-\sqrt{\rho}) = 0$, as required in (23.14).

In summary, by virtue of these considerations, it can be shown that the complex potentials for two pores under hydrostatic loading are

$$\begin{aligned}\Phi(\zeta) &= A\Big[K(\zeta, e^{i\theta}/\sqrt{\rho}) + K(\zeta, e^{-i\theta}/\sqrt{\rho}) - K(-\sqrt{\rho}, e^{i\theta}/\sqrt{\rho}) - K(-\sqrt{\rho}, e^{-i\theta}/\sqrt{\rho})\Big],\\ \Psi(\zeta) &= -\overline{f}(1/\zeta) - \overline{\Phi}(1/\zeta) - \overline{f}(1/\zeta)\frac{d\Phi/d\zeta}{df/d\zeta} + \overline{\gamma}.\end{aligned} \tag{23.24}$$

For more details of this construction, and a fuller analysis of the solutions, the reader is referred to [5], where other boundary conditions are also treated.

It should be clear from this doubly connected example how quadrature domains of higher connectivity—that is, elastic regions with more than two pores—will also provide a route to analytical solutions to this, and other, boundary value problems in plane elasticity.

An inverse problem: Cherepanov's problem. Consider the problem of $M + 1$ holes in an unbounded elastic medium, where $M \geq 0$, on which two of the surface tractions are specified to be constant, i.e.,

$$\sigma_{nn} = p, \qquad \sigma_{nt} = \tau, \tag{23.25}$$

where

$$\sigma_{nn} = n_i \sigma_{ij} n_j, \qquad \sigma_{tn} = t_i \sigma_{ij} n_j, \qquad \sigma_{tt} = t_i \sigma_{ij} t_j, \tag{23.26}$$

t_i denotes the components of the unit tangent vector on the boundary, and the stresses are assumed to vanish in the far field. This problem, for a given set of hole shapes, can be solved; indeed, an explicit example involving two holes or $M = 1$, with $p = 1, \tau = 0$, was just given in the previous section. This is the direct problem, and it is linear. We saw how analytical solutions in terms of the prime function are available when the elastic regions are quadrature domains.

Interestingly, regions akin to quadrature domains also arise quite naturally in a more difficult inverse problem of plane elasticity.

An inverse problem, which is nonlinear, is to ask for which hole shapes will the so-called hoop stress σ_{tt} in the above problem *also* be constant, i.e.,

$$\sigma_{tt} = \sigma = \text{constant}, \qquad \text{on the hole boundaries.} \tag{23.27}$$

This additional condition cannot possibly be satisfied for any choice of the hole shapes since the boundary value problem is now overdetermined. Rather, this is a free boundary problem: we must simultaneously find the shapes of the holes, and the stress distribution around them, such that all three conditions in (23.25) and (23.27) can be satisfied simultaneously. While the hoop stress assumes the constant value σ, it turns out that we will also need to find a consistent value of σ necessary for equilibrium. This problem was studied by Cherepanov [74].

This nonlinear free boundary problem can be solved using the mathematical methods of Part I, as we now show. In fact, there are two different ways to tackle it, and, since it provides an instructive case study of the material in Part I, we will present both solution methods.

First note that the stress components $\{\sigma_{nn}, \sigma_{nt}, \sigma_{tt}\}$ can be expressed in terms of $\{\sigma_{xx}, \sigma_{xy}, \sigma_{yy}\}$ via the relations

$$\sigma_{tt} + \sigma_{nn} = \sigma_{xx} + \sigma_{yy}, \tag{23.28}$$

$$\sigma_{tt} - \sigma_{nn} + 2\mathrm{i}\sigma_{tn} = -\left(\frac{dz}{ds}\right)^2 (\sigma_{yy} - \sigma_{xx} + 2\mathrm{i}\sigma_{xy}). \tag{23.29}$$

Establishing these relations is just an algebraic exercise using the definitions (23.26). Next, since the stresses tend to a constant value in the far field, then as $z \to \infty$,

$$4\mathrm{Re}[\phi'(z)] \to \sigma_{xx}^{\infty} + \sigma_{yy}^{\infty}, \tag{23.30}$$

$$2\left[\overline{z}\phi''(z) + \psi''(z)\right] \to \sigma_{yy}^{\infty} - \sigma_{xx}^{\infty} + 2\mathrm{i}\sigma_{xy}^{\infty}. \tag{23.31}$$

From (23.1) and (23.25), the real part of $\phi'(z)$ is equal to the same constant on the boundary of all the holes and is bounded at infinity. Then, since $\phi'(z)$ must be single-valued so that the stresses and displacements are single-valued around each hole, we must have

$$\phi'(z) = \frac{1}{4}\left(\sigma_{xx}^{\infty} + \sigma_{yy}^{\infty}\right) + \mathrm{i}c, \tag{23.32}$$

where c is some real constant. Clearly, $\phi''(z) = 0$. Also, from (23.28), we must have

$$\sigma_{xx}^{\infty} + \sigma_{yy}^{\infty} = \sigma + p, \tag{23.33}$$

which forces the choice of the hoop stress to be

$$\sigma = \sigma_{xx}^{\infty} + \sigma_{yy}^{\infty} - p. \tag{23.34}$$

Now (23.2) together with (23.25), (23.27), (23.29), and (23.32) imply

$$\sigma - p + 2\mathrm{i}\tau = -2\left(\frac{dz}{ds}\right)^2 \psi''(z), \qquad \text{on each hole boundary.} \tag{23.35}$$

Equivalently this can be written as

$$\psi''(z) = -a\frac{d\overline{z}}{dz}, \qquad a = \frac{\sigma - p}{2} + \mathrm{i}\tau, \qquad \text{on each hole boundary,} \tag{23.36}$$

where we recall that $ds^2 = |dz|^2$. On integration of (23.36) with respect to z,

$$\psi'(z) + a\overline{z} = A_j, \qquad \text{on the boundary of hole } j, \tag{23.37}$$

where A_j is a constant that must be assumed to take a different value on each hole boundary. Moreover, from the far-field stress condition (23.31) we must have

$$\psi''(z) \to \frac{1}{2}\left(\sigma_{yy}^{\infty} - \sigma_{xx}^{\infty}\right) + \mathrm{i}\sigma_{xy}^{\infty} \tag{23.38}$$

or

$$\psi'(z) \to bz + \mathcal{O}(1/z^2), \qquad b \equiv \frac{1}{2}\left(\sigma_{yy}^{\infty} - \sigma_{xx}^{\infty}\right) + \mathrm{i}\sigma_{xy}^{\infty}, \tag{23.39}$$

where we have chosen an arbitrary constant of integration in the definition of $\psi'(z)$ to be zero.

We now describe the two possible solution methods. Both involve the introduction of a conformal mapping function

$$z = \mathcal{Z}(\zeta) \tag{23.40}$$

from a multiply connected circular domain D_ζ to the elastic region exterior to the $M+1$ holes. The circles $\{C_j | j = 0, 1, \ldots, M\}$ are transplanted to the boundaries of the holes in the elastic region. Let some point $\zeta_\infty \neq 0$ be the preimage of the point at infinity (the following can be adapted to the case $\zeta_\infty = 0$) where

$$\mathcal{Z}(\zeta) \to \frac{R}{\zeta - \zeta_\infty} + \ldots \qquad \text{as } \zeta \to \zeta_\infty. \tag{23.41}$$

The first solution method involves taking the real and imaginary part of (23.37):

$$\mathrm{Re}\left[\psi'(z) + a\overline{z}\right] = \mathrm{Re}\left[\psi'(z) + \overline{a}z\right] = \text{constant}, \tag{23.42}$$

$$\mathrm{Im}\left[\psi'(z) + a\overline{z}\right] = \mathrm{Im}\left[\psi'(z) - \overline{a}z\right] = \text{constant}, \tag{23.43}$$

where these conditions hold on each hole boundary. Here we have used the elementary fact that the real parts of complex conjugate quantities are equal while the imaginary parts are the negative of each other.

Let us now introduce the composed function

$$\Psi(\zeta) \equiv \psi'(\mathcal{Z}(\zeta)), \tag{23.44}$$

which must be analytic in D_ζ except for a simple pole at ζ_∞. Relations (23.42) and (23.43) can now be written as

$$\operatorname{Re}[\Psi(\zeta)+\overline{a}\mathcal{Z}(\zeta)] = \text{constant}, \tag{23.45}$$

$$\operatorname{Im}[\Psi(\zeta)-\overline{a}\mathcal{Z}(\zeta)] = \text{constant} \tag{23.46}$$

on each hole boundary. We now recognize that both functions in square brackets in (23.45) and (23.46) are analytic and single-valued everywhere in D_ζ except for simple poles at ζ_∞. Indeed, from (23.39) and (23.41),

$$\Psi(\zeta)+\overline{a}\mathcal{Z}(\zeta) \to \frac{R(b+\overline{a})}{\zeta-\zeta_\infty}+\cdots, \qquad \Psi(\zeta)-\overline{a}\mathcal{Z}(\zeta) \to \frac{R(b-\overline{a})}{\zeta-\zeta_\infty}+\cdots. \tag{23.47}$$

Moreover, conditions (23.45) and (23.46) are precisely those satisfied by parallel slit maps $\phi_\theta(\zeta,\zeta_\infty)$ introduced in Part I with angles $\theta=\pi/2$ and 0, respectively, where, we recall, on the boundaries of D_ζ,

$$\operatorname{Re}[\phi_{\pi/2}(\zeta,\zeta_\infty)] = \text{constant}, \qquad \operatorname{Im}[\phi_0(\zeta,\zeta_\infty)] = \text{constant}. \tag{23.48}$$

Taking real multiples of $\phi_\theta(\zeta,\zeta_\infty)$ will not change these properties, so we deduce that

$$\Psi(\zeta)+\overline{a}\mathcal{Z}(\zeta) = \operatorname{Re}[R(b+\overline{a})]\phi_{\pi/2}(\zeta,\zeta_\infty)+\mathrm{i}\operatorname{Im}[R(b+\overline{a})]\phi_0(\zeta,\zeta_\infty)+c \tag{23.49}$$

and

$$\Psi(\zeta)-\overline{a}\mathcal{Z}(\zeta) = \operatorname{Re}[R(b-\overline{a})]\phi_0(\zeta,\zeta_\infty)+\mathrm{i}\operatorname{Im}[R(b-\overline{a})]\phi_{\pi/2}(\zeta,\zeta_\infty)+d, \tag{23.50}$$

where c and d are constants and where we have used the fact that the parallel slit maps $\phi_\theta(\zeta,\zeta_\infty)$ were constructed to have unit residue at $\zeta=\zeta_\infty$. On addition of (23.49) and (23.50)

$$2\Psi(\zeta) = Rb\left(\phi_{\pi/2}(\zeta,\zeta_\infty)+\phi_0(\zeta,\zeta_\infty)\right)+\overline{R}a\left(\phi_{\pi/2}(\zeta,\zeta_\infty)-\phi_0(\zeta,\zeta_\infty)\right)+c+d, \tag{23.51}$$

and on subtraction of (23.50) from (23.49) we find

$$2\overline{a}\mathcal{Z}(\zeta) = R\overline{a}\left(\phi_{\pi/2}(\zeta,\zeta_\infty)+\phi_0(\zeta,\zeta_\infty)\right)+\overline{Rb}\left(\phi_{\pi/2}(\zeta,\zeta_\infty)-\phi_0(\zeta,\zeta_\infty)\right)+c-d. \tag{23.52}$$

On use of the formulas (6.45) and (6.46) we find

$$\Psi(\zeta) = -\left[\frac{Rb}{\zeta_\infty}\mathcal{K}(\zeta,\zeta_\infty)+\frac{\overline{R}a}{\overline{\zeta_\infty}}\mathcal{K}(\zeta,1/\overline{\zeta_\infty})\right]+\text{constant}, \tag{23.53}$$

$$\mathcal{Z}(\zeta) = -\left[\frac{R}{\zeta_\infty}\mathcal{K}(\zeta,\zeta_\infty)+\frac{\overline{Rb}}{\overline{a}\overline{\zeta_\infty}}\mathcal{K}(\zeta,1/\overline{\zeta_\infty})\right]+\text{constant}. \tag{23.54}$$

This completes the solution of the problem, with (23.53) and (23.54) representing the solution in terms of (derivatives of) the prime function. These are expressions for the conformal mapping describing the shapes of the holes and the function $\Psi(\zeta)$ giving the associated stress field in the elastic medium.

An alternative approach is to think about the Schwarz functions of the boundaries of the holes. Let

$$\mathcal{S}_j(z) = \overline{z} \qquad \text{on the boundary of hole } j \tag{23.55}$$

so that $\mathcal{S}_j(z)$ is the Schwarz function of this hole boundary. (23.37) implies that

$$\mathcal{S}_j(z) = -\frac{\psi'(z)}{a}+\frac{A_j}{a} \qquad \text{on the boundary of hole } j. \tag{23.56}$$

Without loss of generality we can choose $A_0 = 0$ so that $\mathcal{S}_0(z) = -\psi'(z)/a$, and we conclude that

$$\mathcal{S}_j(z) = \mathcal{S}_0(z) + \overline{\gamma_j}, \qquad j = 1, \ldots, M, \tag{23.57}$$

where

$$\overline{\gamma_j} = \frac{A_j}{a}. \tag{23.58}$$

We see that the Schwarz functions of the individual hole boundaries are the same function up to an additive constant. We also know from (23.39) that, as $z \to \infty$,

$$\mathcal{S}_0(z) \sim -\frac{bz}{a} + \cdots \tag{23.59}$$

and, importantly, that there are no other singularities of $\mathcal{S}_0(z)$ in the elastic region. This is reminiscent of the property of the multiply connected quadrature domains considered in Chapter 11, where the Schwarz functions of the individual hole boundaries were all (meromorphic) continuations of each other. The situation is almost the same, but not quite. If we now rewrite the Schwarz functions in terms of the conformal mapping function (23.40), we find

$$\overline{\mathcal{Z}}(\overline{\theta_j}(1/\zeta)) = \overline{\mathcal{Z}}(1/\zeta) + \overline{\gamma_j}, \qquad j = 1, \ldots, M, \tag{23.60}$$

from which we infer, on taking a complex conjugate and setting $1/\overline{\zeta} \mapsto \zeta$, that

$$\mathcal{Z}(\theta_j(\zeta)) = \mathcal{Z}(\zeta) + \gamma_j \qquad j = 1, \ldots, M. \tag{23.61}$$

The conformal mapping $\mathcal{Z}(\zeta)$ is therefore a quasi-automorphic function with respect to the Schottky group Θ generated by $\{\theta_j(\zeta), \theta_j^{-1}(\zeta) | j = 1, \ldots, M\}$ under composition. Recall that, for the quadrature domains of Chapter 11, the associated conformal mappings to them from a preimage circular domain were automorphic functions with respect to Θ.

We can now use the function theory in Part I—in particular that presented in Chapter 9—to construct the required conformal mapping function $\mathcal{Z}(\zeta)$.

From (23.41) we know that $\mathcal{Z}(\zeta)$ has a simple pole at ζ_∞, and then, from (23.59), we deduce that it also has a simple pole at $1/\overline{\zeta_\infty}$. We can therefore deploy the quasi-automorphic function $\mathcal{K}(\zeta, .)$ introduced in Chapter 9 and write

$$\mathcal{Z}(\zeta) = -\left[\frac{R}{\zeta_\infty}\mathcal{K}(\zeta, \zeta_\infty) + \frac{\overline{Rb}}{\overline{a}\overline{\zeta_\infty}}\mathcal{K}(\zeta, 1/\overline{\zeta_\infty})\right] + \text{constant}, \tag{23.62}$$

where the coefficients of these functions have been chosen to ensure that the simple poles at ζ_∞ and $1/\overline{\zeta_\infty}$ have the correct strengths. This coincides with (23.54), derived using the first solution method, and it allows us to determine the shapes of the holes.

From Chapter 11, we know that the conformal maps to quadrature domains are characterized by conformal mappings from circular preimage regions D_ζ that are automorphic with respect to the Schottky group Θ. Domains characterized by quasi-automorphic conformal mapping functions also have an interpretation as generalized quadrature domains. All these matters have been explored by Marshall [98], who was the first to present the second approach to this problem via Schwarz functions.

In summary this chapter has demonstrated how to solve analytically one of the fundamental problems of plane elasticity (the direct problem) in elastic domains given by quadrature domains; expressions for the solutions for the complex potentials governing the stress distribution can be written down concisely in terms of the prime function. We also showed how to solve an inverse problem by considering two methods: one based on slit mappings and another on the use of Schwarz functions. It is interesting that the elastic regions satisfying the inverse problem also emerge as generalized quadrature domains.

The usefulness of the prime function machinery and the relevance of quadrature domain theory to problems in solid mechanics do not appear to have been fully appreciated in the existing literature, and there is much scope for future exploitation of these ideas in this field. Indeed, the author appears to be the first to have pointed out [5] that some of the fundamental problems of plane elasticity can be solved by combining elements of the theory of Part I, in particular the theory of quadrature domains from Chapter 11 and the theory of loxodromic functions from Chapter 8.

Chapter 24

Vortex patch equilibria of the Euler equations

The Euler equations governing the velocity $\mathbf{u}$ of an ideal incompressible inviscid fluid of constant density ρ are

$$\frac{\partial \mathbf{u}}{\partial t} + \mathbf{u}.\nabla \mathbf{u} = -\frac{1}{\rho}\nabla p, \tag{24.1}$$

where $p(x, y, z, t)$ is the fluid pressure. The vorticity is defined as a curl of the velocity field. On taking a curl of equation (24.1), and then restricting consideration to a two-dimensional velocity field $\mathbf{u} = (u(x, y, t), v(x, y, t), 0)$ in an (x, y) plane, we arrive at the equation

$$\frac{\partial \omega(x, y, t)}{\partial t} + \mathbf{u}.\nabla \omega(x, y, t) = 0 \tag{24.2}$$

for the vorticity component in the out-of-plane direction:

$$\omega(x, y, t) \equiv \frac{\partial v}{\partial x} - \frac{\partial u}{\partial y}, \tag{24.3}$$

where we can now think of ∇ as denoting a two-dimensional gradient in the (x, y) plane.

In fluid mechanics it is well known that if (24.3) can be solved for the vorticity distribution, then the fluid velocity can be reconstructed from it using an integral formula of Biot–Savart type [113]. Since the two-dimensional flow is incompressible, so that

$$\nabla.\mathbf{u} = \frac{\partial u}{\partial x} + \frac{\partial v}{\partial y} = 0, \tag{24.4}$$

a streamfunction $\psi(x, y, t)$ can be introduced from which we can derive the velocity and vorticity fields as follows:

$$u = \frac{\partial \psi}{\partial y}, \qquad v = -\frac{\partial \psi}{\partial x}, \qquad \omega = -\nabla^2 \psi. \tag{24.5}$$

An important subclass of solutions to equation (24.2) involves the choice

$$\omega(x, y, t) = \omega_j = \text{constant}, \qquad z \in D_j(t), \tag{24.6}$$

where $D_j(t)$ is a simply connected subregion, or "patch," of finite area and where j labels the subregion or patch. Such a subregion is called a *vortex patch*. Figure 24.1 shows

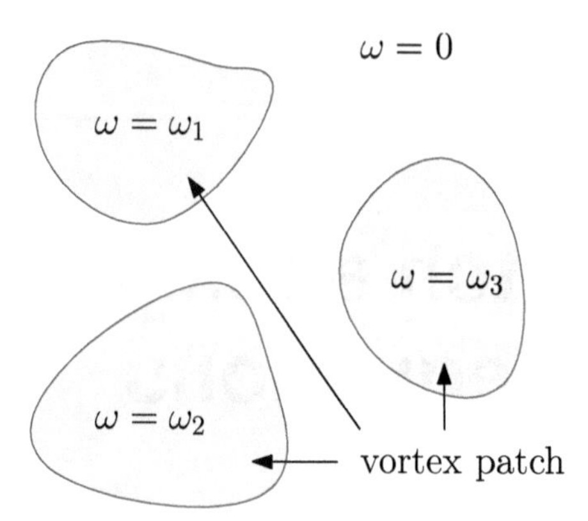

Figure 24.1. *Three regions of uniform vorticity, or "vortex patches," in an otherwise irrotational flow. The shapes of patches in equilibrium must be determined as the solution of a free boundary problem. The region exterior to the patches is triply connected in this example.*

an arrangement of three vortex patches: the vorticity is a piecewise constant field taking possibly different values in different patches. That (24.2) satisfies (24.3) is clear, although the vorticity field $\omega(x, y, t)$ clearly has jumps at the boundaries of the patches, so the main concern is what happens at the patch boundaries. The boundary of a uniform vortex patch is called a *vortex jump*. While there is a discontinuity in the vorticity field, we must ensure that there is no discontinuity in the fluid pressure; any discontinuity would lead to unbounded hydrodynamic forces and must be disallowed. Under the effect of the hydrodynamic forces the vortex patches will evolve in time, and their shape evolution must be determined. This is a nonlinear free boundary problem for the shapes of the vortex patches.

It is known [113] that, at the boundary $\partial D_j(t)$ of a time-evolving vortex patch, the fluid velocity must be continuous across a vortex jump and, as such, is well defined as one approaches the boundary from either inside or outside the patch. Furthermore, a vortex jump must move with the fluid velocity, which, by the continuity of the velocity field at the jump, is well defined there. These two conditions together turn out to guarantee that the fluid pressure is continuous at the vortex jump [113].

There are certain geometrical distributions of vortex patches for which the balance of hydrodynamic forces allows the configuration to remain in steady equilibrium, or in some relative equilibrium where the time dependence of the configuration is trivial. For example, the configuration might translate at some uniform speed or rotate at some constant angular velocity without change in shape. In this way, for a relative equilibrium, by moving to a cotraveling or corotating frame of reference, the geometrical configuration of vortex patches does not change in time.

Finding such equilibria is an important problem in fluid mechanics; they are equilibria (or relative equilibria) of the Euler equations. This amounts to a free boundary problem where the shape of the patches must be found. This is done by satisfying the following two conditions:

(a) The fluid velocity must be continuous across the vortex jump and, as such, is well defined as one approaches the boundary from either inside or outside the patch.

(b) The vortex jump must be a streamline in the laboratory frame (if the configuration is an equilibrium) or in a cotraveling or corotating frame (if the configuration is a relative equilibrium that is either steadily translating or rotating).

The *Rankine vortex* is a well-known radially symmetric equilibrium of the Euler equation. It is a circular vortex patch. The streamfunction associated with it is

$$\psi(r) = \begin{cases} -\frac{\omega_0}{4} r^2 + \text{constant}, & r < a, \\ -\frac{\omega_0 a^2}{2} \log r, & |r| \geq a, \end{cases} \tag{24.7}$$

where ω_0 is the value of the constant vorticity inside the patch and where the patch is taken to have radius a. It can be checked that the vortex jump $r = a$ is a streamline and that the associated velocity field is continuous there. The unspecified additive constant in (24.7) is unimportant since it does not affect the velocity field, but it can be chosen to ensure that ψ is continuous at the vortex jump. The associated vorticity distribution is

$$\omega(r) = \begin{cases} \omega_0, & r < a, \\ 0, & |r| \geq a. \end{cases} \tag{24.8}$$

The *circulation* Γ associated with a vortex patch D is defined to be

$$\Gamma = \int_{\partial D} \mathbf{u}.\mathbf{dx}. \tag{24.9}$$

By the Stokes theorem we can write this as

$$\Gamma = \int_D \omega dA, \tag{24.10}$$

where dA denotes the differential area element. The circulation of the Rankine vortex is $\pi a^2 \omega_0$.

Multipolar vortices: Much experimental and numerical evidence has accrued in recent years concerning the existence of *multipolar equilibria* of the Euler equation. These structures, which have been called *multipolar vortices*, are finite-area regions of vorticity having a core of one signed vorticity surrounded by $N \geq 3$ satellite vortices of opposite signed circulation such that the total circulation associated with the structure is zero; a zero-net-circulation structure is called shielded. Such multipolar vortex equilibria are found to result from a nonlinear destabilization of a monopolar, or radially symmetric, structure with zero net circulation.

The Rankine vortex is a monopolar, or radially symmetric, vortex, but it has a nonzero net circulation $\pi a^2 \omega_0$. A simple way to produce a monopolar structure with *zero* total circulation from the Rankine vortex is to add a point vortex of circulation $-\pi a^2 \omega_0$ at its center; a point vortex is the name for a δ-function distribution of vorticity. Such a point vortex will exactly cancel the circulation $\pi a^2 \omega_0$ of the Rankine vortex. The streamfunction associated with a point vortex of circulation Γ at some point z_α locally takes the form

$$\psi = \text{Im}\left[-\frac{\mathrm{i}\Gamma}{2\pi} \log(z - z_\alpha) \right] + \text{a regular function}. \tag{24.11}$$

The associated complex velocity field, near the vortex, takes the form

$$u - \mathrm{i}v = -\frac{\mathrm{i}\Gamma}{2\pi} \frac{1}{z - z_\alpha} + \tilde{u}_\alpha - \mathrm{i}\tilde{v}_\alpha, \tag{24.12}$$

where $\tilde{u}_\alpha - i\tilde{v}_\alpha$ is regular at z_α. The condition that the point vortex at z_α is in equilibrium is

$$\tilde{u}_\alpha - i\tilde{v}_\alpha \Big|_{z_\alpha} = 0. \tag{24.13}$$

That is, the velocity field at the vortex position once the singular contribution from the vortex itself has been subtracted must vanish. This dynamical condition can be derived by insisting that the point vortex be free of net force [113].

The streamfunction associated with such a *shielded Rankine vortex*—this term was coined by the author in [34]—is, in plane polar coordinates,

$$\psi(r) = \begin{cases} -\frac{\omega_0}{4}\left[r^2 - a^2 \log r^2\right] + \text{constant}, & r < a, \\ 0, & |r| \geq a. \end{cases} \tag{24.14}$$

This can be written, in terms of the complex variable $z = x + iy$, as

$$\psi(z, \overline{z}) = \begin{cases} -\frac{1}{4}\left[z\overline{z} - \int^z S(z')dz' - \overline{\int^z S(z')dz'}\right], & z \in D, \\ \\ 0, & z \notin D, \end{cases} \tag{24.15}$$

where

$$S(z) = \frac{a^2}{z} \tag{24.16}$$

is the Schwarz function of the circle $|z| = a$ [76].

An important observation is that (24.15) has the functional form of a *modified Schwarz potential* as introduced in §11.14. In an attempt to find generalized equilibria having a more geometrically complicated structure, and in view of this observation, it is natural to consider posing modified Schwarz potentials as possible streamfunctions for other steady equilibria of the Euler equations.

We focus on this possibility for a simply connected patch first.

Suppose D is some simply connected vortex patch, and consider the streamfunction defined by

$$\psi(z, \overline{z}) = \begin{cases} -\frac{1}{4}\left[z\overline{z} - \int^z S(z')dz' - \overline{\int^z S(z')dz'}\right], & z \in D, \\ \\ 0, & z \notin D, \end{cases} \tag{24.17}$$

where $S(z)$ is the Schwarz function of the patch boundary ∂D [76]. We want to examine if such an arrangement can be a possible steady equilibrium. This means we must examine if conditions (a) and (b) listed above can be satisfied.

The velocity field in D is

$$u - iv = 2i\frac{\partial \psi}{\partial z} = -\frac{i}{2}\left[\overline{z} - S(z)\right]. \tag{24.18}$$

Hence, by the properties of the Schwarz function $S(z)$, the velocity field is continuous with the quiescent flow outside the patch. This means that condition (a) is satisfied by (24.17).

It is also clear that, on ∂D,

$$d\psi = \frac{\partial \psi}{\partial z}dz + \frac{\partial \psi}{\partial \overline{z}}d\overline{z} = -\frac{1}{4}\left(\overline{z} - S(z)\right)dz - \frac{1}{4}\left(z - \overline{S(z)}\right)d\overline{z} = 0. \tag{24.19}$$

This means that ∂D is also a streamline, which is required by condition (b).

That we have been able to satisfy both boundary conditions (a) and (b) on the patch boundary without even specifying the shape of the patch should be of concern. It suggests that *any* vortex patch is an equilibrium solution of the Euler equation! But this is clearly not true.

This conundrum is resolved by recognizing that the Schwarz function of any closed analytic curve is singular inside the curve. Thus, for a general domain D, the Schwarz function $S(z)$ will be singular in the patch and the streamfunction (24.17) will in general have unphysical singularities inside the patch and will not therefore represent a physically admissible form of a streamfunction.

Nevertheless, since we are seeking multipolar vortex structures with N satellite vortices surrounding a central core, it is natural to ask if the class of vortex shapes D can be restricted to those whose Schwarz functions $S(z)$ are meromorphic inside D with $N+1$ simple pole singularities of purely real residues. Such singularities of $S(z)$, when they appear in (24.17) as singularities of a streamfunction, are physically admissible since they correspond to point vortices. Even so, there remains the matter of ensuring that any such distribution of point vortices is *steady* so that the global vortex structure will be in equilibrium. This means that condition (24.13) must be enforced at all the singularities of $S(z)$. Only if all point vortices can be rendered stationary does an equilibrium of the Euler equations exist.

It turns out that a class of exact solutions of the Euler equations within this class can indeed be found. The shape of the equilibrium patches is given explicitly by the conformal map from the unit disc $|\zeta| < 1$,

$$z = f(\zeta) = R(a,N)\zeta\left[1 + \frac{b(a,N)}{\zeta^N - a^N}\right], \tag{24.20}$$

where $N \geq 3$, $a > 1$ is a real parameter, and $R(a,N)$ is a real normalization parameter that simply sets the size of the vortex structure. The Schwarz function $S(z)$ associated with the map (24.20) is

$$S(z) = \overline{z}(1/\zeta) = \frac{R(a,N)}{\zeta}\left[1 + \frac{b(a,N)\zeta^N}{1 - \zeta^N a^N}\right], \tag{24.21}$$

which has simple poles at $\zeta = 0$ and at the N-fold rotationally symmetric points

$$\zeta_j = \frac{\omega_j}{a}, \qquad \omega_j = e^{2\pi i j/N}, \qquad j = 0, 1, \ldots, N-1. \tag{24.22}$$

On use of (24.18) we see that simple poles of $S(z)$ inside the patch D yield simple poles of the complex velocity field, which correspond physically to point vortices if the residues of $S(z)$ at these poles are purely real. In this case it can be seen that the point vortex locations in the physical plane are at the origin and at

$$z_j = f(1/a)e^{2\pi i j/N}, \qquad j = 0, 1, \ldots, N-1. \tag{24.23}$$

The point vortex at $z = 0$ will be in equilibrium for all choices of the parameters owing to the N-fold rotational symmetry of the global vorticity distribution. However, for a given N and a the value of $b(a,N)$ must be chosen to ensure that one (and hence all, by rotational

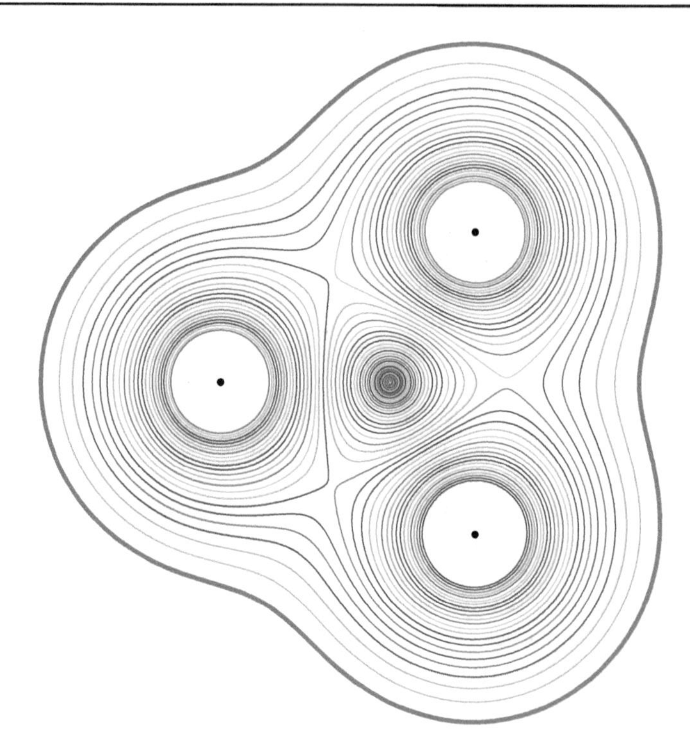

Figure 24.2. *Streamlines for a quadrupolar vortex structure. The region bounded by the blue line is a vortex patch which is a simply connected quadrature domain described by a conformal mapping (24.20) from the disc with $N = 3$ and $a = 2$ and with an associated streamfunction (24.17) which is a modified Schwarz potential. The patch has four superposed point vortices. This vortex structure is surrounded by irrotational, quiescent flow.*

symmetry) of the satellite vortices are in equilibrium. An analytical expression for the parameter $b(a, N)$ can be found explicitly. The form of this explicit expression can be found in [35]. With the steady state boundary conditions on the vortex jump satisfied, and the condition satisfied that all point vortices are in equilibrium, we have therefore derived a global equilibrium of the two-dimensional Euler equation.

Figure 24.2 shows the streamline distribution calculated from (24.20) and (24.17) for a typical quadrupolar vortex comprising a simply connected vortex patch with four superposed point vortices: one at the center of the patch, and three satellite vortices symmetrically disposed about the patch center. At all points where streamlines are shown, the vorticity, away from the point vortex locations shown as black dots, is a uniform constant. Outside the patch boundary, the flow is quiescent.

Since the Schwarz function of the boundary ∂D of the class of exact multipolar vortices just found is meromorphic inside D, we know from Chapter 11 that D is a simply connected quadrature domain.

Equilibria involving multiply connected vortex patches: It is natural to ask if more geometrically complex vortex patch equilibria, perhaps involving multiply connected vortex patches, can be constructed from modified Schwarz potentials in a similar way.

The answer turns out to be yes. In order both to compute the shapes of the associated equilibrium vortex patches and to ensure that any point vortices superposed on the vortex are in equilibrium, it is convenient to use the function theory described in Part I, in particular that outlined in Chapter 11. This section is based on ideas originally presented in [30].

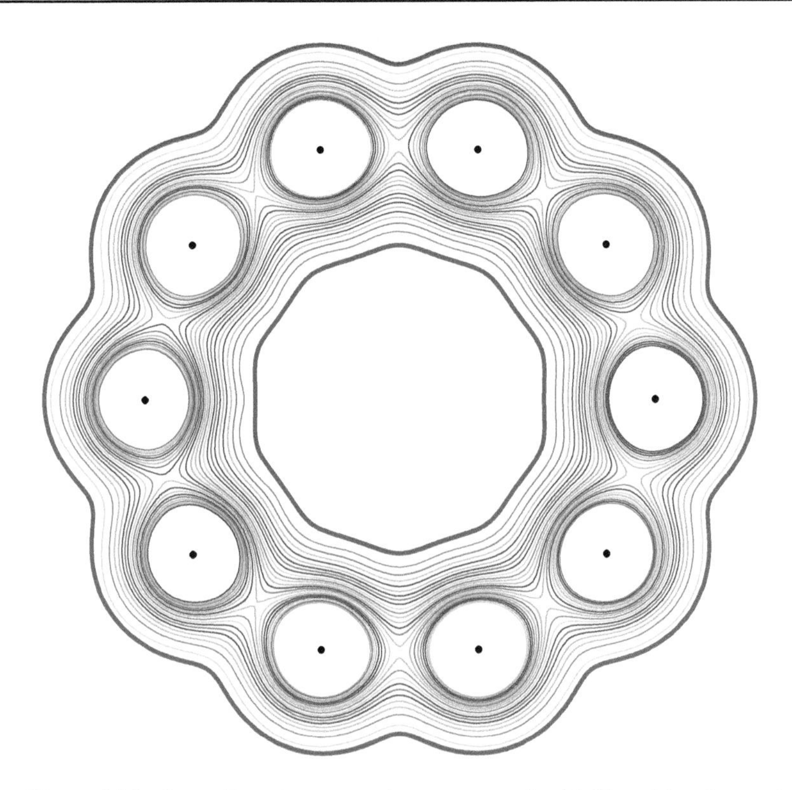

Figure 24.3. *Streamlines for an annular vortex patch with $N = 10$ point vortices. The vortex patch here, bounded by the blue lines, is a doubly connected quadrature domain and the associated streamfunction is a modified Schwarz potential (24.17). There is irrotational, quiescent flow both surrounding the vortex structure and inside the enclosed hole. The mapping from a concentric annulus is given by (24.24).*

As an example, consider the conformal mapping from the concentric annulus $\rho < |\zeta| < 1$ to a doubly connected vortex patch given by

$$z = f(\zeta) = A\zeta \prod_{k=1}^{N} \left[\frac{\omega(\zeta, \omega^{k-1}a/\rho^{2/N})}{\omega(\zeta, \omega^{k-1}a)} \right], \tag{24.24}$$

$$1 < a < 1/\rho, \qquad \omega = e^{2\pi \mathrm{i}/N}, \qquad a, A \in \mathbb{R},$$

where $N \geq 3$. This is the same class of domains considered in (11.90) of Chapter 11. This mapping provides a parametrization of the boundaries of a doubly connected quadrature domain of nontrivial shape and with N-fold rotational symmetry.

In a manner similar to the simply connected patch just described, provided the parameters ρ and a are appropriately chosen, the modified Schwarz potential (24.17) defined over these quadrature domains provides the equilibrium streamfunction for an annular, doubly connected vortex patch with N superposed point vortices symmetrically disposed around the patch, as shown in Figure 24.3. This figure shows a typical equilibrium shape and streamline distribution for $N = 10$, where the parameters a and ρ are chosen to ensure that condition (24.13) is satisfied at all the point vortex locations. Full details can be found in [30].

Steadily rotating vortex structures: All the solutions just described are equilibria of the steady Euler equations. But streamfunctions in the form of modified Schwarz potentials for relative equilibrium comprising steadily rotating vortex structures can also be found.

We therefore now turn our attention to relative equilibria that rotate at some constant angular velocity. Multivortex patch configurations will be considered so that the irrotational region exterior to the collection of patches is multiply connected. The case of two vortex patches will be considered in detail. The following analysis was first presented by Crowdy and Marshall in [52].

Suppose some configuration of two vortex patches, called D_1 and D_2, each of vorticity ω_0, are in pure solid-body rotation with constant angular velocity $\omega_0/2$. The irrotational fluid region exterior to the two vortices will be called D; it is a doubly connected region. The streamfunction associated with the flow is

$$\psi(z,\overline{z}) = \begin{cases} -\dfrac{\omega_0 z\overline{z}}{4}, & z \notin D_1, \\ -\dfrac{\omega_0}{4}\mathrm{Re}[C(z)], & z \in D, \\ -\dfrac{\omega_0 z\overline{z}}{4}, & z \notin D_2, \end{cases} \tag{24.25}$$

where $C(z)$ is some complex potential characterizing the irrotational flow exterior to the vortices. The condition that the vortex patches be in pure solid-body rotation is a special assumption; vortex patches of uniform vorticity need not necessarily be in solid-body rotation. This assumption appears to be the key to unlocking equilibrium solutions that can be represented explicitly in terms of the prime function.

Consider the special class of unbounded fluid regions exterior to two vortex patches described by conformal maps from the annulus $\rho < |\zeta| < 1$ having the form

$$z = f(\zeta) = R\left[\frac{\omega(\zeta,-\sqrt{\rho})\omega(\zeta,-1/\sqrt{\rho})\omega(\zeta,1/\sqrt{\rho})}{\omega(\zeta,\sqrt{\rho})\omega(\zeta,\mathrm{e}^{\mathrm{i}\theta}/\sqrt{\rho})\omega(\zeta,\mathrm{e}^{-\mathrm{i}\theta}/\sqrt{\rho})}\right], \tag{24.26}$$

where R, θ, and ρ are real constants and where $\omega(.,.)$ is the prime function for the concentric annulus $\rho < |\zeta| < 1$. This is the same function considered in Exercise 11.7: it has three simple zeros at $-\sqrt{\rho}, \pm\sqrt{\rho}^{-1}$ and three simple poles at $\sqrt{\rho}, \sqrt{\rho}^{-1}\mathrm{e}^{\pm\mathrm{i}\theta}$. Under the mapping (24.26) the circles C_0 and C_1 are transplanted to the patch boundaries, the point $\zeta = \sqrt{\rho}$ maps to $z = \infty$, and $\zeta = -\sqrt{\rho}$ maps to the origin $z = 0$. (24.26) satisfies the functional relations

$$f(\rho^2\zeta) = f(\zeta), \qquad f(\rho/\zeta) = -f(\zeta), \qquad \overline{f}(\zeta) = f(\zeta). \tag{24.27}$$

These relations can be used to verify that the Schwarz functions of the two boundaries ∂D_1 and ∂D_2 are meromorphic continuations of each other through the domain D; in particular, the Schwarz function $S_0(z)$ of the image of C_0 under the mapping (24.26) is

$$S_0(z) = \overline{z} = \overline{f(\zeta)} = \overline{f}(1/\zeta) = f(1/\zeta), \tag{24.28}$$

and the Schwarz function $S_1(z)$ of the image of C_1 under the mapping (24.26) is

$$S_1(z) = \overline{z} = \overline{f(\zeta)} = \overline{f}(\rho^2/\zeta) = f(1/\zeta) = S_0(z). \tag{24.29}$$

Suppose we move to a frame of reference corotating with the vortices and, in addition, choose the function $C(z)$ in (24.25) such that the streamfunction in D is a modified Schwarz potential. This is possible precisely because, for this class of vortex patch shapes, we have

$$S(z) = S_0(z) = S_1(z). \tag{24.30}$$

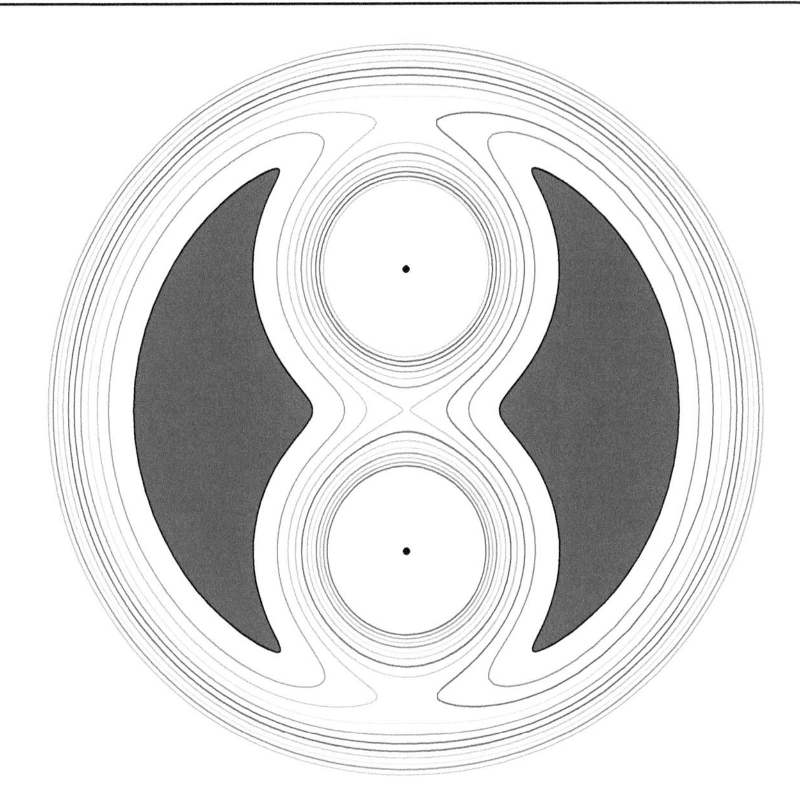

Figure 24.4. *Streamlines, in a rotating frame, around two corotating vortex patches, shown in blue, and a rotating vortex pair, shown as black dots. This solution has* $\rho = 0.15$ *and* $\theta = 2\pi/3$.

In this frame, we propose that a possible streamfunction for an equilibrium configuration is given by

$$\psi(z, \overline{z}) = \begin{cases} 0, & z \notin D_1, \\ \dfrac{\omega_0}{4}\left[z\overline{z} - \displaystyle\int^z S(z')dz' - \overline{\displaystyle\int^z S(z')dz'}\right], & z \in D, \\ 0, & z \notin D_2. \end{cases} \tag{24.31}$$

In this corotating frame it is easily checked, on use of (24.30), that the velocity field associated with this streamfunction satisfies condition (a) that it is continuous on the two vortex jumps. Moreover, condition (b) is also satisfied: the two vortex jumps are streamlines.

As in the construction of the multipolar vortices, the function $S(z)$ is singular in D, and, in general, this will induce unphysical singularities in the streamfunction (24.31). However the choice of vortex patch shapes encoded in (24.26) has the special property that the associated common Schwarz function $S_0(z) = S_1(z) = S(z)$ has only simple pole singularities in D with real residues. In fact, there are two of these with preimages in the annulus at

$$\sqrt{\rho}e^{\pm i\theta}. \tag{24.32}$$

These correspond to the preimages of two point vortices in the streamfunction (24.31) at

$$z_1 = f(\sqrt{\rho}e^{i\theta}), \qquad z_2 = f(\sqrt{\rho}e^{-i\theta}), \tag{24.33}$$

and these are physically admissible singularities. In order for the entire vortex configuration to be in equilibrium, however, it remains to ensure that these two vortices are steady. This can be done by an appropriate choice of the parameters appearing in the mapping function (24.26). See [52] for more details, where, in addition, an interpretation of the construction just described as "growing" vortex patches at the two stagnation points of a corotating point vortex pair is described.

It should be clear that the construction outlined above extends to *any number* of corotating vortex patches, and we know from Chapter 11 that the shapes will be described by conformal maps that are automorphic functions with respect to Schottky groups with more generators.

The construction of such multipatch solutions was carried out by Crowdy and Marshall [51], and the reader is referred there for more details. It provides an interesting case study in use of the prime function in constructing automorphic functions relevant to a physical application.

The message of this chapter is that modified Schwarz potentials provide a fertile source of equilibria, and relative equilibria, of the two-dimensional Euler equation [52, 30, 35, 29, 51, 33]. The vortex patch regions have interpretations as quadrature domains, and the function theory based on the prime function can be used to realize their boundary shapes using automorphic function conformal mappings.

Chapter 25

Free surface Euler flow

Free streamline theory is the name given to the study of irrotational flows of ideal fluids involving free boundaries between the ideal fluid and, most commonly, air or some other inviscid fluid of different density assumed to be at constant pressure. While the boundary value problem for the associated velocity potential is linear, the fact that the domain in which one solves it is unknown renders it a nonlinear free boundary problem: the shape of the fluid domain, and hence of any boundary streamlines, must be determined as part of the solution [85].

Of the many possible problems within this field that might be studied, we focus on *hollow vortices* [113, 107, 41, 38, 39]. A hollow vortex is a finite-area, constant-pressure region, or "bubble," in a two-dimensional irrotational incompressible flow of an ideal fluid with nonzero circulation around it; it is this nonzero circulation that explains why we use the designation "vortex." In the fluid exterior to such a vortex, supposed to be of density ν, there is an irrotational flow $\mathbf{u} = (u, v)$ that is the gradient of a scalar velocity potential ϕ:

$$\mathbf{u} = \nabla\phi. \tag{25.1}$$

In an incompressible two-dimensional flow there is an associated streamfunction ψ with velocity components related to ϕ and ψ via

$$u = \frac{\partial\phi}{\partial x} = \frac{\partial\psi}{\partial y}, \qquad v = \frac{\partial\phi}{\partial y} = -\frac{\partial\psi}{\partial x}. \tag{25.2}$$

Thus ϕ and ψ satisfy the Cauchy–Riemann equations, and the combination

$$h(z,t) \equiv \phi + \mathrm{i}\psi \tag{25.3}$$

constitutes a complex potential that, for an unsteady flow, is a generally time dependent analytic function of $z = x + \mathrm{i}y$.

The shape of a hollow vortex will generally evolve in time, and the flow itself will be unsteady. However it is natural to ask about equilibrium configurations in which the flow and the shape of the vortex are steady.

According to the steady version of Bernoulli's theorem [65] for irrotational flow, the fluid pressure p is related to the velocity potential via

$$\frac{p}{\nu} + \frac{1}{2}|\nabla\phi|^2 = H, \tag{25.4}$$

which holds everywhere in the fluid region. H is sometimes called the Bernoulli constant. In the absence of any singular force distributions, such as surface tension, on the boundary of the vortex, the fluid pressure there must be constant in order that the fluid pressure is continuous with the assumed constant pressure region inside the vortex. This means, on use of (25.4), that, on the boundary of the hollow vortex, we must have

$$|\nabla\phi| = \text{constant}. \tag{25.5}$$

Since the boundary of a hollow vortex moves with the fluid, and since we seek an equilibrium configuration where the shape of the vortex does not change, we also require that the boundary of the vortex is a streamline, namely,

$$\psi = \text{constant}. \tag{25.6}$$

Together this constitutes a nonlinear free boundary problem in which both the shape of the hollow vortex and the associated flow around it must be found simultaneously.

The velocity (u, v) is given in complex form in terms of the complex potential (25.3)—now a function of z but not t in this steady situation—as

$$\frac{dh}{dz} = u - \mathrm{i}v, \tag{25.7}$$

and (25.5) is equivalent to

$$\left|\frac{dh}{dz}\right| = \text{constant}. \tag{25.8}$$

The streamline condition (25.6) means that on the vortex boundary we must also have

$$\text{Im}[h(z)] = \text{constant}. \tag{25.9}$$

The particular situation we focus on is important in applications: a staggered hollow vortex street comprising an L-periodic row of hollow vortices of circulation $+\Gamma$ located next to another L-periodic row of hollow vortices of circulation $-\Gamma$, as depicted schematically in Figure 25.1. The analysis for a street of *point vortices*, where the vorticity has no spatial extent but is concentrated into a delta function distribution, is classical and due to Von Kármán [113, 65]. From that classical analysis we expect to find relative equilibrium configurations of hollow vortices traveling, without change of form, at a constant speed U in the x direction. A hollow vortex can be thought of as a desingularization of a point vortex. Vortex street structures are commonly observed in the bluff-body wakes on a variety of length scales.

It is natural to move to a frame of reference moving with all the vortices at this speed. The flow, and the geometrical configuration of vortices, will be steady in this cotraveling frame, and, in the far field as $|y| \to \infty$, the fluid velocity will tend to $(-U, 0)$.

We will seek a special class of staggered vortex street equilibria in which the shapes of all vortices in the configuration are identical, as indicated in Figure 25.1. The analysis to follow was originally presented in [38], with further developments to the compressible case given in [39].

Let $h(z)$ be the steady complex potential associated with the flow in this cotraveling reference frame. As $y \to \infty^{\pm}$, which we use to denote the top, $y \to \infty$, and bottom, $y \to -\infty$, of the period window, as shown in Figure 25.1, we require that

$$\frac{dh}{dz} \to -U. \tag{25.10}$$

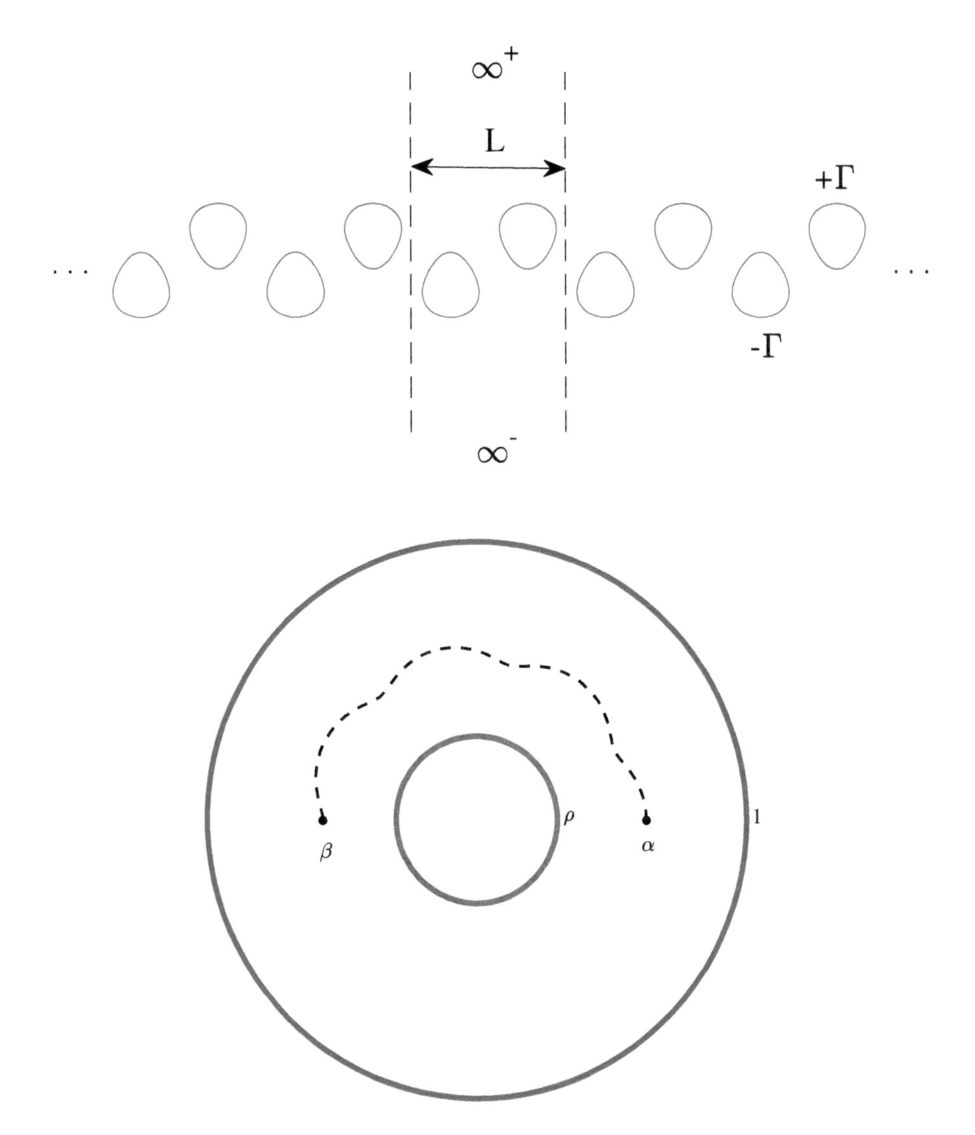

Figure 25.1. *Conformal mapping from a concentric annulus with two logarithmic branch points at α and β to a single period of a staggered periodic hollow vortex street in which all vortices have the same shape. The two sides of a branch cut joining α and β correspond to the two sides of a period window. Circles C_0 and C_1 map to the boundaries of the two vortices in each period window.*

If we let ∂D_+ denote the boundary of the vortex with circulation $+\Gamma$ and let ∂D_- denote the boundary of the vortex with circulation $-\Gamma$, then we require that

$$\int_{\partial D_\pm} dh = \pm\Gamma. \tag{25.11}$$

Since the shapes of the vortices are unknown, it is natural to introduce a conformal map to solve this problem. In this way, finding the unknown shape of the hollow vortices is equivalent to finding the functional form of the conformal mapping.

By the L-periodicity of the geometrical arrangement, it is enough to find a conformal mapping from some canonical region to a single representative period window of the flow.

There are two hollow vortices in each period window, so the fluid region in this period window exterior to the two vortices is triply connected. Our first instinct is to introduce a conformal mapping from a canonical triply connected circular region comprising the unit disc with two excised circular discs. The required mapping might then be taken to have two logarithmic branch points on one of the boundary circles corresponding to the two points $\infty^{\pm}$. The two segments of the unit circle between these branch points will be transplanted to the two edges of the period window. This is certainly a possible approach.

But it is not the only approach. Since the edges of the period window are not "boundaries" in the same sense as the actual boundaries of the vortices—indeed, the shape of the edges of the period window is arbitrary (they have been drawn as straight lines in Figure 25.1), and usually of no particular interest—in the analysis they can be treated differently from the hollow vortex boundaries.

More concretely, consider the mapping

$$z = f(\zeta) \tag{25.12}$$

from a *doubly* connected concentric annulus in a parametric ζ plane

$$\rho < |\zeta| < 1 \tag{25.13}$$

having two logarithmic singularities at $\zeta = \alpha, \beta$ strictly inside this annulus. These internal branch points will be taken to be the preimages of the points $\infty^{\pm}$. Indeed, let α be the preimage of ∞^{+} and let β be the preimage of ∞^{-}. Near $\zeta = \alpha$ it is easy to see that we require that the mapping have the local form

$$f(\zeta) = -\frac{\mathrm{i}L}{2\pi}\log(\zeta - \alpha) + \text{a locally analytic function,} \tag{25.14}$$

and, near $\zeta = \beta$, we require

$$f(\zeta) = +\frac{\mathrm{i}L}{2\pi}\log(\zeta - \beta) + \text{a locally analytic function.} \tag{25.15}$$

It is worth remarking that the approach we are adopting here has much in common with that used in §7.7 to derive formulas for S–C mappings to periodic domains with polygonal boundaries. There we also introduced two logarithmic branch points inside the preimage region to be the preimages of the two points $\infty^{\pm}$ in each period window. The basic idea is the same, although, of course, we are no longer constructing conformal mappings to periodic polygonal domains. Instead the structure of the conformal mapping will be dictated by the equilibrium conditions imposed by the Euler equations.

It is a noteworthy feature that, owing to the periodicity of the arrangement, a problem involving an ostensibly triply connected period window can in fact be tackled using the function theory associated with a doubly connected preimage domain.

To solve the free boundary problem for the equilibrium shapes of the hollow vortices, we must determine the functional form of the complex potential $h(z)$ and the conformal mapping $f(\zeta)$; these will tell us the velocity of the fluid and the shape of the equilibrium vortices.

To do so, we adopt the main ideas of free streamline theory, which are summarized as follows. Suppose that it is possible to find both the complex potential

$$H(\zeta) \equiv h(f(\zeta)) \tag{25.16}$$

as a function of the parametric variable ζ and the complex velocity function

$$\frac{dh}{dz} = u - \mathrm{i}v, \tag{25.17}$$

also as a function of ζ. Since, by the chain rule,

$$\frac{dh}{dz} = \frac{dH/d\zeta}{df/d\zeta}, \tag{25.18}$$

the derivative of the required mapping function $df/d\zeta$ is given by the known ratio

$$\frac{df}{d\zeta} = \frac{dh/dz}{dH/d\zeta}. \tag{25.19}$$

The mapping $f(\zeta)$ then follows by integration.

While $h(z)$ and $f(\zeta)$ are the functions of interest, it is sufficient to find instead the functions $H(\zeta)$ and $h'(z)$. This is convenient since, as we now show, the latter two functions can be constructed, up to a finite set of parameters, using several of the conformal slit mappings presented in Chapter 5 of Part I.

The calculus for potential theory expounded in Chapter 15 is useful in the construction of $H(\zeta)$. Condition (25.10) implies that, as $y \to \infty^{\pm}$,

$$h(z) \to -Uz, \tag{25.20}$$

and, together with (25.14) and (25.15), we deduce that near $\zeta = \alpha$,

$$H(\zeta) = \frac{\mathrm{i}LU}{2\pi} \log(\zeta - \alpha) + \text{a locally analytic function}, \tag{25.21}$$

and, near $\zeta = \beta$,

$$H(\zeta) = -\frac{\mathrm{i}LU}{2\pi} \log(\zeta - \beta) + \text{a locally analytic function}. \tag{25.22}$$

Since $H(\zeta)$ has two logarithmic singularities, of opposite sign and purely imaginary strength, at α, β and also satisfies, from (25.6),

$$\mathrm{Im}[H(\zeta)] = \text{constant}, \qquad |\zeta| = \rho, 1, \tag{25.23}$$

we are reminded of the Green's functions $\mathcal{G}_0(\zeta, \alpha)$ and $\mathcal{G}_0(\zeta, \beta)$. Indeed, the function

$$-LU\mathcal{G}_0(\zeta, \alpha) + LU\mathcal{G}_0(\zeta, \beta) \tag{25.24}$$

is readily confirmed to have all the properties (25.21)–(25.23) listed above. On use of the properties of $\mathcal{G}_0$, it also satisfies

$$\int_{\partial D_{\pm}} d\left[-LU\mathcal{G}_0(\zeta, \alpha) + LU\mathcal{G}_0(\zeta, \beta)\right] = 0, \tag{25.25}$$

which means that this candidate function does not have the property (25.11) needed to ensure that the vortices have the required circulations. This can be fixed by adding the term

$$\Gamma v_1(\zeta) \tag{25.26}$$

to our candidate function since the latter satisfies

$$\int_{C_j} d\,[\Gamma v_1(\zeta)] = \Gamma, \qquad j = 0, 1. \tag{25.27}$$

This implies that the circulations in the physical plane will be $\pm\Gamma$ since, owing to the enclosed logarithmic singularities, the image of C_0 under the mapping $z = f(\zeta)$ is traversed in a clockwise direction if C_0 is traversed anticlockwise. Since (25.26) also has constant imaginary part on C_0 and C_1, we deduce that the required complex potential is

$$H(\zeta) = \frac{\mathrm{i}LU}{2\pi} \log\left[\frac{|\beta|\omega(\zeta,\alpha)\omega(\zeta,1/\overline{\beta})}{|\alpha|\omega(\zeta,\beta)\omega(\zeta,1/\overline{\alpha})}\right] - \frac{\mathrm{i}\Gamma}{2\pi}\log\zeta. \tag{25.28}$$

We still have to find dh/dz. This function is analytic everywhere in the period window and, when viewed as an analytic function of ζ, is therefore analytic everywhere in the annulus $\rho < |\zeta| < 1$, including the two boundary circles. Moreover condition (25.8) tells us that, in this reference frame where the vortex is steady, the modulus of dh/dz is constant on the boundaries of both vortices; hence, when viewed as a function of ζ, dh/dz must have constant modulus on C_0 and C_1. We are thus reminded of the two bounded circular slit maps associated with this doubly connected domain,

$$\eta_0(\zeta,\gamma) = e^{2\pi\mathrm{i}\mathcal{G}_0(z,a)} = \frac{\omega(\zeta,\gamma)}{|\gamma|\omega(\zeta,1/\overline{\gamma})}, \qquad \eta_1(\zeta,\gamma) = e^{2\pi\mathrm{i}\mathcal{G}_1(z,a)}, \tag{25.29}$$

introduced in §5.6 of Chapter 5. Both have constant modulus on the circles C_0 and C_1. Each of these functions has a simple zero at γ and a simple pole at $1/\overline{\gamma}$. Recall that, as a conformal mapping, the image of C_0 under the map η_0 is closed but the image of C_1 is a circular slit; on the other hand, the image of C_0 under the map η_1 is a circular slit but the image of C_1 is a closed circle.

It is important now to anticipate the expected *stagnation points* of the flow in the steady frame of reference; a stagnation point is defined to be a point where the velocity (u, v) vanishes. From the analysis of the analogous configuration of point vortices we know that there are two stagnation points of the flow in each period window in the cotraveling frame of reference. We expect the same to be true for a street of hollow vortices. This is because a small hollow vortex with vanishing area is expected to have a boundary that is close to a circle and will locally generate the same flow around it as a point vortex of the same circulation located at its center.

Suppose the preimages of these two stagnation points in the annulus $\rho < |\zeta| < 1$ are at $\zeta = \gamma_1, \gamma_2$. Since $df/d\zeta$ must not vanish in the annulus (if the conformal map is to be one-to-one, as we assume), we must have

$$\left.\frac{dH}{d\zeta}\right|_{\gamma_1,\gamma_2} = 0, \tag{25.30}$$

which provides relations between γ_1 and γ_2 and the other parameters in the problem. Since

$$\zeta\frac{dH}{d\zeta} = \frac{\mathrm{i}LU}{2\pi}\left[K(\zeta,\alpha) - K(\zeta,\beta) + K(\zeta,1/\overline{\beta}) - K(\zeta,1/\overline{\alpha}) - \frac{\Gamma}{U}\right], \tag{25.31}$$

where we have used the function $K(.,.)$ introduced in §5.12 of Chapter 5, it follows that the parameters $\{\alpha, \beta, \gamma_1, \gamma_2\}$ must satisfy

$$K(\gamma_1,\alpha) - K(\gamma_1,\beta) + K(\gamma_1,1/\overline{\beta}) - K(\gamma_1,1/\overline{\alpha}) - \frac{\Gamma}{U} = 0, \tag{25.32}$$

$$K(\gamma_2,\alpha) - K(\gamma_2,\beta) + K(\gamma_2,1/\overline{\beta}) - K(\gamma_2,1/\overline{\alpha}) - \frac{\Gamma}{U} = 0. \tag{25.33}$$

Consider the function

$$\eta_0(\zeta,\gamma_1)\eta_1(\zeta,\gamma_2), \tag{25.34}$$

involving the functions (25.29). It is analytic everywhere in the annulus and clearly has two simple zeros at γ_1 and γ_2. By using both η_0 and η_1 to introduce these zeros, rather than, say, $\eta_0(\zeta,\gamma_1)\eta_0(\zeta,\gamma_2)$ or $\eta_1(\zeta,\gamma_1)\eta_1(\zeta,\gamma_2)$, which both have the required zeros and boundary properties, we have chosen a velocity field that is consistent with the nonzero circulations around both vortices (these two alternative choices can be discounted not least because, intuitively, they clearly favor one vortex over the other). We therefore pick dh/dz, the complex conjugate velocity field, to be a multiple of the function (25.34):

$$u - \mathrm{i}v = \frac{dh}{dz} = -\frac{U\rho}{\sqrt{\chi}|\gamma_1|^2|\gamma_2|^2\zeta}\left[\frac{\omega(\zeta,\gamma_1)\omega(\zeta,\gamma_2)}{\omega(\zeta,1/\overline{\gamma_1})\omega(\zeta,1/\overline{\gamma_2})}\right], \tag{25.35}$$

where the constant $\sqrt{\chi}$, which is introduced here for convenience later on, is chosen to ensure that $u - \mathrm{i}v \to -U$ as $\zeta \to \alpha, \beta$:

$$\sqrt{\chi} = \frac{\rho}{\alpha|\gamma_1\gamma_2|^2}\frac{\omega(\alpha,\gamma_1)\omega(\alpha,\gamma_2)}{\omega(\alpha,1/\overline{\gamma_1})\omega(\alpha,1/\overline{\gamma_2})} = \frac{\rho}{\beta|\gamma_1\gamma_2|^2}\frac{\omega(\beta,\gamma_1)\omega(\beta,\gamma_2)}{\omega(\beta,1/\overline{\gamma_1})\omega(\beta,1/\overline{\gamma_2})}. \tag{25.36}$$

These conditions ensure that the fluid velocity as $y \to \infty^{\pm}$ tends to $-U$, as required.

In this way, we have used the prime function of the annulus, and the conformal slit maps expressible in terms of it, to find the explicit expressions (25.28) and (25.35) for the functions $H(\zeta)$ and dh/dz that we seek. These expressions are given in terms of a finite set of mathematical parameters ρ, α, β, γ_1, and γ_2 that can be related to the physical parameters U, Γ, and L of the problem via (25.32), (25.33), and (25.36). It is also necessary to ensure that the conformal mapping function $f(\zeta)$ obtained from integration of (25.19) is single-valued in the annulus, and this provides another constraint on the parameters.

For a staggered street of symmetric vortices, where the shapes of all vortices in the street are identical, it is found [38, 39] that an admissible set of parameters α, β, γ_1, and γ_2 are all real and satisfy

$$\alpha\beta = \gamma_1\gamma_2 = -\rho. \tag{25.37}$$

To find the conformal mapping $f(\zeta)$, and hence the vortex shapes, it is useful to observe that while we have found the expression (25.28) for $H(\zeta)$, it is easy to verify from the boundary conditions (25.6) that the function $\zeta dH/d\zeta$ is a loxodromic function. As such, we know from Chapter 8 that it must have an alternative representation of the form

$$\zeta\frac{dH}{d\zeta} = R\frac{\omega(\zeta,\gamma_1)\omega(\zeta,\gamma_2)\omega(\zeta,1/\gamma_1)\omega(\zeta,1/\gamma_2)}{\omega(\zeta,\alpha)\omega(\zeta,\beta)\omega(\zeta,1/\alpha)\omega(\zeta,1/\beta)} \tag{25.38}$$

for some constant R. This is a ratio of products of prime functions and has the advantage that the poles and zeros of the function are obvious from the representation—a useful feature since, on substitution of (25.35) and (25.38) into (25.19), several cancellations occur and we find

$$\frac{df}{d\zeta} = S\frac{\omega(\zeta,1/\gamma_1)^2\omega(\zeta,1/\gamma_2)^2}{\omega(\zeta,\alpha)\omega(\zeta,\beta)\omega(\zeta,1/\alpha)\omega(\zeta,1/\beta)} \tag{25.39}$$

for some constant S. The required conformal mapping function is therefore given by the indefinite integral

$$f(\zeta) = S\int^{\zeta}\frac{\omega(\zeta',1/\gamma_1)^2\omega(\zeta',1/\gamma_2)^2}{\omega(\zeta',\alpha)\omega(\zeta',\beta)\omega(\zeta',1/\alpha)\omega(\zeta',1/\beta)}d\zeta'. \tag{25.40}$$

Figure 25.2. *Streamline distribution for a typical hollow vortex street equilibrium plotted on the basis of formulas (25.28) and (25.41).*

This is a perfectly acceptable form of the solution, being an explicit integral depending on a finite set of parameters and easy to evaluate by quadrature. However it turns out to be possible to evaluate the integral in closed form using the theory of loxodromic functions presented in Chapter 8. There it was shown that a loxodromic function can be expressed either as a product of ratios of prime functions or as a sum of parametric logarithmic derivatives of the prime function. We notice that the integrand in (25.40) has four simple pole singularities at α, $1/\alpha$, β, and $1/\beta$. If we rewrite this integrand using the function $K(.,.)$ introduced in §5.12, the form of the integral then becomes obvious and it is found [39] that

$$z = f(\zeta) = -\frac{\mathrm{i}L}{2\pi}\left[\log\left[\frac{\beta\omega(\zeta,\alpha)}{\alpha\omega(\zeta,\beta)}\right] - \chi\log\left[\frac{\alpha\omega(\zeta,1/\alpha)}{\beta\omega(\zeta,1/\beta)}\right]\right] + \frac{\mathrm{i}L}{4\pi}\log(\beta/\alpha). \quad (25.41)$$

This calculation is the subject of Exercise 8.10.

The streamlines associated with one of these equilibrium solutions are shown in Figure 25.2. They were plotted on the basis of formulas (25.28) and (25.41).

We close this chapter by pointing out that a classical solution for a cotraveling hollow vortex pair found in 1895 by Pocklington [107] using the theory of elliptic functions has been rederived using the prime function framework in a concentric annulus as set out in Chapter 5. Readers are referred to [41] for a further example of how the function theory on the annulus of Chapter 5 and the theory of loxodromic functions of Chapter 8 can be harnessed to tackle this classical problem involving a doubly connected domain.

There are also fascinating connections between this hollow vortex free boundary problem and other areas of mathematics, such as the theory of minimal surfaces [121]. Our framework is therefore likely to be of use there too.

Chapter 26

Laplacian growth and Hele-Shaw flow

The term "Laplacian growth" refers to a collection of free boundary problems in which a harmonic potential satisfying a boundary value problem of Dirichlet type in some domain, simply or multiply connected, gives the normal displacement of the domain boundary as a function of the Neumann data for that potential. This is a class of nonlinear free boundary problems where, in the unsteady problem, the challenge is to track the evolution of a time-evolving domain $D_z(t)$ given the initial domain $D_z(0)$. The Hele-Shaw problem [114], which refers to the evolution of a planar region of viscous fluid sandwiched between two flat plates known as a Hele-Shaw cell, is one manifestation of such a Laplacian growth problem.

We have picked three different problems involving Hele-Shaw flow where the mathematical techniques of Part I can be used to some advantage. There is a vast literature on Laplacian growth and the Hele-Shaw problem which we do not survey here. Our choice of illustrative problems is highly selective and guided purely by our intent to show in action the mathematical techniques of Part I.

We start with a steady free boundary problem and then move on to unsteady problems driven by two different mechanisms: injection due to a finite array of point sources and rotation of the Hele-Shaw cell.

Multiple steady bubbles in a Hele-Shaw cell: The problem of finding the steady shapes of a finite collection of cotraveling bubbles in a Hele-Shaw cell is one in which the conformal slit maps introduced in Part I play a useful role.

Suppose $M + 1$ air bubbles, for $M \geq 0$, each of finite area and of constant pressure, are trapped in the viscous fluid in a Hele-Shaw cell well away from any boundary walls of the cell. The analysis of the single bubble case $M = 0$ case dates back to Taylor and Saffman [120], who also considered the effect of the walls of the cell.

Let D_z denote the region of viscous fluid. We seek solutions for relative equilibria in which the bubbles all travel at some speed U parallel to the real axis in an ambient fluid traveling, due to some external agency, at uniform speed U_0 parallel to the real axis. By a choice of nondimensionalization we can set $U_0 = 1$.

The analysis to follow is based on work originally presented by the author in [17], and the reader is referred there for more details and additional background.

In the Hele-Shaw approximation the Navier–Stokes equations governing the motion of a viscous fluid are depth-averaged over the width of the Hele-Shaw cell. The resulting mathematical problem is to solve for a velocity potential ϕ, which turns out to be proportional to the pressure in the viscous fluid, satisfying

$$\nabla^2\phi = 0, \qquad z \in D_z, \tag{26.1}$$

with

$$\phi \to x, \qquad |z| \to \infty. \tag{26.2}$$

To solve this problem we can introduce a complex potential

$$h(z) = \phi + \mathrm{i}\psi, \tag{26.3}$$

where ψ, which corresponds to a streamfunction for the flow exterior to the bubbles, is the harmonic conjugate to ϕ. As $|z| \to \infty$, it follows from (26.2) that

$$h(z) \to z + \text{constant}. \tag{26.4}$$

It is expedient to move to a frame of reference cotraveling at speed U with the bubbles. A natural way to find the shape of the bubbles is to consider a conformal map from a conformally equivalent circular preimage region D_ζ to the region D_z exterior to the bubbles in this cotraveling frame. As usual we take D_ζ to be a canonical circular domain: the unit ζ disc with M smaller circular discs excised with centers $\{\delta_j | j = 1, \ldots, M\}$ and radii $\{q_j | j = 1, \ldots, M\}$. Let the map from D_ζ to D_z be denoted by

$$z = f(\zeta), \tag{26.5}$$

where, since D_z is unbounded, there must be some point a inside D_ζ where

$$f(\zeta) = \frac{A}{\zeta - a} + \text{a locally analytic function}, \tag{26.6}$$

and where, by using the rotational degree of freedom in the mapping theorem, we can insist that A is real and positive. The point a is the preimage of the point at infinity in the Hele-Shaw cell. We do not yet know the form of $f(\zeta)$. It must be found as part of the solution.

Now introduce the complex potential composed with the conformal mapping function, i.e.,

$$H(\zeta) \equiv h(f(\zeta)). \tag{26.7}$$

From (26.4) and (26.6), it follows that

$$H(\zeta) = \frac{A}{\zeta - a} + \text{a locally analytic function}. \tag{26.8}$$

Since ϕ is proportional to the pressure in the viscous fluid, and since the bubbles are regions of constant pressure and we assume that there are no singular force distributions, such as surface tension, active on the bubble boundaries, a second condition on each bubble boundary is

$$\phi = \mathrm{Re}[H(\zeta)] = \begin{cases} 0, & \zeta \in C_0, \\ d_j, & \zeta \in C_j, \; j = 1, \ldots, M, \end{cases} \tag{26.9}$$

where $\{d_j | j = 1, \ldots, M\}$ is a set of real constants. Without loss of generality we have set the pressure inside the bubble whose boundary is the image of C_0 to zero.

In a frame of reference cotraveling with the steady bubbles, their boundaries must all be streamlines. Since the complex potential for uniform flow with speed U is Uz, the streamfunction in the cotraveling frame is the imaginary part of $w(z)-Uz$, so we introduce the new function

$$S(\zeta) \equiv H(\zeta) - Uf(\zeta). \tag{26.10}$$

As $|z| \to \infty$, it is easy to check that

$$S(\zeta) \to \frac{(1-U)A}{\zeta - a} + \text{a locally analytic function.} \tag{26.11}$$

The streamline conditions on all bubble boundaries can then be stated as

$$\mathrm{Im}[S(\zeta)] = \begin{cases} 0, & \zeta \in C_0, \\ c_j, & \zeta \in C_j, \ j = 1, \ldots, M, \end{cases} \tag{26.12}$$

where $\{c_j | j = 1, \ldots, M\}$ is a set of real constants.

To solve the free boundary problem we must find the two functions $H(\zeta)$, which will give the flow in D, and $f(\zeta)$, which will give the shape of the bubbles.

On inspection of (26.8), (26.9), (26.11), and (26.12) it is clear that the two functions $H(\zeta)$ and $S(\zeta)$ each have a simple pole at $\zeta = a$ inside D_ζ and, moreover, satisfy conditions that remind us of those satisfied by the parallel slit maps $\phi_\theta(\zeta, a)$ encountered in Chapters 1, 5, and 6 of Part I (in the simply, doubly, and higher connected cases, respectively). With this observation, the solution is now immediate:

$$H(\zeta) = A[\phi_{\pi/2}(\zeta, a) - \phi_{\pi/2}(1, a)], \tag{26.13}$$
$$S(\zeta) = (1-U)A[\phi_0(\zeta, a) - \phi_0(1, a)], \tag{26.14}$$

where we have chosen the angles θ in the parallel slit maps to suit the boundary conditions (26.9) and (26.12) and constants have been added to fix the conditions on C_0. It is now a simple matter to find $f(\zeta)$, and hence the bubble shapes, given $H(\zeta)$ and $S(\zeta)$, since on rearrangement of (26.10),

$$f(\zeta) = \frac{1}{U}\left[H(\zeta) - S(\zeta)\right]. \tag{26.15}$$

Figure 26.1 shows typical bubble shapes, for the two-bubble case $M = 1$, computed using (26.13)–(26.15), for different values of the speed U and where all bubbles have the same area.

Figure 26.2 shows a typical streamline distribution around one of these equilibrium solutions.

It is a straightforward matter to use the prime functions in (26.13)–(26.15) corresponding to higher connected circular preimage regions D_ζ to compute solutions for larger numbers of cotraveling bubbles in equilibrium. Solutions involving more than two bubbles can be found in [17].

It is instructive to remark that this same class of Hele-Shaw problems can alternatively be tackled using the ideas described in Chapter 7 to derive the multiply connected S–C mapping formulas. This line of inquiry has been explored by Green and Vasconcelos [84].

Time-dependent motion driven by sources: An unsteady free boundary problem involving a Hele-Shaw cell is related to the industrial process of injection molding, where viscous fluid is injected into the cell at multiple fixed points by a distribution of fluid sources (including sinks, where fluid might be sucked out of the cell). The simplest mathematical model of this process is as follows.[56]

[56]Work on this problem in multiply connected domains has been done by Richardson [109, 110], although he does not make use of the prime function as we advocate here.

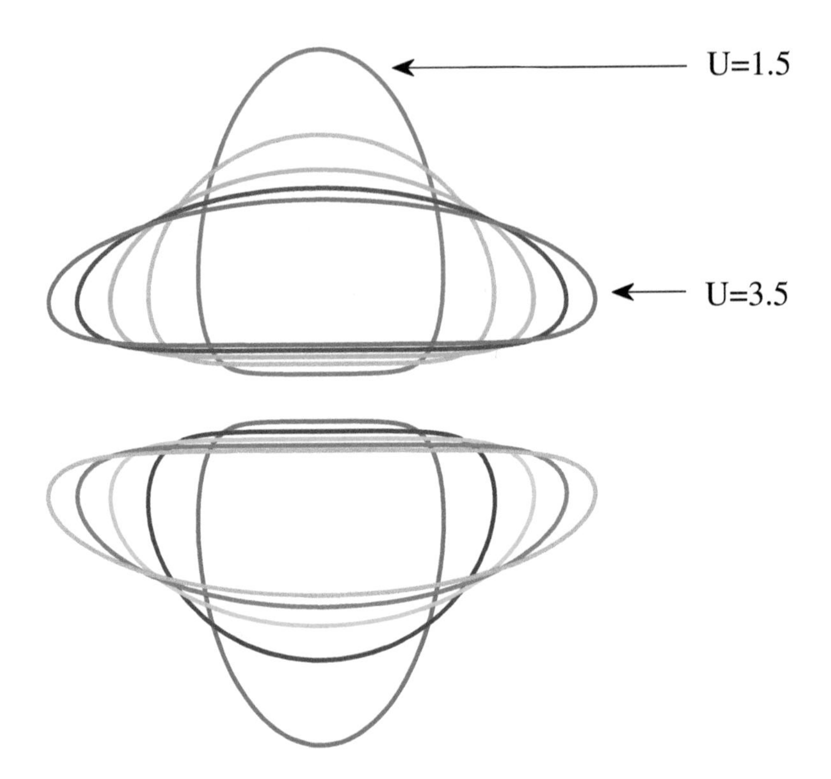

Figure 26.1. *Superposition of several pairs of symmetric bubbles traveling with different speeds* $U = 1.5$, 2, 2.5, 3, *and* 3.5 *parallel to the real axis. The solutions correspond to the choice* $\rho = 0.4$, *and the bubbles all have the same area.*

The velocity potential ϕ at each instant satisfies the following Dirichlet problem in the time-dependent multiply connected domain $D(t)$:

$$\nabla^2\phi = \sum_{k=1}^{N} m_k(t)\delta(z - z_k), \qquad z \in D(t), \tag{26.16}$$

where $\{m_k(t)|k = 1, \dots, N\}$ is the set of injection rates at the different sources/sinks. These injection rates can vary in time. These N sources are located at positions $\{z_k|k = 1, \dots, N\}$, assumed to be fixed inside the time-evolving fluid domain $D(t)$. On the boundary of the viscous fluid region $D(t)$ a Dirichlet boundary condition holds:

$$\phi = \Phi_j(t), \qquad z \in \partial D_j(t), \tag{26.17}$$

where $\Phi_j(t)$ is independent of position and $\partial D_j(t)$ denotes the jth component of the boundary of a multiply connected region $D(t)$. Without loss of generality, we can set

$$\Phi_0(t) = 0. \tag{26.18}$$

The field ϕ is a velocity potential for a two-dimensional velocity field

$$\mathbf{u} = \nabla\phi. \tag{26.19}$$

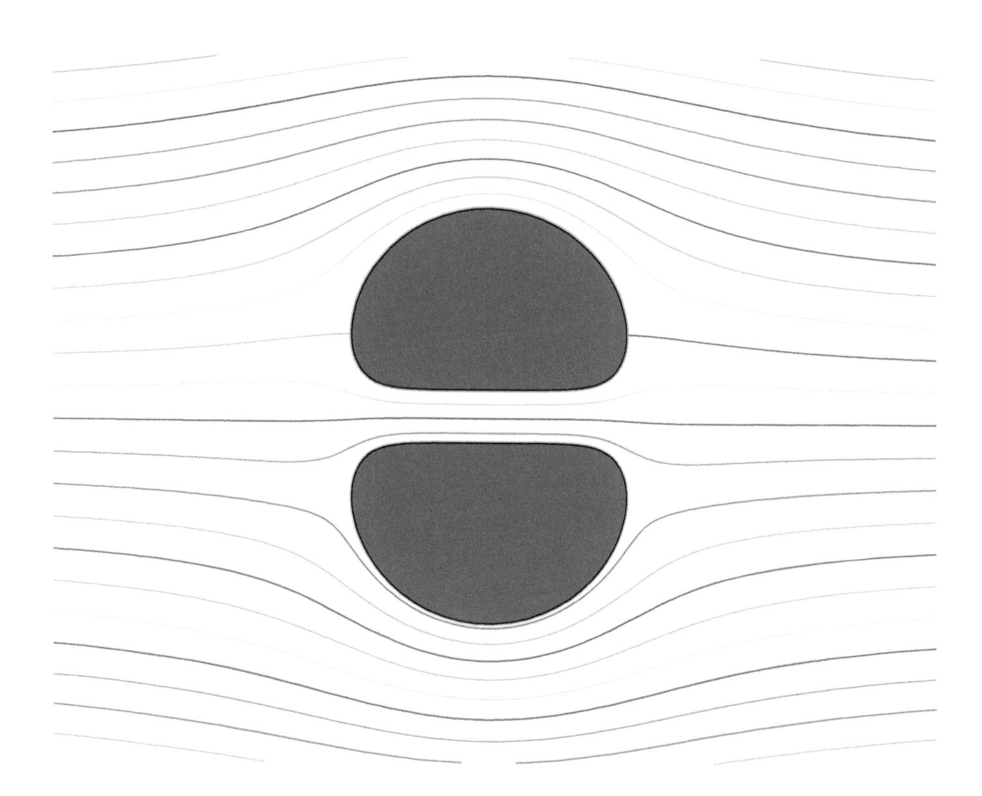

Figure 26.2. *Streamlines around a pair of symmetric bubbles traveling with speed* $U = 2$ *in a Hele-Shaw cell with* $\rho = 0.4$.

The evolution of the boundary is given by the following law for its normal velocity V_n:

$$V_n = \mathbf{u}.\mathbf{n}. \tag{26.20}$$

We introduce a complex potential $h(z,t)$, analytic in D, whose real part is ϕ:

$$h(z,t) = \phi + \mathrm{i}\psi, \tag{26.21}$$

where we have introduced ψ as the harmonic conjugate of ϕ.

A useful way to tackle this free boundary problem is to ask how the Cauchy transform of the time-evolving domain $D(t)$ evolves. For this we need to compute the following differential on the boundary $\partial D(t)$:

$$\mathrm{Im}\left[z_t d\overline{z}\right], \tag{26.22}$$

where, as in Chapter 12, we use the notation z_t to denote the time derivative of points on the boundary. This quantity is determined by the boundary conditions. Condition (26.20) can be written as

$$\mathrm{Re}\left[z_t\left(\mathrm{i}\frac{d\overline{z}}{ds}\right)\right] = \mathrm{Re}\left[\frac{\partial h}{\partial z}\left(\mathrm{i}\frac{dz}{ds}\right)\right] = \mathrm{Re}\left[\mathrm{i}\frac{\partial \overline{h}}{\partial s}\right] = -\mathrm{Re}\left[\mathrm{i}\frac{\partial h}{\partial s}\right], \tag{26.23}$$

where, in the last equality, we have used the fact that the real part of a complex number equals the real part of its complex conjugate. The differential form of (26.23) is

$$\mathrm{Im}\left[z_t d\overline{z}\right] = -\mathrm{Im}[dh]. \tag{26.24}$$

Now

$$\mathrm{Im}[dh] = d\left(\frac{h(z,t) - \overline{h(z,t)}}{2\mathrm{i}}\right) = \frac{dh(z,t)}{\mathrm{i}}, \tag{26.25}$$

where we have used the boundary condition (26.17), which implies that

$$\overline{h(z,t)} = -h(z,t) + 2\Phi_j, \qquad z \in \partial D_j(t). \tag{26.26}$$

Hence we deduce that

$$\frac{\partial C_j(z,t)}{\partial t} = \frac{1}{\pi \mathrm{i}} \int_{\partial D(t)} \frac{h'(z',t)}{z'-z} dz', \tag{26.27}$$

where we have introduced the notation $h'(z,t)$ to denote the quantity $\partial h/\partial z$.

From (26.16) we deduce that $h(z,t)$ must have the form

$$h(z,t) = \sum_{k=1}^{N} \frac{m_k(t)}{2\pi} \log(z - z_k) + \tilde{h}(z,t), \tag{26.28}$$

where $\tilde{h}(z,t)$ is analytic in $D(t)$. On substitution into (26.27) we find

$$\frac{\partial C_j(z,t)}{\partial t} = \frac{1}{\pi \mathrm{i}} \int_{\partial D(t)} \frac{1}{z'-z} \left[\sum_{k=1}^{N} \frac{m_k(t)}{2\pi} \frac{1}{(z'-z_k)} + \tilde{h}'(z)\right] dz' \tag{26.29}$$

$$= \sum_{k=1}^{N} \frac{m_k(t)}{\pi} \frac{1}{z_k - z}. \tag{26.30}$$

This can be integrated with respect to time to give

$$C_j(z,t) = -\sum_{k=1}^{N} \frac{Q_k(t)}{\pi} \frac{1}{z - z_k} + C_j(z,0), \tag{26.31}$$

where

$$Q_k(t) \equiv \int_0^t m_k(t')dt'. \tag{26.32}$$

The parameter $Q_k(t)$ is the total area of fluid that has been injected at the point z_k from the initial instant until time t.

Equation (26.31) shows that if the initial domain $D(0)$ is chosen to be in the class such that all the functions $\{C_j(z,0)\}$ are the same, then this feature will remain true for $t > 0$, so that all the functions $\{C_j(z,t)|j = 0,1,\ldots,M\}$ will also be equal. Since we know that, on any boundary curve of D,

$$\overline{z} = S_j(z,t) = C_I(z,t) - C_j(z,t), \qquad j = 0,1,\ldots,M, \tag{26.33}$$

where $S_j(z,t)$ is the Schwarz function of the jth boundary component of D. But then having all the $\{C_j(z,t)|j = 0,1,\ldots,M\}$ equal means that all the Schwarz functions $\{S_j(z,t)|j = 0,1,\ldots,M\}$ are equal to the same rational function.

We know from Chapter 12 that another way of stating all this is that if $D(0)$ is initially a quadrature domain, then $D(t)$ remains a quadrature domain under evolution with time-evolving quadrature data.

Let the common Schwarz function be denoted by $S(z,t)$, i.e.,

$$S_j(z,t) = S(z,t) \qquad \forall j = 0,1,\ldots,M, \tag{26.34}$$

and let the common Cauchy transform be denoted by $C(z,t)$, i.e.,

$$C_j(z,t) = C(z,t) \qquad \forall j = 0,1,\ldots,M. \tag{26.35}$$

Then from (26.31)

$$C(z,t) = -\sum_{k=1}^{N} \frac{Q_k(t)}{\pi} \frac{1}{z - z_k} + C(z,0). \tag{26.36}$$

Now (26.33) implies

$$S(z,t) = C_I(z,t) - C(z,t), \qquad j = 0,1,\ldots,M. \tag{26.37}$$

Equations (26.36) and (26.37) together provide expressions for the strengths of the poles of the Schwarz function $S(z,t)$ inside D.

There is another way to derive the same result. Suppose the jth boundary curve ∂D_j of D has Schwarz function $S_j(z,t)$:

$$\overline{z} = S_j(z,t) \qquad \text{on } \partial D_j. \tag{26.38}$$

On taking a partial derivative with respect to time, we get

$$\overline{z_t} = S_j'(z,t) z_t + \frac{\partial S_j}{\partial t} \qquad \text{on } \partial D_j, \tag{26.39}$$

where we again use $S_j'(z,t)$ to denote $\partial S_j(z,t)/\partial z$. Now on the boundary ∂D_j for $j = 0,1,\ldots,M$,

$$S_j'(z,t) = \frac{d\overline{z}}{dz}; \tag{26.40}$$

hence (26.39) is equivalent to

$$\overline{z_t} dz - z_t d\overline{z} = \frac{\partial S_j}{\partial t} dz \qquad \text{on } \partial D_j. \tag{26.41}$$

The kinematic condition on the interface is

$$\text{Re}\left[\overline{z_t}\left(-\text{i}\frac{dz}{ds}\right)\right] = \text{Re}\left[h'(z,t)\left(-\text{i}\frac{dz}{ds}\right)\right] \tag{26.42}$$

or

$$\text{Im}\left[\overline{z_t} dz\right] = \text{Im}\left[h'(z,t) dz\right]. \tag{26.43}$$

But we also know that, on the boundary ∂D_j,

$$h(z,t) + \overline{h(z,t)} = 2\Phi_j(t); \tag{26.44}$$

hence

$$h'(z,t) dz = -\overline{h'(z,t) dz}. \tag{26.45}$$

Now (26.43) becomes

$$\text{Im}\left[\overline{z_t} dz\right] = \frac{1}{2\text{i}} 2h'(z,t) dz. \tag{26.46}$$

On comparison with (26.41) we deduce

$$\frac{\partial S_j}{\partial t} = 2h'(z,t), \tag{26.47}$$

where, we notice, the right-hand side is independent of j. We can write

$$S_j(z,t) = \int^t 2h'(z,\tilde{t})d\tilde{t} + S_j(z,0) \tag{26.48}$$

and conclude, again, that if all the Schwarz functions of the boundaries of the initial $D(0)$ are equal—that is, if $D(0)$ is a quadrature domain—then $D(t)$ will also be a quadrature domain. Equation (26.48) provides the strengths of the poles of the Schwarz function inside D.

A natural way to proceed is to introduce a time-evolving conformal map

$$z = f(\zeta,t) \tag{26.49}$$

from a time-evolving preimage circular region D_ζ to the domain $D(t)$. For the initial conditions allowing quadrature domain solutions we know that the function $f(\zeta,t)$ will be, at each instant, an automorphic function. It is then natural to use the prime function associated with D_ζ to construct analytical expressions for such mappings. This will allow the boundary evolution to be tracked in a convenient fashion.

However, for a multiply connected fluid domain, we know from Chapters 11 and 12 that knowledge of the evolution of the Cauchy transform alone does not provide enough information to determine the boundary evolution and, hence, the functional form of $f(\zeta,t)$. Additional physical information about the behavior of the holes is needed. This seems natural when you look back at how we just found the evolution of the Cauchy transform: the precise values $\{\Phi_j | j = 1,\ldots,M\}$ of the fluid pressures on the boundaries did not appear in the derivation. But the values of these pressures must, of course, affect the evolution in some way.

In fact, the values of $\{\Phi_j | j = 1,\ldots,M\}$ are in direct correspondence with the rate of change of the areas of the M holes. There is an easy way to see this using Green's second identity.

Let $\mathcal{A}_j(t)$ be the area of the jth hole. Let $\tilde{\sigma}_j(z,t)$ denote the harmonic function associated with the jth boundary component ∂D_j satisfying, for $j = 1,\ldots,M$,

$$\tilde{\sigma}_j(z,t) = \begin{cases} 0, & z \in \partial D_0, \\ 1, & z \in \partial D_j, \\ 0, & z \in \partial D_k,\ k \neq j, \end{cases} \tag{26.50}$$

and let the period matrix be

$$\tilde{\mathcal{P}}_{jk} = -\int_{\partial D_k} \frac{\partial \tilde{\sigma}_j}{\partial n_z} ds_z. \tag{26.51}$$

It is emphasized that the harmonic functions $\{\tilde{\sigma}_j(z,t) | j = 1,\ldots,M\}$ and the period matrix $\tilde{\mathcal{P}}_{jk}$ are defined for the domain $D(t)$ in the physical z plane, and not in the preimage ζ plane. This is why we have added tildes to make this distinction from the harmonic functions introduced in Chapter 2.

By Green's second identity we have, for each $j = 1,\ldots,M$, and at each instant,

$$\int\int_{D(t)} [\phi \nabla^2 \tilde{\sigma}_j - \tilde{\sigma}_j \nabla^2 \phi]\, dA_z = \int_{\partial D(t)} \left[\phi \frac{\partial \tilde{\sigma}_j}{\partial n_z} - \tilde{\sigma}_j \frac{\partial \phi}{\partial n_z} \right] ds_z. \tag{26.52}$$

On use of (26.16), (26.50), and (26.51), this gives

$$-\sum_{k=1}^{N} m_k(t)\tilde{\sigma}_j(z_k) = \sum_{k=1}^{M} \tilde{\mathcal{P}}_{jk}\Phi_k + \frac{d\mathcal{A}_j}{dt}, \qquad j = 1, \ldots, M, \tag{26.53}$$

where $\mathcal{A}_j(t)$ is the area of the jth hole in the domain. We have used the fact that

$$\frac{d\mathcal{A}_j}{dt} = -\int_{\partial D_j(t)} \frac{\partial \phi}{\partial n_z} ds_z, \tag{26.54}$$

since n_z is the outward normal to the fluid region.

The system (26.53) is a set of M linear equations relating the set of M boundary values $\{\Phi_j | j = 1, \ldots, M\}$ to the rate of change of the areas $\{\mathcal{A}_j | j = 1, \ldots, M\}$ of the M holes. The equations (26.53) depend on the harmonic measures $\{\tilde{\sigma}_j(z,t) | j = 1, \ldots, M\}$ and the period matrix $\tilde{\mathcal{P}}_{jk}$ of $D(t)$.

Recall, however, from Exercise 2.7 that the boundary value problem defining the harmonic functions in each domain is conformally invariant, meaning that

$$\tilde{\sigma}_j(z,t) = \sigma_j(\zeta,t), \;\; j = 1, \ldots, M, \qquad \tilde{\mathcal{P}}_{jk} = \mathcal{P}_{jk}. \tag{26.55}$$

This means that (26.53) provides a useful set of relations between the conformal mapping parameters, the boundary pressures $\{\Phi_j | j = 1, \ldots, M\}$, and the rates of change of the hole areas $\{\mathcal{A}_j | j = 1, \ldots, M\}$ that can be used to calculate the mapping (26.49).

There is an alternative, and more direct, way to proceed that involves finding an explicit expression for the potential $h(z,t)$ governing the flow; this can be done using the machinery of the prime function. The ideas are best illustrated in the context of a specific example, which allows us to compare the two methods.

Consider injection of fluid into the quarter plane bounded by impenetrable walls $y = \pm x$ from a point source situated at $z = 1$; see Figure 26.3. By the symmetry of this arrangement, it should be clear that this problem is equivalent to finding the flow in the unbounded plane produced by four equal point sources located at $\pm 1, \pm \mathrm{i}$. Assuming that there is initially no fluid in the corner region, and that the mass flux $m(t) = 1$, the free surface of the fluid will be an expanding circle until it hits the two walls. Because the source is placed on the real axis $y = 0$, the free boundary will meet both walls at the same instant, immediately forming a region of trapped air near the corner. A natural assumption is to take the fluid pressure in the trapped region to be equal to the fluid pressure in the air far from the corner, which is assumed to be zero.

A little thought shows that the free boundary evolution after the fluid hits the walls is equivalent to the problem of the evolution of a fluid annulus, with four injection points at the points $\pm 1, \pm \mathrm{i}$, with equal fluid pressures outside the annulus and in the single enclosed hole. Since there was no fluid initially, the Cauchy transform of the disconnected fluid domain is the rational function

$$C(z,t) = -\frac{Q}{\pi}\frac{1}{z-1} - \frac{Q}{\pi}\frac{1}{z-i} - \frac{Q}{\pi}\frac{1}{z+1} - \frac{Q}{\pi}\frac{1}{z+i}, \tag{26.56}$$

where

$$Q = \int_0^t m(t')dt' = t. \tag{26.57}$$

After the fluid hits the walls or, equivalently, after the four expanding circular discs of fluid intersect, it is usual to invoke the continuity in time of the Cauchy transform and suppose

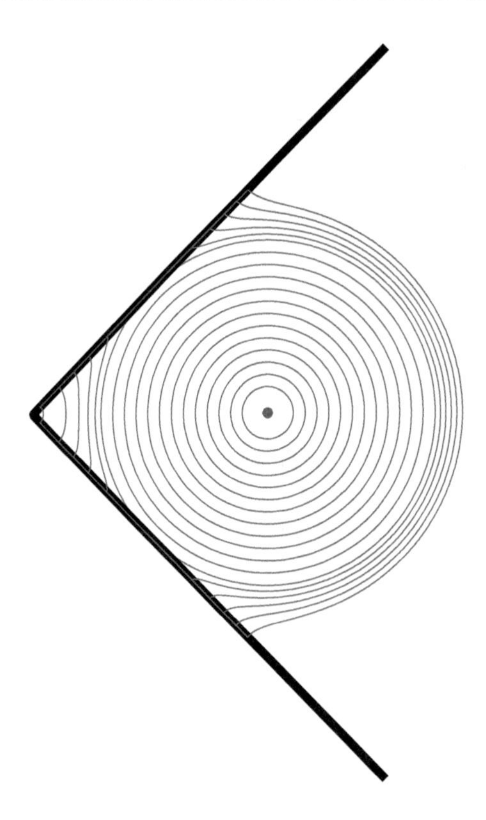

Figure 26.3. *Injection by a source in a quarter plane. The free boundary is circular until the time it simultaneously reaches the two sides of the wedge. After that time the free boundary splits into two parts, whose shape is described by (26.59).*

that the Cauchy transform for z both in the trapped region and far from the annular fluid domain continues to be given by (26.56). This means that the Schwarz functions of both the inner and outer boundaries are the same; equivalently, the domain will evolve as a quadrature domain.

This means that, after the fluid has hit the walls, the conformal map from the annulus D_ζ given by

$$\rho(t) < |\zeta| < 1 \tag{26.58}$$

to the annular fluid region will be a loxodromic function, as considered in Chapter 8. Indeed, the required function is

$$z = f(\zeta, t) = R\zeta \prod_{k=1}^{4} \frac{\omega(\zeta, a\Omega_k/\sqrt{\rho})}{\omega(\zeta, a\Omega_k)}, \qquad \Omega_k = e^{2\pi \mathrm{i}(k-1)/4}, \tag{26.59}$$

where $1 < a < 1/\rho$ and R are real parameters. This function depends on three real parameters: ρ, a, and R. We must find the evolution equations for these.

Two conditions relating these parameters arise from insisting that the singularity of the Cauchy transform is at $z = 1$ and of strength $-Q/\pi$ in accordance with (26.56). But a third condition is still to be found.

One option is to use (26.53), which, given that $M = 1$ in this doubly connected case, gives one real condition. Indeed, on use of the facts that

$$\sigma_1(\zeta) = \frac{\log|\zeta|}{\log\rho}, \qquad \mathcal{P}_{11} = -\frac{2\pi}{\log\rho}, \tag{26.60}$$

if the fluid pressure on the inner boundary is $\Phi_1 = 0$, then (26.53) gives

$$\frac{d\mathcal{A}}{dt} = -\mathcal{P}_{11}\Phi_1 + 4\frac{\log a}{\log\rho} = 4\frac{\log a}{\log\rho}. \tag{26.61}$$

This is the required third equation determining the evolution of ρ, a, and A.

As just suggested above, there is an alternative approach, which deploys other aspects of the framework of Part I. The idea is to introduce

$$H(\zeta,t) \equiv h(f(\zeta,t),t). \tag{26.62}$$

On use of the functional identities (5.4)–(5.6) associated with the prime function of the concentric annulus, it can be shown that the required potential is given by

$$H(\zeta,t) = \frac{1}{2\pi}\sum_{k=1}^{4}\log\left[\frac{1}{|a|}\frac{\omega(\zeta,\Omega_k/a)}{\omega(\zeta,\Omega_k a)}\right] + \left[\Phi_1 + \frac{2\log|a|}{\log\rho}\right]\frac{\log\zeta}{\log\rho}. \tag{26.63}$$

Moreover, on use of the facts that

$$\frac{dz}{ds} = \begin{cases} +\dfrac{\mathrm{i}\zeta f'(\zeta,t)}{|f'(\zeta,t)|}, & \zeta \in C_0, \\[2ex] -\dfrac{\mathrm{i}\zeta f'(\zeta,t)}{\rho|f'(\zeta,t)|}, & \zeta \in C_1, \end{cases} \tag{26.64}$$

and that, on C_1,

$$\frac{dz}{dt} = \frac{\dot\rho}{\rho}f'(\zeta,t) + \dot f(\zeta,t), \tag{26.65}$$

where we use the shorthand notation

$$f'(\zeta,t) = \frac{\partial f(\zeta,t)}{\partial\zeta}, \qquad \dot f(\zeta,t) = \frac{\partial f(\zeta,t)}{\partial t}, \tag{26.66}$$

the boundary condition (26.43) takes the form

$$\mathrm{Re}\left[\frac{\dot f(\zeta,t)}{\zeta f'(\zeta,t)}\right] = \begin{cases} \mathrm{Re}\left[\dfrac{\zeta H'(\zeta,t)}{|f'(\zeta,t)|^2}\right], & \zeta \in C_0, \\[2ex] \mathrm{Re}\left[\dfrac{\zeta H'(\zeta,t)}{\rho^2|f'(\zeta,t)|^2}\right] - \dfrac{\dot\rho}{\rho}, & \zeta \in C_1. \end{cases} \tag{26.67}$$

The function in square brackets on the left-hand side of (26.67) is analytic in D_z and single-valued there, meaning that (26.67) constitutes a modified Schwarz problem in the annulus, as considered in Chapter 13. As such, there is a compatibility condition on the data on the right-hand side of (26.67). In this case, it takes the form

$$\frac{1}{2\pi\mathrm{i}}\int_{C_0}\mathrm{Re}\left[\frac{\zeta H'(\zeta,t)}{|f'(\zeta,t)|^2}\right]\frac{d\zeta}{\zeta} = \frac{1}{2\pi\mathrm{i}}\int_{C_1}\left[\mathrm{Re}\left[\frac{\zeta H'(\zeta,t)}{\rho^2|f'(\zeta,t)|^2}\right] - \frac{\dot\rho}{\rho}\right]\frac{d\zeta}{\zeta}. \tag{26.68}$$

This equation provides a third evolution equation to complete the determination of ρ, a, and A. Indeed, it provides that

$$\dot{\rho} = \frac{\rho}{2\pi \mathrm{i}} \left[\frac{1}{2\pi \mathrm{i}} \int_{C_1} \mathrm{Re} \left[\frac{\zeta H'(\zeta, t)}{\rho^2 |f'(\zeta, t)|^2} \right] \frac{d\zeta}{\zeta} - \int_{C_0} \mathrm{Re} \left[\frac{\zeta H'(\zeta, t)}{|f'(\zeta, t)|^2} \right] \frac{d\zeta}{\zeta} \right]. \tag{26.69}$$

The right-hand side is readily computable since $H(\zeta, t)$ is known, in terms of the prime function of the annulus, from (26.63).

We have therefore given two distinct ways to find the missing evolution equation, namely, the one not given by the evolution of the Cauchy transform. It can be verified numerically that use of (26.68) is equivalent to use of (26.61). The equation (26.61) has the advantage that it does not require the potential $H(\zeta, t)$ to be determined.

The two methods just described for determining this third condition associated with the assumed physics of the trapped air both appear to be new, and, to the best of our knowledge, neither has previously been used in existing studies of singularity-driven Hele-Shaw flows (although a variant of the second method was used by Crowdy and Tanveer [57] to study pressure-driven Hele-Shaw flow of a finite blob of fluid in a Hele-Shaw channel). Most prior studies make use of yet another, quite different, method introduced by Richardson [111] based on supplementing the evolution of the Cauchy transform with additional data obtained from certain integrals in the physical z plane involving logarithms in the integrands.

Figure 26.3 shows the evolution of the free boundary of a blob of fluid injected at a fixed point along the bisector of a right-angled corner. The evolution is shown up to the point where the fluid touches the apex where the two solid walls meet. The blob evolution beyond this can be tracked using a conformal mapping from a simple unit disc.[57]

Flow in a rotating Hele-Shaw cell: As a final example of a time-evolving Hele-Shaw flow, we remark on a problem with a different driving mechanism. We suppose that a fluid blob $D(t)$ is situated in a Hele-Shaw cell that is in solid-body rotation. The velocity potential ϕ (or pressure) is now harmonic in some time-dependent multiply connected domain $D(t)$:

$$\nabla^2 \phi = 0, \qquad z \in D(t). \tag{26.70}$$

There are now no sources or sinks driving the motion of the fluid, although these can easily be incorporated. Instead the free boundary evolution is driven by the rotation of the cell.

The boundary condition (26.17) on the harmonic velocity potential ϕ becomes

$$\phi = \frac{\Omega}{2} z\overline{z} + \Phi_j(t), \qquad z \in \partial D_j(t), \tag{26.71}$$

where Ω is a specified parameter that depends on the angular velocity of the rotating cell. Without loss of generality, because of an additive degree of freedom in specifying the potential, we can set

$$\Phi_0(t) = 0. \tag{26.72}$$

The evolution of the boundary is given by the following law for its normal velocity V_n:

$$V_n = \mathbf{u}.\mathbf{n} = \nabla \phi.\mathbf{n}. \tag{26.73}$$

We introduce the complex potential

$$h(z) = \phi + \mathrm{i}\psi, \tag{26.74}$$

[57] For an alternative approach to this same problem using elliptic function theory rather than the prime function, see Richardson [111].

whose dependence on time is suppressed in our notation for brevity. This function must be analytic in D_z. As before,

$$\mathrm{Im}\left[\frac{\partial z}{\partial t}d\overline{z}\right] = -\mathrm{Im}[dh], \tag{26.75}$$

but now

$$\mathrm{Im}[dh] = \left(\frac{dh(z) - \overline{dh(z)}}{2\mathrm{i}}\right) = \frac{2dh(z) - \Omega d(z\overline{z})}{2\mathrm{i}}, \tag{26.76}$$

where we have used the boundary condition (26.71), which implies that

$$\overline{h(z)} = -h(z) + 2\Phi_j + \Omega z\overline{z} \qquad \text{on } \partial D_j(t). \tag{26.77}$$

We will again make use of considerations based on the Cauchy transform of the time-evolving domain. Suppose that the point z sits in the jth hole of D_z, and consider the analytic function $C_j(z,t)$ defined by the Cauchy transform integral for z in this hole. On use of the general expression (12.53) from Chapter 12 its evolution is given by

$$\frac{\partial C_j(z,t)}{\partial t} = \frac{1}{2\pi\mathrm{i}}\int_{\partial D(t)}\left[\frac{2h'(z')}{z'-z}dz' - \frac{\Omega d(z'\overline{z'})}{z'-z}\right]. \tag{26.78}$$

Since $h(z)$ is analytic in $D(t)$—we have assumed the motion is driven purely by rotation and that there are no sources or sinks—the first integral on the right-hand side vanishes by Cauchy's theorem. Integration by parts of the second integral yields

$$\frac{\partial C_j(z,t)}{\partial t} = \frac{1}{2\pi\mathrm{i}}\int_{\partial D(t)} -\frac{\Omega z'\overline{z'}dz'}{(z'-z)^2} = -\frac{1}{\pi}\frac{\partial}{\partial z}\int\!\!\int_{D(t)}\frac{\Omega z' dA_{z'}}{z'-z}. \tag{26.79}$$

The Cauchy transform for z in the jth hole therefore satisfies a partial differential equation of the form

$$\frac{\partial C_j(z,t)}{\partial t} + \frac{\partial I_j(z,t)}{\partial z} = 0, \tag{26.80}$$

where

$$I_j(z,t) = \frac{1}{\pi}\int\!\!\int_{D(t)}\frac{\sigma(z',t)dA_{z'}}{z'-z} \tag{26.81}$$

and where

$$\sigma(z,t) = \Omega z \tag{26.82}$$

is analytic in $D(t)$.

The evolution of a blob of fluid in a rotating Hele-Shaw cell is therefore a free boundary problem of the special form described in §12.7 of Chapter 12. This means that we can immediately infer from the considerations featured there that an initial domain that is a quadrature domain, and therefore for which all boundaries share a common Schwarz function that is meromorphic in D_z, will remain a quadrature domain under evolution. In such a case,

$$C_j(z,t) = C(z,t), \qquad I_j(z,t) = I(z,t) \qquad \forall j = 0,1,\ldots,M, \tag{26.83}$$

with

$$\frac{\partial C(z,t)}{\partial t} + \frac{\partial I(z,t)}{\partial z} = 0, \tag{26.84}$$

where

$$I(z,t) = \frac{1}{\pi}\int\!\!\int_{D(t)}\frac{\sigma(z',t)\,dA_{z'}}{z'-z}. \tag{26.85}$$

In fact it was the result of Exercise 12.4 that, for this particular choice of $\sigma(z,t)$, the solution of the partial differential equation (26.84) for an initial quadrature domain is given by

$$C(z,t) = e^{-\Omega t} C_0(z e^{-\Omega t}). \tag{26.86}$$

This immediately provides the time evolution of the Cauchy transform given a rational function $C_0(z) = C(z,0)$ encoding the shape of the initial domain.

It was seen in the previously considered problem of source-driven Hele-Shaw flows that, for multiply connected domains, the evolution of the Cauchy transform of the fluid domain is not enough to uniquely determine its evolution. Additional information on the physical conditions associated with the holes is needed. It therefore remains to specify conditions associated with the evolution of the holes.

This can be done by generalizing a method already advocated for in the context of source-driven flows. By Green's second identity we have, for each $j = 1, \ldots, M$,

$$\int\int_{D(t)} [\phi \nabla^2 \tilde{\sigma}_j - \tilde{\sigma}_j \nabla^2 \phi] \, dA_z = \int_{\partial D(t)} \left[\phi \frac{\partial \tilde{\sigma}_j}{\partial n_z} - \tilde{\sigma}_j \frac{\partial \phi}{\partial n_z} \right] ds_z. \tag{26.87}$$

On use of (26.70), (26.50), (26.51), and (26.71) this gives

$$-\frac{\Omega}{2} \sum_{k=1}^{M} \int_{C_k} z\bar{z} \frac{\partial \tilde{\sigma}_j}{\partial n_z} ds_z = \sum_{k=1}^{M} \tilde{\mathcal{P}}_{jk} \Phi_k - \frac{d\mathcal{A}_j}{dt}, \qquad j = 1, \ldots, M, \tag{26.88}$$

where $\mathcal{A}_j(t)$ is the area of the jth hole in the domain. (26.88) is a set of M equations that provide an explicit relationship between the set of M boundary values $\{\Phi_j | j = 1, \ldots, M\}$ and the rate of change of the areas $\{\mathcal{A}_j | j = 1, \ldots, M\}$ of the M holes. To complete the specification of the evolution of D_z, one can, for example, specify all the hole pressures $\{\Phi_j | j = 1, \ldots, M\}$ or all the hole areas $\{\mathcal{A}_j | j = 1, \ldots, M\}$.

More details on the problem of an evolving fluid annulus in a Hele-Shaw cell can be found in [31].

Chapter 27

Free surface Stokes flow

Chapter 22 illustrated how problems of Stokes flow, or the low-Reynolds-number motion of very viscous fluids, can be solved using the mathematical methods of Part I. Those problems involved fixed domains with given boundaries. *Free surface* Stokes flow problems are more difficult. For these, it is necessary to find a streamfunction satisfying

$$\nabla^4 \psi = 0, \qquad \nabla^2 = \frac{\partial^2}{\partial x^2} + \frac{\partial^2}{\partial y^2} \tag{27.1}$$

in some fluid domain having free surfaces which must be found as part of the solution. The fact that the shape of the boundary is unknown a priori makes such problems inherently nonlinear.

As in Chapter 26 a free boundary problem can be either steady or unsteady. We again consider example problems of each kind, but now for slow viscous Stokes flow. The steady problem we have chosen to use as an example has the additional complication that it is a free boundary problem of mixed type akin to those studied in Chapter 21. Consequently many varied ingredients from the theory of Part I come into play, making it a valuable case study.

The general solution of (27.1) for the streamfunction ψ associated with a steady Stokes flow can be written in terms of two analytic functions, called $f(z)$ and $g(z)$, i.e.,

$$\psi = \mathrm{Im}[\overline{z} f(z) + g(z)], \qquad z = x + \mathrm{i}y. \tag{27.2}$$

Suppose a free capillary surface surrounds a constant-pressure bubble in a two-dimensional fluid; this means that surface tension is active on the bubble boundary. The assumption of constant pressure means that we do not need to resolve any fluid motion inside the bubble. It will be assumed that there is a surface tension $T(s)$ which depends on a surfactant concentration $\Gamma(s)$ which is a function of position, or arclength s, along the interface. It is usual to assume an equation of state for this dependence of surface tension on surfactant concentration which, here, we suppose has the linear form

$$T(s) = T_0(1 - \beta\Gamma(s)/\Gamma_0), \tag{27.3}$$

where T_0 is the surface tension of the clean interface, Γ_0 is the uniform concentration of surfactant in the absence of flow, and β is an adjustable parameter determining the sensitivity of the interfacial surface tension to changes in the surfactant concentration. The case $\beta = 0$ corresponds to the "clean flow" problem where there is no surfactant present and where the surface tension is uniform along the interface.

In general, there is a dynamical evolution equation governing the kinetics of the surfactant between the interface and the fluid. This couples the surface concentration $\Gamma(s)$ with the flow field, as well as a bulk concentration of surfactant. For the purposes of this chapter, we will only need to invoke the fact that in steady state conditions where it is assumed that surface diffusion of surfactant is negligible, this evolution equation simply tells us that, at any point on the interface, either

$$\Gamma = 0 \qquad \text{or} \qquad \mathbf{u}.\mathbf{t} = 0. \tag{27.4}$$

The associated dynamic boundary condition, with n_i and t_i denoting components of the unit normal and tangent vectors along the boundary, respectively, is

$$\sigma_{ij} n_j = T(s)\kappa n_i - \frac{dT}{ds} t_i, \tag{27.5}$$

where κ is the surface curvature. (27.5) is a statement of the fact that the normal fluid stress on the free surface is being balanced by a normal force associated with surface tension, while the tangential fluid stress is balanced by a Marangoni stress associated with a nonuniform distribution of surfactant along the interface. In the simpler case where the surface tension is uniform, (27.5) simplifies to

$$\sigma_{ij} n_j = T_0 \kappa n_i. \tag{27.6}$$

We will study both situations in this chapter.

If the free boundary problem is steady, involving the determination of some equilibrium free surface shape, then there is an additional requirement on the free surface that it be a streamline, i.e.,

$$\psi = \text{constant} \tag{27.7}$$

on the surface. In such a case the normal velocity of the fluid at the boundary, which is the same as the normal velocity of the interface, vanishes and the free surface remains of fixed shape.

On the other hand, if the free boundary evolves in time, then it is usual to adopt the assumption of quasi-steadiness, which means that once the streamfunction ψ is solved at each instant according, say, to the above-stated boundary value problem for the fluid stress, then the free boundary evolves according to the kinematic requirement that the normal speed of the boundary, denoted by V_n, is given by the normal fluid velocity

$$V_n = \mathbf{u}.\mathbf{n}, \tag{27.8}$$

where $\mathbf{u} = (u, v)$.

We consider both steady and quasi-steady problems in this chapter. We will suppress the dependence of the functions on time t, except where it is needed for emphasis.

In terms of the Goursat functions $f(z)$ and $g'(z)$ we can write the velocity field (u, v) in the complex form

$$u + \mathrm{i}v = -f(z) + z\overline{f'(z)} + \overline{g'(z)}. \tag{27.9}$$

The complex form of the stress tensor $\sigma_{ij} n_j$ is

$$2\mu\mathrm{i}\frac{dH(z,\overline{z})}{ds}, \tag{27.10}$$

where

$$H(z,\overline{z}) = f(z) + z\overline{f'(z)} + \overline{g'(z)}. \tag{27.11}$$

The curvature $\kappa = \theta_s$, where

$$\frac{dz}{ds} = e^{\mathrm{i}\theta(s)}, \tag{27.12}$$

so that $\theta(s)$ denotes the angle made by the boundary tangent vector to the positive x axis as a function of the arclength s, can be expressed as

$$\kappa = \theta_s = -\mathrm{i}\frac{d^2z/ds^2}{dz/ds}. \tag{27.13}$$

On use of this, the boundary condition (27.5) can be written, after an integration with respect to s, as

$$H(z,\overline{z}) = f(z) + z\overline{f'(z)} + \overline{g'(z)} = -\frac{\mathrm{i}T(s)}{2\mu}\frac{dz}{ds} + A_j, \tag{27.14}$$

where A_j is independent of position on the jth interface. Without loss of generality we can set

$$A_0 = 0. \tag{27.15}$$

However $\{A_j | j = 1, \dots, M\}$ must be determined from the physical conditions imposed on the boundaries.

Steady stagnant-cap surfactant-laden bubble in strain: Consider the situation in a complex $z = x + \mathrm{i}y$ plane in which a single bubble is placed in an ambient irrotational linear straining flow (u, v) having the far-field form

$$(u,v) \sim -Q(x,-y) \qquad \text{as } |z| \to \infty, \tag{27.16}$$

where $Q > 0$ is the far-field strain rate of the flow. We will also assume that there is no diffusion of surfactant off the interface. This problem was first solved analytically by Siegel [119] using Riemann–Hilbert methods. It was later reappraised and generalized using the techniques of this monograph in [6].

The corresponding far-field forms of the functions $f(z)$ and $g(z)$ are easily shown from (27.9) to be

$$f(z) \sim \frac{p_\infty}{4\mu}z + f_\infty + \mathcal{O}(1/z), \qquad g(z) \to -\frac{Q}{2}z^2 + g_\infty + \mathcal{O}(1/z), \tag{27.17}$$

where p_∞ is the far-field fluid pressure and f_∞, g_∞ are constants. The boundary conditions on the interface are that

$$\mathrm{Im}[\overline{z}f(z) + g(z)] = 0, \tag{27.18}$$

which is equivalent to (27.7), with the constant set to zero without loss of generality, and

$$H(z,\overline{z}) = f(z) + z\overline{f'(z)} + \overline{g'(z)} = -\frac{\mathrm{i}T(s)}{2\mu}\frac{dz}{ds}, \tag{27.19}$$

where we have used (27.15). The challenge is to use the two conditions (27.18)–(27.19) to determine the two functions $f(z)$ and $g(z)$ as well as the shape of the free surface of the bubble.

To make progress we first note that it is possible to use (27.18) and (27.19) to show that, on the free surface,

$$\overline{z}f(z) + g(z) = 0. \tag{27.20}$$

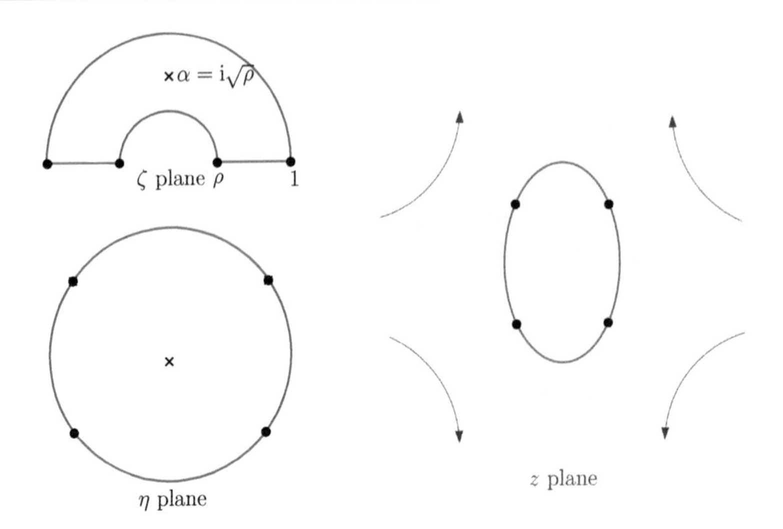

Figure 27.1. *If a surfactant-laden bubble is situated in a linear straining flow (27.16) that is sufficiently strong, the surfactant is swept to the bubble tips where stagnant caps can form (shown in red) leaving clean zones (shown in blue).*

The imaginary part of the left-hand side of (27.20) is known immediately to be zero from (27.18), but, on use of (27.19), it follows that its real part also vanishes.

Since, on the interface, we know that

$$\overline{z} = S(z), \tag{27.21}$$

where $S(z)$ is the Schwarz function of the interface [76], it can be argued from (27.20) that

$$S(z) = -\frac{g(z)}{f(z)} \tag{27.22}$$

and hence, from (27.17), that $S(z)$ has a simple pole as $|z| \to \infty$.

An effective way to find the shape of the free surface is to introduce a conformal map from the interior of a unit disc in a complex η plane, say, to the unbounded simply connected fluid region exterior to the bubble. Equivalently, we can map the exterior $|\eta| > 1$ of the unit η disc to the fluid region. By what has been deduced above about the Schwarz function of the free surface, the mapping from the exterior of the unit η disc given by

$$z = f(\eta) = a\eta + \frac{b}{\eta}, \qquad a, b \in \mathbb{R}, \tag{27.23}$$

is easily shown to provide a Schwarz function with the required simple pole at $z \to \infty$. The bubble shape is an ellipse with semiaxes $a \pm b$.

Suppose, however, that we seek a solution for a surfactant-laden bubble where the strain rate Q is sufficiently strong that the surfactant on the bubble surface has been swept to the two tips of the bubble at the points of maximum elongation. Since we also assume, as discussed earlier, that there is no diffusion of surfactant off the surface, this implies, on use of (27.4), that there will be two localized regions on the surface, occupied by surfactant, and where

$$u - \mathrm{i}v = 0. \tag{27.24}$$

This is because both the normal and tangential components of the velocity must vanish at any point on the boundary where surfactant is present. Between these two no-slip zones will be two regions of uniform surface tension where there is no surfactant, i.e., $\Gamma(s) = 0$. On the latter portions of the surface, the boundary condition (27.14) with $T(s) = T_0$ will hold. Henceforth we set $T_0 = 1$.

This is now a free boundary value problem that is also of *mixed* type akin to those considered in Chapter 21.

We now illustrate the perhaps counterintuitive fact that although the fluid region is simply connected, the mixed character of the boundary value problem to be solved is such that the function theory associated with an annulus

$$\rho < |\zeta| < 1 \tag{27.25}$$

in a parametric ζ plane can be useful. This is because we seek solutions with two stagnant caps.

Consider the composition of conformal mappings given by

$$z = \mathcal{Z}(\eta) = a\eta(\zeta) + \frac{b}{\eta(\zeta)}, \qquad \eta(\zeta) = \frac{\omega(\zeta, \overline{\alpha})\omega(\zeta, 1/\overline{\alpha})}{\omega(\zeta, \alpha)\omega(\zeta, 1/\alpha)}, \qquad \alpha = \mathrm{i}\sqrt{\rho}. \tag{27.26}$$

The mapping $\eta(\zeta)$ here is precisely the special case of an unbounded circular slit map from a concentric annulus featured in (5.38) of Chapter 5. The point

$$\zeta = \alpha = \mathrm{i}\sqrt{\rho} \tag{27.27}$$

is the preimage of $z = \infty$; moreover, the upper half annulus

$$\rho < |\zeta| < 1, \qquad \mathrm{Im}[\zeta] > 0, \tag{27.28}$$

is the preimage of the fluid region exterior to the bubble. We will suppose that the upper half circle

$$|\zeta| = 1, \qquad \mathrm{Im}[\zeta] > 0, \tag{27.29}$$

is transplanted to one of the surfactant-free zones on the bubble surface and that the upper half circle

$$|\zeta| = \rho, \qquad \mathrm{Im}[\zeta] > 0, \tag{27.30}$$

is transplanted to the other one. The two portions of the real ζ axis given by

$$-1 < \zeta < -\rho, \qquad \rho < \zeta < 1 \tag{27.31}$$

will be the preimages of the two no-slip portions of the bubble surface.

To see why this construction is valuable, note that because of (27.20), the velocity field on the interface can be written as

$$u - \mathrm{i}v = -\overline{f(z)} - \frac{d\overline{z}}{dz} f(z) \tag{27.32}$$

so that on the no-slip portions of the interface we have

$$\mathrm{Re}\left[\frac{f(z)}{dz/ds}\right] = 0. \tag{27.33}$$

On the portions of the interface that are free of surfactant, the complex conjugate of boundary condition (27.14) with $T_0 = 1$ implies, on use of (27.20), that

$$\overline{H} = \overline{f(z)} - \frac{d\overline{z}}{dz} f(z) = \frac{\mathrm{i}}{2} \frac{d\overline{z}}{ds} \tag{27.34}$$

or

$$\text{Im}\left[\frac{f(z)}{dz/ds}\right] = -\frac{1}{4}. \tag{27.35}$$

On the real ζ axis where $\overline{\zeta} = \zeta$ we have

$$\frac{dz}{ds} = -\frac{\mathcal{Z}'(\zeta)}{|\mathcal{Z}'(\zeta)|}, \tag{27.36}$$

where we use the prime notation to denote derivatives with respect to ζ, while on $|\zeta| = 1$,

$$\frac{dz}{ds} = \frac{\mathrm{i}\zeta\mathcal{Z}'(\zeta)}{|\mathcal{Z}'(\zeta)|}, \qquad \zeta \in C_0, \tag{27.37}$$

and on $|\zeta| = \rho$,

$$\frac{dz}{ds} = -\frac{\mathrm{i}\zeta\mathcal{Z}'(\zeta)}{\rho|\mathcal{Z}'(\zeta)|}, \qquad \zeta \in C_1. \tag{27.38}$$

On substitution of (27.36) into (27.33) we find that

$$\text{Re}\left[\frac{F(\zeta)}{\mathcal{Z}'(\zeta)}\right] = 0, \tag{27.39}$$

or, equivalently, on division by ζ,

$$\text{Re}\left[\frac{F(\zeta)}{\zeta\mathcal{Z}'(\zeta)}\right] = 0, \tag{27.40}$$

where we have introduced the function

$$F(\zeta) \equiv f(\mathcal{Z}(\zeta)). \tag{27.41}$$

Similarly, on substitution of (27.37) and (27.38) into (27.35), we find

$$\text{Re}\left[\frac{F(\zeta)}{\zeta\mathcal{Z}'(\zeta)}\right] = \begin{cases} +\dfrac{1}{4|\mathcal{Z}'(\zeta)|}, & \zeta \in C_0^+, \\[2ex] -\dfrac{1}{4\rho|\mathcal{Z}'(\zeta)|}, & \zeta \in C_1^+, \end{cases} \tag{27.42}$$

where we use C_0^+ and C_1^+ to denote the semicircular portions of C_0 and C_1 that are in the upper half ζ plane.

Since there are singularities of known type on the bubble boundary, it is now expedient to introduce the auxiliary function

$$M(\zeta) = \frac{\mathrm{i}}{\rho^2\zeta}\left[\frac{\omega(\zeta,1)\omega(\zeta,-1)\omega(\zeta,\rho)\omega(\zeta,-\rho)}{\omega(\zeta,\alpha)\omega(\zeta,1/\alpha)\omega(\zeta,\overline{\alpha})\omega(\zeta,1/\overline{\alpha})}\right], \tag{27.43}$$

where $\omega(.,.)$ is the prime function associated with the concentric annulus $\rho < |\zeta| < 1$. On use of the properties (5.4)–(5.6) of this prime function it can be verified that

$$\overline{M}(\zeta) = -M(\zeta), \qquad \overline{M}(1/\zeta) = M(\zeta), \qquad M(\rho^2\zeta) = M(\zeta). \tag{27.44}$$

These functional relations can be used to show that $M(\zeta)$ is real on the circles C_0 and C_1. Notice that $M(\zeta)$ is an example of a loxodromic function as studied in Chapter 8. It has been introduced to explicitly account for known singularities in the function in square brackets on the left-hand side of (27.42) in order that it can be found as the solution of a modified Schwarz problem in the following way. What follows is another example in the spirit of the ideas presented in §13.8, where boundary value problems of Schwarz type with singularities on the boundaries are discussed.

Armed with $M(\zeta)$ we can introduce the function

$$W(\zeta) \equiv \frac{F(\zeta)M(\zeta)}{\zeta \mathcal{Z}'(\zeta)}. \tag{27.45}$$

It can be verified that this function is analytic in the upper half annulus

$$\rho < |\zeta| < 1, \qquad \mathrm{Im}[\zeta] > 0, \tag{27.46}$$

since the zeros of $\mathcal{Z}'(\zeta)$ at $\zeta = \pm\rho, \pm 1$ are removed on multiplication by $M(\zeta)$, while the second order pole of $F(\zeta)M(\zeta)$ at $\zeta = \alpha$ is removed on division by $\mathcal{Z}'(\zeta)$, which also has a second order pole there. Moreover it can be shown that

$$\overline{W}(\zeta) = W(\zeta). \tag{27.47}$$

This implies, by the Schwarz reflection principle, that $W(\zeta)$ is also analytic in the lower half annulus

$$\rho < |\zeta| < 1, \qquad \mathrm{Im}[\zeta] < 0, \tag{27.48}$$

and, consequently, in the entire annulus $\rho < |\zeta| < 1$. Moreover, since $M(\zeta)$ is real on both C_0 and C_1, we know from (27.42) that

$$\mathrm{Re}\left[W(\zeta)\right] = \begin{cases} \dfrac{M(\zeta)}{4|\mathcal{Z}'(\zeta)|}, & \zeta \in C_0^+, \\[2ex] -\dfrac{M(\zeta)}{4\rho|\mathcal{Z}'(\zeta)|}, & \zeta \in C_1^+. \end{cases} \tag{27.49}$$

Now suppose that ζ is a point on the boundary of the lower half annulus; then $\overline{\zeta}$ will be on the boundary of the upper half annulus and, by (27.47),

$$W(\zeta) = \overline{W}(\zeta) = \overline{W(\overline{\zeta})} \tag{27.50}$$

so that, on taking real parts,

$$\mathrm{Re}[W(\zeta)] = \mathrm{Re}[\overline{W(\overline{\zeta})}] = \mathrm{Re}[W(\overline{\zeta})]. \tag{27.51}$$

In the last equality, we have used the fact that the real parts of complex conjugate quantities are equal. But the quantity on the right-hand side of (27.51) is known from (27.49) since $\overline{\zeta}$ is on the boundary of the upper half annulus. The reader may recall that we used similar arguments in Chapter 18.

As a result of all these considerations we know that $W(\zeta)$ is analytic in the entire annulus $\rho < |\zeta| < 1$, including its boundaries, and we know its real part on both boundary circles C_0 and C_1. We also know that $W(\zeta)$ is single-valued in the annulus since $F(\zeta)$ must be single-valued (this is because the bubble must be free of net force).

In this way, the boundary value problem for $W(\zeta)$ is a modified Schwarz problem in the annulus $\rho < |\zeta| < 1$ as treated in Chapter 13. It can be solved using the integral representation of the solution in terms of the prime function.

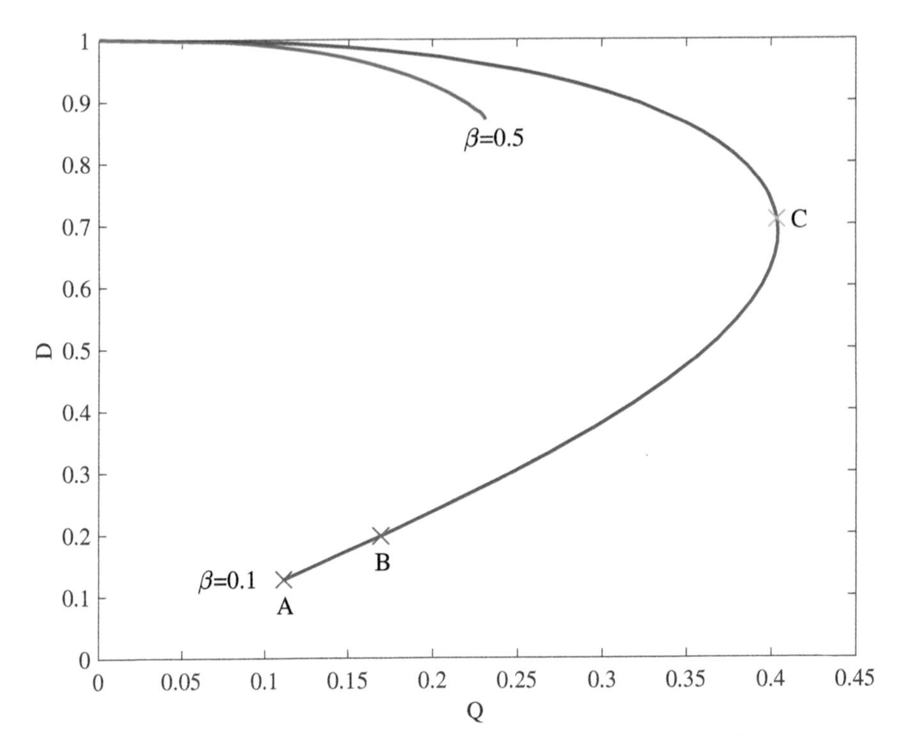

Figure 27.2. *Graphs of the deformation D of bubbles with stagnant caps as a function of the strain rate Q for $\beta = 0.1$ and 0.5. The ends of the solution branches correspond to a bubble fully coated with surfactant.*

Armed with this explicit integral expression for $W(\zeta)$ the function $F(\zeta)$ can be found and, then, the second function $g'(z)$. Any quantities of physical interest are then readily calculated.

Figure 27.2 shows graphs, for two different values of β, 0.1 and 0.5, of the deformation parameter D of the ellipse defined by

$$D = \frac{|a| - |b|}{|a| + |b|} \tag{27.52}$$

as a function of the strain rate parameter Q. On the $\beta = 0.1$ graph three points are marked: A, B, and C. For each of these points, Figure 27.3 shows graphs of the surfactant distribution as a function of $\arg[\eta]$, which is closed related to the arclength around the bubble boundary.

It is worth remarking on the many theoretical aspects of Part I that have been combined to develop the solution just described: we used the special class of unbounded circular slit mappings from an annulus described in §5.7, a loxodromic function constructed using ideas from Chapter 8, and an integral representation of the generalized Poisson kernel for the concentric annulus as derived in Chapter 13.

All these ingredients rely on use of the very same prime function, thereby underlining its central importance. And while all these aspects of our framework were explained separately, and in different chapters, this example shows the potential for them all coming together to provide a powerful suite of solution techniques.

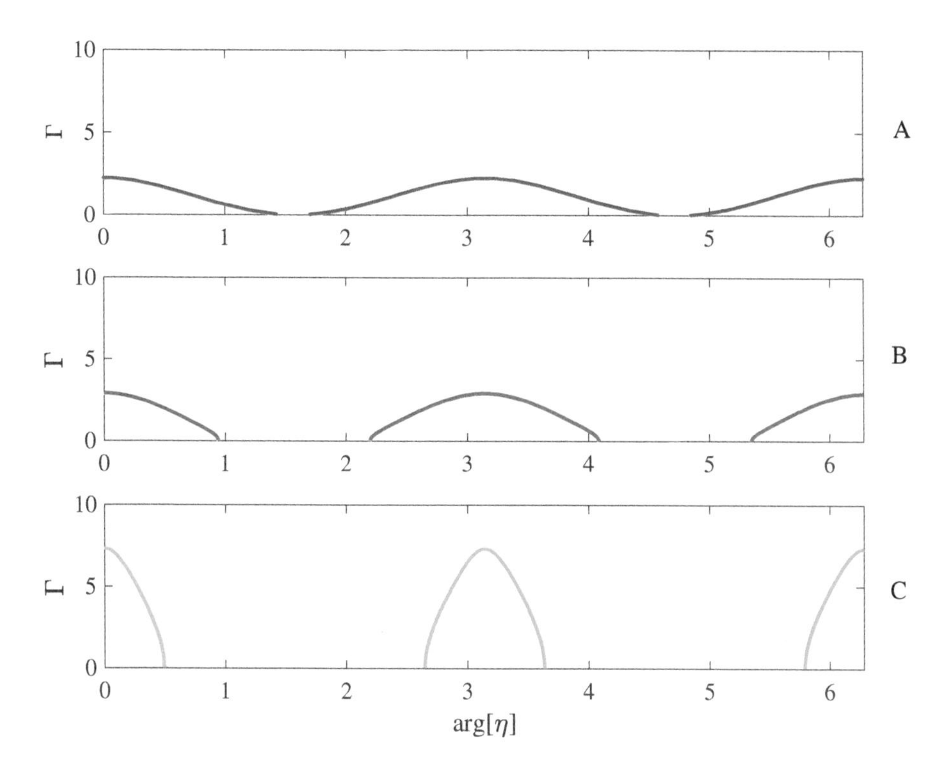

Figure 27.3. *Surfactant distribution* Γ *as a function of* $\arg[\eta]$ *for* $\beta = 0.1$ *(shifted so that* $\arg[\eta] = 0, \pi$ *are the points where the stagnant caps begin to develop). The graphs correspond to the crosses marked A, B, and C in Figure 27.2. The top graph, corresponding to point A, shows a bubble almost totally covered in surfactant.*

The analysis above was first presented by the author in [6], where more details of the underlying physical problem can be found.

Quasi-steady droplet motion driven by surface tension: In this final section we turn from an isolated bubble in an unbounded region of fluid to a finite region, or "blob," of fluid in the plane with surface tension active on its boundaries but with no surfactants present. The surface tension is therefore uniform over the boundary. In this case, however, it is not assumed that the blobs are in equilibrium. Rather, the shape of the blob of fluid will change in time, driven by the surface tension effects on the boundary. Surface tension will act to smear out any regions of high curvature in the boundaries of the fluid blob and make them smoother. Engineers call this physical process of mass transfer driven by surface energy effects *sintering*.

Motivated by the considerations of Chapter 12, and in view of the analysis of the quite different free boundary problem given in Chapter 26, it is useful in this unsteady problem to study the evolution of the Cauchy transform of the fluid domain.

From Chapter 12 we know that the evolution of the Cauchy transform $C_0(z,t)$ for $z \notin D$ is given by

$$\frac{\partial}{\partial t} C_0(z,t) = -\frac{1}{\pi} \int_{\partial D(t)} \frac{\mathrm{Im}[z_t' d\overline{z'}]}{z' - z}, \tag{27.53}$$

where, as in Chapter 12, we adopt the shorthand notation z_t to denote the time derivative of boundary points. Therefore to compute its evolution we need to find the quantity

$$\mathrm{Im}[z_t d\overline{z}] \tag{27.54}$$

on the boundaries of the fluid domain. This is determined from the boundary conditions applied there.

To exemplify the mathematical approach we focus first on the evolution of a simply connected region of fluid. The extension later to a doubly connected region will be a natural extension of it.

If there is a single blob of fluid, then, without loss of generality, we can pick $A_0 = 0$ on this boundary so that the boundary condition (27.14) becomes

$$H(z,\overline{z},t) = f(z,t) + z\overline{f'(z,t)} + \overline{g'(z,t)} = -\frac{\mathrm{i}T_0}{2\mu}\frac{dz}{ds}. \tag{27.55}$$

The kinematic condition (27.8) becomes

$$\mathrm{Re}\left[z_t\left(\mathrm{i}\frac{d\overline{z}}{ds}\right)\right] = \mathrm{Re}\left[(u+\mathrm{i}v)\left(\mathrm{i}\frac{d\overline{z}}{ds}\right)\right]. \tag{27.56}$$

On the free surface, we have

$$u + \mathrm{i}v = -2f(z,t) - \frac{\mathrm{i}T_0}{2\mu}\frac{dz}{ds}. \tag{27.57}$$

Hence (27.56) is

$$\mathrm{Re}\left[z_t\left(\mathrm{i}\frac{d\overline{z}}{ds}\right)\right] = \mathrm{Re}\left[\left(-2f(z,t) - \frac{\mathrm{i}T_0}{2\mu}\frac{dz}{ds}\right)\left(\mathrm{i}\frac{d\overline{z}}{ds}\right)\right] \tag{27.58}$$

or

$$\mathrm{Im}[z_t d\overline{z}] = -\mathrm{Im}[2f(z,t)d\overline{z}] - \frac{T_0}{2\mu}ds. \tag{27.59}$$

On substitution of this quantity into (27.53) we find

$$\frac{\partial}{\partial t}C_0(z,t) = \frac{1}{\pi}\int_{\partial D(t)} \frac{\mathrm{Im}[2f(z',t)d\overline{z'}] + T_0/(2\mu)ds'}{z'-z} \tag{27.60}$$

$$= \frac{1}{2\pi\mathrm{i}}\int_{\partial D(t)} \frac{2f(z',t)d\overline{z'} - 2\overline{f(z',t)}dz' + (\mathrm{i}T_0/\mu)ds'}{z'-z}. \tag{27.61}$$

Use of integration by parts on the first term on the right-hand side leads to

$$\frac{\partial}{\partial t}C_0(z,t) = \frac{1}{2\pi\mathrm{i}}\int_{\partial D(t)} \frac{-2f'(z',t)\overline{z'}dz' - 2\overline{f(z',t)}dz' + (\mathrm{i}T_0/\mu)ds'}{z'-z}$$
$$+\frac{1}{2\pi\mathrm{i}}\int_{\partial D(t)} \frac{2f'(z',t)\overline{z'}dz'}{(z'-z)^2}. \tag{27.62}$$

The complex conjugate form of (27.55) is

$$\overline{f(z,t)} + \overline{z}f'(z,t) + g'(z,t) = \frac{\mathrm{i}T_0}{2\mu}\frac{d\overline{z}}{ds} \tag{27.63}$$

or, in differential form,

$$2\overline{f(z,t)}dz + 2\bar{z}df(z,t) + 2dg(z,t) = \frac{\mathrm{i}T_0}{\mu}ds. \tag{27.64}$$

This can be used in the first line of (27.62) to give

$$\frac{\partial}{\partial t}C_0(z,t) = \frac{1}{2\pi\mathrm{i}}\int_{\partial D(t)} \frac{2dg(z',t)}{z'-z} + \frac{1}{2\pi\mathrm{i}}\int_{\partial D(t)} \frac{2f'(z',t)\overline{z'}dz'}{(z'-z)^2}. \tag{27.65}$$

The first integral on the right-hand side is zero by Cauchy's theorem since $g(z,t)$, and hence $g'(z,t)$, are analytic in $D(t)$ and $z \notin D(t)$. We arrive at

$$\frac{\partial}{\partial t}C_0(z,t) = \frac{\partial}{\partial z}\frac{1}{2\pi\mathrm{i}}\int_{\partial D(t)} \frac{2f'(z',t)\overline{z'}dz'}{(z'-z)}. \tag{27.66}$$

This can be written as

$$\frac{\partial C_0}{\partial t} + \frac{\partial I_0}{\partial z} = 0, \tag{27.67}$$

where

$$I_0(z,t) = \frac{1}{\pi}\int\int_{D(t)} \frac{-2f(z',t)}{z'-z}dA_{z'}. \tag{27.68}$$

This partial differential equation for the evolution of the Cauchy transform is of exactly the special kind considered in §12.7 of Chapter 12 and where we make the association

$$\sigma(z,t) = -2f(z,t) \tag{27.69}$$

in the class of free boundary problems studied there [32]. It was shown that, for this special class of problems, an initially rational Cauchy transform having the form

$$C_0(z,0) = \sum_{k=1}^{N} \frac{R_n(0)}{z - z_n(0)}, \tag{27.70}$$

say, will remain rational under evolution and take the form

$$C_0(z,t) = \sum_{k=1}^{N} \frac{R_n(t)}{z - z_n(t)} \tag{27.71}$$

with

$$R_n(t) = R_n(0), \qquad \frac{dz_n}{dt} = -2f(z_n(t),t), \qquad n = 1,\ldots,N. \tag{27.72}$$

Given that $f(z,t)$ can be determined at each instant from the current configuration $D(t)$, then (27.72) provides a set of $2N$ ordinary differential equations for the evolution of the Cauchy transform $C_0(z,t)$. If $C_0(z,t)$ is a known rational function, then the (simply connected) fluid domain at each instant can, in principle, be reconstructed from it.

As it turns out, it is not necessary to find $f(z,t)$ at each instant. To see why this can be avoided we introduce a time-dependent conformal map to describe the shape of the evolving interface. Now from (27.59) we can write

$$\mathrm{Im}\left[\frac{z_t + 2f(z,t)}{dz/ds}\right] = -\frac{T_0}{2\mu}, \tag{27.73}$$

where we have divided by ds and used $(dz/ds)(d\bar{z}/ds) = 1$. Let

$$z = Z(\zeta, t) \tag{27.74}$$

be a conformal mapping from the unit disc in a parametric ζ plane to $D(t)$. Now it is easy to show, using the chain rule, that

$$\frac{dz}{ds} = \frac{\mathrm{i}\zeta Z'(\zeta, t)}{|Z'(\zeta, t)|}, \tag{27.75}$$

where we use the notation

$$Z'(\zeta, t) = \frac{\partial Z}{\partial \zeta}(\zeta, t). \tag{27.76}$$

On substitution into (27.73) it transforms to

$$\mathrm{Re}\left[\frac{\partial Z/\partial t + 2f(z, t)}{\zeta Z'(\zeta, t)}\right] = \frac{T_0}{2\mu|Z'|}. \tag{27.77}$$

Provided we pick the constant of integration in $f(z, t)$ so that

$$f(0, t) = 0 \qquad \forall t \geq 0, \tag{27.78}$$

which can always be done using an additive degree of freedom in the definition of $f(z, t)$ at each instant, the function in square brackets on the left-hand side is analytic in $|\zeta| < 1$ and the Poisson integral formula of Chapter 13 can be used to deduce

$$\frac{\partial Z}{\partial t} + 2f(z, t) = \zeta Z'(\zeta, t) I(\zeta, t), \tag{27.79}$$

where

$$I(\zeta, t) = \frac{1}{2\pi\mathrm{i}} \int_{|\zeta'|=1} \frac{\zeta' + \zeta}{\zeta' - \zeta} \frac{T_0}{2\mu|Z'(\zeta', t)|} \frac{d\zeta'}{\zeta'}, \tag{27.80}$$

where a purely imaginary constant has been set equal to zero to preserve symmetry.

As an example consider the case of two sintering viscous blobs of fluid with Cauchy transform

$$C(z, t) = \frac{R_1}{z - z_a} + \frac{R_1}{z + z_a}. \tag{27.81}$$

The boundary of the corresponding fluid domain can be described by the rational function conformal map

$$z = Z(\zeta, t) = \frac{R(t)\zeta}{\zeta^2 - a(t)^2}. \tag{27.82}$$

The time evolution, or sintering, of two equal near-circular touching blobs of fluid is shown in Figure 27.4. This solution was first found, using more direct methods, by Hopper [91]. It was later reappraised in terms of Cauchy transforms in [32].

Now we turn to the analogous evolution problem for a doubly connected region of fluid. We will study the case of rotationally symmetric domains such as those resulting from the surface tension–driven coalescence of a rotationally symmetric array of near-circular touching blobs of fluid. In this case we can argue [28, 112] that, at each instant in the evolution, the rotational symmetry is preserved and hence that

$$A_0(t) = A_1(t) = 0. \tag{27.83}$$

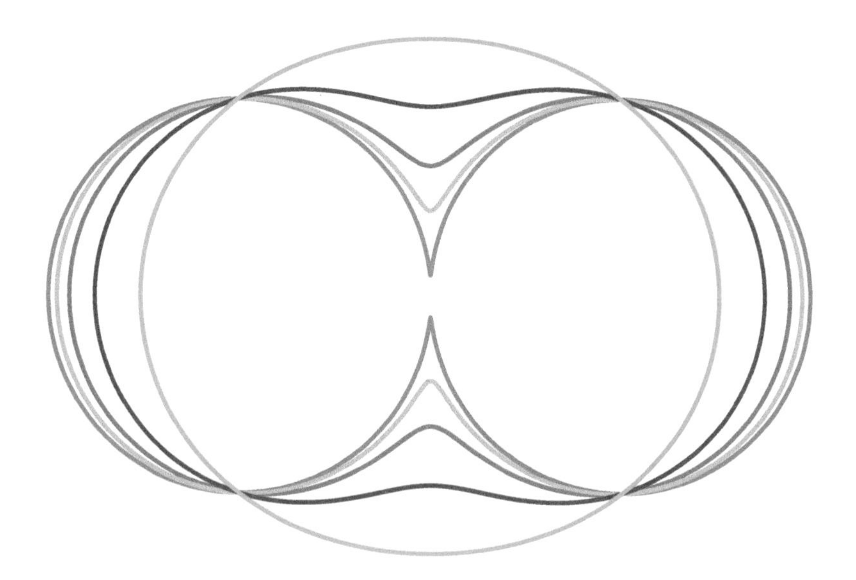

Figure 27.4. *Viscous sintering of two equal near-circular touching blobs of fluid evolving under surface tension. The domain boundaries at different times are superposed. Each domain is a quadrature domain.*

In this case, since the boundary conditions on both inner and outer boundaries of the annular region of the fluid are identical, the same manipulations used to derive the partial differential equation (27.67) carry over. This is an important observation since it shows that a fluid domain that is initially a doubly connected quadrature domain will remain a quadrature domain under evolution.[58]

We assume the Cauchy transform of an initial domain with $N \geq 3$ equal-sized, near-circular blobs of fluid in an annular array has the form

$$C(z,0) = \sum_{k=1}^{N} \frac{R}{z - z_a \Omega_N^{k-1}}, \tag{27.84}$$

where Ω_N is an Nth root of unity. The functional form of $C(z,t)$ under evolution will be

$$C(z,t) = \sum_{k=1}^{N} \frac{R}{z - z_a(t) \Omega_N^{k-1}}, \tag{27.85}$$

where

$$\frac{dz_a}{dt} = -2f(z_a, t). \tag{27.86}$$

To actually compute the shapes of the evolving interfaces we introduce a time-dependent conformal map

$$z = Z(\zeta, t) \tag{27.87}$$

from a time-evolving annulus

$$\rho(t) < |\zeta| < 1 \tag{27.88}$$

[58]Condition (27.83) is a requirement needed to guarantee the preservation of the time-evolving time quadrature domains. It holds for the rotationally symmetric domains considered here. It is an open question whether this condition can be relaxed and generalized classes of solutions found.

to the fluid region. We know from the considerations of Chapter 11 that the functional form of such maps can be written in terms of the prime function $\omega(.,.)$ of the annulus as

$$Z(\zeta,t) = A(t)\zeta \prod_{k=1}^{N} \frac{\omega(\zeta, \Omega_N^{k-1} a(t)/\rho^{2/N})}{\omega(\zeta, \Omega_N^{k-1} a(t))}. \tag{27.89}$$

This function depends on just three real parameters, $\rho(t)$, $a(t)$, and $A(t)$, whose evolution in time must be determined.

Remark: It should be noted that, since $\rho(t)$ is evolving in time, so is the preimage annulus $\rho(t) < |\zeta| < 1$ and hence there is time dependence in the prime function $\omega(.,.)$, too, even though this is not evident from our notation.

As for the simply connected fluid region just discussed, the need to calculate the Goursat function $f(z,t)$ in (27.86) at each instant can be avoided by making use of the generalized Poisson kernel relevant to this situation. We introduce a time-dependent conformal map

$$z = Z(\zeta,t) \tag{27.90}$$

from a time-evolving annulus $\rho(t) < |\zeta| < 1$ to the fluid region. From (27.59), on each boundary, we can write

$$\mathrm{Im}\left[\frac{z_t + 2f(z,t)}{dz/ds}\right] = -\frac{T_0}{2\mu}, \tag{27.91}$$

where we have divided by ds and used $(dz/ds)(d\bar{z}/ds) = 1$. On use of the chain rule, it can be shown that

$$\frac{dz}{ds} = \begin{cases} \dfrac{\mathrm{i}\zeta Z'(\zeta,t)}{|Z'(\zeta,t)|}, & \zeta \in C_0, \\ -\dfrac{\mathrm{i}\zeta Z'(\zeta,t)}{\rho|Z'(\zeta,t)|}, & \zeta \in C_1. \end{cases} \tag{27.92}$$

Also, on C_1, whose radius is evolving in time,

$$z_t = \frac{\partial Z}{\partial t} + \frac{d\zeta}{dt}\frac{\partial Z}{\partial \zeta}(\zeta,t) = \frac{\partial Z}{\partial t} + \frac{1}{\rho}\frac{d\rho}{dt}\zeta Z'(\zeta,t), \tag{27.93}$$

where we have used the fact that $\zeta = \rho(t)e^{\mathrm{i}\arg[\zeta]}$ on C_1. On substitution into (27.73) it transforms to

$$\mathrm{Re}\left[\frac{\partial Z/\partial t + 2f(z,t)}{\zeta Z'(\zeta,t)}\right] = \begin{cases} \dfrac{T_0}{2\mu|Z'|}, & \zeta \in C_0, \\ -\dfrac{T_0}{2\mu\rho|Z'|} - \dfrac{1}{\rho}\dfrac{d\rho}{dt}, & \zeta \in C_1. \end{cases} \tag{27.94}$$

Owing to the fact that the surface tension forces around the enclosed hole sum to zero, meaning that there is no net force on the enclosed air pocket, it follows that the period of the function $f(z,t)$ around the air pocket must vanish. This physical constraint means that the function in square brackets on the left-hand side is analytic and single-valued in the annulus $\rho < |\zeta| < 1$. Now the integral representation for the solution of the modified Schwarz problem in the annulus can be used to deduce

$$\frac{\partial Z}{\partial t} + 2f(z,t) = \zeta Z'(\zeta,t) I(\zeta,t), \tag{27.95}$$

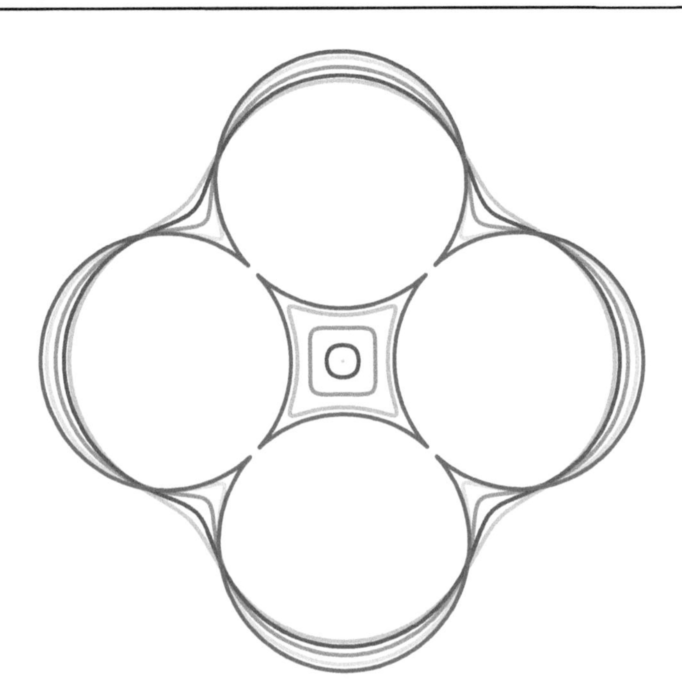

Figure 27.5. *Viscous sintering of four equal near-circular touching blobs of fluid, in a doubly connected configuration, evolving under surface tension to a single circular blob. The blue curves are the initial boundaries of the doubly connected fluid region. Its subsequent shapes at four later times are shown in different colors. The calculation is terminated at the point of disappearance of the enclosed air hole. Further details concerning this calculation can be found in [28].*

where

$$I(\zeta,t) = \frac{1}{2\pi \mathrm{i}}\int_{C_0} \frac{T_0}{2\mu|Z'(\zeta',t)|}\left[2d\log\omega(\zeta',\zeta) - d\log\zeta'\right] \tag{27.96}$$

$$-\frac{1}{2\pi \mathrm{i}}\int_{C_1}\left[-\frac{T_0}{2\mu\rho|Z'(\zeta',t)|} - \frac{1}{\rho}\frac{d\rho}{dt}\right]\left[2d\log\omega(\zeta',\zeta)\right], \tag{27.97}$$

where $\omega(.,.)$ is the prime function of the annulus and where a purely imaginary constant has been set equal to zero to preserve symmetry. For the concentric annulus, as clarified in §13.7 of Chapter 13, this result is equivalent to the classical Villat formula. In order for the solution to the modified Schwarz problem to exist, the following compatibility condition must be satisfied by the data on the right-hand side of (27.94):

$$\frac{1}{2\pi \mathrm{i}}\int_{C_0} \frac{T_0}{2\mu|Z'(\zeta,t)|}\frac{d\zeta}{\zeta} = \frac{1}{2\pi \mathrm{i}}\int_{C_1}\left[-\frac{T_0}{2\mu\rho|Z'(\zeta,t)|} - \frac{1}{\rho}\frac{d\rho}{dt}\right]\frac{d\zeta}{\zeta}. \tag{27.98}$$

From this we deduce that

$$\frac{1}{\rho}\frac{d\rho}{dt} = -\frac{1}{2\pi \mathrm{i}}\int_{C_0} \frac{T_0}{2\mu|Z'(\zeta,t)|}\frac{d\zeta}{\zeta} - \frac{1}{2\pi \mathrm{i}}\int_{C_1}\left[\frac{T_0}{2\mu\rho|Z'(\zeta,t)|}\right]\frac{d\zeta}{\zeta}, \tag{27.99}$$

which provides the evolution equation for $\rho(t)$. From the relations (27.86) and (27.95) and the fact that

$$z_a = Z(1/a,t), \tag{27.100}$$

we deduce the following ordinary differential equation for $a(t)$:

$$\frac{da}{dt} = aI(1/a, t). \tag{27.101}$$

Finally, the third equation needed to find the evolution of $A(t)$ follows from the fact that the (identical) residues of the simple poles of the Cauchy transform (27.85) are constant in time.

A typical calculation of the sintering of four near-circular blobs of viscous fluid—the case $N = 4$—under the effects of surface tension is shown in Figure 27.5. The calculation is terminated at the critical time where the interior air hole vanishes and the doubly connected fluid domain becomes simply connected. In principle, the calculation can be continued beyond this time using rational function conformal maps to describe simply connected time-evolving quadrature domains.

The theory just presented for studying fluid regions evolving (sintering) under the effects of surface tension is a summary of the approach based on use of the prime function originally developed in [58, 28, 32, 56], where additional discussion and theoretical background can be found.[59]

These examples have demonstrated the remarkable fact that the problem of unsteady two-dimensional viscous sintering dynamically preserves certain initial planar quadrature domains. The review article [27] surveys how quadrature domains arise in various guises throughout the field of fluid dynamics.

[59] A different approach to the same sintering problem of doubly connected domains using standard elliptic function theory is presented in [112].

Chapter 28
Epilogue

But now I leave my cetological System standing thus unfinished, even as the great Cathedral of Cologne was left, with the crane still standing upon the top of the uncompleted tower. For small erections may be finished by their first architects; grand ones, true ones, ever leave the copestone to posterity. God keep me from ever completing anything. This whole book is but a draught—nay, but the draft of a draft.

Moby Dick
Herman Melville

We end this book by summarizing what has been done and outlining all the many things that remain to be done.

A framework for solving a broad class of problems in multiply connected domains has been set out, one which naturally extends concepts that are familiar in the simply connected case. The prime function has been the theoretical linchpin. An alternative title for the book might have been *The Prime Function in Multiply Connected Domains*. But this was rejected because so few applied scientists are yet familiar with the concept of the prime function. Perhaps this book will help change this status quo.

Starting with multiply connected circular domains—an important canonical class—and the first-type Green's function associated with them, Part I developed the theory of the prime function associated with those domains. Using the prime function, a number of fundamental results were established. Among them were formulas for basic canonical slit mappings; formulas for multiply connected S–C mappings, which are the natural extension of slit domains to the broader class of polygonal domains; the form of the ordinary differential equations satisfied by conformal mappings to multiply connected polycircular arc domains, which are the next natural domain class beyond polygons; formulas for loxodromic and automorphic functions, including those parametrizing the boundaries of multiply connected quadrature domains; and expressions for the Bergman kernel and modified Bergman kernel functions and expressions for the analogues of the Poisson kernel for multiply connected circular domains. Many other ideas have been treated in passing, mostly in the exercises: the connections of our slit maps to Jacobi elliptic functions, the link between our function theory in an annulus to the Weierstrass elliptic functions, how the *method of images* familiar to applied scientists fits in, links between the Poisson and Villat kernels using circular slit maps, and how the prime function connects with the theta functions of Jacobi and Riemann, to name just a few. Novel strategies for the numerical computation of the prime function have been described and open source codes advertised.

Part II surveyed a variety of problems in applications that can be tackled using the framework set out in Part I. Used in combination these results form a formidable toolbox for constructive solution of a variety of applied problems. On a practical level, all that is needed is a small set of codes to compute the prime function in a given multiply connected circular domain, and all the rest follows from the formulas developed here.

With the main tenets of our vision now laid out, it is appropriate to point out some subsidiary mathematical ideas and to touch on a few other application areas not discussed in the main text. It was a conscious and deliberate decision, in planning the writing of this book, to suppress any mention of the following ideas in the main text. While these more technical settings of the ideas are familiar and attractive to many mathematicians, they are alien to many applied scientists.

Part I could have been written instead in the language of compact Riemann surface theory: the circles $\{C_j | j = 1, \ldots, M\}$ and their Möbius-identified reflections $\{C'_j | j = 1, \ldots, M\}$ can be viewed as a canonical homology basis of a cycles, with the barriers $\{(L_j, L_j^+) | j = 1, \ldots, M\}$ introduced in Chapter 5 as a basis of b cycles. The role of the a and b cycles can also be reversed, swapped, or interchanged, as discussed in [44]. Indeed we showed in the main text that this feature can be used to some advantage in applications (see Exercise 6.13 on swapping the correspondences and Chapter 17, for example).

The underlying compact Riemann surface we have modeled here within a Schottky model is the Schottky double of the planar domain [86, 87]. The Schottky model uses the complex plane quotiented by the action of a Schottky group to keep things a function of a single variable, rather than moving to a multidimensional Jacobi variety [80, 79]. The Schottky doubles relevant here are examples of so-called M-curves, and we have similarly avoided mentioning any connections with algebraic curves, or indeed algebraic geometry (except for Exercise 11.6, where we pointed out that the boundaries of quadrature domains are algebraic curves). The methods here can, of course, be used for constructive algebraic geometry [42]. The set of functions $\{\sigma_j(z) | j = 1, \ldots, M\}$ is sometimes called the set of harmonic measures, and their analytic extensions, here denoted by $\{\hat{v}_j(z) | j = 1, \ldots, M\}$, are a set of first-kind Abelian integrals on the Schottky double [116, 75]; the differential $d\Pi^{z,w}(a)$ that was crucial in the construction of the prime function is a third-kind Abelian differential on the surface with simple poles at z and w [64, 80, 79]. Many other objects we have introduced can be given alternative designations and interpretations. By being concrete with our model from the start we have avoided, as far as possible, the language of differential geometry and differential calculus on manifolds.

With its eponymous focus on multiply connected planar domains, the text has presented a distillation of the relevant theory of compact Riemann surfaces, and algebraic geometry, needed to constructively solve problems in multiply connected domains and in a way that we hope will appeal to applied scientists. The underlying Riemann surfaces relevant for us have been special, so many results for more general surfaces are simply not needed. At the same time, these surfaces have special symmetries—the M-curve symmetry just mentioned—not enjoyed by more general surfaces, and, as such, the associated function theory inherits useful features not usually appearing in more general treatments. For example, the identity

$$\overline{\omega}(1/z, 1/a) = -\frac{1}{za}\omega(z, a)$$

satisfied by all the prime functions featured here has been critical to the theoretical development, yet it is not a property of prime functions on more general surfaces.

There are many other areas not explicitly mentioned here where the framework we have developed, and extensions of it, are likely to be useful.

Perhaps most important of all: many of the problems considered here can be rephrased as Riemann–Hilbert problems [61, 102, 99], although we have deliberately avoided viewing them as such. The Riemann–Hilbert approach, which has enjoyed a renaissance in recent decades, represents a huge area of endeavor, which we expect to be a fertile area for deploying the techniques discussed herein. The author has elsewhere [1, 16] pointed out the possibility of using the prime function framework in the solution of certain classes of Riemann–Hilbert problems, but much more is possible in this direction and remains to be developed. This is the topic of a monograph all to itself!

The mathematical framework is also likely to find use in the closely related theory of isomonodromic deformations of Fuchsian differential equations; indeed, some relevant mathematical ideas appear in Chapter 10 in the context of constructing multiply connected polycircular arc mappings. The classical uniformization theorem has found new relevance to modern ideas in quantum field theory, e.g., Liouville field theory. Although it has not been written from this point of view, in its focus on constructive conformal mapping this book provides a foundation for constructive uniformization challenges in this and many other areas.

In recent decades, the methods of algebraic geometry have been useful in finite-gap integration of integrable systems (a topic intertwined with the Riemann–Hilbert method mentioned above). Historically, it has been most common to use theta functions in this regard [67], although the benefits of the Schottky model over this approach have been pointed out at least once before [69], especially when one is interested in calculating solutions or plotting waveforms, that is, in constructive methods that "effectivize" the theory and which have been the primary focus of this monograph. Even so, use of the prime function—as opposed to use of *Poincaré series*, for example—within the Schottky model for finite-gap integration of integrable systems has not really been explored and deserves further attention. The author [26, 24] has touched on these possibilities by pointing out, for example, how to use the prime function formalism to conveniently parametrize finite-gap solutions of the Benney hierarchy.

An area in which the prime function has been used, to a certain extent, is the theory of extremal functions defined over multiple intervals [95], and much more would seem possible here. Other related areas where the theory based on the prime function is likely to find further application is in the theory of orthogonal polynomials, random matrix theory, statistical physics, and spectral theory [4].

Far from being, in the words of Ishmael, the "copestone" on this mathematical edifice, this monograph, despite its flaws and imperfections, will perhaps provide a foundation that will inspire readers to reinforce the structure, and build it up further. *Floreat domus.*

Bibliography

[1] Y. A. ANTIPOV & D. G. CROWDY, Riemann-Hilbert problem for automorphic functions and the Schottky-Klein prime function, *Complex Anal. Oper. Theory*, **1**(3), 317–334 (2007). (Cited on p. 423)

[2] P. J. BADDOO & D. G. CROWDY, Periodic Schwarz-Christoffel mappings with multiple boundaries per period, *Proc. Roy. Soc. A*, **475**, 20190225 (2019). (Cited on p. 141)

[3] D. G. CROWDY, Perturbation analysis of subphase gas and meniscus curvature effects for longitudinal flows over superhydrophobic surfaces, *J. Fluid Mech.*, **822**, 307–326 (2017). (Cited on p. 351)

[4] D. G. CROWDY, Finite-gap Jacobi matrices and the Schottky–Klein prime function, *Comput. Methods Funct. Theory*, 1–23 (2016). (Cited on p. 423)

[5] D. G. CROWDY, Stress fields around two pores in an elastic body: Exact quadrature domain solutions, *Proc. Roy. Soc. A*, **471**, 20150240 (2015). (Cited on pp. 213, 364, 366, 371)

[6] D. G. CROWDY, Surfactant-induced stagnant zones in the Jeong-Moffatt free surface Stokes flow problem, *Phys. Fluids.*, **25**, 092104 (2013). (Cited on pp. 407, 413)

[7] D. G. CROWDY, Analytical formulae for source and sink flows in multiply connected domains, *Theoret. Comput. Fluid Dyn.*, **27**, 1–19 (2013). (Cited on p. 302)

[8] D. G. CROWDY, Exact solutions for cylindrical "slip-stick" Janus swimmers in Stokes flow, *J. Fluid Mech.*, **719**, R2 (2013). (Cited on p. 258)

[9] D. G. CROWDY, Wall effects on self-diffusiophoretic Janus particles: A theoretical study, *J. Fluid Mech.*, **735**, 473–498 (2013). (Cited on pp. 258, 259)

[10] D. G. CROWDY, Conformal slit maps in applied mathematics, *ANZIAM J.*, **53**(3), 171–189 (2012). (Cited on pp. xvi, 351)

[11] D. G. CROWDY, Frictional slip lengths and blockage coefficients, *Phys. Fluids*, **23**, 091703 (2011). (Cited on pp. 308, 349)

[12] D. G. CROWDY, Frictional slip lengths for unidirectional superhydrophobic grooved surfaces, *Phys. Fluids*, **23**, 072001 (2011). (Cited on pp. 349, 350, 351)

[13] D. G. CROWDY, The Schottky-Klein prime function on the Schottky double of planar domains, *Comput. Methods Funct. Theory*, **10**(2), 501–517 (2010). (Cited on p. xvi)

[14] D. G. CROWDY, A new calculus for two dimensional vortex dynamics, *Theoret. Comput. Fluid Dyn.*, **24**, 9–24 (2010). (Cited on pp. 289, 298)

[15] D. G. CROWDY, The spreading phase in Lighthill's model of the Weis-Fogh lift mechanism, *J. Fluid Mech.* **641**, 195–204 (2009). (Cited on p. 96)

[16] D. G. CROWDY, Explicit solution of a class of Riemann-Hilbert problems, *Ann. Univ. Paed. Cracov (Studia Math.)*, **8**, 5–18 (2009). (Cited on pp. 156, 423)

[17] D. G. CROWDY, Multiple steady bubbles in a Hele-Shaw cell, *Proc. Roy. Soc. A.*, **465**, 421–435 (2009). (Cited on pp. 391, 393)

[18] D. G. CROWDY, Explicit solution for the potential flow due to an assembly of stirrers in an inviscid fluid, *J. Eng. Math.*, **62**(4), 333–344 (2008). (Cited on p. 303)

[19] D. G. CROWDY, Geometric function theory: A modern view of a classical subject, *Nonlinearity*, **21** (10), T205–T219 (2008). (Cited on p. xvi)

[20] D. G. CROWDY, The Schwarz problem in multiply connected domains and the Schottky-Klein prime function, *Complex Var. Elliptic Equations*, **53**(3), 1–16 (2008). (Cited on pp. 251, 254)

[21] D. G. CROWDY, Schwarz-Christoffel mappings to unbounded multiply connected polygonal regions, *Math. Proc. Cambridge Philos. Soc.*, **142**, 319–339 (2007). (Cited on pp. 127, 140, 153)

[22] D. G. CROWDY, Analytical solutions for uniform potential flow past multiple cylinders, *Eur. J. Mech. B/Fluids*, **25**(4), 459–470 (2006). (Cited on p. 295)

[23] D. G. CROWDY, Calculating the lift on a finite stack of cylindrical aerofoils, *Proc. Roy. Soc. A*, **462**, 1387–1407 (2006). (Cited on p. 302)

[24] D. G. CROWDY, Genus-N algebraic reductions of the Benney hierarchy within a Schottky model, *J. Phys. A: Math. Gen.*, **38**, 10917–10934 (2005). (Cited on p. 423)

[25] D. G. CROWDY, The Schwarz-Christoffel mapping to bounded multiply connected polygonal domains, *Proc. Roy. Soc. A*, **461**, 2653–2678 (2005) (Cited on pp. 127, 135, 136, 139, 140)

[26] D. G. CROWDY, The Benney hierarchy and the Dirichlet boundary problem in two dimensions, *Phys. Lett. A*, **319–329** (2005). (Cited on p. 423)

[27] D. G. CROWDY, *Quadrature domains and fluid dynamics: Quadrature domains and applications, a Harold Shapiro Anniversary volume* (Eds. Ebenfelt, Gustafsson, Khavinson, Putinar), Birkhäuser (2005). (Cited on pp. 213, 420)

[28] D. G. CROWDY, Viscous sintering of unimodal and bimodal cylindrical packings with shrinking pores, *Eur. J. Appl. Math.*, **14**, 421–445 (2003). (Cited on pp. 228, 416, 419, 420)

[29] D. G. CROWDY, Exact solutions for rotating vortex arrays with finite-area cores, *J. Fluid Mech.*, **469**, 209–235 (2002). (Cited on pp. 230, 382)

[30] D. G. CROWDY, On the construction of exact multipolar equilibria of the 2D Euler equations *Phys. Fluids*, **14**(1), 257–267 (2002). (Cited on pp. 230, 378, 379, 382)

[31] D. G. CROWDY, Theory of exact solutions for the evolution of a fluid annulus in a rotating Hele-Shaw cell, *Quart. Appl. Math.*, **60**(1), 11–36 (2002). (Cited on pp. 228, 229, 248, 404)

[32] D. G. CROWDY, On a class of geometry-driven free boundary problems, *SIAM J. Appl. Math.*, **62**(2), 945–954 (2002). (Cited on pp. 415, 416, 420)

[33] D. G. CROWDY, Multipolar vortices and algebraic curves, *Proc. Roy. Soc. A*, **457**, 2337–2359 (2001). (Cited on pp. 230, 231, 382)

[34] D. G. CROWDY, A class of exact multipolar vortices *Phys. Fluids*, **11**(9), 2556–2564 (1999). (Cited on pp. 230, 376)

[35] D. G. CROWDY & M. CLOKE, Stability analysis of a class of two-dimensional multipolar vortex equilibria, *Phys. Fluids*, **14**(6), 1862–1876 (2002). (Cited on pp. 378, 382)

[36] D. G. CROWDY & A. S. FOKAS, Conformal mappings to a doubly connected polycircular arc domain, *Proc. Roy. Soc. A*, **463**, 1885–1907 (2007). (Cited on pp. 168, 196)

[37] D. G. CROWDY, A. S. FOKAS, & C. C. GREEN, Conformal mappings to multiply connected polycircular arc domains, *Comput. Methods Funct. Theory*, **11**(2), 685–706 (2011). (Cited on pp. 193, 196)

[38] D. G. CROWDY & C. C. GREEN, Analytical solutions for von Kármán streets of hollow vortices, *Phys. Fluids*, **23**, 126602 (2011). (Cited on pp. 383, 384, 389)

[39] D. G. CROWDY & V. S. KRISHNAMURTHY, The effect of core size on the speed of compressible hollow vortex streets, *J. Fluid Mech.*, **836**, 797–827 (2017). (Cited on pp. 170, 383, 384, 389, 390)

[40] D. G. CROWDY, E. H. KROPF, C. C. GREEN, & M. M. S. NASSER, The Schottky–Klein prime function: A theoretical and computational tool for applications, *IMA J. Appl. Math.*, **81**(3), 589–628 (2016). (Cited on pp. 266, 275, 276, 277)

[41] D. G. CROWDY, S. LLEWELLYN SMITH, & D. FREILICH, Translating hollow vortex pairs, *Eur. J. Mech. B/Fluids*, **37**, 180–186 (2012). (Cited on pp. 383, 390)

[42] D. G. CROWDY & J. S. MARSHALL, Uniformizing real hyperelliptic M-curves using the Schottky-Klein prime function, in *Computational approaches to Riemann surfaces*, Lecture Notes in Mathematics, Springer (2013). (Cited on p. 422)

[43] D. G. CROWDY & J. S. MARSHALL, Multiply connected quadrature domains and the Bergman kernel function, *Complex Anal. Oper. Theory*, **3**(2), 379–397 (2009). (Cited on pp. 226, 228)

[44] D. G. CROWDY & J. S. MARSHALL Uniformizing the boundaries of multiply connected quadrature domains using Fuchsian groups, *Physica D*, **235**, 82–89 (2007). (Cited on pp. 322, 422)

[45] D. G. CROWDY & J. S. MARSHALL, Computing the Schottky-Klein prime function on the Schottky double of planar domains, *Comput. Methods Funct. Theory*, **7**(1), 293–308 (2007). (Cited on pp. 266, 268, 275, 276)

[46] D. G. CROWDY & J. S. MARSHALL, Green's functions for Laplace's equation in multiply connected domains, *IMA J. Appl. Math.*, **72**, 278–301 (2007). (Cited on p. 66)

[47] D. G. CROWDY & J. S. MARSHALL, Conformal mappings between canonical multiply connected domains, *Comput. Methods Funct. Theory*, **6**(1), 59–76 (2006). (Cited on p. 120)

[48] D. G. CROWDY & J. S. MARSHALL, Analytical formulae for the Kirchhoff-Routh path function in multiply connected domains, *Proc. Roy. Soc. A*, **461**, 2477–2501 (2005). (Cited on p. 309)

[49] D. G. CROWDY & J. S. MARSHALL, The motion of a point vortex around multiple circular islands, *Phys. Fluids*, **17**, 056602 (2005). (Cited on pp. 309, 313)

[50] D. G. CROWDY & J. S. MARSHALL, The motion of a point vortex through gaps in walls, *J. Fluid Mech.*, **541**, 231–261 (2005). (Cited on pp. 95, 309, 313)

[51] D. G. CROWDY & J. S. MARSHALL, Analytical solutions for rotating vortex arrays involving multiple vortex patches, *J. Fluid Mech.*, **523**, 307–338 (2005). (Cited on pp. 230, 382)

[52] D. G. CROWDY & J. S. MARSHALL, Growing vortex patches, *Phys. Fluids*, **16**, 3122–3129 (2004). (Cited on pp. 230, 380, 382)

[53] D. G. CROWDY & J. S. MARSHALL, Constructing multiply connected quadrature domains, *SIAM J. Appl. Math.*, **64**, 1334–1359 (2004). (Cited on pp. 217, 222)

[54] D. G. CROWDY & J. ROENBY, Hollow vortices, capillary water waves and double quadrature domains, *Fluid Dyn. Res.*, **46**, 031424 (2014). (Cited on p. 233)

[55] D. G. CROWDY & O. SAMSON, Stokes flow past gaps in a wall, *Proc. Roy. Soc. A*, **466**, 2727–2746 (2010). (Cited on p. 355)

[56] D. G. CROWDY & M. SIEGEL, Exact solutions for the evolution of a bubble in Stokes flow: A Cauchy transform approach, *SIAM J. Appl. Math.*, **65**, 941–963 (2005). (Cited on p. 420)

[57] D. G CROWDY & S. TANVEER, The effect of finiteness in the Saffman Taylor viscous fingering problem, *J. Stat. Phys.*, **114**, 1501–1536 (2004). (Cited on p. 402)

[58] D. G. CROWDY & S. TANVEER, A theory of exact solutions for annular viscous blobs, *J. Nonlinear Sci.*, **8**(4), 375–400 (1998). (Cited on pp. 228, 420)

[59] M. HODES, T. KIRK, & D. G. CROWDY, Spreading and contact resistance formulae capturing boundary curvature and contact distribution effects, *J. Heat Transfer*, **140**, 104503 (2018). (Cited on p. 351)

Further reading

[60] D. A. ABANIN & L. S. LEVITOV, Conformal invariance and shape-dependent conductance of graphene samples, *Phys. Rev. B*, **78**, 035416 (2008). (Cited on p. 319)

[61] M. ABLOWITZ & A. S. FOKAS, *Complex variables: Introduction and applications*, Cambridge University Press (2011). (Cited on pp. 7, 9, 10, 13, 40, 215, 239, 256, 264, 423)

[62] T. AKAZA, Singular sets of some Kleinian groups, *Nagoya Math J.*, **827**, 127–143 (1966). (Cited on p. 268)

[63] T. AKAZA & K. INOUE, Limit sets of geometrically finite free Kleinian groups, *Tohoku Math. J.*, **36**, 1–16 (1984). (Cited on p. 268)

[64] H. BAKER, *Abelian functions: Abel's theorem and the allied theory of theta functions*, Cambridge University Press (1996). (Cited on pp. 64, 265, 266, 270, 422)

[65] G. K. BATCHELOR, *An introduction to fluid mechanics*, Cambridge University Press (2012). (Cited on pp. 323, 383, 384)

[66] S. R. BELL, *The Cauchy transform, potential theory and conformal mapping*, Chapman and Hall/CRC (2015). (Cited on pp. 225, 237, 251)

[67] E. D. BELOKOLOS, A. I. BOBENKO, V. Z. ENOL'SKII, & A. R. ITS, Algebro-geometric approach to nonlinear integrable equations, *Springer Series in Nonlinear Dynamics*, Springer Verlag (1994). (Cited on p. 423)

[68] D. BETSAKOS, K. SAMUELSSON, & M. VUORINEN, The computation of capacity of planar condensers, *Publications de L'Institut Mathématique*, **89**, 233–252 (2004). (Cited on p. 145)

[69] A. I. BOBENKO, Schottky uniformization and finite-gap integration, *Soviet Math. Dokl.*, **36**(1), 38–42 (1988). (Cited on p. 423)

[70] A. B. BOGATYREV, Prime form and Schottky model, *Comput. Methods Funct. Theory*, **9**, 47–55 (2009). (Cited on p. 64)

[71] U. BOTTAZZINI & J. GRAY, Hidden harmony—geometric fantasies: The rise of complex function theory, *Sources and Studies in the History of Mathematics and Physical Sciences*, Springer (2013). (Cited on p. xv)

[72] W. BURNSIDE, On a class of automorphic functions, *Proc. Lond. Math. Soc.*, **23**, 49–88 (1891). (Cited on p. 64)

[73] W. BURNSIDE, Further note on automorphic functions, *Proc. Lond. Math. Soc.*, **23**, 281–295 (1892). (Cited on p. 64)

[74] G. P. CHEREPANOV, Inverse problems of the plane theory of elasticity, *J. Appl. Math. Mech.*, **38**, 915–931 (1974). (Cited on p. 367)

[75] R. COURANT, *Dirichlet's principle, conformal mapping and minimal surfaces*, Dover Publications (2005). (Cited on pp. 29, 422)

[76] P. J. DAVIS, The Schwarz function and its applications, *Carus Mathematical Monographs*, Mathematical Association of America (1974). (Cited on pp. 17, 40, 376, 408)

[77] T. K. DELILLO, A. R. ELCRAT, & J. A. PFALTZGRAFF, Schwarz-Christoffel mapping of multiply connected domains, *J. Anal. Math.*, **94**, 17–47 (2004). (Cited on p. 140)

[78] T. K. DELILLO, Schwarz-Christoffel mapping of bounded, multiply connected domains, *Comput. Methods Funct. Theory*, **6**, 275–300 (2006). (Cited on p. 140)

[79] H. M. FARKAS & I. KRA, *Riemann surfaces*, Springer Verlag (1980). (Cited on p. 422)

[80] J. FAY, Theta functions on Riemann surfaces, *Lecture Notes in Mathematics*, **352**, Springer Verlag (1973). (Cited on pp. 270, 422)

[81] L. R. FORD, *Automorphic functions*, AMS Chelsea Publishing (2006). (Cited on pp. 160, 177, 179)

[82] G. M. GOLUZIN, *Geometric theory of functions of a complex variable*, American Mathematical Society (1969). (Cited on pp. 27, 28, 127)

[83] J. N. GOODIER, An analogy between the slow motions of a viscous fluid in two dimensions, and systems of plane stress, *The London, Edinburgh, and Dublin Philosophical Magazine and Journal of Science*, Series 7, **17**(113), 554–576 (1934). (Cited on p. 363)

[84] C. C. GREEN & G. VASCONCELOS, Multiple steadily translating bubbles in a Hele-Shaw channel, *Proc. Roy. Soc. A*, **470**, 20130698 (2014). (Cited on p. 393)

[85] M. I. GUREVICH, *The theory of jets in an ideal fluid*, Pergamon Press (1966). (Cited on p. 383)

[86] B. GUSTAFSSON, Quadrature domains and the Schottky double, *Acta. Appl. Math.*, **1**, 209–240 (1983). (Cited on pp. 213, 215, 229, 231, 422)

[87] B. GUSTAFSSON & A. SEBBAR, Critical points of Green's function and geometric function theory, *Ind. Univ. Math J.*, **61**(3), 939–1017 (2012). (Cited on p. 422)

[88] B. GUSTAFSSON & M. PUTINAR, Hyponormal quantization of planar domains, *Lecture Notes in Mathematics*, **2199**, Springer International Publishing (2017). (Cited on p. 237)

[89] H. HASIMOTO, On the flow of a viscous fluid past a thin screen at small Reynolds number, *J. Phys. Soc. Japan*, **13**, 633–639 (1958). (Cited on p. 354)

[90] D. A. HEJHAL, *Theta functions, kernel functions, and Abelian integrals*, American Mathematical Society (1972). (Cited on pp. xv, 225)

[91] R. W. HOPPER, Plane Stokes flow driven by capillarity on a free surface, *J. Fluid Mech.*, **213**, 349–375 (1990). (Cited on p. 416)

[92] M. H. HU & Y. P. CHANG, Optimization of finned tubes for heat transfer in laminar flow, *J. Heat Transfer*, **95**(3), 332–338 (1973). (Cited on p. 329)

[93] S. G. KRANTZ, *Geometric function theory: Explorations in complex analysis*, Birkhäuser (2006). (Cited on p. 4)

[94] C. C. LIN, On the motion of vortices in two dimensions. I. Existence of the Kirchhoff–Routh function, *Proc. Nat. Acad. Sci.*, **27**, 570–575 (1941); On the motion of vortices in two dimensions. II. Some further investigations on the Kirchhoff–Routh function, *Proc. Nat. Acad. Sci.*, **27**, 575–577 (1941). (Cited on pp. 309, 313)

[95] A. LUKASHOV, On Chebyshev-Markov rational functions over several intervals, *J. Approx. Theory*, **95**(3), 333–352 (1998). (Cited on p. 423)

[96] J. S. MARSHALL, On the construction of multiply connected arc integral quadrature domains, *Comput. Methods Funct. Theory*, **14** (1), 107–138 (2014). (Cited on p. 226)

[97] J. S. MARSHALL, Analytical solutions for an escape problem in a disc with an arbitrary distribution of exit holes along its boundary, *J. Stat. Phys.*, **165**, 920–952 (2016). (Cited on p. 351)

[98] J. S. MARSHALL, On sets of multiple equally strong holes in an infinite elastic plate: Parameterization and existence, *SIAM J. Appl. Math.*, **79**, 2288–2312 (2019). (Cited on p. 370)

[99] V. MITYUSHEV & S. A. ROGOSIN, Constructive methods for linear and nonlinear boundary value problems for analytic functions, *Monographs and Surveys in Pure and Applied Mathematics*, Chapman & Hall CRC (1999). (Cited on p. 423)

[100] B. YA. MOIZHES, Averaged electrostatic boundary conditions for metallic meshes, *Zh. Tekh. Fiz.*, **25**, 167–176 (1955). (Cited on p. 351)

[101] D. MUMFORD, C. SERIES, & D. WRIGHT, *Indra's pearls: The vision of Felix Klein*, Cambridge University Press (2015). (Cited on pp. 48, 186)

[102] N. I. MUSKHELISHVILI, *Singular integral equations*, Springer (1958). (Cited on p. 423)

[103] T. NEEDHAM, *Visual complex analysis*, Oxford University Press (1999). (Cited on pp. 6, 13)

[104] Z. NEHARI, *Conformal mapping*, Dover Publications (2003). (Cited on pp. 4, 7)

[105] R. B. NELSON, B. PROTAS, & T. SAKAJO, Linear feedback stabilization of point-vortex equilibria near a Kasper wing, *J. Fluid Mech.*, **46**, 121–154 (2017). (Cited on p. 313)

[106] J. R. PHILIP, Flows satisfying mixed no-slip and no-shear conditions, *J. Appl. Math. Phys. (ZAMP)*, **23**, 353–372 (1972). (Cited on p. 351)

[107] H. C. POCKLINGTON, The configuration of a pair of equal and opposite hollow straight vortices of finite cross-section, moving steadily through fluid, *Proc. Cambridge Philos. Soc.*, **8**, 178–187 (1895). (Cited on pp. 383, 390)

[108] R. W. RENDELL & S. M. GIRVIN, Hall voltage dependence on inversion-layer geometry in the quantum Hall-effect regime, *Phys. Rev. B*, **23**, 6610–6614 (1981). (Cited on p. 319)

[109] S. RICHARDSON, Hele-Shaw flows with time-dependent free boundaries in which the fluid occupies a multiply connected region, *Eur. J. Appl. Math.*, **5**, 97–122 (1994). (Cited on p. 393)

[110] S. RICHARDSON, Hele-Shaw flows with time-dependent free boundaries involving a multiply connected fluid region, *Eur. J. Appl. Math.*, **12**, 571–599 (2001). (Cited on p. 393)

[111] S. RICHARDSON, Hele-Shaw flows with free boundaries in a corner or around a wedge. Part II: Air at the vertex, *Eur. J. Appl. Math.*, **12**, 677–688 (2001). (Cited on p. 402)

[112] S. RICHARDSON, Plane Stokes flows with time-dependent free boundaries in which the fluid occupies a doubly connected region, *Eur. J. Appl. Math.*, **11**, 249–269 (2000). (Cited on pp. 416, 420)

[113] P. G. SAFFMAN, *Vortex dynamics*, Cambridge University Press (1992). (Cited on pp. 298, 309, 373, 374, 376, 383, 384)

[114] P. G. SAFFMAN & G. I. TAYLOR, The penetration of a fluid into a medium or Hele-Shaw cell containing a more viscous liquid, *Proc. Roy. Soc. A*, **245**, 312–329 (1958). (Cited on p. 391)

[115] T. SAKAJO, Numerical construction of potential flows in multiply connected channel domains. *Comput. Methods Funct. Theory*, **11**, 415–438 (2012). (Cited on p. 313)

[116] M. SCHIFFER & D. C. SPENCER, *Functionals of finite Riemann surfaces*, Princeton University Press (1954). (Cited on p. 422)

[117] M. V. SCHNEIDER, Microstrip lines for microwave integrated circuits, *Bell Syst. Tech. J.*, **48**, 1421–1444 (1969). (Cited on p. 95)

[118] R. K. SHAH & A. L. LONDON, *Laminar flow forced convection in ducts*, Academic Press (1978). (Cited on p. 335)

[119] M. SIEGEL, Influence of surfactant on rounded and pointed bubbles in two-dimensional Stokes flow, *SIAM J. Appl. Math.*, **59**, 1998–2027 (1999). (Cited on p. 407)

[120] G. I. TAYLOR & P. G. SAFFMAN, A note on the motion of bubbles in a Hele-Shaw cell and porous medium, *Quart. J. Mech. Appl. Math.*, **12**, 265–279 (1959). (Cited on p. 391)

[121] M. TRAIZET, Hollow vortices and minimal surfaces, *J. Math. Phys.*, **56**, 083101 (2015). (Cited on p. 390)

[122] G. VALIRON, *Théorie des fonctions*, Masson (1955). (Cited on pp. 160, 256)

[123] G. VASCONCELOS, Generalization of the Schwarz-Christoffel mapping to multiply connected polygonal domains, *Proc. Roy. Soc. A*, **470**, 20130848 (2014). (Cited on p. 140)

[124] C. T. WANG, *Applied elasticity*, McGraw-Hill (1953). (Cited on p. 331)

[125] R. WICK, Solution of the field problem of the germanium gyrator, *J. Appl. Phys.*, **25**, 741 (1954). (Cited on p. 318)

Index